CHAPMAN & HALL/CRC APPLIED MATHEMATICS
AND NONLINEAR SCIENCE SERIES

Mixed Boundary Value Problems

CHAPMAN & HALL/CRC APPLIED MATHEMATICS AND NONLINEAR SCIENCE SERIES

Series Editors *Goong Chen and Thomas J. Bridges*

Published Titles

Computing with hp-ADAPTIVE FINITE ELEMENTS, Volume 1, One and Two Dimensional Elliptic and Maxwell Problems, Leszek Demkowicz

Computing with hp-ADAPTIVE FINITE ELEMENTS, Volume 2, Frontiers: Three Dimensional Elliptic and Maxwell Problems with Applications, Leszek Demkowicz, Jason Kurtz, David Pardo, Maciej Paszyński, Waldemar Rachowicz, and Adam Zdunek

CRC Standard Curves and Surfaces with Mathematica®*: Second Edition,* David H. von Seggern

Exact Solutions and Invariant Subspaces of Nonlinear Partial Differential Equations in Mechanics and Physics, Victor A. Galaktionov and Sergey R. Svirshchevskii

Geometric Sturmian Theory of Nonlinear Parabolic Equations and Applications, Victor A. Galaktionov

Introduction to Fuzzy Systems, Guanrong Chen and Trung Tat Pham

Introduction to non-Kerr Law Optical Solitons, Anjan Biswas and Swapan Konar

Introduction to Partial Differential Equations with MATLAB®, Matthew P. Coleman

Introduction to Quantum Control and Dynamics, Domenico D'Alessandro

Mathematical Methods in Physics and Engineering with Mathematica, Ferdinand F. Cap

Mathematics of Quantum Computation and Quantum Technology, Goong Chen, Louis Kauffman, and Samuel J. Lomonaco

Mixed Boundary Value Problems, Dean G. Duffy

Optimal Estimation of Dynamic Systems, John L. Crassidis and John L. Junkins

Quantum Computing Devices: Principles, Designs, and Analysis, Goong Chen, David A. Church, Berthold-Georg Englert, Carsten Henkel, Bernd Rohwedder, Marlan O. Scully, and M. Suhail Zubairy

Stochastic Partial Differential Equations, Pao-Liu Chow

Forthcoming Titles

Mathematical Theory of Quantum Computation, Goong Chen and Zijian Diao

Multi-Resolution Methods for Modeling and Control of Dynamical Systems, John L. Junkins and Puneet Singla

CHAPMAN & HALL/CRC APPLIED MATHEMATICS
AND NONLINEAR SCIENCE SERIES

Mixed Boundary Value Problems

Dean G. Duffy

CRC Press
Taylor & Francis Group
Boca Raton London New York

CRC Press is an imprint of the
Taylor & Francis Group, an **informa** business
A CHAPMAN & HALL BOOK

CRC Press
Taylor & Francis Group
6000 Broken Sound Parkway NW, Suite 300
Boca Raton, FL 33487-2742

First issued in paperback 2019

ISBN-13: 978-1-58488-579-5 (hbk)
ISBN-13: 978-0-367-38758-7 (pbk)

Library of Congress Cataloging-in-Publication Data

Duffy, Dean G.
Mixed boundary value problems / Dean G. Duffy.
p. cm. -- (Chapman & Hall/CRC applied mathematics & nonlinear science series ; 15)
Includes bibliographical references and index.
ISBN 978-1-58488-579-5 (alk. paper)
1. Boundary value problems--Numerical solutions. 2. Boundary element methods. I. Title. II. Series.

QA379.G78 2008
515'.35--dc22 2007049899

Visit the Taylor & Francis Web site at
http://www.taylorandfrancis.com

and the CRC Press Web site at
http://www.crcpress.com

Dedicated to Dr. Stephen Teoh

Contents

Acknowledgments

I am indebted to R. S. Daniels and M. A. Truesdale of the Defense College of Management and Technology for their aid in obtaining the portrait of Prof. Tranter. My appreciation goes to all the authors and publishers who allowed me the use of their material from the scientific and engineering literature. Finally, many of the plots and calculations were done using MATLAB®.

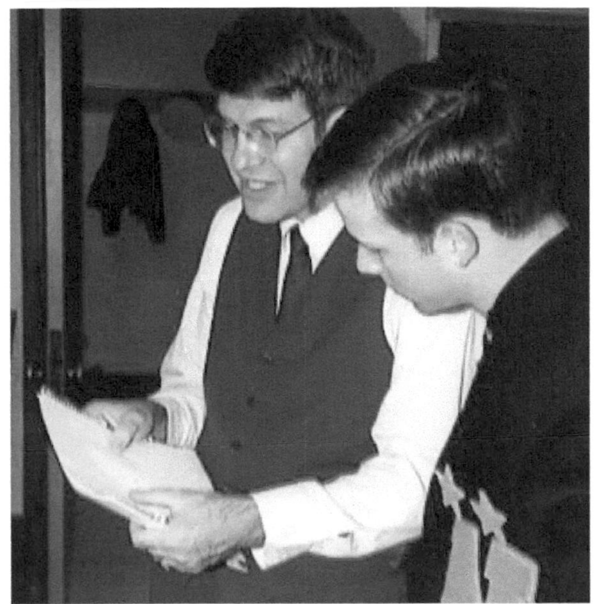

Author

Dean G. Duffy received his bachelor of science in geophysics from Case Institute of Technology (Cleveland, Ohio) and his doctorate of science in meteorology from the Massachusetts Institute of Technology (Cambridge, Massachusetts). He served in the United States Air Force from September 1975 to December 1979 as a numerical weather prediction officer. After his military service, he began a twenty-five year (1980 to 2005) association with NASA at the Goddard Space Flight Center (Greenbelt, Maryland) where he focused on numerical weather prediction, oceanic wave modeling and dynamical meteorology. He also wrote papers in the areas of Laplace transforms, antenna theory and mechanical engineering. In addition to his NASA duties he taught engineering mathematics, differential equations and calculus at the United States Naval Academy (Annapolis, Maryland) and the United States Military Academy (West Point, New York). Drawing from his teaching experience, he has written several books on transform methods, engineering mathematics and Green's functions. This present volume is his fourth book for Chapman & Hall/CRC Press.

Introduction

Purpose. This book was conceived while I was revising my engineering mathematics textbook. I noticed that in many engineering and scientific problems the nature of the boundary condition changes, say from a Dirichlet to a Neumann condition, along a particular boundary. Although these mixed boundary value problems appear in such diverse fields as elasticity and biomechanics, there are only two books (by Sneddon[1] and Fabrikant[2]) that address this problem and they are restricted to the potential equation. The purpose of this book is to give an updated treatment of this subject.

The solution of mixed boundary value problems requires considerable mathematical skill. Although the analytic solution begins using a conventional technique such as separation of variables or transform methods, the mixed boundary condition eventually leads to a system of equations, involving series or integrals, that must be solved. The solution of these equations often yields a Fredholm integral equation of the second kind. Because these integral equations usually have no closed form solution, numerical methods must be employed. Indeed, this book is just as much about solving integral equations as it involves mixed boundary value problems.

Prerequisites. The book assumes that the reader is familiar with the conventional methods of mathematical physics: generalized Fourier series, transform methods, Green's functions and conformal mapping.

---

[1] Sneddon, I. N., 1966: *Mixed Boundary Value Problems in Potential Theory.* North Holland, 283 pp.

[2] Fabrikant, V. I., 1991: *Mixed Boundary Value Problems of Potential Theory and Their Applications in Engineering.* Kluwer Academic, 451 pp.

Audience. This book may be used as either a textbook or a reference book for anyone in the physical sciences, engineering, or applied mathematics.

Chapter Overview. The purpose of Chapter 1 is twofold. The first section provides examples of what constitutes a mixed boundary value problem and how their solution differs from commonly encountered boundary value problems. The second part provides the mathematical background on integral equations and special functions that the reader might not know.

Chapter 2 presents mixed boundary value problems in their historical context. Classic problems from mathematical physics are used to illustrate how mixed boundary value problems arose and some of the mathematical techniques that were developed to handle them.

Chapters 3 and 4 are the heart of the book. Most mixed boundary value problems are solved using separation of variables if the domain is of limited extent or transform methods if the domain is of infinite or semi-infinite extent. For example, transform methods lead to the problem of solving dual or triple Fourier or Bessel integral equations. We then have a separate section for each of these integral equations.

Chapters 5 through 7 are devoted to additional techniques that are sometimes used to solve mixed boundary value problems. Here each technique is presented according to the nature of the partial differential or the domain for which it is most commonly employed or some other special technique.

Numerical methods play an important role in this book. Most integral equations here require numerical solution. All of this is done using MATLAB and the appropriate code is included. MATLAB is also used to illustrate the solutions.

We have essentially ignored brute force numerical integration of mixed boundary value problems. In most instances conventional numerical methods are simply applied to these problems. Because the solution is usually discontinuous along the boundary that contains the mixed boundary condition, analytic techniques are particularly attractive.

An important question in writing any book is what material to include or exclude. This is especially true here because many examples become very cumbersome because of the nature of governing equations. Consequently we include only those problems that highlight the mathematical techniques in a straightforward manner. The literature includes many more problems that involve mixed boundary value problems but are too complicated to be included here.

Features. Although this book should be viewed primarily as a source book on solving mixed boundary value problems, I have included problems for those who truly wish to master the material. As in my earlier books, I have included intermediate results so that the reader has confidence that he or she is on the right track.

List of Definitions

Function	Definition
$\delta(t-a)$	$= \begin{cases} \infty, & t = a, \\ 0, & t \neq a, \end{cases} \qquad \int_{-\infty}^{\infty} \delta(t-a)\, dt = 1$
$\Gamma(x)$	gamma function
$H(t-a)$	$= \begin{cases} 1, & t > a, \\ 0, & t < a. \end{cases}$
$H_n^{(1)}(x), H_n^{(2)}(x)$	Hankel functions of first and second kind and of order n
$I_n(x)$	modified Bessel function of the first kind and order n
$J_n(x)$	Bessel function of the first kind and order n
$K_n(x)$	modified Bessel function of the second kind and order n
$P_n(x)$	Legendre polynomial of order n
$\operatorname{sgn}(t-a)$	$= \begin{cases} -1, & t < a, \\ 1, & t > a. \end{cases}$
$Y_n(x)$	Bessel function of the second kind and order n

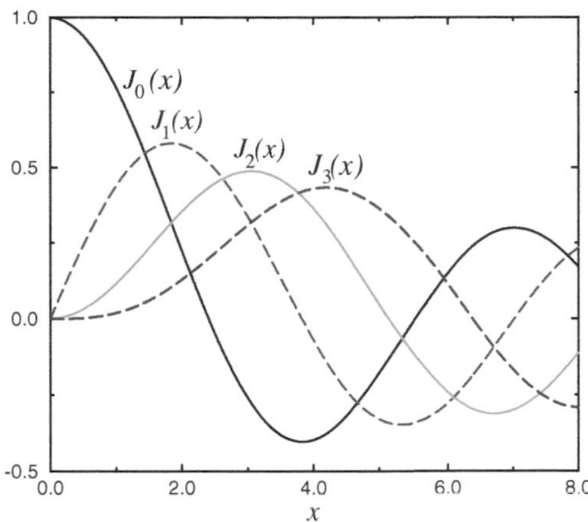

Chapter 1
Overview

In the solution of differential equations, an important class of problems involves satisfying boundary conditions either at end points or along a boundary. As undergraduates, we learn that there are three types of boundary conditions: 1) the solution has some particular value at the end point or along a boundary (Dirichlet condition), 2) the derivative of the solution equals a particular value at the end point or in the normal direction along a boundary (Neumann condition), or 3) a linear combination of Dirichlet and Neumann conditions, commonly called a "Robin condition." In the case of partial differential equations, the nature of the boundary condition can change along a particular boundary, say from a Dirichlet condition to a Neumann condition. The purpose of this book is to show how to solve these *mixed boundary value* problems.

1.1 EXAMPLES OF MIXED BOUNDARY VALUE PROBLEMS

Before we plunge into the details of how to solve a mixed boundary value problem, let us examine the origins of these problems and the challenges to their solution.

- **Example 1.1.1: Separation of variables**

Mixed boundary value problems arise during the solution of Laplace's equation within a specified region. A simple example[1] is

$$\frac{\partial^2 u}{\partial x^2} + \frac{\partial^2 u}{\partial y^2} = 0, \qquad 0 < x < \pi, \quad 0 < y < \infty, \tag{1.1.1}$$

subject to the boundary conditions

$$u_x(0, y) = u(\pi, y) = 0, \qquad 0 < y < \infty, \tag{1.1.2}$$

$$\lim_{y\to\infty} u(x, y) \to 0, \qquad 0 < x < \pi, \tag{1.1.3}$$

and

$$\begin{cases} u(x, 0) = 1, & 0 \le x < c, \\ u_y(x, 0) = 0, & c < x \le \pi. \end{cases} \tag{1.1.4}$$

The interesting aspect of this problem is the boundary condition given by Equation 1.1.4. For x between 0 and c, it satisfies a Dirichlet condition which becomes a Neumann condition as x runs between c and π.

The problem posed by Equation 1.1.1 to Equation 1.1.4 is very similar to those solved in an elementary course on partial differential equations. For that reason, let us try and apply the method of separation variables to solve it. Assuming that $u(x, y) = X(x)Y(y)$, we obtain

$$\frac{X''}{X} = -\frac{Y''}{Y} = -k^2, \tag{1.1.5}$$

with

$$X'(0) = X(\pi) = 0, \qquad \text{and} \qquad \lim_{y\to\infty} Y(y) \to 0. \tag{1.1.6}$$

Particular solutions that satisfy Equation 1.1.5 and Equation 1.1.6 are

$$u_p(x, y) = B_n \exp\left[-\left(n - \tfrac{1}{2}\right) y\right] \cos\left[\left(n - \tfrac{1}{2}\right) x\right], \tag{1.1.7}$$

with $n = 1, 2, 3, \ldots$. Because the most general solution to our problem consists of a superposition of these particular solutions, we have that

$$u(x, y) = \sum_{n=1}^{\infty} \frac{A_n}{n - \frac{1}{2}} \exp\left[-\left(n - \tfrac{1}{2}\right) y\right] \cos\left[\left(n - \tfrac{1}{2}\right) x\right]. \tag{1.1.8}$$

---

[1] See, for example, Mill, P. L., S. S. Lai, and M. P. Duduković, 1985: Solution methods for problems with discontinuous boundary conditions in heat conduction and diffusion with reaction. *Indust. Eng. Chem. Fund.*, **24**, 64–77.

Substituting this general solution into the boundary condition given by Equation 1.1.4, we obtain

$$\sum_{n=1}^{\infty} \frac{A_n}{n-\frac{1}{2}} \cos\left[\left(n-\tfrac{1}{2}\right)x\right] = 1, \qquad 0 \le x < c, \tag{1.1.9}$$

and

$$\sum_{n=1}^{\infty} A_n \cos\left[\left(n-\tfrac{1}{2}\right)x\right] = 0, \qquad c < x \le \pi. \tag{1.1.10}$$

Both Equations 1.1.9 and 1.1.10 have the form of a Fourier series except that there are two of them! Clearly the challenge raised by the boundary condition along $y = 0$ is the solution of this *dual Fourier cosine series* given by Equation 1.1.9 and Equation 1.1.10. This solution of these dual Fourier series will be addressed in Chapter 3.

- **Example 1.1.2: Transform methods**

In the previous problem, we saw that we could apply the classic method of separation of variables to solve mixed boundary value problems where the nature of the boundary condition changes along a boundary of finite length. How do we solve problems when the boundary becomes infinite or semi-infinite in length? The answer is transform methods.

Let us solve Laplace's equation[2]

$$\frac{\partial^2 u}{\partial x^2} + \frac{\partial^2 u}{\partial y^2} = 0, \qquad 0 < x < \infty, \quad 0 < y < h, \tag{1.1.11}$$

subject to the boundary conditions

$$u_x(0,y) = 0 \quad \text{and} \quad \lim_{x\to\infty} u(x,y) \to 0, \qquad 0 < y < h, \tag{1.1.12}$$

$$\begin{cases} u_y(x,0) = 1/h, & 0 \le x < 1, \\ u(x,0) = 0, & 1 < x < \infty, \end{cases} \tag{1.1.13}$$

and

$$u(x,h) = 0, \qquad 0 \le x < \infty. \tag{1.1.14}$$

The interesting aspect of this problem is the boundary condition given by Equation 1.1.13. It changes from a Neumann condition to a Dirichlet condition along the boundary $x = 1$.

To solve this boundary value problem, let us introduce the Fourier cosine transform

$$u(x,y) = \frac{2}{\pi}\int_0^\infty U(k,y)\cos(kx)\,dk, \tag{1.1.15}$$

[2] See Chen, H., and J. C. M. Li, 2000: Anodic metal matrix removal rate in electrolytic in-process dressing. I: Two-dimensional modeling. *J. Appl. Phys.*, **87**, 3151–3158.

which automatically fulfills the boundary condition given by Equation 1.1.12. Then, the differential equation given by Equation 1.1.11 and boundary condition given by Equation 1.1.14 become

$$\frac{d^2U(k,y)}{dy^2} - k^2U(k,y) = 0, \qquad 0 < y < h, \tag{1.1.16}$$

with $U(k,h) = 0$. Solving Equation 1.1.16 and inverting the transform, we find that

$$u(x,y) = \frac{2}{\pi}\int_0^\infty A(k)\frac{\sinh[k(h-y)]}{\sinh(kh)}\cos(kx)\,dk. \tag{1.1.17}$$

Substituting Equation 1.1.17 into Equation 1.1.13, we obtain

$$-\frac{2}{\pi}\int_0^\infty k\coth(kh)A(k)\cos(kx)\,dk = 1/h, \qquad 0 \le x < 1, \tag{1.1.18}$$

and

$$\frac{2}{\pi}\int_0^\infty A(k)\cos(kx)\,dk = 0, \qquad 1 < x < \infty. \tag{1.1.19}$$

Equation 1.1.18 and Equation 1.1.19 are a set of *dual integral equations* where $A(k)$ is the unknown. In Chapter 4 we will show how to solve this kind of integral equation.

- **Example 1.1.3: Wiener-Hopf technique**

In the previous example we showed how mixed boundary value problems can be solved using transform methods. Although we have not addressed the question of how to solve the resulting integral equations, the analysis leading up to that point is quite straightforward. To show that this is not always true, consider the following problem:[3]

$$\frac{\partial^2 u}{\partial x^2} + \frac{\partial^2 u}{\partial y^2} - \alpha^2 u = 0, \qquad -\infty < x < \infty, \quad 0 < y. \tag{1.1.20}$$

At infinity, we have that $\lim_{y\to\infty} u(x,y) \to 0$ while along $y = 0$,

$$\begin{cases} u(x,0) = 1, & x < 0, \\ u(x,0) = 1 + \lambda u_y(x,0), & x > 0, \end{cases} \tag{1.1.21}$$

where $0 < \alpha, \lambda$.

[3] A considerably simplified version of a problem solved by Dawson, T. W., and J. T. Weaver, 1979: *H*-polarization induction in two thin half-sheets. *Geophys. J. R. Astr. Soc.*, **56**, 419–438.

Before tackling the general problem, let us find the solution at large $|x|$. In these regions, the solution becomes essentially independent of x and Equation 1.1.20 becomes an ordinary differential equation in y. The solution as $x \to -\infty$ is

$$u(x, y) = e^{-\alpha y}, \qquad (\mathbf{1.1.22})$$

while for $x \to \infty$ the solution approaches

$$u(x, y) = \frac{e^{-\alpha y}}{1 + \alpha\lambda}. \qquad (\mathbf{1.1.23})$$

Note how these solutions satisfy the differential equation and boundary conditions as $y \to \infty$ and at $y = 0$.

Why are these limiting cases useful? If we wish to use Fourier transforms to solve the general problem, then $u(x, y)$ must tend to zero as $|x| \to \infty$ so that the Fourier transform exists. Does that occur here? No, because $u(x, y)$ tends to constant, nonzero values as $x \to -\infty$ *and* $x \to \infty$. Therefore, the use of the conventional Fourier transform is not justified.

Let us now introduce the intermediate dependent variable $v(x, y)$ so that

$$u(x, y) = \frac{e^{-\alpha y}}{1 + \alpha\lambda} + v(x, y). \qquad (\mathbf{1.1.24})$$

Substituting Equation 1.1.24 into Equation 1.1.20, we obtain

$$\frac{\partial^2 v}{\partial x^2} + \frac{\partial^2 v}{\partial y^2} - \alpha^2 v = 0, \qquad -\infty < x < \infty, \quad 0 < y. \qquad (\mathbf{1.1.25})$$

The boundary condition at infinity now reads $\lim_{y\to\infty} v(x, y) \to 0$ while

$$\begin{cases} v(x, 0) = \alpha\lambda/(1 + \alpha\lambda), & x < 0, \\ v(x, 0) = \lambda v_y(x, 0), & x > 0. \end{cases} \qquad (\mathbf{1.1.26})$$

This substitution therefore yields a $v(x, y)$ that tends to zero as $x \to \infty$. Unfortunately, $v(x, y)$ does not tend to zero as $x \to -\infty$. Consequently, once again we cannot use the conventional Fourier transform to solve this mixed boundary value problem; we appear no better off than before. In Chapter 5, we show that is not true and how the Wiener-Hopf technique allows us to solve these cases analytically.

• Example 1.1.4: Green's function

In Example 1.1.2 we used transform methods to solve a mixed boundary value problem that eventually lead to integral equations that we must solve. An alternative method of solving this problem involves Green's functions as the following example shows.

Consider the problem[4]

$$\frac{\partial^2 u}{\partial x^2} + \frac{\partial^2 u}{\partial y^2} = 0, \qquad -\infty < x < \infty, \quad 0 < y < L, \tag{1.1.27}$$

subject to the boundary conditions

$$\begin{cases} u_y(x,0) = -h(x), & |x| < 1, \\ u(x,0) = 0, & |x| > 1, \end{cases} \tag{1.1.28}$$

$$u(x,L) = 0, \qquad -\infty < x < \infty, \tag{1.1.29}$$

and

$$\lim_{|x|\to\infty} u(x,y) \to 0, \qquad 0 < y < L. \tag{1.1.30}$$

From the theory of Green's functions, the solution to Equation 1.1.27 through Equation 1.1.30 is

$$u(x,y) = \int_{-1}^{1} f(\xi) \frac{\partial g(x,y|\xi,0)}{\partial \eta}\, d\xi, \tag{1.1.31}$$

where $g(x,y|\xi,\eta)$ is the Green's function defined by

$$\frac{\partial^2 g}{\partial x^2} + \frac{\partial^2 g}{\partial y^2} = -\delta(x-\xi)\delta(y-\eta), \qquad -\infty < x,\xi < \infty, \quad 0 < y,\eta < L, \tag{1.1.32}$$

and the boundary conditions

$$g(x,0|\xi,\eta) = g(x,L|\xi,\eta) = 0, \qquad -\infty < x < \infty, \tag{1.1.33}$$

and

$$\lim_{|x|\to\infty} g(x,y|\xi,\eta) \to 0, \qquad 0 < y < L, \tag{1.1.34}$$

and $f(\xi)$ is an unknown function such that $u(\xi,0) = f(\xi)$ if $|\xi| \le 1$.

To find $g(x,y|\xi,\eta)$, we first take the Fourier transform of Equation 1.1.32 through Equation 1.1.34 with respect to x. This yields

$$\frac{d^2 G}{dy^2} - k^2 G = -\frac{e^{-ik\xi}}{L}\delta(y-\eta), \tag{1.1.35}$$

with the boundary conditions $G(k,0|\xi,\eta) = G(k,L|\xi,\eta) = 0$. Because

$$\delta(y-\eta) = \frac{2}{L}\sum_{n=1}^{\infty} \sin\left(\frac{n\pi\eta}{L}\right)\sin\left(\frac{n\pi y}{L}\right) \tag{1.1.36}$$

[4] Taken with permission from Yang, F., V. Prasad, and I. Kao, 1999: The thermal constriction resistance of a strip contact spot on a thin film. *J. Phys. D. Appl. Phys.*, **32**, 930–936. Published by IOP Publishing Ltd.

and assuming

$$G(k, y|\xi, \eta) = \sum_{n=1}^{\infty} A_n \sin\left(\frac{n\pi y}{L}\right), \tag{1.1.37}$$

direct substitution gives

$$\left(\frac{n^2\pi^2}{L^2} + k^2\right) A_n = \frac{2}{L} e^{-ik\xi} \sin\left(\frac{n\pi\eta}{L}\right), \tag{1.1.38}$$

and

$$G(k, y|\xi, \eta) = \frac{2}{L} e^{-ik\xi} \sum_{n=1}^{\infty} \frac{\sin(n\pi\eta/L)\sin(n\pi y/L)}{k^2 + n^2\pi^2/L^2}. \tag{1.1.39}$$

Inverting the Fourier transforms in Equation 1.1.39 term by term, we obtain

$$g(x, y|\xi, \eta) = \frac{1}{\pi} \sum_{n=1}^{\infty} \frac{1}{n} \exp\left(-\frac{n\pi}{L}|x - \xi|\right) \sin\left(\frac{n\pi\eta}{L}\right) \sin\left(\frac{n\pi y}{L}\right) \tag{1.1.40}$$

$$= \frac{1}{2\pi} \sum_{n=1}^{\infty} \frac{1}{n} \exp\left(-\frac{n\pi}{L}|x - \xi|\right) \times \left\{\cos\left[\frac{n\pi(y-\eta)}{L}\right] - \cos\left[\frac{n\pi(y+\eta)}{L}\right]\right\}. \tag{1.1.41}$$

Because

$$\sum_{n=1}^{\infty} \frac{q^n}{n} \cos(n\alpha) = -\ln\left[\sqrt{1 - 2q\cos(\alpha) + q^2}\right], \tag{1.1.42}$$

provided that $|q| < 1$ and $0 \le \alpha \le 2\pi$, we have that

$$g(x, y|\xi, \eta) = -\frac{1}{4\pi} \ln\left\{\frac{\cosh[\pi(x-\xi)/L] - \cos[\pi(y-\eta)/L]}{\cosh[\pi(x-\xi)/L] - \cos[\pi(y+\eta)/L]}\right\}. \tag{1.1.43}$$

Substituting Equation 1.1.43 into Equation 1.1.31 and simplifying, we have

$$u(x, y) = \frac{1}{2L} \int_{-1}^{1} \frac{f(\xi)\sin(\pi y/L)}{\cosh[\pi(x-\xi)/L] - \cos(\pi y/L)}\, d\xi. \tag{1.1.44}$$

Finally, computing $u_y(x, 0)$ and using Equation 1.1.28, we obtain the integral equation

$$\frac{\pi}{4L^2} \int_{-1}^{1} \frac{f(\xi)}{\sinh^2[\pi(x-\xi)/(2L)]}\, d\xi = -h(x), \qquad |x| < 1. \tag{1.1.45}$$

Integrating by parts with $f(1) = f(-1) = 0$, an equivalent integral equation is

$$\frac{1}{2L} \int_{-1}^{1} f'(\xi) \coth[\pi(x-\xi)/(2L)]\, d\xi = h(x), \qquad |x| < 1. \tag{1.1.46}$$

Once again, we have reduced the mixed boundary value problem to finding the solution of an integral equation. Chapter 6 is devoted to solving Equation 1.1.46 as well as other mixed boundary value problems via Green's function.

• Example 1.1.5: Conformal mapping

Conformal mapping is a mathematical technique involving two complex variables: $z = x + iy$ and $t = r + is$. Given an analytic function $t = g(z)$, the domain over which Laplace's equation holds in the z-plane is mapped into some portion of the t-plane, such as an upper half-plane, rectangle or circle. It is readily shown that Laplace's equation and the Dirichlet and/or Neumann conditions in the z-plane also apply in the t-plane. For this method to be useful, the solution of Laplace's equation in the t-plane must be easier than in the z-plane.

For us the interest in conformal mapping lies in the fact that a solution to Laplace's equation in the xy-plane is also a solution to Laplace's equation in the rs-plane. Of equal importance, if the solution along a boundary in the xy-plane is constant, it is also constant along the corresponding boundary in the rs-plane. For these reasons, conformal mapping has been a powerful method for solving Laplace's equation since the nineteenth century. Let us see how we can use this technique to solve a mixed boundary value problem.

Let us solve Laplace's equation[5]

$$\frac{\partial^2 u}{\partial x^2} + \frac{\partial^2 u}{\partial y^2} = 0, \qquad 0 < x < a, \quad 0 < y < \infty, \tag{1.1.47}$$

subject to the boundary conditions

$$u(0, y) = 0, \qquad 0 < y < \infty, \tag{1.1.48}$$

$$\begin{cases} u(a, y) = 1, & 0 < y < c, \\ u_x(a, y) = 0, & c < y < \infty, \end{cases} \tag{1.1.49}$$

and

$$u_y(x, 0) = 0, \quad \lim_{y\to\infty} u(x, y) \to 0, \qquad 0 < x < a. \tag{1.1.50}$$

Consider the transformation $t = -\cos(\pi z/a)$. As Figure 1.1.1 shows, this transformation maps the strip $0 < x < a$, $0 < y < \infty$ into the half-plane $0 < \Im(t)$: The boundary $x = a$, $y > 0$ is mapped into $\Re(t) > \cosh(\pi c/a)$, $\Im(t) = 0$ while the x-axis lies along $-1 < \Re(t) < 1$, $\Im(t) = 0$.

Consider next, the fractional linear transformation

$$s = \frac{\alpha t + \beta}{\gamma t + \delta}, \tag{1.1.51}$$

[5] Laporte, O., and R. G. Fowler, 1966: Resistance of a plasma slab between juxtaposed disk electrodes. *Phys. Rev.*, **148**, 170–175.

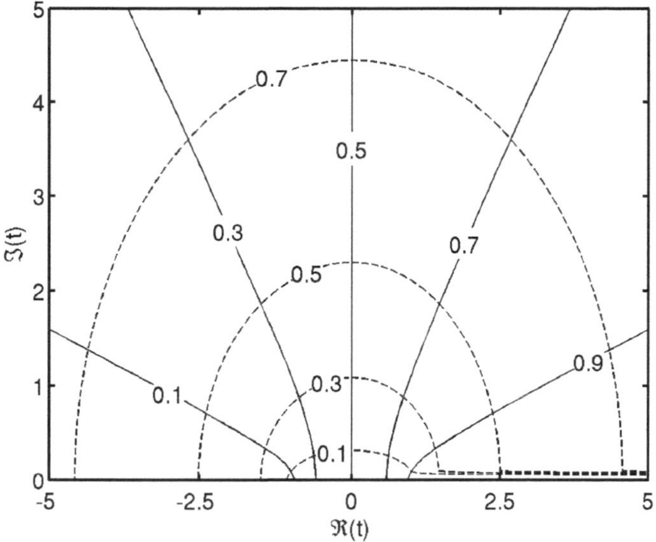

Figure 1.1.1: Lines of constant values of of x/a (solid line) and y/a (dashed lines) in the t-plane given by the conformal mapping $t = -\cos(\pi z/a)$.

where

$$\alpha = D/k, \qquad \gamma = D, \tag{1.1.52}$$

$$\beta - (D/2)[(1 - 1/k)\psi - (1 + 1/k)], \tag{1.1.53}$$

$$\delta = (D/2)[(1/k - 1)\psi - (1 + 1/k)], \tag{1.1.54}$$

D is a free parameter, and

$$\frac{1}{k} = \frac{\psi + 3 + \sqrt{8\psi + 8}}{\psi - 1}; \qquad \psi = \cosh(\pi c/a). \tag{1.1.55}$$

We illustrate this conformal mapping in Figure 1.1.2 when $c/a = 1$ and $D = 1$.

Finally, we introduce the conformal mapping $s = \text{sn}(\zeta, k)$, where $\text{sn}(\cdot, k)$ denotes one of the Jacobian elliptic functions.[6] This maps the half-plane $\Im(s) > 0$ into a rectangular box with vertices at $(K, 0)$, (K, K'), $(-K, K')$ and $(-K, 0)$, where K and K' are the real and imaginary quarter-periods, respectively. We show this conformal mapping in Figure 1.1.3 when $c/a = 1$ and $D = 1$.

Why have we introduced these three conformal mappings? After applying these three mappings, our original problem, Equation 1.1.47 through Equation

---

[6] See Milne-Thomson, L. M., 1965: Jacobian elliptic functions and theta functions. *Handbook of Mathematical Functions*, M. Abromowitz and I. A. Stegun, Eds., Dover, 567–586.

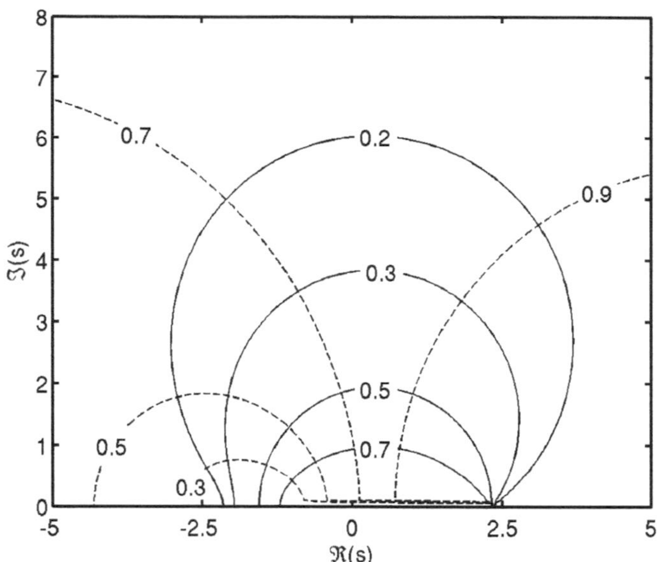

Figure 1.1.2: Same as Figure 1.1.1 except that we have the additional mapping given by Equation 1.1.51 with $c/a = 1$ or $k = 0.430$ by Equation 1.1.55. If $D = 1$, $\alpha = 2.325$, $\beta = -9.344$, $\gamma = 1$, and $\delta = 6.018$.

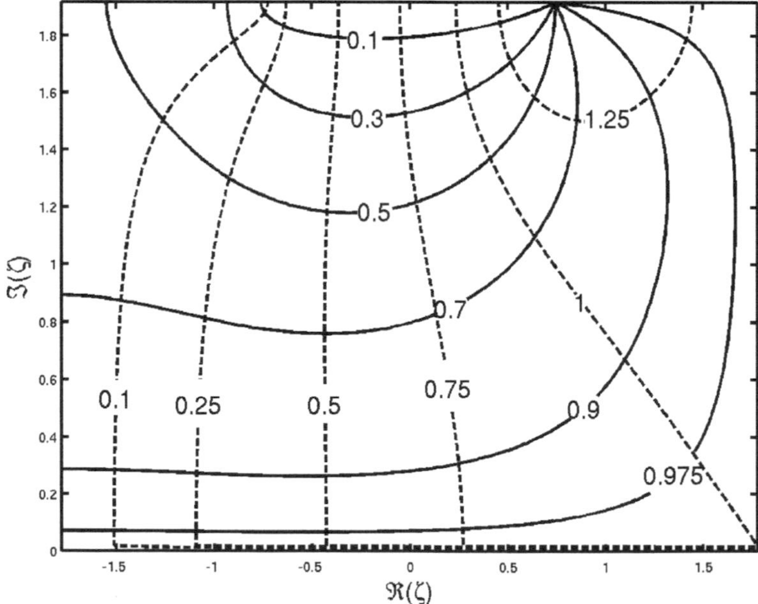

Figure 1.1.3: Same as Figure 1.1.2 except that we have the additional mapping $s = \text{sn}(\zeta, k)$, where $K = 1.779$ and $K' = 1.918$.

1.1.50, becomes

$$\frac{\partial^2 u}{\partial \xi^2} + \frac{\partial^2 u}{\partial \eta^2} = 0, \qquad -K < \xi < K, \quad 0 < \eta < K', \tag{1.1.56}$$

subject to the boundary conditions

$$u(\xi, 0) = 1, \quad u(\xi, K') = 0, \qquad -K < \xi < K, \tag{1.1.57}$$

and

$$u_\xi(-K, \eta) = u_\xi(K, \eta) = 0, \tag{1.1.58}$$

where $\zeta = \xi + i\eta$. The solution to Equation 1.1.56 through Equation 1.1.58 is

$$u(\xi, \eta) = 1 - \eta/K'. \tag{1.1.59}$$

By using Equation 1.1.59 and the three conformal mappings, we can solve Equation 1.1.47 through Equation 1.1.50 as follows: For a rectangular grid over the domain $-K < \xi < K$, $0 < \eta < K'$, we compute the values of $u(\xi, \eta)$. Next, using the MATLAB® procedure `ellipj`, we find the values of the Jacobian elliptic functions to compute s. Because ζ is complex, we use the relationship

$$\begin{aligned} \operatorname{sn}(x + iy, k) = &\frac{\operatorname{sn}(x,k)\operatorname{dn}(y,k')}{\operatorname{cn}^2(y,k') + k\ \operatorname{sn}^2(x,k)\operatorname{sn}^2(y,k')} \\ &+ i\frac{\operatorname{sn}(x,k)\operatorname{dn}(x,k)\operatorname{sn}(y,k')\operatorname{dn}(y,k')}{\operatorname{cn}^2(y,k') + k\ \operatorname{sn}^2(x,k)\operatorname{sn}^2(y,k')}, \end{aligned} \tag{1.1.60}$$

where $k' = 1 - k$. Next, we use Equation 1.1.51 to compute t given s. Finally, $z = \arccos(-t)/\pi$. Thus, for a particular value of x and y, we have $u(x, y)$. Figure 1.1.4 illustrates this solution. In Chapter 7, we will explore this technique further.

• Example 1.1.6: Numerical methods

Numerical methods are necessary in certain instances because the geometry may be simply too complicated for analytic techniques. These techniques are similar to those applied to solve most partial differential equations. However, because most of the solutions are discontinuous along the boundary, a few papers have examined the application of finite differences to mixed boundary value problems.[7]

---

[7] Greenspan, D., 1964: On the numerical solution of problems allowing mixed boundary conditions. *J. Franklin Inst.*, **277**, 11–30; Bramble, J. H., and B. E. Hubbard, 1965: Approximation of solutions of mixed boundary value problems for Poisson's equation by finite differences. *J. Assoc. Comput. Mach.*, **12**, 114–123; Thuraisamy, V., 1969: Approximate solutions for mixed boundary value problems by finite-difference methods. *Math. Comput.*, **23**, 373–386.

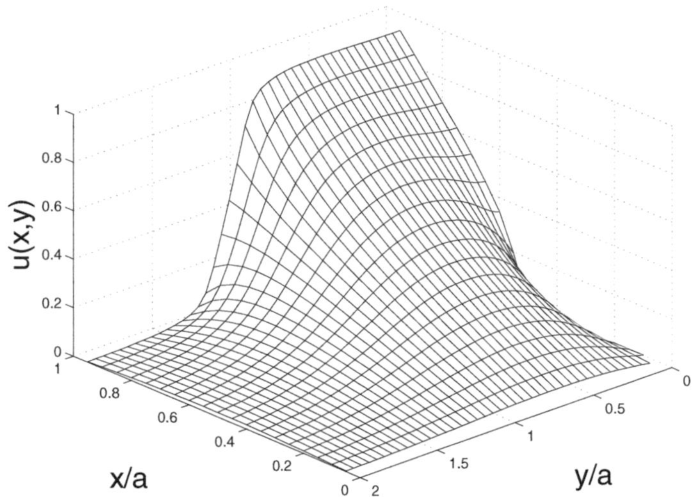

Figure 1.1.4: The solution to the mixed boundary value problem governed by Equation 1.1.47 to Equation 1.1.50.

Consider the mixed boundary value problem

$$\frac{\partial^2 u}{\partial r^2} + \frac{1}{r}\frac{\partial u}{\partial r} + \frac{\partial^2 u}{\partial z^2} = 0, \qquad 0 \le r < \infty, \qquad -\infty < z < \infty, \tag{1.1.61}$$

subject to the boundary conditions

$$\lim_{r\to 0} |u(r,z)| < \infty, \qquad \lim_{r\to\infty} u(r,z) \to 0, \qquad \lim_{|z|\to\infty} u(r,z) \to 0, \tag{1.1.62}$$

and

$$u(r, \pm h) = \pm 1, \qquad 0 < r < a. \tag{1.1.63}$$

Due to symmetry, we can reduce the domain to the half-space $z > 0$ by requiring that $u(r, 0) = 0$ for $0 \le r < \infty$.

Although there is an exact solution[8] to this problem, we will act as if there is none and solve it purely by numerical methods. Introducing a grid with nodal points located at $r_n = n\Delta r$ and $z_m = m\Delta z$, where $n = 0, 1, 2, \ldots, N$ and $m = 0, 1, 2, \ldots, M$, and applying simple second-order, finite differences to represent the partial derivatives, Equation 1.1.61 can be approximated by

$$\frac{u_{n+1,m} - 2u_{n,m} + u_{n-1,m}}{(\Delta r)^2} + \frac{u_{n+1,m} - u_{n-1,m}}{2n(\Delta r)^2} + \frac{u_{n,m+1} - 2u_{n,m} + u_{n,m-1}}{(\Delta z)^2} = 0, \tag{1.1.64}$$

[8] Bartlett, D. F., and T. R. Corle, 1985: The circular parallel plate capacitor: A numerical solution for the potential. *J. Phys. A*, **18**, 1337–1342. See also Schwarzbek, S. M., and S. T. Ruggiero, 1986: The effect of fringing fields on the resistance of a conducting film. *IEEE Microwave Theory Tech.*, **MTT-34**, 977–981.

where $n = 1, 2, \ldots, N-1$ and $m = 1, 2, \ldots, M-1$. Axial symmetry yields for $n = 0$:

$$\frac{2u_{1,m} - 2u_{0,m}}{(\Delta r)^2} + \frac{u_{0,m+1} - 2u_{0,m} + u_{0,m-1}}{(\Delta z)^2} = 0, \qquad \textbf{(1.1.65)}$$

with $m = 1, 2, \ldots, M-1$. Finally,

$$u_{n,0} = u_{n,M} = 0, \qquad n = 0, 1, 2, \ldots, N, \qquad \textbf{(1.1.66)}$$

$$u_{N,m} = 0, \qquad m = 0, 1, 2, \ldots, M, \qquad \textbf{(1.1.67)}$$

and

$$u_{n,H} = 1, \qquad n = 0, 1, 2, \ldots, I, \qquad \textbf{(1.1.68)}$$

where $a = I\Delta r$ and $h = H\Delta z$. In Equation 1.1.64 through Equation 1.1.68, we have denoted $u(r_n, z_m)$ simply by $u_{n,m}$.

Although this system of equations could be solved using techniques from linear algebra, that would be rather inefficient; in general, these equations form a sparse matrix. For this reason, an iterative method is best. A simple one is to solve for $u_{n,m}$ in Equation 1.1.64. Assuming $\Delta r = \Delta z$, we obtain

$$u_{n,m}^{i+1} = \tfrac{1}{4}\left[u_{n+1,m}^{i} + u_{n-1,m}^{i+1} + u_{n,m+1}^{i} + u_{n,m-1}^{i+1} + \left(u_{n+1,m}^{i} - u_{n-1,m}^{i+1}\right)/(2n)\right], \qquad \textbf{(1.1.69)}$$

where $n = 1, 2, \ldots, N-1$ and $m = 1, 2, \ldots, M-1$. Similarly,

$$u_{0,m}^{i+1} = \tfrac{1}{4}\left(2u_{1,m}^{i} + u_{0,m+1}^{i} + u_{0,m-1}^{i+1}\right), \qquad \textbf{(1.1.70)}$$

where $m = 1, 2, \ldots, M-1$. Here, we denote the value of $u_{n,m}$ during the ith iteration with the subscript i. This iterative scheme is an example of the Gauss-Seidel method. It is particularly efficient because $u_{n-1,m}$ and $u_{n,m-1}$ have already been updated.

Figure 1.1.5 illustrates this numerical solution by showing $u_{n,m}$ at various points during the iterative process. Initially, there is dramatic change in the solution, followed by slower change as i becomes large.

1.2 INTEGRAL EQUATIONS

An *integral equation* is any equation in which the unknown appears in the integrand. Let $\varphi(t)$ denote the unknown function, $f(x)$ is a known function, and $K(x,t)$ is a known integral kernel, then a wide class of integral equations can be written as

$$f(x) = \int_a^b K(x,t)\varphi(t)\,dt, \qquad \textbf{(1.2.1)}$$

or

$$\varphi(x) = f(x) + \int_a^b K(x,t)\varphi(t)\,dt. \qquad \textbf{(1.2.2)}$$

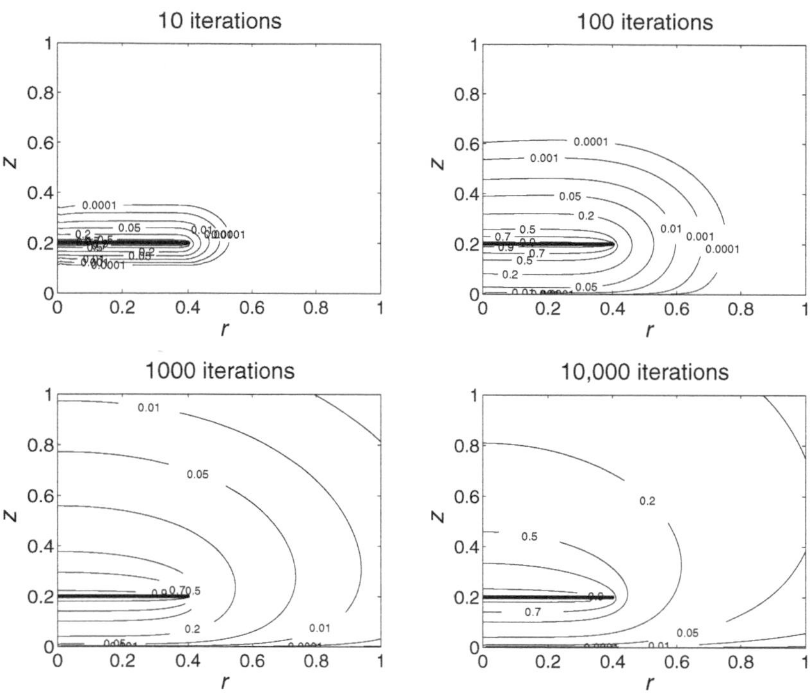

Figure 1.1.5: A portion of the numerical solution of Equation 1.1.61 through Equation 1.1.63 using the Gauss-Seidel scheme to solve the finite differenced equations. The parameters used in this example are $N = M = 200$, $\Delta r = \Delta z = 0.01$, $a = 0.4$, and $h = 0.2$.

A common property of these integral equations is the fixed limits in the integral. These integral equations are collectively called "Fredholm integral equations," named after the Swedish mathematician Erik Ivar Fredholm (1866–1927) who first studied them. Equation 1.2.1 is referred to as a Fredholm integral equation of the first kind while Equation 1.2.2 is a Fredholm integral equation of the second kind; integral equations of the second kind differ from integral equations of the first kind in the appearance of the unknown outside of the integral. In an analogous manner, integral equations of the form

$$f(x) = \int_a^x K(x,t)\varphi(t)\,dt, \tag{1.2.3}$$

and

$$\varphi(x) = f(x) + \int_a^x K(x,t)\varphi(t)\,dt, \tag{1.2.4}$$

are referred to as Volterra integrals of the first and second kind, respectively. Named after the Italian mathematician Vito Volterra (1860–1940), they have in common the property that one of the limits is a variable.

- **Example 1.2.1**

An example of Volterra equation of the first kind that we will often encounter is

$$\int_a^x \frac{f(t)}{[h(x)-h(t)]^\alpha}\,dt = g(x), \qquad a < x < b, \tag{1.2.5}$$

where $0 < \alpha < 1$ and $h(t)$ is a strictly monotonically increasing function in (a, b). Srivastav[9] showed that integral equations of the form of Equation 1.2.5 have the solution

$$f(t) = \frac{\sin(\pi\alpha)}{\pi}\frac{d}{dt}\left\{\int_a^t \frac{h'(\tau)g(\tau)}{[h(t)-h(\tau)]^{1-\alpha}}\,d\tau\right\}, \qquad a < t < b. \tag{1.2.6}$$

In a similar manner, the integral equation

$$\int_x^b \frac{f(t)}{[h(t)-h(x)]^\alpha}\,dt = g(x), \qquad a < x < b, \tag{1.2.7}$$

where $0 < \alpha < 1$ and $h(t)$ is a strictly monotonically increasing function in (a, b) has the solution

$$f(t) = -\frac{\sin(\pi\alpha)}{\pi}\frac{d}{dt}\left\{\int_t^b \frac{h'(\tau)g(\tau)}{[h(\tau)-h(t)]^{1-\alpha}}\,d\tau\right\}, \qquad a < t < b. \tag{1.2.8}$$

To illustrate Equation 1.2.5, let us choose $h(\tau) = 1 - \cos(\tau)$ and $\alpha = \frac{1}{2}$. In this case, it now reads

$$\int_a^x \frac{f(t)}{\sqrt{\cos(t)-\cos(x)}}\,dt = g(x), \qquad 0 \le a < x < b \le \pi, \tag{1.2.9}$$

has the solution

$$f(t) = \frac{1}{\pi}\frac{d}{dt}\left[\int_a^t \frac{\sin(\tau)g(\tau)}{\sqrt{\cos(\tau)-\cos(t)}}\,d\tau\right], \qquad a < t < b. \tag{1.2.10}$$

Similarly,

$$\int_x^b \frac{f(t)}{\sqrt{\cos(x)-\cos(t)}}\,dt = g(x), \qquad 0 \le a < x < b \le \pi, \tag{1.2.11}$$

has the solution

$$f(t) = -\frac{1}{\pi}\frac{d}{dt}\left[\int_t^b \frac{\sin(\tau)g(\tau)}{\sqrt{\cos(t)-\cos(\tau)}}\,d\tau\right], \qquad a < t < b. \tag{1.2.12}$$

---

[9] Srivastav, R. P., 1963: A note on certain integral equations of Abel-type. *Proc. Edinburgh Math. Soc., Ser. 2*, **13**, 271–272.

A second application occurs if we set $h(\tau) = \tau^2$. In this case we have the integral equation (Abel-type integral equation)

$$\int_a^x \frac{f(t)}{(x^2 - t^2)^\alpha}\,dt = g(x), \qquad 0 < \alpha < 1, \quad a < x < b, \tag{1.2.13}$$

which has the solution

$$f(t) = \frac{2\sin(\pi\alpha)}{\pi}\frac{d}{dt}\left[\int_a^t \frac{\tau\, g(\tau)}{(t^2 - \tau^2)^{1-\alpha}}\,d\tau\right], \qquad a < t < b. \tag{1.2.14}$$

Taken together, Equation 1.2.13 and Equation 1.2.14 are known as an interior Abel transform. Similarly,

$$\int_x^b \frac{f(t)}{(t^2 - x^2)^\alpha}\,dt = g(x), \qquad 0 < \alpha < 1, \quad a < x < b, \tag{1.2.15}$$

has the solution

$$f(t) = -\frac{2\sin(\pi\alpha)}{\pi}\frac{d}{dt}\left[\int_t^b \frac{\tau\, g(\tau)}{(\tau^2 - t^2)^{1-\alpha}}\,d\tau\right], \qquad a < t < b. \tag{1.2.16}$$

Equation 1.2.15 and Equation 1.2.16 are known as an exterior Abel transform.

Finally, if $\alpha = \frac{1}{2}$, $f(t) = F'(t)$, and $h(t) = t$, Equation 1.2.5 and Equation 1.2.6 yield

$$F'(t) = \frac{1}{\pi}\frac{d}{dt}\left[\int_a^t \frac{g(\eta)}{\sqrt{t-\eta}}\,d\eta\right], \tag{1.2.17}$$

where

$$g(\eta) = \int_a^\eta \frac{F'(\xi)}{\sqrt{\eta - \xi}}\,d\xi. \tag{1.2.18}$$

Combining Equation 1.2.17 and Equation 1.2.18,

$$F'(t) = \frac{1}{\pi}\frac{d}{dt}\left\{\int_a^t \frac{1}{\sqrt{t-\eta}}\left[\int_a^\eta \frac{F'(\xi)}{\sqrt{\eta-\xi}}\,d\xi\right]\,d\eta\right\}. \tag{1.2.19}$$

Integrating both sides of Equation 1.2.19 from a to x, we obtain

$$\int_a^x \frac{1}{\sqrt{x-\eta}}\left[\int_a^\eta \frac{F'(\xi)}{\sqrt{\eta-\xi}}\,d\xi\right]\,d\eta = \pi[F(x) - F(a)]. \tag{1.2.20}$$

In a similar manner, Equation 1.2.7 and Equation 1.2.8 can be combined to yield

$$\int_x^a \frac{1}{\sqrt{x-\eta}}\left[\int_a^\eta \frac{F'(\xi)}{\sqrt{\eta-\xi}}\,d\xi\right]\,d\eta = \pi[F(a) - F(x)]. \tag{1.2.21}$$

Hafen[10] derived Equation 1.2.20 and Equation 1.2.21 in 1910.

• **Example 1.2.2**

Consider the integral equation of the form

$$\int_0^x K(x^2 - t^2) f(t)\, dt = g(x), \qquad 0 < x, \tag{1.2.22}$$

where $K(\cdot)$ is known. We can solve equations of this type by reducing them to

$$\int_0^\xi K(\xi - \tau) F(\tau)\, d\tau = G(\xi), \qquad 0 < \xi, \tag{1.2.23}$$

via the substitutions $x = \sqrt{\xi}$, $t = \sqrt{\tau}$, $F(\tau) = f(\sqrt{\tau}\,) \,/\, (2\sqrt{\tau}\,)$, $G(\xi) = g\big(\sqrt{\xi}\,\big)$. Taking the Laplace transform of both sides of Equation 1.2.23, we have by the convolution theorem

$$\mathcal{L}[K(t)]\mathcal{L}[F(t)] = \mathcal{L}[G(t)]. \tag{1.2.24}$$

Defining $\mathcal{L}[L(t)] = 1/\{s\mathcal{L}[K(t)]\}$, we have by the convolution theorem

$$F(\xi) = \frac{d}{d\xi}\left[\int_0^\xi L(\xi - \tau) G(\tau)\, d\tau\right], \tag{1.2.25}$$

or

$$f(x) = 2\frac{d}{dx}\left[\int_0^x t g(t) L(x^2 - t^2)\, dt\right]. \tag{1.2.26}$$

To illustrate this method, we choose $K(t) = \cos\big(k\sqrt{t}\,\big) \,/\sqrt{t}$ which has the Laplace transform $\mathcal{L}[K(t)] = \sqrt{\pi} e^{-k^2/(4s)}/\sqrt{s}$. Then,

$$L(t) = \mathcal{L}^{-1}\left[\frac{e^{k^2/(4s)}}{\sqrt{\pi s}}\right] = \frac{\cosh\big(k\sqrt{t}\,\big)}{\pi\sqrt{t}}. \tag{1.2.27}$$

Therefore, the integral equation

$$\int_0^x \frac{\cosh\big(k\sqrt{x^2 - t^2}\,\big)}{\sqrt{x^2 - t^2}} f(t)\, dt = g(x), \qquad 0 < x, \tag{1.2.28}$$

has the solution

$$f(x) = \frac{2}{\pi}\frac{d}{dx}\left[\int_0^x \frac{\cosh\big(k\sqrt{x^2 - t^2}\,\big)}{\sqrt{x^2 - t^2}} t g(t)\, dt\right]. \tag{1.2.29}$$

---

[10] Hafen, M., 1910: Studien über einige Probleme der Potentialtheorie. *Math. Ann.*, **69**, 517–537.

In particular, if $g(0) = 0$, then

$$f(x) = \frac{2x}{\pi} \int_0^x \frac{\cosh\left(k\sqrt{x^2 - t^2}\right)}{\sqrt{x^2 - t^2}} g'(t)\, dt. \tag{1.2.30}$$

• **Example 1.2.3**

In 1970, Cooke[11] proved that the solution to the integral equation

$$\int_0^1 \ln\left|\frac{x+t}{x-t}\right| h(t)\, dt = \pi f(x), \qquad 0 < x < 1, \tag{1.2.31}$$

is

$$h(t) = -\frac{2}{\pi}\frac{d}{dt}\left[\int_t^1 \frac{\alpha\, S(\alpha)}{\sqrt{\alpha^2 - t^2}}\, d\alpha\right] + \frac{2f(0^+)}{\pi t\sqrt{1 - t^2}}, \tag{1.2.32}$$

where

$$S(\alpha) = \int_0^\alpha \frac{f'(\xi)}{\sqrt{\alpha^2 - \xi^2}}\, d\xi. \tag{1.2.33}$$

We will use this result several times in this book. For example, at the beginning of Chapter 4, we must solve the integral equation

$$-\frac{1}{\pi}\int_0^1 g(t) \ln\left|\frac{\tanh(\beta x) + \tanh(\beta t)}{\tanh(\beta x) - \tanh(\beta t)}\right| dt = \frac{x}{h}, \qquad 0 \le x < 1, \tag{1.2.34}$$

where $2h\beta = \pi$. How does Equation 1.2.31 help us here? If we introduce the variables $\tanh(\beta t) = \tanh(\beta)T$ and $\tanh(\beta x) = \tanh(\beta)X$, then Equation 1.2.34 transforms into an integral equation of the form Equation 1.2.31. Substituting back into the original variables, we find that

$$h(t) = \frac{1}{\pi^2}\frac{d}{dt}\left[\int_t^1 \frac{\tanh(\beta\alpha)S(\alpha)}{\cosh^2(\beta\alpha)\sqrt{\tanh^2(\beta\alpha) - \tanh^2(\beta t)}}\, d\alpha\right], \tag{1.2.35}$$

where

$$S(\alpha) = \int_0^\alpha \frac{d\xi}{\sqrt{\tanh^2(\beta\alpha) - \tanh^2(\beta\xi)}}. \tag{1.2.36}$$

Another useful result derived by Cooke is that the solution to the integral equation

$$\int_0^1 g(y) \log\left(\frac{|x^2 - y^2|}{y^2}\right) dy = \pi f(x), \qquad 0 < x < 1, \tag{1.2.37}$$

---

[11] Cooke, J. C., 1970: The solution of some integral equations and their connection with dual integral equations and series. *Glasgow Math. J.*, **11**, 9–20.

is

$$g(y) = \frac{2}{\pi\sqrt{1-y^2}} \int_0^1 \frac{x\sqrt{1-x^2}}{x^2-y^2} f'(x)\,dx + \frac{C}{\sqrt{1-y^2}}, \tag{1.2.38}$$

or

$$g(y) = \frac{2}{\pi y}\frac{d}{dy}\left[\frac{\tau\, S(\tau)}{\sqrt{\tau^2-y^2}}\,d\tau\right] + \frac{C}{\sqrt{1-y^2}}, \tag{1.2.39}$$

where

$$S(\tau) = \tau\frac{d}{d\tau}\left[\int_0^\tau \frac{f(x)}{\sqrt{\tau^2-x^2}}\,dx\right] = \int_0^\tau \frac{x\, f'(x)}{\sqrt{\tau^2-x^2}}\,dx. \tag{1.2.40}$$

If $f(x)$ is a constant, the equation has no solution. We will use Equation 1.2.39 and Equation 1.2.40 in Chapter 6.

• Example 1.2.4

Many solutions to dual integrals hinge on various improper integrals that contain Bessel functions. For example, consider the known result[12] that

$$\int_0^\infty J_\nu(\beta x)\frac{J_\mu\big(\alpha\sqrt{x^2+z^2}\,\big)}{\sqrt{(x^2+z^2)^\mu}}x^{\nu+1}\,dx \tag{1.2.41}$$

$$= \frac{\beta^\nu}{\alpha^\mu}\left(\frac{\sqrt{\alpha^2-\beta^2}}{z}\right)^{\mu-\nu-1} J_{\mu-\nu-1}\Big(z\sqrt{\alpha^2-\beta^2}\,\Big)\,H(\alpha-\beta),$$

if $\Re(\mu) > \Re(\nu) > -1$. Akhiezer[13] used Equation 1.2.41 along with the result[14] that the integral equation

$$g(x) = \int_0^x f(t)\left(\frac{\sqrt{x^2-t^2}}{ik}\right)^{-p} J_{-p}\Big(ik\sqrt{x^2-t^2}\,\Big)\,dt \tag{1.2.42}$$

has the solution

$$f(x) = \frac{d}{dx}\left[\int_0^x t\, g(t)\left(\frac{\sqrt{x^2-t^2}}{k}\right)^{-q} J_{-q}\Big(k\sqrt{x^2-t^2}\,\Big)\,dt\right], \tag{1.2.43}$$

---

[12] Gradshteyn, I. S., and I. M. Ryzhik, 1965: *Table of Integrals, Series, and Products*. Academic Press, Formula 6.596.6.

[13] Akhiezer, N. I., 1954: On some coupled integral equations (in Russian). *Dokl. Akad. Nauk USSR*, **98**, 333–336.

[14] Polyanin, A. D., and A. V. Manzhirov, 1998: *Handbook of Integral Equations*. CRC Press, Formula 1.8.71.

where $p > 0$, $q > 0$, and $p + q = 1$.

Let us use these results to find the solution to the dual integral equations

$$\int_\alpha^\infty C(k) J_\nu(kx)\, dk = f(x), \qquad 0 < x < 1, \tag{1.2.44}$$

and

$$\int_0^\infty C(k) J_{-\nu}(kx)\, dk = 0, \qquad 1 < x < \infty, \tag{1.2.45}$$

where $\alpha \geq 0$ and $0 < \nu^2 < 1$.

We begin by introducing

$$C(k) = k^{1-\nu} \int_0^1 h(t) J_0\left(t\sqrt{k^2-\alpha^2}\right) dt. \tag{1.2.46}$$

Then,

$$\begin{aligned}
\int_0^\infty & C(k) J_{-\nu}(kx)\, dk \\
&= \int_0^1 h(t) \left[\int_0^\infty k^{1-\nu} J_{-\nu}(kx) J_0\left(t\sqrt{k^2-\alpha^2}\right) dk\right] dt && (1.2.47)\\
&= \int_0^1 \frac{h(t)}{x^\nu} \left(\frac{\sqrt{t^2-x^2}}{i\alpha}\right)^{\nu-1} J_{\nu-1}\left(i\alpha\sqrt{t^2-x^2}\right) H(t-x)\, dt && (1.2.48)\\
&= 0, && (1.2.49)
\end{aligned}$$

because $0 \leq t \leq 1 < x < \infty$. Therefore, the introduction of the integral definition for $C(k)$ results in Equation 1.2.45 being satisfied identically. Turning now to Equation 1.2.44,

$$\int_0^1 h(t) \left[\int_\alpha^\infty k^{1-\nu} J_\nu(kx) J_0\left(t\sqrt{k^2-\alpha^2}\right) dk\right] dt = f(x) \tag{1.2.50}$$

for $0 < x < 1$. Now, if we introduce $\eta^2 = k^2 - \alpha^2$,

$$\begin{aligned}
\int_\alpha^\infty & k^{1-\nu} J_\nu(kx) J_0\left(t\sqrt{k^2-\alpha^2}\right) dk \\
&= \int_0^\infty J_\nu\left(x\sqrt{\eta^2+\alpha^2}\right) J_0(t\eta) \frac{\eta}{\sqrt{(\eta^2+\alpha^2)^\nu}}\, d\eta && (1.2.51)\\
&= \frac{1}{x^\nu} \left(\frac{\sqrt{x^2-t^2}}{-\alpha}\right)^{\nu-1} J_{\nu-1}\left(-\alpha\sqrt{x^2-t^2}\right) H(x-t), && (1.2.52)
\end{aligned}$$

upon applying Equation 1.2.41. Therefore, Equation 1.2.50 simplifies to

$$\int_0^x \frac{h(t)}{x^\nu} \left(\frac{\sqrt{x^2-t^2}}{-\alpha}\right)^{\nu-1} J_{\nu-1}\left(-\alpha\sqrt{x^2-t^2}\right) dt = f(x) \tag{1.2.53}$$

with $0 < x < 1$. Therefore, the solution to the dual integral equations, Equation 1.2.44 and Equation 1.2.45, consists of Equation 1.2.46 and

$$h(t) = \frac{d}{dt}\left[\int_0^t f(x)x^{\nu+1}\left(\frac{\sqrt{t^2-x^2}}{i\alpha}\right)^{-\nu} J_{-\nu}\left(i\alpha\sqrt{t^2-x^2}\right)dx\right] \quad \textbf{(1.2.54)}$$

if $0 < \nu < 1$. This solution also applies to the dual integral equation:

$$\int_\alpha^\infty C(k)J_{1+\nu}(kx)k\,dk = -x^\nu \frac{d}{dx}\left[x^{-\nu}f(x)\right], \qquad 0 < x < 1, \quad \textbf{(1.2.55)}$$

and

$$\int_0^\infty C(k)J_{-1-\nu}(kx)k\,dk = 0, \qquad 1 < x < \infty, \quad \textbf{(1.2.56)}$$

when $-1 < \nu < 0$.

In a similar manner, let us show the dual integral equation

$$\int_\alpha^\infty S(k)J_0(kx)k\,dk = f(x), \qquad 0 < x < 1, \quad \textbf{(1.2.57)}$$

and

$$\int_0^\infty S(k)J_0(kx)(k^2-\alpha^2)^p k\,dk = 0, \qquad 1 < x < \infty, \quad \textbf{(1.2.58)}$$

where $\alpha \geq 0$ and $0 < p^2 < 1$.

We begin by introducing

$$S(k) = \int_0^1 \frac{h(t)}{\left(\sqrt{k^2-\alpha^2}\,\right)^p} J_{-p}\left(t\sqrt{k^2-\alpha^2}\right)dt. \quad \textbf{(1.2.59)}$$

It is straightforward to show that this choice for $S(k)$ satisfies Equation 1.2.58 identically. When we perform an analysis similar to Equation 1.2.50 through Equation 1.2.53, we find that

$$f(x) = \int_0^x \frac{h(t)}{t^p}\left(\frac{\sqrt{x^2-t^2}}{\alpha}\right)^{p-1} J_{p-1}\left(\alpha\sqrt{x^2-t^2}\right)dt \quad \textbf{(1.2.60)}$$

if $0 < p < 1$; or

$$f(x) = \frac{1}{x}\frac{d}{dx}\left[\int_0^x \frac{h(t)}{t^p}\left(\frac{\sqrt{x^2-t^2}}{\alpha}\right)^p J_p\left(\alpha\sqrt{x^2-t^2}\right)dt\right] \quad \textbf{(1.2.61)}$$

if $-1 < p < 0$. To obtain Equation 1.2.61, we integrated Equation 1.2.57 with respect to x which yields

$$\frac{1}{x}\int_0^x f(\xi)\xi\,d\xi = \int_\alpha^\infty S(k)J_1(kx)\,dx. \quad \textbf{(1.2.62)}$$

Applying Equation 1.2.42 and Equation 1.2.43, we have that

$$h(t) = t^p \frac{d}{dt}\left[\int_0^\infty x\, f(x) \left(\frac{\sqrt{t^2 - x^2}}{i\alpha}\right)^{-p} J_{-p}\left(i\alpha\sqrt{t^2-\alpha^2}\right)\, dx\right] \tag{1.2.63}$$

for $0 < p < 1$; and

$$h(t) = t^{1+p} \int_0^t x\, f(x) \left(\frac{\sqrt{t^2 - x^2}}{i\alpha}\right)^{-1-p} J_{-1-p}\left(i\alpha\sqrt{t^2-\alpha^2}\right)\, dx \tag{1.2.64}$$

for $-1 < p < 0$. Therefore, the solution to the dual integral equations Equation 1.2.57 and Equation 1.2.58 has the solution Equation 1.2.59 along with Equation 1.2.63 or Equation 1.2.64 depending on the value of p.

1.3 LEGENDRE POLYNOMIALS

In this book we will encounter special functions whose properties will be repeatedly used to derive important results. This section focuses on Legendre polynomials.

Legendre polynomials[15] are defined by the power series:

$$P_n(x) = \sum_{k=0}^{m} (-1)^k \frac{(2n-2k)!}{2^n k!(n-k)!(n-2k)!} x^{n-2k}, \tag{1.3.1}$$

where $m = n/2$, or $m = (n-1)/2$, depending upon which is an integer. Figure 1.3.1 illustrates the first four Legendre polynomials.

Legendre polynomials were originally developed to satisfy the differential equation

$$(1-x^2)\frac{d^2y}{dx^2} - 2x\frac{dy}{dx} + n(n+1)y = 0, \tag{1.3.2}$$

or

$$\frac{d}{dx}\left[(1-x^2)\frac{dy}{dx}\right] + n(n+1)y = 0, \tag{1.3.3}$$

that arose in the separation-of-variables solution of partial differential equations in spherical coordinates. Several of their properties are given in Table 1.3.1.

---

[15] Legendre, A. M., 1785: Sur l'attraction des sphéroïdes homogènes. *Mém. math. phys. présentés à l'Acad. Sci. pars divers savants*, **10**, 411–434. The best reference on Legendre polynomials is Hobson, E. W., 1965: *The Theory of Spherical and Ellipsoidal Harmonics.* Chelsea Publishing Co., 500 pp.

Table 1.3.1: Some Useful Relationships Involving Legendre Polynomials

---

Rodrigues's formula

$$P_n(x) = \frac{1}{2^n n!}\frac{d^n}{dx^n}(x^2-1)^n$$

Recurrence formulas

$$(n+1)P_{n+1}(x) - (2n+1)xP_n(x) + nP_{n-1}(x) = 0, \qquad n = 1, 2, 3, \ldots$$

$$P'_{n+1}(x) - P'_{n-1}(x) = (2n+1)P_n(x), \qquad n = 1, 2, 3, \ldots$$

Orthogonality condition

$$\int_{-1}^{1} P_n(x)P_m(x)\,dx = \begin{cases} 0, & m \neq n, \\ \dfrac{2}{2n+1}, & m = n. \end{cases}$$

For us, the usefulness in Legendre polynomials arises from the integral representation

$$P_n[\cos(\theta)] = \frac{2}{\pi}\int_0^{\theta} \frac{\cos[(n+\frac{1}{2})x]}{\sqrt{2[\cos(x)-\cos(\theta)]}}\,dx \tag{1.3.4}$$

$$= \frac{2}{\pi}\int_{\theta}^{\pi} \frac{\sin[(n+\frac{1}{2})x]}{\sqrt{2[\cos(\theta)-\cos(x)]}}\,dx. \tag{1.3.5}$$

Equation 1.3.4 is known as the *Mehler formula*.[16]

• **Example 1.3.1**

Let us simplify

$$\frac{1}{\sqrt{2}}\sum_{n=1}^{\infty}[1-\cos(nx)]\left\{P_{n-1}[\cos(t)] - P_n[\cos(t)]\right\},$$

where $0 < x, t < \pi$.

---

[16] Mehler, F. G., 1881: Ueber eine mit den Kugel- und Cylinderfunctionen verwandte Function und ihre Anwendung in der Theorie der Elektricitätsvertheilung. *Math. Ann.*, **18**, 161–194.

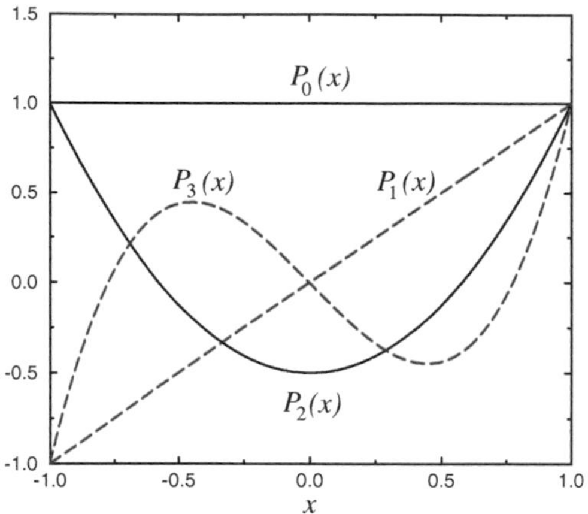

Figure 1.3.1: The first four Legendre functions of the first kind.

We begin by noting that

$$\frac{1}{\sqrt{2}}\sum_{n=1}^{\infty}[1-\cos(nx)]\left\{P_{n-1}[\cos(t)]-P_n[\cos(t)]\right\}$$

$$= \frac{1}{\sqrt{2}}\sum_{n=0}^{\infty} P_n[\cos(t)] - \cos[(n+1)x]P_n[\cos(t)]$$

$$- \frac{1}{\sqrt{2}}\sum_{n=0}^{\infty} P_n[\cos(t)] - \cos(nx)P_n[\cos(t)] \tag{1.3.6}$$

$$= \frac{1}{\sqrt{2}}\sum_{n=0}^{\infty} P_n[\cos(t)]\left\{\cos(nx) - \cos[(n+1)x]\right\} \tag{1.3.7}$$

$$= \sqrt{2}\,\sin\left(\frac{x}{2}\right)\sum_{n=0}^{\infty} P_n[\cos(t)]\sin\left[\frac{(2n+1)x}{2}\right]. \tag{1.3.8}$$

Applying the results from Problem 1, we finally obtain

$$\frac{1}{\sqrt{2}}\sum_{n=1}^{\infty}[1-\cos(nx)]\left\{P_{n-1}[\cos(t)]-P_n[\cos(t)]\right\} = \frac{\sin(x/2)H(x-t)}{\sqrt{\cos(t)-\cos(x)}}. \tag{1.3.9}$$

• Example 1.3.2

Tranter and Cooke[17] used Legendre polynomials to prove that

$$\sum_{n=0}^{\infty} J_{2n+1}(z)\cos\left[\left(n+\tfrac{1}{2}\right)x\right] = \frac{z}{4\sqrt{2}}\int_x^{\pi}\frac{\sin(\eta)J_0[z\sin(\eta/2)]}{\sqrt{\cos(x)-\cos(\eta)}}\,d\eta. \tag{1.3.10}$$

---

[17] Tranter, C. J., and J. C. Cooke, 1973: A Fourier-Neumann series and its application to the reduction of triple cosine series. *Glasgow Math. J.*, **14**, 198–201.

They began by using the formula[18]

$$\sqrt{2}\sum_{n=0}^{\infty} \cos\left[\left(n+\tfrac{1}{2}\right)x\right] P_n[\cos(\eta)] = \begin{cases} 1/\sqrt{\cos(x)-\cos(\eta)}, & 0 \le x < \eta < \pi, \\ 0, & 0 < \eta < x < \pi. \end{cases} \tag{1.3.11}$$

Now,[19]

$$\tfrac{1}{2} z J_0[z\sin(\eta/2)] = \sum_{n=0}^{\infty} (2n+1) P_n[\cos(\eta)] J_{2n+1}(z). \tag{1.3.12}$$

Here the hypergeometric function in Watson's formula is replaced with Legendre polynomials. Multiplying Equation 1.3.12 by $\sin(\eta)/\sqrt{\cos(x)-\cos(\eta)}$, integrating with respect to η from x to π and using the results from Problem 2, we obtain Equation 1.3.10.

In a similar manner, we also have

$$\sum_{n=0}^{\infty} J_{2n+1}(z) \sin\left[\left(n+\tfrac{1}{2}\right)x\right] = \frac{z}{4\sqrt{2}} \int_0^x \frac{\sin(\eta) J_0[z\sin(\eta/2)]}{\sqrt{\cos(\eta)-\cos(x)}}\, d\eta. \tag{1.3.13}$$

Problems

1. Using Equation 1.3.4 and Equation 1.3.5, show that the following generalized Fourier series hold:

$$\frac{H(\theta-t)}{\sqrt{2\cos(t)-2\cos(\theta)}} = \sum_{n=0}^{\infty} P_n[\cos(\theta)] \cos\left[\left(n+\tfrac{1}{2}\right)t\right], \quad 0 \le t, \theta \le \pi,$$

if we use the eigenfunction $y_n(x) = \cos\left[\left(n+\frac{1}{2}\right)x\right]$, $0 < x < \pi$, and $r(x) = 1$; and

$$\frac{H(t-\theta)}{\sqrt{2\cos(\theta)-2\cos(t)}} = \sum_{n=0}^{\infty} P_n[\cos(\theta)] \sin\left[\left(n+\tfrac{1}{2}\right)t\right], \quad 0 \le \theta, t \le \pi,$$

if we use the eigenfunction $y_n(x) = \sin\left[\left(n+\frac{1}{2}\right)x\right]$, $0 < x < \pi$, and $r(x) = 1$.

2. The series given in Problem 1 are also expansions in Legendre polynomials. In that light, show that

$$\int_0^t \frac{P_n[\cos(\theta)]\, \sin(\theta)}{\sqrt{2\cos(\theta)-2\cos(t)}}\, d\theta = \frac{\sin\left[\left(n+\frac{1}{2}\right)t\right]}{n+\frac{1}{2}},$$

---

18 Gradshteyn and Ryzhik, op. cit., Formula 8.927.

19 Watson, G. N., 1966: *A Treatise on the Theory of Bessel Functions.* Cambridge University Press, 804 pp. See Equation (3) in Section 5.21.

and

$$\int_t^\pi \frac{P_n[\cos(\theta)]\sin(\theta)}{\sqrt{2\cos(t)-2\cos(\theta)}}\,d\theta = \frac{\cos\left[\left(n+\frac{1}{2}\right)t\right]}{n+\frac{1}{2}},$$

where $0 < t < \pi$.

3. Using

$$\frac{1}{\sqrt{2}}\sum_{n=0}^{\infty}\{P_{n-1}[\cos(t)] - P_n[\cos(t)]\}\sin(nx)$$
$$= \frac{1}{\sqrt{2}}\sum_{n=0}^{\infty}P_n[\cos(t)]\,\{\sin[(n+1)x] - \sin(nx)\}$$

and the results from Problem 1, show that

$$\frac{1}{\sqrt{2}}\sum_{n=1}^{\infty}\{P_{n-1}[\cos(t)] - P_n[\cos(t)]\}\sin(nx) = \frac{\sin(x/2)H(t-x)}{\sqrt{\cos(x)-\cos(t)}},$$

and

$$\frac{1}{\sqrt{2}}\sum_{n=1}^{\infty}\{P_{n-1}[\cos(t)] + P_n[\cos(t)]\}\sin(nx) = \frac{\cos(x/2)H(x-t)}{\sqrt{\cos(t)-\cos(x)}},$$

provided $0 < x, t < \pi$.

4. The generating function for Legendre polynomials is

$$\left[1 - 2\xi\cos(\theta) + \xi^2\right]^{-1/2} = \sum_{n=0}^{\infty}P_n[\cos(\theta)]\xi^n, \qquad |\xi| < 1.$$

Setting $\xi = h$, then $\xi = -h$, and finally adding and subtracting the resulting equations, show[20] that

$$2\sum_{n=0}^{\infty}P_{2n+m}[\cos(\theta)]\,h^{2n+m} = \left[1 - 2h\cos(\theta) + h^2\right]^{-1/2}$$
$$+ (-1)^m\left[1 + 2h\cos(\theta) + h^2\right]^{-1/2}$$

for $m = 0$ and 1.

---

[20] For its use, see Minkov, I. M., 1963: Electrostatic field of a sectional spherical capacitor. *Sov. Tech. Phys.*, **7**, 1041–1043.

Multiplying the previous equation by $\sqrt{h}$, setting $h = e^{it}$, and then separating the real and imaginary parts, show that

$$2\sum_{n=0}^{\infty} P_{2n+m}[\cos(\theta)] \sin\left[\left(2n+m+\tfrac{1}{2}\right)t\right] = \frac{H(t-\theta)}{\sqrt{2\left[\cos(\theta)-\cos(t)\right]}},$$

and

$$2\sum_{n=0}^{\infty} P_{2n+m}[\cos(\theta)] \cos\left[\left(2n+m+\tfrac{1}{2}\right)t\right] = \frac{H(\theta-t)}{\sqrt{2\left[\cos(t)-\cos(\theta)\right]}} + \frac{(-1)^m}{\sqrt{2\left[\cos(\theta)+\cos(t)\right]}},$$

where $0 < t, \theta < \pi/2$.

1.4 BESSEL FUNCTIONS

The solution to the classic differential equation

$$r^2\frac{d^2y}{dr^2} + r\frac{dy}{dr} + (\lambda^2 r^2 - n^2)y = 0, \qquad \textbf{(1.4.1)}$$

commonly known as *Bessel's equation of order n with a parameter λ*, is

$$y(r) = c_1 J_n(\lambda r) + c_2 Y_n(\lambda r), \qquad \textbf{(1.4.2)}$$

where $J_n(\cdot)$ and $Y_n(\cdot)$ are nth order Bessel functions of the first and second kind, respectively. Figure 1.4.1 illustrates $J_0(x)$, $J_1(x)$, $J_2(x)$, and $J_3(x)$ while in Figure 1.4.2 $Y_0(x)$, $Y_1(x)$, $Y_2(x)$, and $Y_3(x)$ are graphed. Bessel functions have been exhaustively studied and a vast literature now exists on them.[21] The Bessel function $J_n(z)$ is an entire function, has no complex zeros, and has an infinite number of real zeros symmetrically located with respect to the point $z = 0$, which is itself a zero if $n > 0$. All of the zeros are simple, except the point $z = 0$, which is a zero of order n if $n > 0$. On the other hand, $Y_n(z)$ is analytic in the complex plane with a branch cut along the segment $(-\infty, 0]$ and becomes infinite as $z \to 0$.

Considerable insight into the nature of Bessel functions is gained from their asymptotic expansions. These expansions are

$$J_n(z) \sim \left(\frac{2}{\pi z}\right)^{1/2} \cos\left(z - \tfrac{1}{2}n\pi - \tfrac{1}{4}\pi\right), \quad |\arg(z)| \le \pi - \epsilon, \quad |z| \to \infty, \quad \textbf{(1.4.3)}$$

[21] *The* standard reference is Watson, op. cit.

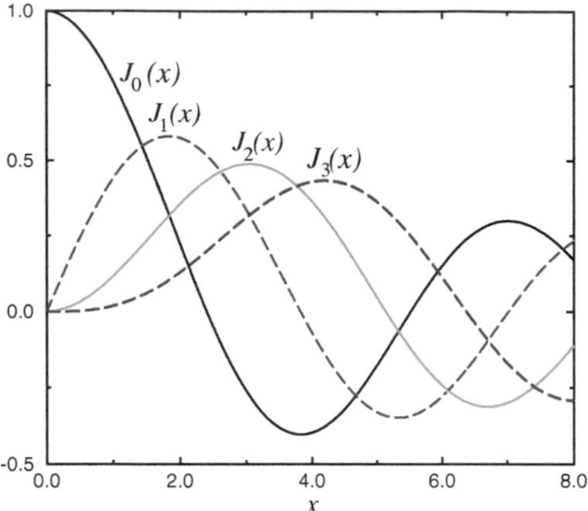

Figure 1.4.1: The first four Bessel functions of the first kind over $0 \leq x \leq 8$.

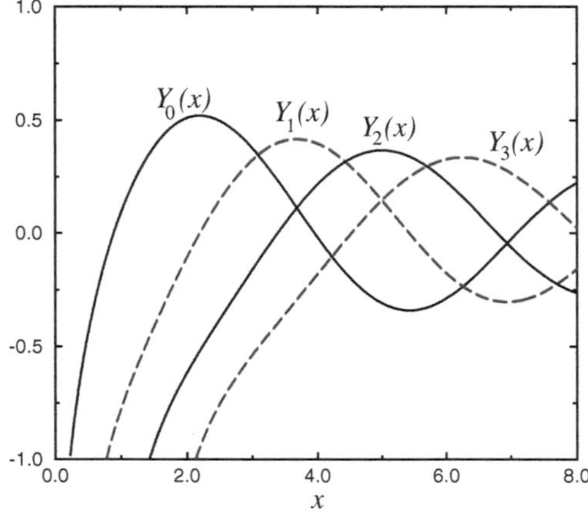

Figure 1.4.2: The first four Bessel functions of the second kind over $0 \leq x \leq 8$.

and

$$Y_n(z) \sim \left(\frac{2}{\pi z}\right)^{1/2} \sin\left(z - \tfrac{1}{2}n\pi - \tfrac{1}{4}\pi\right), \quad |\arg(z)| \leq \pi - \epsilon, \quad |z| \to \infty, \quad \textbf{(1.4.4)}$$

where ϵ denotes an arbitrarily small positive number. Therefore, Bessel functions are sinusoidal in nature and decay as $z^{-1/2}$.

A closely related differential equation is the *modified Bessel equation*

$$r^2 \frac{d^2 y}{dr^2} + r\frac{dy}{dr} - (\lambda^2 r^2 + \nu^2) y = 0. \qquad \textbf{(1.4.5)}$$

Its general solution is

$$y(r) = c_1 I_\nu(\lambda r) + c_2 K_\nu(\lambda r), \qquad \textbf{(1.4.6)}$$

where $I_\nu(\cdot)$ and $K_\nu(\cdot)$ are ν-th order, modified Bessel functions of the first and second kind, respectively. Both $I_\nu(\cdot)$ and $K_\nu(\cdot)$ are analytic functions

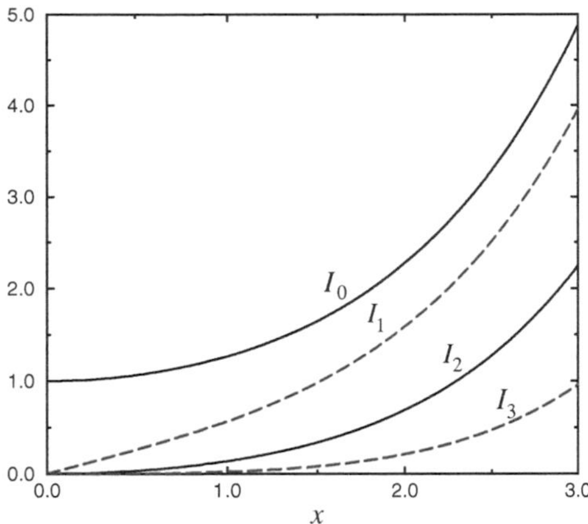

Figure 1.4.3: The first four modified Bessel functions of the first kind over $0 \le x \le 3$.

in the complex z-plane provided that we introduce a branch cut along the segment $(-\infty, 0]$. As $z \to 0$, $K_\nu(z)$ becomes infinite. Figure 1.4.3 illustrates $I_0(x)$, $I_1(x)$, $I_2(x)$, and $I_3(x)$ while in Figure 1.4.4 $K_0(x)$, $K_1(x)$, $K_2(x)$, and $K_3(x)$ are graphed.

Turning to the zeros, $I_\nu(z)$ has zeros that are purely imaginary for $\nu > -1$. On the other hand, $K_\nu(z)$ has no zeros in the region $|\arg(z)| \le \pi/2$. In the remaining portion of the cut z-plane, it has a finite number of zeros.

The modified Bessel functions also have asymptotic representations:

$$I_\nu(z) \sim \frac{e^z}{\sqrt{2\pi z}} + \frac{e^{-z\pm\pi(\nu+\frac{1}{2})}}{\sqrt{2\pi z}}, \quad |\arg(z)| \le \pi - \epsilon, \quad |z| \to \infty, \tag{1.4.7}$$

and

$$K_n(z) \sim \frac{\pi e^{-z}}{\sqrt{2\pi z}}, \quad |\arg(z)| \le \pi - \epsilon, \quad |z| \to \infty, \tag{1.4.8}$$

where we chose the plus sign if $\Im(z) > 0$, and the minus sign if $\Im(z) < 0$. Note that $K_n(z)$ decrease exponentially as $x \to \infty$, while $I_n(z)$ increases exponentially as $x \to \infty$ *and* $x \to -\infty$.

Having introduced Bessel functions, we now turn to some of their useful properties. Repeatedly in the following chapters, we will encounter them in improper integrals. Here, we list some of the most common ones:[22]

$$\int_0^t \frac{x\, J_0(kx)}{\sqrt{t^2 - x^2}}\, dx = \frac{\sin(kt)}{k}, \qquad k > 0, \tag{1.4.9}$$

$$\int_0^\infty e^{-\alpha x} J_\nu(\beta x)\, x^\nu\, dx = \frac{(2\beta)^\nu \Gamma\left(\nu + \frac{1}{2}\right)}{\sqrt{\pi}\,(\alpha^2 + \beta^2)^{\nu+\frac{1}{2}}}, \quad \Re(\nu) > -\tfrac{1}{2},\ \Re(\alpha) > |\Im(\beta)|, \tag{1.4.10}$$

[22] Gradshteyn and Ryzhik, op. cit., Formulas 6.554.2, 6.623.1, 6.623.2, 6.671.1, 6.671.2, and 6.699.8.

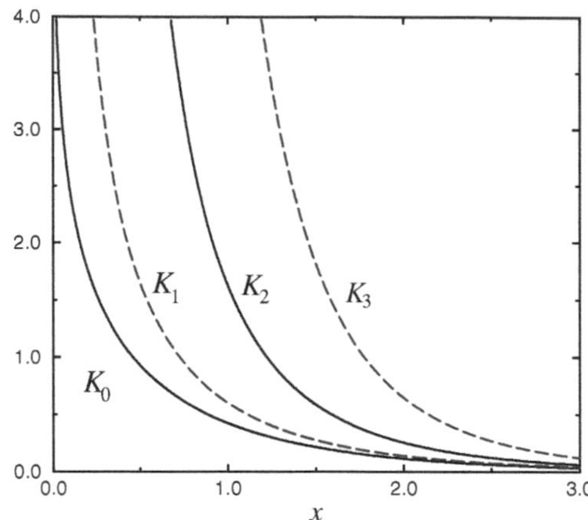

Figure 1.4.4: The first four modified Bessel functions of the second kind over $0 \le x \le 3$.

$$\int_0^\infty e^{-\alpha x} J_\nu(\beta x)\, x^\nu \, dx = \frac{(2\beta)^\nu \Gamma\left(\nu + \frac{1}{2}\right)}{\sqrt{\pi}\,(\alpha^2 + \beta^2)^{\nu + \frac{1}{2}}}, \quad \Re(\nu) > -\tfrac{1}{2},\ \Re(\alpha) > |\Im(\beta)|, \tag{1.4.11}$$

$$\int_0^\infty e^{-\alpha x} J_\nu(\beta x)\, x^{\nu+1} \, dx = \frac{2\alpha(2\beta)^\nu \Gamma\left(\nu + \frac{3}{2}\right)}{\sqrt{\pi}\,(\alpha^2 + \beta^2)^{\nu + \frac{3}{2}}}, \quad \Re(\nu) > -1,\ \Re(\alpha) > |\Im(\beta)|, \tag{1.4.12}$$

$$\int_0^\infty \sin(bx) J_\nu(ax)\, dx = \begin{cases} \dfrac{\sin\left[\nu \arcsin(b/a)\right]}{\sqrt{a^2 - b^2}}, & b < a, \\ \infty \text{ or } 0, & b = a, \quad \Re(\nu) > -2, \\ \dfrac{a^\nu \cos(\nu\pi/2)}{\sqrt{b^2 - a^2}\left(b + \sqrt{b^2 - a^2}\right)^\nu}, & b > a, \end{cases} \tag{1.4.13}$$

and

$$\int_0^\infty \cos(bx) J_\nu(ax)\, dx = \begin{cases} \dfrac{\cos\left[\nu \arcsin(b/a)\right]}{\sqrt{a^2 - b^2}}, & b < a, \\ \infty \text{ or } 0, & b = a, \quad \Re(\nu) > -1. \\ \dfrac{-a^\nu \sin(\nu\pi/2)}{\sqrt{b^2 - a^2}\left(b + \sqrt{b^2 - a^2}\right)^\nu}, & b > a, \end{cases} \tag{1.4.14}$$

Just as sines and cosines can be used in a Fourier series to reexpress an arbitrary function $f(x)$, Bessel functions can also be used to create the Fourier-Bessel series

$$f(x) = \sum_{k=1}^\infty A_k J_n(\mu_k x), \qquad 0 < x < L, \tag{1.4.15}$$

Table 1.4.1: Some Useful Relationships Involving Bessel Functions of Integer Order

$$J_{n-1}(z) + J_{n+1}(z) = \frac{2n}{z} J_n(z), \qquad n = 1, 2, 3, \ldots$$

$$J_{n-1}(z) - J_{n+1}(z) = 2J_n'(z), \; n = 1, 2, 3, \ldots; \quad J_0'(z) = -J_1(z)$$

$$\frac{d}{dz}\left[z^n J_n(z)\right] = z^n J_{n-1}(z), \qquad n = 1, 2, 3, \ldots$$

$$\frac{d}{dz}\left[z^{-n} J_n(z)\right] = -z^{-n} J_{n+1}(z), \qquad n = 0, 1, 2, 3, \ldots$$

$$I_{n-1}(z) - I_{n+1}(z) = \frac{2n}{z} I_n(z), \qquad n = 1, 2, 3, \ldots$$

$$I_{n-1}(z) + I_{n+1}(z) = 2I_n'(z), \; n = 1, 2, 3, \ldots; \quad I_0'(z) = I_1(z)$$

$$K_{n-1}(z) - K_{n+1}(z) = -\frac{2n}{z} K_n(z), \qquad n = 1, 2, 3, \ldots$$

$$K_{n-1}(z) + K_{n+1}(z) = -2K_n'(z), \; n = 1, 2, 3, \ldots; \quad K_0'(z) = -K_1(z)$$

$$J_n(ze^{m\pi i}) = e^{nm\pi i} J_n(z)$$

$$I_n(ze^{m\pi i}) = e^{nm\pi i} I_n(z)$$

$$K_n(ze^{m\pi i}) = e^{-mn\pi i} K_n(z) - m\pi i \frac{\cos(mn\pi)}{\cos(n\pi)} I_n(z)$$

$$I_n(z) = e^{-n\pi i/2} J_n(ze^{\pi i/2}), \qquad -\pi < \arg(z) \le \pi/2$$

$$I_n(z) = e^{3n\pi i/2} J_n(ze^{-3\pi i/2}), \qquad \pi/2 < \arg(z) \le \pi$$

where

$$A_k = \frac{1}{C_k} \int_0^L x f(x) J_n(\mu_k x)\, dx. \tag{1.4.16}$$

The values of μ_k and C_k depend on the condition at $x = L$. If $J_n(\mu_k L) = 0$, then

$$C_k = \tfrac{1}{2} L^2 J_{n+1}^2(\mu_k L). \tag{1.4.17}$$

On the other hand, if $J_n'(\mu_k L) = 0$, then

$$C_k = \frac{\mu_k^2 L^2 - n^2}{2\mu_k^2} J_n^2(\mu_k L). \tag{1.4.18}$$

Finally, if $\mu_k J_n'(\mu_k L) = -hJ_n(\mu_k L)$,

$$C_k = \frac{\mu_k^2 L^2 - n^2 + h^2 L^2}{2\mu_k^2} J_n^2(\mu_k L). \quad \textbf{(1.4.19)}$$

All of the preceding results must be slightly modified when $n = 0$ and the boundary condition is $J_0'(\mu_k L) = 0$ or $\mu_k J_1(\mu_k L) = 0$. For this case, Equation 1.4.15 now reads

$$f(x) = A_0 + \sum_{k=1}^{\infty} A_k J_0(\mu_k x), \quad \textbf{(1.4.20)}$$

where the equation for finding A_0 is

$$A_0 = \frac{2}{L^2} \int_0^L f(x)\, x\, dx, \quad \textbf{(1.4.21)}$$

and Equation 1.4.16 and Equation 1.4.18 with $n = 0$ give the remaining coefficients.

• **Example 1.4.1**

Let us expand $f(x) = x$, $0 < x < 1$, in the series

$$f(x) = \sum_{k=1}^{\infty} A_k J_1(\mu_k x), \quad \textbf{(1.4.22)}$$

where μ_k denotes the kth zero of $J_1(\mu)$. From Equation 1.4.16 and Equation 1.4.18,

$$A_k = \frac{2}{J_2^2(\mu_k)} \int_0^1 x^2 J_1(\mu_k x)\, dx. \quad \textbf{(1.4.23)}$$

However, from the third line of Table 1.4.1,

$$\frac{d}{dx}\left[x^2 J_2(x)\right] = x^2 J_1(x), \quad \textbf{(1.4.24)}$$

if $n = 2$. Therefore, Equation 1.4.23 becomes

$$A_k = \left.\frac{2x^2 J_2(x)}{\mu_k^3 J_2^2(\mu_k)}\right|_0^{\mu_k} = \frac{2}{\mu_k J_2(\mu_k)}, \quad \textbf{(1.4.25)}$$

and the resulting expansion is

$$x = 2\sum_{k=1}^{\infty} \frac{J_1(\mu_k x)}{\mu_k J_2(\mu_k)}, \qquad 0 \le x < 1. \quad \textbf{(1.4.26)}$$

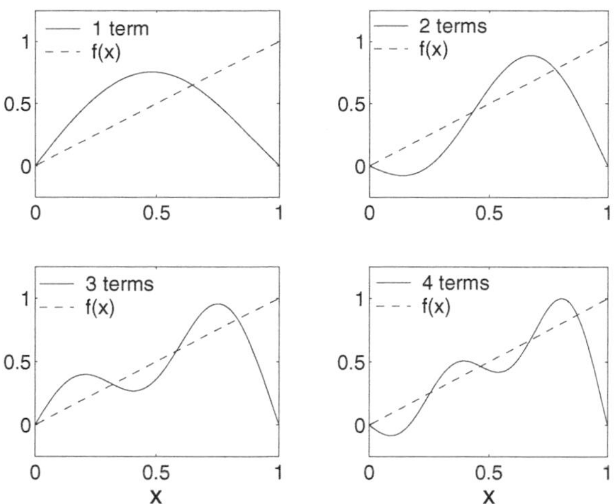

Figure 1.4.5: The Fourier-Bessel series representation, Equation 1.4.26, for $f(x) = x$, $0 < x < 1$, when we truncate the series so that it includes only the first, first two, first three, and first four terms.

Figure 1.4.5 shows the Fourier-Bessel expansion of $f(x) = x$ in truncated form when we only include one, two, three, and four terms.

• **Example 1.4.2**

Let us expand the function $f(x) = x^2$, $0 < x < 1$, in the series

$$f(x) = \sum_{k=1}^{\infty} A_k J_0(\mu_k x), \tag{1.4.27}$$

where μ_k denotes the kth positive zero of $J_0(\mu)$. From Equation 1.4.16 and Equation 1.4.17,

$$A_k = \frac{2}{J_1^2(\mu_k)} \int_0^1 x^3 J_0(\mu_k x)\, dx. \tag{1.4.28}$$

If we let $t = \mu_k x$, the integration Equation 1.4.28 becomes

$$A_k = \frac{2}{\mu_k^4 J_1^2(\mu_k)} \int_0^{\mu_k} t^3 J_0(t)\, dt. \tag{1.4.29}$$

We now let $u = t^2$ and $dv = tJ_0(t)\, dt$ so that integration by parts results in

$$A_k = \frac{2}{\mu_k^4 J_1^2(\mu_k)} \left[t^3 J_1(t)\Big|_0^{\mu_k} - 2\int_0^{\mu_k} t^2 J_1(t)\, dt \right] \tag{1.4.30}$$

$$= \frac{2}{\mu_k^4 J_1^2(\mu_k)} \left[\mu_k^3 J_1(\mu_k) - 2\int_0^{\mu_k} t^2 J_1(t)\, dt \right], \tag{1.4.31}$$

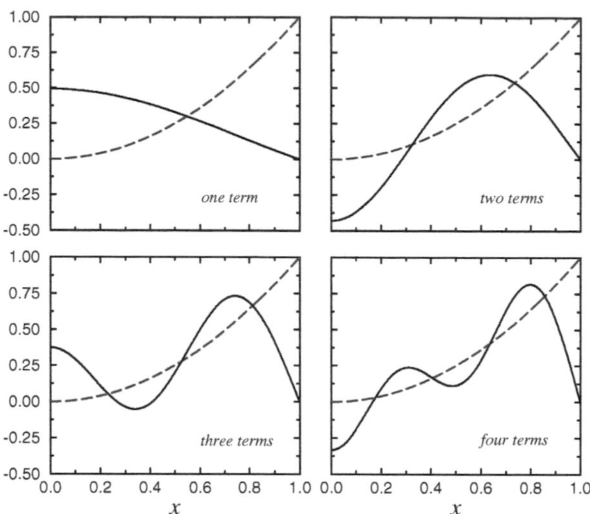

Figure 1.4.6: The Fourier-Bessel series representation, Equation 1.4.37, for $f(x) = x^2$, $0 < x < 1$, when we truncate the series so that it includes only the first, first two, first three, and first four terms.

because $v = tJ_1(t)$ from the fourth line of Table 1.4.1. If we integrate by parts once more, we find that

$$A_k = \frac{2}{\mu_k^4 J_1^2(\mu_k)} \left[\mu_k^3 J_1(\mu_k) - 2\mu_k^2 J_2(\mu_k) \right] \tag{1.4.32}$$

$$= \frac{2}{J_1^2(\mu_k)} \left[\frac{J_1(\mu_k)}{\mu_k} - \frac{2J_2(\mu_k)}{\mu_k^2} \right]. \tag{1.4.33}$$

However, from the first line of Table 1.4.1 with $n = 1$,

$$J_1(\mu_k) = \tfrac{1}{2}\mu_k \left[J_2(\mu_k) + J_0(\mu_k) \right], \tag{1.4.34}$$

or

$$J_2(\mu_k) = \frac{2J_1(\mu_k)}{\mu_k}, \tag{1.4.35}$$

because $J_0(\mu_k) = 0$. Therefore,

$$A_k = \frac{2(\mu_k^2 - 4)J_1(\mu_k)}{\mu_k^3 J_1^2(\mu_k)}, \tag{1.4.36}$$

and

$$x^2 = 2\sum_{k=1}^{\infty} \frac{(\mu_k^2 - 4)J_0(\mu_k x)}{\mu_k^3 J_1(\mu_k)}, \qquad 0 < x < 1. \tag{1.4.37}$$

Figure 1.4.6 shows the representation of x^2 by the Fourier-Bessel series given by Equation 1.4.37 when we truncate it so that it includes only one, two, three, or four terms.

We have repeatedly noted how Bessel functions are very similar in nature to sine and cosine. This suggests that for axisymmetric problems we could develop an alternative transform based on Bessel functions. To examine this possibility, let us write the two-dimensional Fourier transform pair as

$$f(x,y) = \frac{1}{2\pi}\int_{-\infty}^{\infty}\int_{-\infty}^{\infty} F(k,\ell)\, e^{i(kx+\ell y)}\, dk\, d\ell, \tag{1.4.38}$$

where

$$F(k,\ell) = \frac{1}{2\pi}\int_{-\infty}^{\infty}\int_{-\infty}^{\infty} f(x,y)\, e^{-i(kx+\ell y)}\, dx\, dy. \tag{1.4.39}$$

Consider now the special case where $f(x,y)$ is only a function of $r = \sqrt{x^2+y^2}$, so that $f(x,y) = g(r)$. Then, changing to polar coordinates through the substitution $x = r\cos(\theta)$, $y = r\sin(\theta)$, $k = \rho\cos(\varphi)$ and $\ell = \rho\sin(\varphi)$, we have that

$$kx + \ell y = r\rho[\cos(\theta)\cos(\varphi) + \sin(\theta)\sin(\varphi)] = r\rho\cos(\theta - \varphi) \tag{1.4.40}$$

and

$$dA = dx\, dy = r\, dr\, d\theta. \tag{1.4.41}$$

Therefore, the integral in Equation 1.4.39 becomes

$$F(k,\ell) = \frac{1}{2\pi}\int_0^{\infty}\int_0^{2\pi} g(r)\, e^{-ir\rho\cos(\theta-\varphi)} r\, dr\, d\theta \tag{1.4.42}$$

$$= \frac{1}{2\pi}\int_0^{\infty} r\, g(r)\left[\int_0^{2\pi} e^{-ir\rho\cos(\theta-\varphi)}\, d\theta\right] dr. \tag{1.4.43}$$

If we introduce $\lambda = \theta - \varphi$, the integral within the square brackets can be evaluated as follows:

$$\int_0^{2\pi} e^{-ir\rho\cos(\theta-\varphi)}\, d\theta = \int_{-\varphi}^{2\pi-\varphi} e^{-ir\rho\cos(\lambda)}\, d\lambda \tag{1.4.44}$$

$$= \int_0^{2\pi} e^{-ir\rho\cos(\lambda)}\, d\lambda \tag{1.4.45}$$

$$= 2\pi J_0(\rho r). \tag{1.4.46}$$

Equation 1.4.45 is equivalent to Equation 1.4.44 because the integral of a periodic function over one full period is the same regardless of where the integration begins. Equation 1.4.46 follows from the integral definition of the Bessel function.[23] Therefore,

$$F(k,\ell) = \int_0^{\infty} r\, g(r)\, J_0(\rho r)\, dr. \tag{1.4.47}$$

[23] Watson, op. cit., Section 2.2, Equation 5.

Finally, because Equation 1.4.47 is clearly a function of $\rho = \sqrt{k^2+\ell^2}$, $F(k,\ell) = G(\rho)$ and

$$G(\rho) = \int_0^\infty r\, g(r)\, J_0(\rho r)\, dr. \tag{1.4.48}$$

Conversely, if we begin with Equation 1.4.38, make the same substitution, and integrate over the $k\ell$-plane, we have

$$f(x,y) = g(r) = \frac{1}{2\pi}\int_0^\infty \int_0^{2\pi} F(k,\ell)\, e^{ir\rho\cos(\theta-\varphi)}\rho\, d\rho\, d\varphi \tag{1.4.49}$$

$$= \frac{1}{2\pi}\int_0^\infty \rho\, G(\rho)\left[\int_0^{2\pi} e^{ir\rho\cos(\theta-\varphi)}\, d\varphi\right] d\rho \tag{1.4.50}$$

$$= \int_0^\infty \rho\, G(\rho)\, J_0(\rho r)\, d\rho. \tag{1.4.51}$$

Thus, we obtain the result that if $\int_0^\infty |F(r)|\, dr$ exists, then

$$g(r) = \int_0^\infty \rho\, G(\rho)\, J_0(\rho r)\, d\rho, \tag{1.4.52}$$

where

$$G(\rho) = \int_0^\infty r\, g(r)\, J_0(\rho r)\, dr. \tag{1.4.53}$$

Taken together, Equation 1.4.52 and Equation 1.4.53 constitute the *Hankel transform pair for Bessel function of order 0*, named after the German mathematician Hermann Hankel (1839–1873). The function $G(\rho)$ is called "the Hankel transform of $g(r)$."

For asymmetric problems, we can generalize our results to Hankel transforms of order ν

$$F(k) = \int_0^\infty f(r) J_\nu(kr)\, r\, dr, \qquad -\tfrac{1}{2} < \nu, \tag{1.4.54}$$

and its inverse[24]

$$f(r) = \int_0^\infty F(k) J_\nu(kr)\, k\, dk. \tag{1.4.55}$$

Finally, it is well known that $\sin(\theta)$ and $\cos(\theta)$ can be expressed in terms of the complex exponential $e^{\theta i}$ and $e^{-\theta i}$. In the case of Bessel functions $J_\nu(z)$ and $Y_\nu(z)$, the corresponding representations are called "Hankel functions" (or Bessel functions of the third kind)

$$H_\nu^{(1)}(z) = J_\nu(z) + iY_\nu(z) \quad \text{and} \quad H_\nu^{(2)}(z) = J_\nu(z) - iY_\nu(z), \tag{1.4.56}$$

---

[24] Ibid., Section 14.4. See also MacRobert, T. M., 1931: Fourier integrals. *Proc. R. Soc. Edinburgh, Ser. A*, **51**, 116–126.

Table 1.4.2: Some Useful Recurrence Relations for Hankel Functions

$$\frac{d}{dx}\left[x^n H_n^{(p)}(x)\right] = x^n H_{n-1}^{(p)}(x),\ n=1,2,\ldots;\ \frac{d}{dx}\left[H_0^{(p)}(x)\right] = -H_1^{(p)}(x)$$

$$\frac{d}{dx}\left[x^{-n} H_n^{(p)}(x)\right] = -x^{-n} H_{n+1}^{(p)}(x), \qquad n=0,1,2,3,\ldots$$

$$H_{n-1}^{(p)}(x) + H_{n+1}^{(p)}(x) = \frac{2n}{x} H_n^{(p)}(x), \qquad n=1,2,3,\ldots$$

$$H_{n-1}^{(p)}(x) - H_{n+1}^{(p)}(x) = 2\frac{dH_n^{(p)}(x)}{dx}, \qquad n=1,2,3,\ldots$$

where ν is arbitrary and z is any point in the z-plane cut along the segment $(-\infty, 0]$. The analogy is most clearly seen in the asymptotic expansions for these functions:

$$H_\nu^{(1)}(z) = \sqrt{\frac{2}{\pi z}}e^{i(z-\nu\pi/2-\pi/4)} \quad \text{and} \quad H_\nu^{(2)}(z) = \sqrt{\frac{2}{\pi z}}e^{-i(z-\nu\pi/2-\pi/4)} \tag{1.4.57}$$

for $|z| \to \infty$ with $|\arg(z)| \le \pi - \epsilon$, where ϵ is an arbitrarily small positive number. These functions are linearly independent solutions of

$$\frac{d^2u}{dz^2} + \frac{1}{z}\frac{du}{dz} + \left(1 - \frac{\nu^2}{z^2}\right)u = 0. \tag{1.4.58}$$

Table 1.4.2 gives additional relationships involving Hankel functions.

Problems

1. Show that

$$1 = 2\sum_{k=1}^{\infty} \frac{J_0(\mu_k x)}{\mu_k J_1(\mu_k)}, \qquad 0 \le x < 1,$$

where μ_k is the kth positive root of $J_0(\mu) = 0$.

2. Show that

$$\frac{1-x^2}{8} = \sum_{k=1}^{\infty} \frac{J_0(\mu_k x)}{\mu_k^3 J_1(\mu_k)}, \qquad 0 \le x \le 1,$$

where μ_k is the kth positive root of $J_0(\mu) = 0$.

3. Show that

$$4x - x^3 = -16 \sum_{k=1}^{\infty} \frac{J_1(\mu_k x)}{\mu_k^3 J_0(2\mu_k)}, \qquad 0 \le x \le 2,$$

where μ_k is the kth positive root of $J_1(2\mu) = 0$.

4. Show that

$$x^3 = 2 \sum_{k=1}^{\infty} \frac{(\mu_k^2 - 8) J_1(\mu_k x)}{\mu_k^3 J_2(\mu_k)}, \qquad 0 \le x \le 1,$$

where μ_k is the kth positive root of $J_1(\mu) = 0$.

5. Show that

$$x = 2 \sum_{k=1}^{\infty} \frac{\mu_k J_2(\mu_k) J_1(\mu_k x)}{(\mu_k^2 - 1) J_1^2(\mu_k)}, \qquad 0 \le x \le 1,$$

where μ_k is the kth positive root of $J_1'(\mu) = 0$.

6. Show that

$$1 - x^4 = 32 \sum_{k=1}^{\infty} \frac{(\mu_k^2 - 4) J_0(\mu_k x)}{\mu_k^5 J_1(\mu_k)}, \qquad 0 \le x \le 1,$$

where μ_k is the kth positive root of $J_0(\mu) = 0$.

7. Show that

$$1 = 2\alpha L \sum_{k=1}^{\infty} \frac{J_0(\mu_k x/L)}{(\mu_k^2 + \alpha^2 L^2) J_0(\mu_k)}, \qquad 0 \le x \le L,$$

where μ_k is the kth positive root of $\mu J_1(\mu) = \alpha L J_0(\mu)$.

8. Using the relationship[25]

$$\int_0^a J_\nu(\alpha r) J_\nu(\beta r)\, r\, dr = \frac{a\beta J_\nu(\alpha a) J_\nu'(\beta a) - a\alpha J_\nu(\beta a) J_\nu'(\alpha a)}{\alpha^2 - \beta^2},$$

show that

$$\frac{J_0(bx) - J_0(ba)}{J_0(ba)} = \frac{2b^2}{a} \sum_{k=1}^{\infty} \frac{J_0(\mu_k x)}{\mu_k(\mu_k^2 - b^2) J_1(\mu_k a)}, \qquad 0 \le x \le a,$$

[25] Ibid., Section 5.11, Equation 8.

where μ_k is the kth positive root of $J_0(\mu a) = 0$ and b is a constant.

9. Using Equation 1.4.9, show that

$$\frac{H(t-x)}{\sqrt{t^2-x^2}} = 2\sum_{k=1}^{\infty} \frac{\sin(\mu_k t) J_0(\mu_k x)}{\mu_k J_1^2(\mu_k)}, \qquad 0 < x < 1, \quad 0 < t \le 1,$$

where μ_k is the kth positive root of $J_0(\mu) = 0$.

10. Using Equation 1.4.9, show[26] that

$$\frac{H(a-x)}{\sqrt{a^2-x^2}} = \frac{2}{b}\sum_{n=1}^{\infty} \frac{\sin(\mu_n a/b) J_0(\mu_n x/b)}{\mu_n J_0^2(\mu_n)}, \qquad 0 \le x < b,$$

where $a < b$ and μ_n is the nth positive root of $J_0'(\mu) = -J_1(\mu) = 0$.

11. Given the definite integral[27]

$$\int_0^a \cos(cx)\, J_0\left(b\sqrt{a^2-x^2}\right) dx = \frac{\sin\left(a\sqrt{b^2+c^2}\right)}{\sqrt{b^2+c^2}}, \qquad 0 < b,$$

show that

$$\frac{\cosh\left(b\sqrt{t^2-x^2}\right)}{\sqrt{t^2-x^2}} H(t-x) = \frac{2}{a^2}\sum_{k=1}^{\infty} \frac{\sin\left(t\sqrt{\mu_k^2-b^2}\right) J_0(\mu_k x)}{\sqrt{\mu_k^2-b^2}\, J_1^2(\mu_k a)},$$

where $0 < x < a$ and μ_k is the kth positive root of $J_0(\mu a) = 0$.

12. Using the integral definition of the Bessel function[28] for $J_1(z)$:

$$J_1(z) = \frac{2}{\pi}\int_0^1 \frac{t\,\sin(zt)}{\sqrt{1-t^2}}\, dt, \qquad 0 < z,$$

show that

$$\frac{x}{t\sqrt{t^2-x^2}} H(t-x) = \frac{\pi}{L}\sum_{n=1}^{\infty} J_1\left(\frac{n\pi t}{L}\right) \sin\left(\frac{n\pi x}{L}\right), \qquad 0 \le x < L.$$

---

[26] For an application of this result, see Wei, X. X., and K. T. Chau, 2000: Finite solid circular cylinders subjected to arbitrary surface load. Part II–Application to double-punch test. *Int. J. Solids Struct.*, **37**, 5733–5744.

[27] Gradshteyn and Ryzhik, op. cit., Formula 6.677.6.

[28] Ibid., Formula 3.753.5.

Hint: Treat this as a Fourier half-range sine expansion.

13. Show that

$$\delta(x-b) = \frac{2b}{a^2} \sum_{k=1}^{\infty} \frac{J_0(\mu_k b/a) J_0(\mu_k x/a)}{J_1^2(\mu_k)}, \qquad 0 \le x, b < a,$$

where μ_k is the kth positive root of $J_0(\mu) = 0$.

14. Show that

$$\frac{\delta(x)}{2\pi x} = \frac{1}{\pi a^2} \sum_{k=1}^{\infty} \frac{J_0(\mu_k x/a)}{J_1^2(\mu_k)}, \qquad 0 \le x < a,$$

where μ_k is the kth positive root of $J_0(\mu) = 0$.

15. Using integral tables,[29] show[30] that

$$u(r,z) = \frac{2Ar}{\pi} \int_0^\infty e^{-kz} J_0(kr) \sin(ka) \frac{dk}{k} + \frac{2Ar}{\pi} \int_0^\infty e^{-kz} J_2(kr) \sin(ka) \frac{dk}{k} - \frac{4Aa}{\pi} \int_0^\infty e^{-kz} J_1(kr) \cos(ka) \frac{dk}{k}$$

satisfies the partial differential equation

$$\frac{\partial^2 u}{\partial r^2} + \frac{1}{r}\frac{\partial u}{\partial r} - \frac{u}{r^2} + \frac{\partial^2 u}{\partial z^2} = 0, \qquad 0 \le r < \infty, \quad 0 < z < \infty,$$

with the mixed boundary conditions

$$\lim_{r \to 0} |u(r,z)| < \infty, \quad \lim_{r \to \infty} u(r,z) \to 0, \qquad 0 < z < \infty,$$

$$\lim_{z \to \infty} u(r,z) \to 0, \qquad 0 \le r < \infty,$$

$$\begin{cases} u(r,0) = Ar, & 0 \le r < a, \\ u_z(r,0) = 0, & a < r < \infty. \end{cases}$$

[29] Ibid., Formulas 6.671.1, 6.671.2, 6.693.1, and 6.693.2.

[30] Ray, M., 1936: Application of Bessel functions to the solution of problem of motion of a circular disk in viscous liquid. *Philos. Mag., Ser. 7*, **21**, 546–564.

Chapter 2
Historical Background

Mixed boundary value problems arose, as did other boundary value problems, during the development of mathematical physics in the nineteenth and twentieth centuries. In this chapter we highlight its historical development by examining several of the classic problems.

2.1 NOBILI'S RINGS

Our story begins in 1824 when the Italian physicists Leopoldo Nobili (1784–1835) experimented with the chemical reactions that occur in voltaic cells. In a series of papers,[1] he described the appearance of a series of rainbow colored rings on a positively charged silver plate coated with a thin electrolytic solution when a negatively charge platinum wire was introduced into the solution. Although Riemann[2] formulated a mathematical theory of

[1] Nobili, L., 1827: Ueber ein neue Klasse von electro-chemischen Erscheinungen. *Ann. Phys., Folge 2*, **9**, 183–184; Nobili, L., 1827: Ueber ein neue Klasse von electro-chemischen Erscheinungen. *Ann. Phys., Folge 2*, **10**, 392–424.

[2] Riemann, G. F. B., 1855: Zur Theorie der Nobili'schen Farbenringe. *Ann. Phys., Folge 2*, **95**, 130–139. A more accessible copy of this paper can be found in Riemann, B., 1953: *Gesammelte Mathematische Werke.* Dover, 558 pp. See pp. 55–66. Archibald (Archibald, T., 1991: Riemann and the theory of electrical phenomena: Nobili's ring. *Centaurus*, **34**, 247–271.) has given the background, as well as an analysis, of Riemann's work.

Figure 2.2.1: Despite his short life, (Georg Friedrich) Bernhard Riemann's (1826–1866) mathematical work contained many imaginative and profound concepts. It was in his doctoral thesis on complex function theory (1851) that he introduced the Cauchy-Riemann differential equations. Riemann's later work dealt with the definition of the integral and the foundations of geometry and non-Euclidean (elliptic) geometry. (Portrait courtesy of Photo AKG, London.)

this phenomenon, it is the formulation[3] by Weber that has drawn the greater attention. Both Riemann and Weber formulated the problem as the solution of Laplace's equation over an infinite strip of thickness $2a$:

$$\frac{1}{r}\frac{\partial}{\partial r}\left(r\frac{\partial u}{\partial r}\right) + \frac{\partial^2 u}{\partial z^2} = 0, \qquad 0 \le r < \infty, \quad -a < z < a, \tag{2.1.1}$$

subject to the boundary conditions

$$\lim_{r\to 0} |u(r,z)| < \infty, \qquad \lim_{r\to\infty} u(r,z) \to 0, \qquad -a < z < a, \tag{2.1.2}$$

and

$$\begin{cases} u(r,\pm a) = \pm U_0, & r < c, \\ u_z(r,\pm a) = 0, & r > c. \end{cases} \tag{2.1.3}$$

Although Weber's explanation of Nobili's rings was essentially accepted for a century, its correct solution is relatively recent.[4] Using separation of

---

[3] Weber, H., 1873: Ueber die Besselschen Functionen und ihre Anwendung auf die Theorie der elektrischen Ströme. *J. Reine Angew. Math.*, **65**, 75–105. See Section 6.

[4] Laporte, O., and R. G. Fowler, 1967: Weber's mixed boundary value problem in electrodynamics. *J. Math. Phys.*, **8**, 518–522.

variables or transform methods, the solution to Equation 2.1.1 and Equation 2.1.2 can be written

$$u(r,z) = \frac{2}{\pi}\int_0^\infty A(k)\frac{\sinh(kz)}{\cosh(ka)}J_0(kr)\,dk. \tag{2.1.4}$$

Substituting Equation 2.1.4 into Equation 2.1.3, we obtain the dual integral equations

$$\frac{2}{\pi}\int_0^\infty \tanh(ka)A(k)J_0(kr)\,dk = U_0, \qquad 0 \le r < c, \tag{2.1.5}$$

and

$$\frac{2}{\pi}\int_0^\infty kA(k)J_0(kr)\,dk = 0, \qquad c < r < \infty. \tag{2.1.6}$$

Following Laporte and Fowler, we set

$$A(k) = \int_0^c f(\xi)\cos(k\xi)\,d\xi. \tag{2.1.7}$$

Why have we introduced this definition of $A(k)$? If we substitute Equation 2.1.7 into the condition for $r > c$ in Equation 2.1.3, we have that

$$u_z(r,a) = \frac{2}{\pi}\int_0^\infty k\left[\int_0^c f(\xi)k\cos(k\xi)\,d\xi\right]J_0(kr)\,dk, \tag{2.1.8}$$

or

$$u_z(r,a) = \frac{2}{\pi}\int_0^\infty \left[f(c)\sin(kc) - \int_0^c f'(\xi)\sin(k\xi)\,d\xi\right]J_0(kr)\,dk. \tag{2.1.9}$$

Because

$$\int_0^\infty \sin(kt)J_0(kr)\,dk = \frac{H(t-r)}{\sqrt{t^2-r^2}}, \tag{2.1.10}$$

$$u_z(r,a) = \frac{2}{\pi}\int_0^c f'(\xi)\left[\int_0^\infty \sin(k\xi)J_0(kr)\,dk\right]d\xi, \tag{2.1.11}$$

since $r > c$. Finally, by applying Equation 2.1.10 to the integral within the square brackets of Equation 2.1.11, we see that $u_z(r,a) = 0$ if $r > c$. Thus, $A(k)$, defined by Equation 2.1.7, identically satisfies the boundary condition along $z = \pm a$ and $r > c$.

Turning to the boundary condition for $0 \le r < c$, we have that

$$\int_0^\infty \tanh(ka)\left[\int_0^c f(\xi)\cos(k\xi)\,d\xi\right]J_0(kr)\,dk = \frac{\pi U_0}{2}, \tag{2.1.12}$$

or

$$\int_0^c f(\xi)\left[\int_0^\infty \left(1 - \frac{2}{1+e^{2ak}}\right)\cos(k\xi)J_0(k\rho)\,dk\right]d\xi = \frac{\pi U_0}{2}. \tag{2.1.13}$$

We now multiply both sides of Equation 2.1.13 by $\rho\,d\rho/\sqrt{r^2-\rho^2}$ and integrate from 0 to r. Interchanging the order of integration, we obtain that

$$\begin{aligned}\int_0^c f(\xi)&\left\{\int_0^\infty \cos(k\xi)\left[\int_0^r \frac{\rho\,J_0(k\rho)}{\sqrt{r^2-\rho^2}}\,d\rho\right]dk\right\}d\xi\\ &-2\int_0^c f(\xi)\left\{\int_0^\infty \frac{\cos(k\xi)}{1+e^{2ka}}\left[\int_0^r \frac{\rho\,J_0(k\rho)}{\sqrt{r^2-\rho^2}}\,d\rho\right]dk\right\}d\xi\\ &= \frac{\pi U_0}{2}\int_0^r \frac{\rho}{\sqrt{r^2-\rho^2}}\,d\rho.\end{aligned} \tag{2.1.14}$$

Because

$$\int_0^r \frac{\eta\,J_0(k\eta)}{\sqrt{r^2-\eta^2}}\,d\eta = \frac{\sin(kr)}{k}, \tag{2.1.15}$$

Equation 2.1.14 simplifies to

$$\begin{aligned}\int_0^c f(\xi)&\left[\int_0^\infty \cos(k\xi)\sin(kr)\,\frac{dk}{k}\right]d\xi\\ &-2\int_0^c f(\xi)\left[\int_0^\infty \frac{\cos(k\xi)\sin(kr)}{1+e^{2ka}}\,\frac{dk}{k}\right]d\xi = \frac{\pi U_0 r}{2}.\end{aligned} \tag{2.1.16}$$

Upon taking the derivative of Equation 2.1.16 with respect to r, we find that

$$\begin{aligned}\int_0^c f(\xi)&\left[\int_0^\infty \cos(k\xi)\cos(kr)\,dk\right]d\xi\\ &-2\int_0^c f(\xi)\left[\int_0^\infty \frac{\cos(k\xi)\cos(kr)}{1+e^{2ka}}\,dk\right]d\xi = \frac{\pi U_0}{2}.\end{aligned} \tag{2.1.17}$$

Finally, noting that

$$\frac{2}{\pi}\int_0^\infty \cos(k\xi)\cos(kr)\,dk = \delta(\xi - r), \tag{2.1.18}$$

we find that $f(\xi)$ is given by

$$f(\xi) - \frac{4}{\pi}\int_0^c f(\tau)\left[\int_0^\infty \frac{\cos(k\xi)\cos(k\tau)}{1+e^{2ka}}\,dk\right]d\tau = U_0. \tag{2.1.19}$$

In the limit of $c \to 0$, the integral in Equation 2.1.19 vanishes and $f(\xi) = U_0$. From Equation 2.1.7, $A(k) = U_0\sin(kc)/k$ and we recover Weber's solution:

$$u(r,z) = \frac{2}{\pi}\int_0^\infty \sin(kc)\frac{\sinh(kz)}{\cosh(ka)}J_0(kr)\,\frac{dk}{k}. \tag{2.1.20}$$

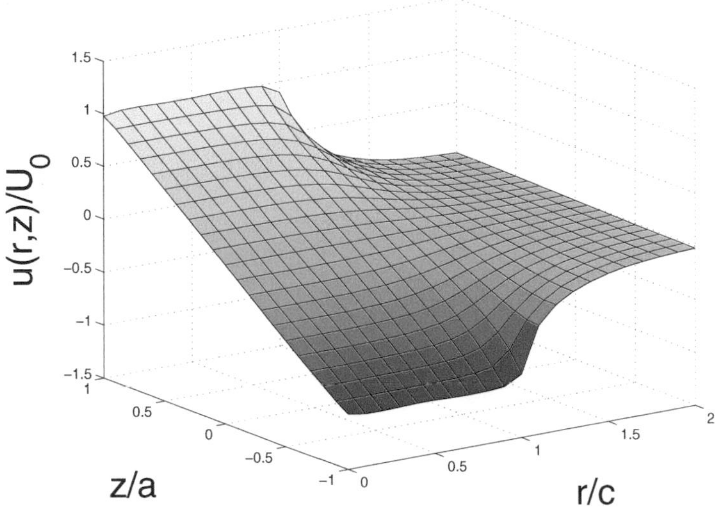

Figure 2.1.2: The solution $u(r,z)/U_0$ to the mixed boundary value problem governed by Equation 2.1.1 through Equation 2.1.3 when $c/a = 2$.

In Chapter 4, we discuss how to numerically solve Equation 2.1.19 for a finite value of c, so that we can evaluate $A(k)$ and then compute $u(r,z)$. Presently, in Fig. 2.1.2, we merely illustrate $u(r,z)$ when $c/a = 2$.

2.2 DISC CAPACITOR

In the previous section we showed that the mathematical explanation of Nobili's rings involved solving Laplace's equation with mixed boundary conditions. The next important problem involving mixed boundary value problems was the capacitance of two oppositely charged, parallel, circular coaxial discs. Kirchhoff[5] used conformal mapping to find the approximate solution for the capacitance of a circular in free space. Since then, a number of authors have refined his calculation. The most general one is by Carlson and Illman[6] and we now present their analysis.

The problem may be expressed mathematically as the potential problem

$$\frac{1}{r}\frac{\partial}{\partial r}\left(r\frac{\partial u}{\partial r}\right) + \frac{\partial^2 u}{\partial z^2} = 0, \qquad 0 \le r < \infty, \quad -\infty < z < \infty, \tag{2.2.1}$$

---

[5] Kirchhoff, G., 1877: Zur Theorie des Kondensators. *Monatsb. Deutsch. Akad. Wiss. Berlin*, 144–162.

[6] Carlson, G. T., and B. L. Illman, 1994: The circular disk parallel plate capacitor. *Am. J. Phys.*, **62**, 1099–1105.

Figure 2.2.1: Gustav Robert Kirchhoff's (1824–1887) most celebrated contributions to physics are the joint founding with Robert Bunsen of the science of spectroscopy, and the discovery of the fundamental law of electromagnetic radiation. Kirchhoff's work on light coincides with his final years as a professor of theoretical physics at Berlin. (Portrait taken from the frontispiece of Kirchhoff, G., 1882: *Gesammelte Abhandlungen*. J. A. Barth, 641 pp.)

subject to the boundary conditions

$$\lim_{r\to 0} |u(r,z)| < \infty, \qquad \lim_{r\to\infty} u(r,z) \to 0, \qquad -\infty < z < \infty, \tag{2.2.2}$$

$$\lim_{|z|\to\infty} u(r,z) \to 0, \qquad 0 \le r < \infty, \tag{2.2.3}$$

$$\begin{cases} u(r,0) = V_1, & 0 \le r < 1, \\ u_z(r,0) = 0, & 1 < r < \infty, \end{cases} \tag{2.2.4}$$

and

$$\begin{cases} u(r,L) = V_2, & 0 \le r < 1, \\ u_z(r,L) = 0, & 1 < r < \infty. \end{cases} \tag{2.2.5}$$

Using separation of variables or transform methods, the solution to Equation 2.2.1 is

$$u(r,z) = \int_0^\infty \left[A(k)e^{-k|z|} + B(k)e^{-k|z-L|}\right] J_0(kr)\,\frac{dk}{k}. \tag{2.2.6}$$

This solution also satisfies the boundary conditions given by Equation 2.2.2 and Equation 2.2.3. Substituting Equation 2.2.6 into Equation 2.2.4 and Equation 2.2.5, we obtain the dual integral equations for $A(k)$ and $B(k)$:

$$\int_0^\infty \left[A(k) + B(k)e^{-kL}\right] J_0(kr)\,\frac{dk}{k} = V_1, \qquad 0 \le r < 1, \tag{2.2.7}$$

$$\int_0^\infty \left[A(k)e^{-kL} + B(k)\right] J_0(kr)\,\frac{dk}{k} = V_2, \qquad 0 \le r < 1, \tag{2.2.8}$$

$$\int_0^\infty A(k)J_0(kr)\,dk = 0, \qquad 1 < r < \infty, \tag{2.2.9}$$

and

$$\int_0^\infty B(k)J_0(kr)\,dk = 0, \qquad 1 < r < \infty. \tag{2.2.10}$$

To solve Equation 2.2.9 and Equation 2.2.10, we introduce

$$A(k) = \frac{2k}{\pi}\int_0^1 f(t)\cos(kt)\,dt, \tag{2.2.11}$$

and

$$B(k) = \frac{2k}{\pi}\int_0^1 g(t)\cos(kt)\,dt. \tag{2.2.12}$$

To show that the $A(k)$ given by Equation 2.2.11 satisfies Equation 2.2.9, we evaluate

$$\int_0^\infty A(k)J_0(kr)\,dk = \frac{2}{\pi}\int_0^\infty \left[\int_0^1 f(t)k\cos(kt)\,dt\right] J_0(kr)\,dk \tag{2.2.13}$$

$$= \frac{2}{\pi}\int_0^\infty f(t)\sin(kt)\Big|_0^1 J_0(kr)\,dk$$

$$- \frac{2}{\pi}\int_0^1 f'(t)\left[\int_0^\infty \sin(kt)J_0(kr)\,dk\right] dt \tag{2.2.14}$$

$$= \frac{2}{\pi}\int_0^\infty f(1)\sin(k)J_0(kr)\,dk$$

$$- \frac{2}{\pi}\int_0^1 f'(t)\left[\int_0^\infty \sin(kt)J_0(kr)\,dk\right] dt \tag{2.2.15}$$

$$= 0, \tag{2.2.16}$$

because $r > 1$ and we used Equation 1.4.13. A similar demonstration holds for $B(k)$.

We now turn to the solution of Equation 2.2.7. Substituting Equation 2.2.11 and Equation 2.2.12 into Equation 2.2.7, we find that

$$\int_0^1 f(t)\left[\int_0^\infty \cos(kt)J_0(kr)\,dk\right]dt + \int_0^1 g(t)\left[\int_0^\infty e^{-kL}\cos(kt)J_0(kr)\,dk\right]dt = \frac{\pi V_1}{2}. \tag{2.2.17}$$

Using Equation 1.4.14, we can evaluate the first integral and obtain

$$h(r) + \int_0^1 g(\tau)\left[\int_0^\infty e^{-kL}\cos(k\tau)J_0(kr)\,dk\right]d\tau = \frac{\pi V_1}{2}, \tag{2.2.18}$$

where

$$h(r) = \int_0^r \frac{f(t)}{\sqrt{r^2 - t^2}}\,dt. \tag{2.2.19}$$

If we now multiply Equation 2.2.18 by $2r\,dr/\left(\pi\sqrt{t^2 - r^2}\right)$, integrate from 0 to t, and then taking the derivative with respect to t, we have

$$\begin{aligned} f(t) + \int_0^1 g(\tau)\left[\frac{2}{\pi}\frac{d}{dt}\left\{\int_0^t \left[\int_0^\infty e^{-kL}\cos(k\tau)J_0(kr)\,dk\right]\frac{r\,dr}{\sqrt{t^2 - r^2}}\right\}\right]d\tau \\ = V_1\frac{d}{dt}\left[\int_0^t \frac{r\,dr}{\sqrt{t^2 - r^2}}\right] = -V_1\frac{d}{dt}\left[\sqrt{t^2 - r^2}\Big|_0^t\right] = V_1, \end{aligned} \tag{2.2.20}$$

since the solution to Equation 2.2.19 is

$$f(t) = \frac{2}{\pi}\frac{d}{dt}\left[\int_0^t \frac{r\,h(r)}{\sqrt{t^2 - r^2}}\,dr\right]. \tag{2.2.21}$$

Defining

$$K(t,\tau) = \frac{2}{\pi}\int_0^\infty e^{-kL}\cos(k\tau)\frac{d}{dt}\left[\int_0^t \frac{rJ_0(kr)}{\sqrt{t^2 - r^2}}\,dr\right]dk, \tag{2.2.22}$$

we substitute Equation 1.4.9 into Equation 2.2.22 and $K(t,\tau)$ simplifies to

$$K(t,\tau) = \frac{2}{\pi}\int_0^\infty e^{-kL}\cos(k\tau)\cos(kt)\,dk \tag{2.2.23}$$

$$= \frac{L}{\pi}\left[\frac{1}{L^2 + (t+\tau)^2} + \frac{1}{L^2 + (t-\tau)^2}\right]. \tag{2.2.24}$$

Therefore, we can write Equation 2.2.20 as

$$f(t) + \int_0^1 K(t,\tau)g(\tau)\,d\tau = V_1. \tag{2.2.25}$$

By using Equation 2.2.8, we can show in a similar manner that

$$g(t) + \int_0^1 K(t,\tau) f(\tau)\, d\tau = V_2. \tag{2.2.26}$$

To evaluate $u(r, z)$ in terms of $f(t)$ and $g(t)$, we substitute Equation 2.2.11 and Equation 2.2.12 into Equation 2.2.6 and find that

$$u(r,z) = \frac{2}{\pi} \int_0^1 \left\{ \int_0^\infty \left[f(t) e^{-k|z|} + g(t) e^{-k|z-L|} \right] \cos(kt) J_0(kr)\, dk \right\} dt \tag{2.2.27}$$

$$= \frac{2}{\pi} \Re \left\{ \int_0^1 \left[\frac{f(t)}{\sqrt{r^2 + (|z| - it)^2}} + \frac{g(t)}{\sqrt{r^2 + (|z-L| + it)^2}} \right] dt \right\}, \tag{2.2.28}$$

where we used Equation 1.4.10.

Figure 2.2.2 illustrates $u(r, z)$ when $L = 1$ and $V_1 = -V_2 = 1$. In Chapter 4 we will discuss the numerical procedure used to solve Equation 2.2.25 and Equation 2.2.26 for specific values of L, V_1, and V_2. We used Simpson's rule to evaluate Equation 2.2.28.

Two special cases of Equation 2.2.25 through Equation 2.2.28 are of historical note. Hafen[7] solved the disc capacitor problem when the electrodes are located at $z = \pm h$ and have a radius of a. The electrode at $z = h$ has the potential of $V_1 = 1$ while the electrode at $z = -h$ has the potential $V_2 = \pm 1$. He obtained

$$u(r,z) = \frac{2}{\pi} \int_0^a \left[\int_0^\infty \left(e^{-k|z-h|} \pm e^{-k|z+h|} \right) \cos(kt) J_0(kr)\, dk \right] f(t)\, dt, \tag{2.2.29}$$

where $f(t)$ is given by

$$f(t) = 1 \mp \frac{2h}{\pi} \int_{-a}^a \frac{f(\tau)}{(t-\tau)^2 + 4h^2}\, d\tau. \tag{2.2.30}$$

Hafen did not present any numerical computations.

The second special case occurs when we set $V_1 = -V_2 = V_0$. For this special case, Equation 2.2.25 and Equation 2.2.26 have the solution $f(t) = -g(t) = V_0 h(t)$, where $h(t)$ is given by

$$h(t) = 1 + \int_0^1 K(t,\tau) h(\tau)\, d\tau. \tag{2.2.31}$$

---

[7] Hafen, M., 1910: Studien über einige Probleme der Potentialtheorie. *Math. Ann.*, **69**, 517–537. See Section 3.

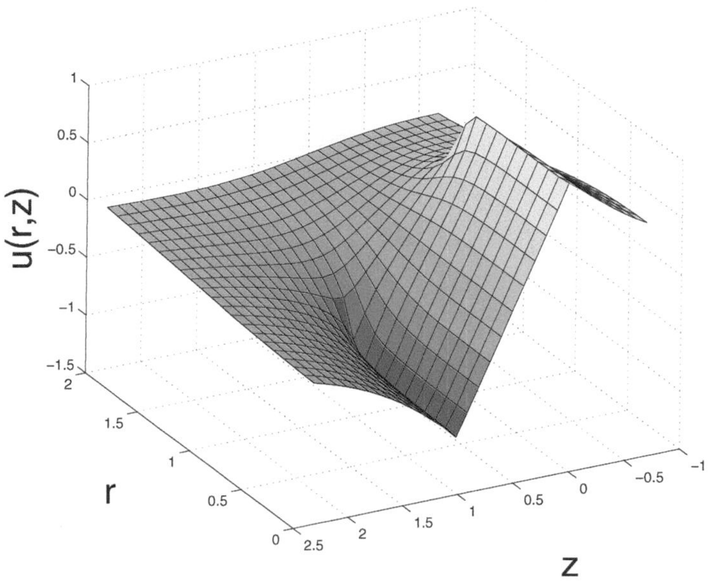

Figure 2.2.2: The solution $u(r,z)$ to the mixed boundary value problem governed by Equation 2.2.1 through Equation 2.2.5 when $L=1$ and $V_1=-V_2=1$.

Love[8] obtained Equation 2.2.31, "Love's equation," by solving Laplace's equation using spheroidal coordinates; Cooke[9] employed Hankel transforms as we did. Regardless of how Equation 2.2.31 is obtained, it requires numerical methods to solve it. During the 1950s, the computational resources were inadequate to handle the case of small L and Cooke[10] developed an approximate method to treat this case. Presently these considerations are of purely historical interest because finite difference methods[11] can be used to find the potential to high accuracy.

2.3 ANOTHER ELECTROSTATIC PROBLEM

The earliest mixed boundary value problem that led to dual Fourier series

---

[8] Love, E. R., 1949: The electrostatic field of two equal circular conducting disks. *Quart. J. Mech. Appl. Math.*, **2**, 428–451.

[9] Cooke, J. C., 1956: A solution of Tranter's dual integral equations problem. *Quart. J. Mech. Appl. Math.*, **9**, 103–110.

[10] Cooke, J. C., 1958: The coaxial circular disc problem. *Z. Angew. Math. Mech.*, **38**, 349–357.

[11] Sloggett, G. J., N. G. Barton, and S. J. Spencer, 1986: Fringing fields in disc capacitors. *J. Phys., Ser. A*, **19**, 2725–2736.

occurred while solving the potential problem:

$$\frac{1}{r}\frac{\partial}{\partial r}\left(r\frac{\partial u}{\partial r}\right) + \frac{1}{r^2}\frac{\partial^2 u}{\partial \theta^2} = 0, \qquad 0 \le r < \infty, \quad 0 < \theta < 2\pi, \tag{2.3.1}$$

subject to the boundary conditions

$$\lim_{r\to 0} |u(r,\theta)| < \infty, \qquad \lim_{r\to\infty} u(r,\theta) \to -Vr\sin(\theta), \qquad 0 < \theta < 2\pi, \tag{2.3.2}$$

$$u(1^-,\theta) = u(1^+,\theta), \qquad 0 < \theta < 2\pi, \tag{2.3.3}$$

and

$$\begin{cases} \kappa u_r(1^-,\theta) = u_r(1^+,\theta), & 0 < \theta < \pi/2, 3\pi/2 < \theta < 2\pi, \\ u(1,\theta) = 0, & \pi/2 < \theta < 3\pi/2, \end{cases} \tag{2.3.4}$$

where 1^- and 1^+ denote points slightly inside and outside of the circle $r = 1$, respectively.

We begin by using the technique of separation of variables. This yields the potential

$$u(r,\theta) = -Vr\sin(\theta) + \sum_{n=1}^{\infty} C_n r^n \sin(n\theta), \qquad 0 < r < 1, \tag{2.3.5}$$

and

$$u(r,\theta) = -Vr\sin(\theta) + \sum_{n=1}^{\infty} C_n r^{-n} \sin(n\theta), \qquad 1 < r < \infty. \tag{2.3.6}$$

This potential satisfies not only Equation 2.3.1, but also the boundary conditions given by Equations 2.3.2 and 2.3.3.

We must next satisfy the mixed boundary value condition given by Equation 2.3.4. Noting the symmetry about the x-axis, we find that we must only consider that $0 < \theta < \pi$. Turning to the $0 < \theta < \pi/2$ case, we substitute Equation 2.3.5 and Equation 2.3.6 into Equation 2.3.4 and obtain

$$(1+\kappa)\sum_{n=1}^{\infty} n\, C_n \sin(n\theta) = (\kappa - 1)V\sin(\theta). \tag{2.3.7}$$

By integrating Equation 2.3.7 with respect to θ, we find that

$$\sum_{n=1}^{\infty} C_n \cos(n\theta) = \frac{\kappa - 1}{\kappa + 1} V\cos(\theta), \qquad 0 < \theta < \pi/2. \tag{2.3.8}$$

Here, C_0 is the constant of integration.

Let us now turn to the boundary condition for $\pi/2 < \theta < \pi$. Substituting Equation 2.3.5 or Equation 2.3.6 into Equation 2.3.4, we find that

$$\sum_{n=1}^{\infty} C_n \sin(n\theta) = V \sin(\theta). \tag{2.3.9}$$

In summary then, we solved our mixed boundary value problem, provided that the C_n's satisfy Equation 2.3.8 and Equation 2.3.9.

It was the dual Fourier series given by Equation 2.3.8 and Equation 2.3.9 that motivated W. M. Shepherd[12] in the 1930s to study dual Fourier series of the form

$$\cos(m\theta) = \sum_{n=0}^{\infty} A_n \cos(n\theta), \qquad 0 < \theta < \pi/2, \tag{2.3.10}$$

and

$$-\sin(m\theta) = \sum_{n=1}^{\infty} A_n \sin(n\theta), \qquad \pi/2 < \theta < \pi, \tag{2.3.11}$$

where m is a positive integer. He considered two cases: If m is an even integer, $2k$, he proved that

$$A_{2n} = (-1)^{n+k} \frac{2k[n][k]}{2k+2n}, \tag{2.3.12}$$

and

$$A_{2n+1} = (-1)^{n+k} \frac{2k[n][k]}{2n-2k+1}, \tag{2.3.13}$$

where

$$[n] = \frac{1 \cdot 3 \cdot 5 \cdots (2n-1)}{2 \cdot 4 \cdot 6 \cdots 2n} \quad \text{and} \quad [0] = 1. \tag{2.3.14}$$

On the other hand, if m is an odd integer, $2k+1$, then

$$A_{2n+1} = (-1)^{n+k+1} \frac{(2k+1)[n][k]}{2k+2n+2}, \tag{2.3.15}$$

and

$$A_{2n} = (-1)^{n+k+1} \frac{(2k+1)[n][k]}{2n-2k-1}. \tag{2.3.16}$$

How can we solve Equation 2.3.7 and Equation 2.3.8 by using Shepherd's results? The difficulty is the $(\kappa-1)/(\kappa+1)$ term in Equation 2.3.8. To

---

[12] Shepherd, W. M., 1937: On trigonometrical series with mixed conditions. *Proc. London Math. Soc., Ser. 2*, **43**, 366–375.

circumvent this difficulty, we use Equation 2.3.10 to rewrite Equation 2.3.8 as follows:

$$\sum_{n=1}^{\infty} C_n \cos(n\theta) = \frac{\kappa V}{\kappa+1}\cos(\theta) - \frac{V}{\kappa+1}\cos(\theta) \tag{2.3.17}$$

$$= \frac{\kappa V}{\kappa+1}\cos(\theta) - \frac{V}{\kappa+1}\sum_{n=1}^{\infty} A_n \cos(n\theta), \tag{2.3.18}$$

or

$$\sum_{n=1}^{\infty}\left[C_n + \frac{V}{\kappa+1}A_n\right]\cos(n\theta) = \frac{\kappa V}{\kappa+1}\cos(\theta). \tag{2.3.19}$$

Similarly, we rewrite Equation 2.3.9 as follows:

$$\sum_{n=1}^{\infty} C_n \sin(n\theta) = \frac{\kappa V}{\kappa+1}\sin(\theta) + \frac{V}{\kappa+1}\sin(\theta) \tag{2.3.20}$$

$$= \frac{\kappa V}{\kappa+1}\sin(\theta) - \frac{V}{\kappa+1}\sum_{n=1}^{\infty} A_n \sin(n\theta), \tag{2.3.21}$$

or

$$\sum_{n=1}^{\infty}\left[C_n + \frac{V}{\kappa+1}A_n\right]\sin(n\theta) = \frac{\kappa V}{\kappa+1}\sin(\theta). \tag{2.3.22}$$

The A_n's are given by Equation 2.3.15 and Equation 2.3.16 with $k = 0$. Using either Equation 2.3.19 or 2.3.22, we equate the coefficients of each harmonic and find that

$$C_1 = \frac{2\kappa+1}{2(\kappa+1)}V, \qquad C_2 = -\frac{V}{2(\kappa+1)}, \tag{2.3.23}$$

and

$$C_{2n-1} = -C_{2n} = (-1)^{n+1}\frac{1\cdot 3\cdots(2n-3)V}{2\cdot 4\cdots 2n}. \tag{2.3.24}$$

Figure 2.3.1 illustrates the solution to Equation 2.3.1 through Equation 2.3.4 when $\kappa = 6$.

2.4 GRIFFITH CRACKS

During the 1920s, A. A. Griffith (1893–1963) sought to explain why a nonductile material, such as glass, ruptures. An important aspect of his work is the assumption that a large number of small cracks exist in the interior of the solid body. Whether these "Griffith cracks" spread depends upon the distribution of the stresses about the crack. Our interest in computing this stress field lies in the fact that many fracture problems contain mixed boundary conditions.

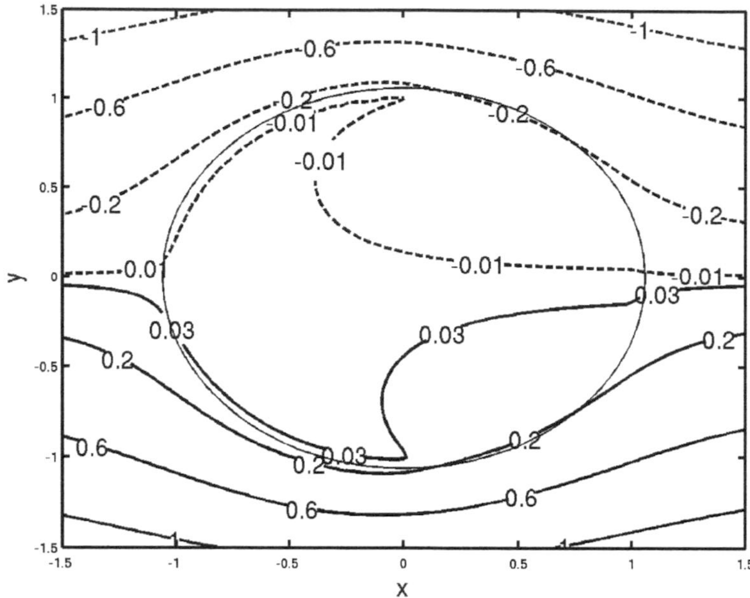

Figure 2.3.1: The solution $u(r,\theta)/V$ to the mixed boundary value problem governed by Equation 2.3.1 through Equation 2.3.4 when $\kappa = 6$.

Modeling Griffith cracks is not easy. Therefore, early models made several assumptions. The first one is that the body is in a state of *plane stress.* Under this condition, the body is idealized to a long right cylinder and then acted upon by external forces which are so arranged that the component of the displacement in the direction of the cylinder vanishes. The remaining components remain constant along the length of the cylinder. Computing the balance of forces on an infinitesimally small element, we find that

$$\frac{\partial \sigma_x}{\partial x} + \frac{\partial \tau_{xy}}{\partial y} + \rho X = 0, \tag{2.4.1}$$

and

$$\frac{\partial \tau_{xy}}{\partial x} + \frac{\partial \sigma_y}{\partial y} + \rho Y = 0, \tag{2.4.2}$$

where σ_x and σ_y are the normal components of the stress while τ_{xy} is the shearing stress. The x- and y-components of the body forces are denoted by X and Y.

These stresses must now be related to the displacements in the x- and y-directions, which we shall denote by $u(x,y)$ and $v(x,y)$, respectively. The stress-strain relations are

$$E\epsilon_x = (1-\sigma^2)\sigma_x - \sigma(1+\sigma)\sigma_y, \tag{2.4.3}$$

$$E\epsilon_y = (1-\sigma^2)\sigma_y - \sigma(1+\sigma)\sigma_x, \tag{2.4.4}$$

and

$$E\gamma_{xy} = 2(1+\sigma)\tau_{xy}, \tag{2.4.5}$$

where

$$\epsilon_x = \frac{\partial u}{\partial x}, \qquad \epsilon_y = \frac{\partial v}{\partial y}, \qquad \gamma_{xy} = \frac{\partial u}{\partial y} + \frac{\partial v}{\partial x}, \tag{2.4.6}$$

and E and σ denote the Young modulus and Poisson ratio of the material, respectively.

The second assumption involves requiring symmetry about the $y = 0$ plane so that our domain can be taken to be the semi-infinite plane $x \geq 0$. Taking the Fourier transform of Equation 2.4.1 through Equation 2.4.6 with $X = Y = 0$, we have that

$$\frac{d\Sigma_x}{dx} + ikT_{xy} = 0, \tag{2.4.7}$$

$$\frac{dT_{xy}}{dx} + ik\Sigma_y = 0, \tag{2.4.8}$$

$$\frac{d^2}{dx^2}\left[\Sigma_y - \sigma(\Sigma_x + \Sigma_y)\right] - k^2\left[\Sigma_x - \sigma(\Sigma_x + \Sigma_y)\right] = 2ik\frac{dT_{xy}}{dx}, \tag{2.4.9}$$

$$\frac{ikE}{1+\sigma}V = \Sigma_y - \sigma(\Sigma_x + \Sigma_y), \tag{2.4.10}$$

and

$$\frac{E}{2(1+\sigma)}\left(\frac{dV}{dx} + ikU\right) = T_{xy}. \tag{2.4.11}$$

Remarkably these five equations can be combined together to yield

$$\left(\frac{d^2}{dx^2} - k^2\right)^2 G(x,k) = 0, \tag{2.4.12}$$

where $G(x,k)$ is one of the quantities Σ_x, Σ_y, T_{xy}, U, or V. The solution of Equation 2.4.12 which tends to zero as $x \to \infty$ is

$$G(x,k) = [A(k) + B(k)x]e^{-|k|x}. \tag{2.4.13}$$

Finally, to evaluate the constants $A(k)$ and $B(k)$, we must state the boundary condition along the crack where $x = 0$ and $|y| < c$. Assuming that the crack occurred because of the external pressure $p(y)$, we have that

$$\tau_{xy}(0,y) = 0, \qquad -\infty < y < \infty, \tag{2.4.14}$$

and

$$\begin{cases} \sigma_x(0,y) = -p(y), & |y| < c, \\ u(0,y) = 0, & |y| > c, \end{cases} \tag{2.4.15}$$

where $p(y)$ is a known even function of y. Because

$$\sigma_x(x,y) = -\frac{1}{2\pi}\int_{-\infty}^{\infty} P(k)\,(1+|k|x)\,e^{-|k|x+iky}\,dk, \tag{2.4.16}$$

$$\sigma_y(x,y) = -\frac{1}{2\pi}\int_{-\infty}^{\infty} P(k)\,(1-|k|x)\,e^{-|k|x+iky}\,dk, \tag{2.4.17}$$

$$\tau_{xy}(x,y) = \frac{ix}{2\pi}\int_{-\infty}^{\infty} kP(k)e^{-|k|x+iky}\,dk, \tag{2.4.18}$$

$$u(x,y) = \frac{1+\sigma}{2\pi E}\int_{-\infty}^{\infty} \frac{P(k)}{|k|}\,[2(1-\sigma)+|k|x\,]\,e^{-|k|x+iky}\,dk, \tag{2.4.19}$$

and

$$v(x,y) = -i\frac{1+\sigma}{2\pi E}\int_{-\infty}^{\infty} \frac{P(k)}{|k|}\,[(1-2\sigma)-|k|x\,]\,e^{-|k|x+iky}\,dk, \tag{2.4.20}$$

Equation 2.4.16 and Equation 2.4.19 yields the dual integral equations

$$\frac{2}{\pi}\int_0^{\infty} P(k)\cos(ky)\,dk = p(y), \qquad 0 \le y < c, \tag{2.4.21}$$

and

$$\int_0^{\infty} P(k)\cos(ky)\frac{dk}{k} = 0, \qquad c < y < \infty, \tag{2.4.22}$$

where

$$P(k) = \int_0^{\infty} p(y)\cos(ky)\,dy. \tag{2.4.23}$$

How do we solve the dual integral equations, Equation 2.4.22 and Equation 2.4.23? We begin by introducing

$$k = \rho/c, \qquad g(\eta) = c\sqrt{\frac{\pi}{2\eta}}\,p(c\eta), \qquad y = c\eta, \tag{2.4.24}$$

and

$$P(\rho/c) = \sqrt{\rho}\,F(\rho), \quad \text{and} \quad \cos(\rho\eta) = \sqrt{\frac{\pi\rho\eta}{2}}\,J_{-\frac{1}{2}}(\rho\eta), \tag{2.4.25}$$

so that Equation 2.4.21 and Equation 2.4.22 become

$$\int_0^{\infty} \rho F(\rho) J_{-\frac{1}{2}}(\rho\eta)\,d\rho = g(\eta), \qquad 0 \le \eta < 1, \tag{2.4.26}$$

Figure 2.4.1: Most of Ian Naismith Sneddon's (1919–2000) life involved the University of Glasgow. Entering at age 16, he graduated with undergraduate degrees in mathematics and physics, returned as a lecturer in physics from 1946 to 1951, and finally accepted the Simon Chair in Mathematics in 1956. In addition to his numerous papers, primarily on elasticity, Sneddon published notable texts on elasticity, mixed boundary value problems, and Fourier transforms. (Portrait from Godfrey Argent Studio, London.)

and

$$\int_0^\infty F(\rho) J_{-\frac{1}{2}}(\rho\eta)\, d\rho = 0, \qquad 1 \le \eta < \infty. \tag{2.4.27}$$

In 1938 Busbridge[13] studied the dual integral equations

$$\int_0^\infty y^\alpha f(y) J_\nu(xy)\, dy = g(x), \qquad 0 \le x < 1, \tag{2.4.28}$$

and

$$\int_0^\infty f(y) J_\nu(xy)\, dy = 0, \qquad 1 \le x < \infty, \tag{2.4.29}$$

where $\alpha > -2$ and $-\nu - 1 < \alpha - \frac{1}{2} < \nu + 1$. He showed that

$$f(x) = \frac{2^{-\alpha/2} x^{-\alpha}}{\Gamma(1+\alpha/2)} \left\{ x^{1+\alpha/2} J_{\nu+\alpha/2}(x) \int_0^1 y^{\nu+1} \left(1 - y^2\right)^{\alpha/2} g(y)\, dy \right.$$

[13] Busbridge, I. W., 1938: Dual integral equations. *Proc. London Math. Soc., Ser. 2*, **44**, 115–125.

$$+\int_0^1 \eta^{\nu+1}\left(1-\eta^2\right)^{\alpha/2}\left[\int_0^1 g(\eta y)(xy)^{2+\alpha/2} J_{\nu+1+\alpha/2}(xy)\,dy\right]d\eta\bigg\}. \tag{2.4.30}$$

If $\alpha > 0$, Sneddon[14] showed that Equation 2.4.30 simplifies to

$$f(x) = \frac{(2x)^{1-\alpha/2}}{\Gamma(\alpha/2)}\int_0^1 \eta^{1+\alpha/2} J_{\nu+\alpha/2}(\eta x)\left[\int_0^1 g(\eta y) y^{1+\nu}\left(1-y^2\right)^{\alpha/2-1}dy\right]d\eta. \tag{2.4.31}$$

In the present case, $\alpha = 1$, $\nu = -\frac{1}{2}$ and Equation 2.4.30 simplifies to

$$F(\rho) = \sqrt{\frac{2\rho}{\pi}}\bigg\{J_0(\rho)\int_0^1 \sqrt{y(1-y^2)}g(y)\,dy \\ +\rho\int_0^1 \sqrt{\eta(1-\eta^2)}\left[\int_0^1 g(\eta y) y^{3/2} J_1(\rho y)\,dy\right]d\eta\bigg\}. \tag{2.4.32}$$

A simple illustration of this solution occurs if $p(y) = p_0$ for all y. Then, $P(y) = p_0 c J_1(ck)$ and $u(0,y) = 2(1-\nu^2)\sqrt{c^2-y^2}/E$. In this case, the crack has the shape of an ellipse with semi-axes of $2(1-\nu^2)p_0/E$ and c.

2.5 THE BOUNDARY VALUE PROBLEM OF REISSNER AND SAGOCI

Mixed boundary value problems often appear in elasticity problems. An early example involved finding the distribution of stress within a semi-infinite elastic medium when a load is applied to the surface $z = 0$. Reissner and Sagoci[15] used separation of variables and spheroidal coordinates. In 1947 Sneddon[16] resolved the static (time-independent) problems applying Hankel transforms. This is the approach that we will highlight here.

If $u(r,z)$ denotes the circumferential displacement, the mathematical theory of elasticity yields the governing equation

$$\frac{\partial^2 u}{\partial r^2} + \frac{1}{r}\frac{\partial u}{\partial r} - \frac{u}{r^2} + \frac{\partial^2 u}{\partial z^2} = 0, \qquad 0 \le r < \infty, \quad 0 < z < \infty, \tag{2.5.1}$$

[14] See Section 12 in Sneddon, I. N., 1995: *Fourier Transforms.* Dover, 542pp.

[15] Reissner, E., and H. F. Sagoci, 1944: Forced torsional oscillation of an elastic half-space. *J. Appl. Phys.*, **15**, 652–654; Sagoci, H. F., 1944: Forced torsional oscillation of an elastic half-space. II. *J. Appl. Phys.*, **15**, 655–662.

[16] Sneddon, I. N., 1947: Note on a boundary value problem of Reissner and Sagoci. *J. Appl. Phys.*, **18**, 130–132; Rahimian, M., A. K. Ghorbani-Tanha, and M. Eskandari-Ghadi, 2006: The Reissner-Sagoci problem for a transversely isotropic half-space. *Int. J. Numer. Anal. Methods Geomech.*, **30**, 1063–1074.

subject to the boundary conditions

$$\lim_{r\to 0} |u(r,z)| < \infty, \quad \lim_{r\to\infty} u(r,z) \to 0, \qquad 0 < z < \infty, \tag{2.5.2}$$

$$\lim_{z\to\infty} u(r,z) \to 0, \qquad 0 \le r < \infty, \tag{2.5.3}$$

and

$$\begin{cases} u(r,0) = r, & 0 \le r \le a, \\ u_z(r,0) = 0, & a \le r < \infty. \end{cases} \tag{2.5.4}$$

Hankel transforms are used to solve Equation 2.5.1 via

$$U(k,z) = \int_0^\infty r\, u(r,z) J_1(kr)\, dr \tag{2.5.5}$$

which transforms the governing partial differential equation into the ordinary differential equation

$$\frac{d^2U(k,z)}{dz^2} - k^2U(k,z) = 0, \qquad 0 < z < \infty. \tag{2.5.6}$$

Taking Equation 2.5.3 into account,

$$U(k,z) = A(k)e^{-kz}. \tag{2.5.7}$$

Therefore, the solution to Equation 2.5.1, Equation 2.5.2, and Equation 2.5.3 is

$$u(r,z) = \int_0^\infty k\, A(k) e^{-kz} J_1(kr)\, dk. \tag{2.5.8}$$

Upon substituting Equation 2.5.8 into Equation 2.5.4, we have that

$$\int_0^\infty k\, A(k) J_1(kr)\, dk = r, \qquad 0 \le r < a, \tag{2.5.9}$$

and

$$\int_0^\infty k^2 A(k) J_1(kr)\, dk = 0, \qquad a < r < \infty. \tag{2.5.10}$$

The dual integral equations, Equation 2.5.9 and Equation 2.5.10, can be solved using the Busbridge results, Equation 2.4.28 through Equation 2.4.30. This yields

$$A(k) = \frac{4a}{\pi k^2}\left[\frac{\sin(ak)}{ak} - \cos(ak)\right], \tag{2.5.11}$$

and

$$u(r,z) = \frac{4a}{\pi}\int_0^\infty \left[\frac{\sin(ak)}{ak} - \cos(ak)\right] e^{-kz} J_1(kr)\, \frac{dk}{k}. \tag{2.5.12}$$

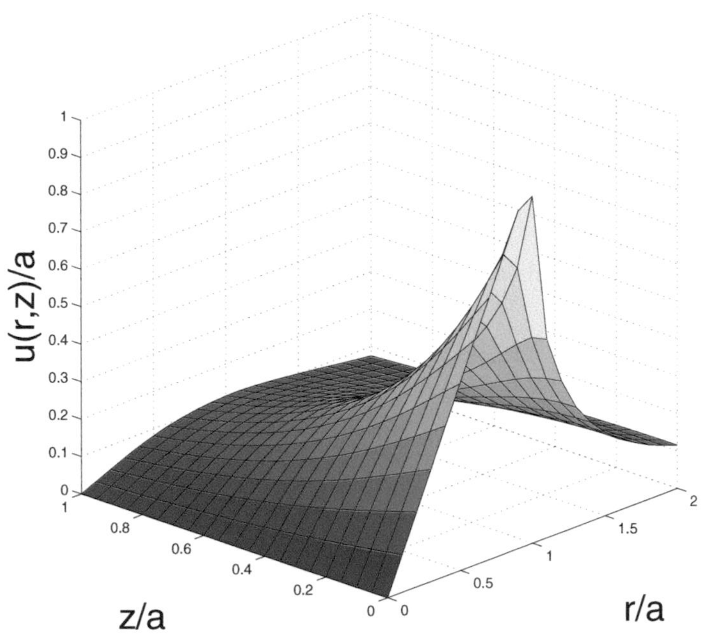

Figure 2.5.1: The solution $u(r,z)/a$ to the mixed boundary value problem governed by Equation 2.5.1 through Equation 2.5.4.

We can evaluate the integral in Equation 2.5.12 and it simplifies to

$$u(r,z) = \frac{2a^2}{r\pi}\left\{\lambda R\sin(\psi+\varphi) - 2R\cos(\varphi) + \frac{r^2}{a^2}\arctan\left[\frac{R\sin(\varphi)+\lambda\sin(\psi)}{R\cos(\varphi)+\lambda\cos(\psi)}\right]\right\}, \tag{2.5.13}$$

where $\lambda^2\sin(2\psi)\tan(\psi)=2$, $\lambda^2 = 1+z^2/a^2$, $z\tan(\psi)=a$,

$$R^4 = \left(\frac{r^2}{a^2}+\frac{z^2}{a^2}-1\right)^2 + \frac{4z^2}{a^2}, \quad \text{and} \quad \frac{2z}{a}\cot(2\varphi) = \frac{r^2}{a^2}+\frac{z^2}{a^2}-1. \tag{2.5.14}$$

We illustrate Equation 2.5.13 in Figure 2.5.1.

We can generalize[17] the original Reissner-Sagoci equation so that it now reads

$$\frac{\partial^2 u}{\partial r^2} + \frac{1}{r}\frac{\partial u}{\partial r} - \frac{u}{r^2} + \frac{\partial^2 u}{\partial z^2} + K\frac{\partial u}{\partial z} + \omega^2 u = 0, \qquad 0 \le r < \infty, \quad 0 < z < \infty, \tag{2.5.15}$$

[17] See Chakraborty, S., D. S. Ray and A. Chakravarty, 1996: A dynamical problem of Reissner-Sagoci type for a non-homogeneous elastic half-space. *Indian J. Pure Appl. Math.*, **27**, 795–806.

subject to the boundary conditions

$$\lim_{r\to 0} |u(r,z)| < \infty, \quad \lim_{r\to\infty} u(r,z) \to 0, \qquad 0 < z < \infty, \tag{2.5.16}$$

$$\lim_{z\to\infty} u(r,z) \to 0, \qquad 0 \le r < \infty, \tag{2.5.17}$$

and

$$\begin{cases} u(r,0) = r, & 0 \le r \le 1, \\ u_z(r,0) = 0, & 1 \le r < \infty. \end{cases} \tag{2.5.18}$$

If we use Hankel transforms, the solution to Equation 2.5.15 through Equation 2.5.17 is

$$u(r,z) = \int_0^\infty A(k) e^{-\kappa(k)z} J_1(kr)\, dk, \tag{2.5.19}$$

where $\kappa(k) = K/2 + \sqrt{k^2 + a^2}$ and $a^2 = K^2/4 - \omega^2$. Upon substituting Equation 2.5.19 into Equation 2.5.18, we have that

$$\int_0^\infty A(k) J_1(kr)\, dk = r, \qquad 0 \le r < 1, \tag{2.5.20}$$

and

$$\int_0^\infty \kappa(k) A(k) J_1(kr)\, dk = 0, \qquad 1 < r < \infty; \tag{2.5.21}$$

or

$$\int_0^\infty B(k)[1 + M(k)] J_1(kr)\, \frac{dk}{k} = 2r, \qquad 0 \le r < 1, \tag{2.5.22}$$

and

$$\int_0^\infty B(k) J_1(kr)\, dk = 0, \qquad 1 < r < \infty, \tag{2.5.23}$$

where $B(k) = 2\kappa(k)A(k)$ and $M(k) = k/\kappa(k) - 1$.

To solve the dual integral equations, Equation 2.5.22 and Equation 2.5.23, we set

$$B(k) = k \int_0^1 h(t) \sin(kt)\, dt. \tag{2.5.24}$$

We have done this because

$$\int_0^\infty B(k) J_1(kr)\, dk = \int_0^1 h(t) \left[\int_0^\infty k \sin(kt) J_1(kr)\, dk \right] dt \tag{2.5.25}$$

$$= -\int_0^1 h(t) \frac{d}{dr} \left[\int_0^\infty \sin(kt) J_0(kr)\, dk \right] dt = 0, \tag{2.5.26}$$

where we used Equation 1.4.13 and $0 \le t \le 1 < r$. Consequently our choice for $B(k)$ satisfies Equation 2.5.23 identically.

Turning to Equation 2.5.22, we now substitute Equation 2.5.24 and interchange the order of integration:

$$\int_0^1 h(t) \left[\int_0^\infty \sin(kt) J_1(kr)\, dk \right] dt + \int_0^1 h(\tau) \left[\int_0^\infty M(k) \sin(k\tau) J_1(kr)\, dk \right] d\tau = 2r. \tag{2.5.27}$$

Using Equation 1.4.13 again, Equation 2.5.27 simplifies to

$$\int_0^t \frac{t\, h(t)}{\sqrt{r^2 - t^2}}\, dt + \int_0^1 h(\tau) \left[\int_0^\infty M(k) \sin(k\tau)\, r J_1(kr)\, dk \right] d\tau = 2r^2. \tag{2.5.28}$$

Applying Equation 1.2.13 and Equation 1.2.14, we have

$$t\, h(t) = \frac{4}{\pi} \frac{d}{dt} \left[\int_0^t \frac{\eta^3}{\sqrt{t^2 - \eta^2}}\, d\eta \right] - \frac{2}{\pi} \int_0^1 h(\tau) \left\{ \int_0^\infty M(k) \sin(k\tau) \frac{d}{dt} \left[\int_0^t \frac{\eta^2 J_1(k\eta)}{\sqrt{t^2 - \eta^2}}\, d\eta \right] dk \right\} d\tau. \tag{2.5.29}$$

Now

$$\frac{d}{dt} \left(\int_0^t \frac{\xi^3}{\sqrt{t^2 - \xi^2}}\, d\xi \right) = 2t^2, \tag{2.5.30}$$

and

$$\frac{d}{dt} \left[\int_0^t \frac{\xi^2 J_1(k\xi)}{\sqrt{t^2 - \xi^2}}\, d\xi \right] = t \sin(kt) \tag{2.5.31}$$

after using integral tables.[18] Substituting Equation 2.5.30 and Equation 2.5.31 into Equation 2.5.29, we finally obtain

$$h(t) + \frac{2}{\pi} \int_0^1 h(\tau) \left[\int_0^\infty M(k) \sin(kt) \sin(k\tau)\, dk \right] d\tau = \frac{8t}{\pi}. \tag{2.5.32}$$

Equation 2.5.32 must be solved numerically. We examine this in detail in Section 4.3. Once $h(t)$ is computed, $B(k)$ and $A(k)$ follow from Equation 2.5.24. Finally Equation 2.5.19 gives the solution $u(r, z)$. We illustrate this solution in Figure 2.5.2.

---

[18] Gradshteyn, I. S., and I. M. Ryzhik, 1965: *Table of Integrals, Series, and Products.* Academic Press, Formula 6.567.1 with $\nu = 1$ and $\mu = -\frac{1}{2}$.

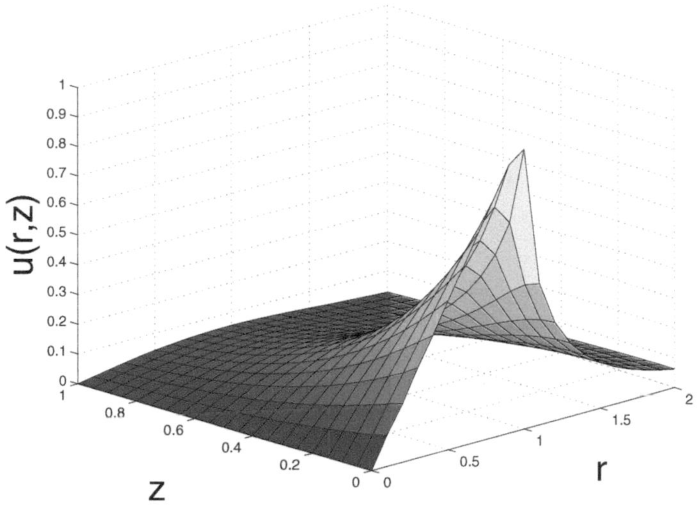

Figure 2.5.2: The solution $u(r, z)$ to the mixed boundary value problem governed by Equation 2.5.15 through Equation 2.5.18.

In 1989 Singh et al.[19] extended the classical Reissner-Sagoci problem so that it now reads

$$\frac{\partial^2 u}{\partial r^2} + \frac{1}{r}\frac{\partial u}{\partial r} - \frac{u}{r^2} + \frac{\partial^2 u}{\partial z^2} = 0, \qquad 0 \le r < \infty, \quad 0 < z < \infty, \tag{2.5.33}$$

subject to the boundary conditions

$$\lim_{r\to 0} |u(r, z)| < \infty, \quad \lim_{r\to\infty} u(r, z) \to 0, \qquad 0 < z < \infty, \tag{2.5.34}$$

$$\lim_{z\to\infty} u(r, z) \to 0, \qquad 0 \le r < \infty, \tag{2.5.35}$$

and

$$\begin{cases} u(r, 0) = r, & 0 \le r < a, \\ u_z(r, 0) = 0, & a < r < b, \\ u(r, 0) = 0, & b < r < \infty. \end{cases} \tag{2.5.36}$$

As we showed earlier, the solution to Equation 2.5.33, Equation 2.5.34 and Equation 2.5.35 is

$$u(r, z) = \int_0^\infty A(k) e^{-kz} J_1(kr)\, dk. \tag{2.5.37}$$

---

[19] Singh, B. M., H. T. Danyluk, and A. P. S. Selvadurai, 1989: The Reissner-Sagoci problem for a half-space with a surface constraint. *Z. Angew. Math. Phys.*, **40**, 762–768. This work was extended in Singh, B. M., H. T. Danyluk, J. Vrbik, J. Rokne, and R. S. Dhaliwal, 2003: The Reissner-Sagoci problem for a non-homogeneous half-space with a surface constraint. *Meccanica*, **38**, 453–465.

Upon substituting Equation 2.5.37 into Equation 2.5.36, we obtain the triple integral equations:

$$\int_0^\infty A(k) J_1(kr)\, dk = r, \qquad 0 \le r < a, \tag{2.5.38}$$

$$\int_0^\infty k\, A(k) J_1(kr)\, dk = 0, \qquad a < r < b, \tag{2.5.39}$$

and

$$\int_0^\infty A(k) J_1(kr)\, dk = 0, \qquad b < r < \infty. \tag{2.5.40}$$

Let us now solve this set of triple integral equations by assuming that

$$\int_0^\infty k\, A(k) J_1(kr)\, dk = \begin{cases} f_1(r), & 0 < r < a, \\ f_2(r), & b < r < \infty. \end{cases} \tag{2.5.41}$$

Taking the inverse of the Hankel transform, we obtain from Equation 2.5.39 and Equation 2.5.41 that

$$A(k) = \int_0^a r f_1(r) J_1(kr)\, dr + \int_b^\infty r f_2(r) J_1(kr)\, dr. \tag{2.5.42}$$

Substituting Equation 2.5.42 into Equation 2.5.38 and Equation 2.5.40, we find that

$$\int_0^a \tau f_1(\tau) L(r,\tau)\, d\tau + \int_b^\infty \tau f_2(\tau) L(r,\tau)\, d\tau = r, \quad 0 < r < a, \tag{2.5.43}$$

and

$$\int_0^a \tau f_1(\tau) L(r,\tau)\, d\tau + \int_b^\infty \tau f_2(\tau) L(r,\tau)\, d\tau = 0, \quad b < r < \infty, \tag{2.5.44}$$

where

$$L(r,\tau) = \int_0^\infty J_1(kr) J_1(k\tau)\, dk. \tag{2.5.45}$$

At this point, we introduce several results by Cooke,[20] namely that

$$L(r,\tau) = \frac{2}{\pi r \tau} \int_0^{\min(r,\tau)} \frac{t^2}{\sqrt{(r^2 - t^2)(\tau^2 - t^2)}}\, dt \tag{2.5.46}$$

$$= \frac{2r\tau}{\pi} \int_{\max(r,\tau)}^\infty \frac{dt}{t^2 \sqrt{(t^2 - r^2)(t^2 - \tau^2)}}, \tag{2.5.47}$$

---

[20] Cooke, J. C., 1963: Triple integral equations problems. *Quart. J. Mech. Appl. Math.*, **16**, 193–201.

$$\int_a^b \int_0^{\min(r,\tau)} (\cdots)\, dt\, d\tau = \int_0^r \int_t^b (\cdots)\, d\tau\, dt + \int_0^a \int_a^b (\cdots)\, d\tau\, dt, \qquad \textbf{(2.5.48)}$$

and

$$\int_a^b \int_{\max(r,\tau)}^\infty (\cdots)\, dt\, d\tau = \int_r^b \int_a^t (\cdots)\, d\tau\, dt + \int_b^\infty \int_a^b (\cdots)\, d\tau\, dt. \qquad \textbf{(2.5.49)}$$

Why have we introduced Equation 2.5.46 through Equation 2.5.49? Applying Equation 2.5.46, we can rewrite Equation 2.5.43 as

$$\int_0^a f_1(\tau) \left[\int_0^{\min(r,\tau)} \frac{t^2}{\sqrt{(r^2-t^2)(\tau^2-t^2)}}\, dt \right] d\tau \qquad \textbf{(2.5.50)}$$
$$+ r^2 \int_b^\infty \tau^2 f_2(\tau) \left[\int_\tau^\infty \frac{dt}{t^2\sqrt{(t^2-r^2)(t^2-\tau^2)}} \right] d\tau = \frac{\pi r^2}{2}$$

for $0 < r < a$. Then, applying Equation 2.5.48 and interchanging the order of integration in the second integral, we obtain

$$\int_0^r \frac{t^2 F_1(t)}{\sqrt{r^2-t^2}}\, dt = \frac{\pi r^2}{2} - r^2 \int_b^\infty \frac{F_2(t)}{t^2\sqrt{t^2-r^2}}\, dt, \qquad 0 < r < a, \qquad \textbf{(2.5.51)}$$

where

$$F_1(t) = \int_t^a \frac{f_1(\tau)}{\sqrt{\tau^2-t^2}}\, d\tau, \qquad 0 < t < a, \qquad \textbf{(2.5.52)}$$

and

$$F_2(t) = \int_b^t \frac{\tau^2 f_2(\tau)}{\sqrt{t^2-\tau^2}}\, d\tau, \qquad b < t < \infty. \qquad \textbf{(2.5.53)}$$

If we regard the right side of Equation 2.5.51 as a known function of r, then it is an integral equation of the Abel type. From Equation 1.2.13 and Equation 1.2.14, its solution is

$$t\, F_1(t) = 2t - \frac{1}{\pi} \int_b^\infty \left(\frac{2\eta t}{\eta^2 - t^2} - \log\left|\frac{\eta - t}{\eta + t}\right| \right) F_2(\eta)\, \frac{d\eta}{\eta^2}, \qquad 0 < t < a, \qquad \textbf{(2.5.54)}$$

where we used the following results:

$$\frac{d}{dt}\left[\int_0^t \frac{r^3}{\sqrt{t^2-r^2}}\, dr \right] = 2t^2, \qquad \textbf{(2.5.55)}$$

and

$$\frac{d}{dt}\left[\int_0^t \frac{r^3}{\sqrt{(t^2-r^2)(\eta^2-r^2)}}\, dr \right] = \frac{t}{2} \left(\frac{2\eta t}{\eta^2-t^2} - \log\left|\frac{\eta-t}{\eta+t}\right| \right). \qquad \textbf{(2.5.56)}$$

Turning to Equation 2.5.44, we employ Equation 2.5.46 and Equation 2.5.47 and find that

$$\int_b^\infty \tau^2 f_2(\tau) \left[\int_{\max(r,\tau)}^\infty \frac{dt}{t^2\sqrt{(t^2-r^2)(t^2-\tau^2)}} \right] d\tau + \frac{1}{r^2} \int_0^a f_1(\tau) \left[\int_0^\tau \frac{t^2}{\sqrt{(r^2-t^2)(\tau^2-t^2)}}\, dt \right] d\tau = 0, \tag{2.5.57}$$

if $b < r < \infty$. If we now apply Equation 2.5.49, interchange the order of integration in the second integral and use Equation 2.5.52 and Equation 2.5.53, we find that

$$\int_r^\infty \frac{F_2(t)}{t^2\sqrt{t^2-r^2}}\, dt = -\frac{1}{r^2} \int_0^a \frac{t^2 F_1(t)}{\sqrt{r^2-t^2}}\, dt, \qquad b < r < \infty. \tag{2.5.58}$$

Solving this integral equation of the Abel type yields

$$\frac{F_2(t)}{t^2} = \frac{2}{\pi} \frac{d}{dt} \left\{ \int_t^\infty \frac{dr}{r\sqrt{r^2-t^2}} \left[\int_0^a \frac{\eta^2 F_1(\eta)}{\sqrt{r^2-\eta^2}}\, d\eta \right] \right\}, \qquad b < t < \infty. \tag{2.5.59}$$

Because

$$\frac{d}{dt} \left[\int_t^\infty \frac{dr}{r\sqrt{(r^2-t^2)(r^2-\eta^2)}} \right] = \frac{1}{2\eta t^2} \log\left| \frac{t-\eta}{t+\eta} \right| - \frac{1}{t(t^2-\eta^2)}, \tag{2.5.60}$$

$$\frac{F_2(t)}{t} = \frac{1}{\pi} \int_0^a \eta F_1(\eta) \left[\frac{1}{t} \log\left| \frac{t-\eta}{t+\eta} \right| - \frac{2\eta}{t^2-\eta^2} \right] d\eta, \qquad b < t < \infty. \tag{2.5.61}$$

Setting $\eta F_1(\eta) = 2aX_1(\eta)$, $F_2(\eta)/\eta = 2aX_2(\eta)$, $c = a/b$, and introducing the variables $\eta = b\eta_1$ and $t = at_1$, we can rewrite Equation 2.5.54 and Equation 2.5.59 as

$$X_1(at_1) = t_1 - \frac{1}{\pi} \int_1^\infty \left[\frac{2ct_1}{\eta_1^2 - c^2t_1^2} - \frac{1}{\eta_1} \log\left| \frac{1-ct_1/\eta_1}{1+ct_1/\eta_1} \right| \right] X_2(b\eta_1)\, d\eta_1, \tag{2.5.62}$$

when $0 < t_1 < 1$; and

$$X_2(bt_1) = \frac{1}{\pi} \int_0^1 \left[\frac{c}{t_1} \log\left| \frac{1-c\eta_1/t_1}{1+c\eta_1/t_1} \right| - \frac{2c^2\eta_1}{t_1^2 - c^2\eta_1^2} \right] X_1(a\eta_1)\, d\eta_1, \tag{2.5.63}$$

when $1 < t_1 < \infty$.

Once we find $X_1(at_1)$ and $X_2(bt_1)$ via Equation 2.5.62 and Equation 2.5.63, we can find $f_1(r)$ and $f_2(r)$ from

$$f_1(r) = -\frac{2}{\pi} \frac{d}{dr} \left[\int_r^a \frac{t\, F_1(t)}{\sqrt{t^2-r^2}}\, dt \right], \qquad 0 < r < a, \tag{2.5.64}$$

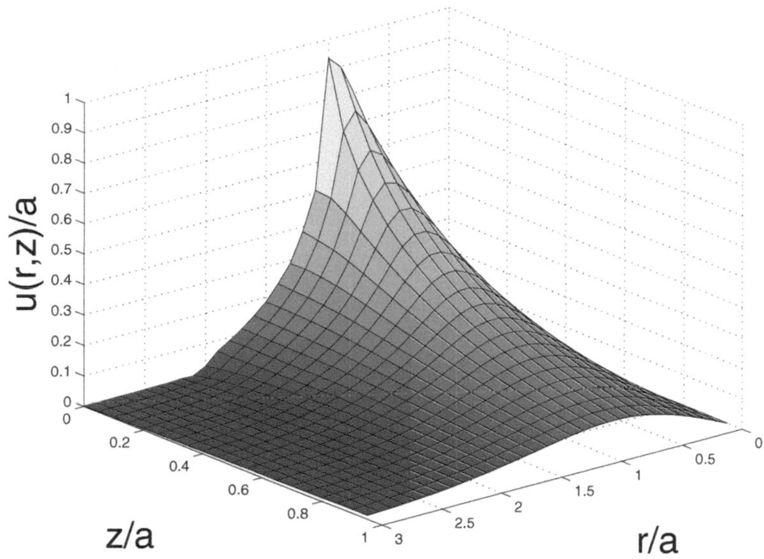

Figure 2.5.3: The solution $u(r,z)/a$ to the mixed boundary value problem governed by Equation 2.5.33 through Equation 2.5.36 when $c = \frac{1}{2}$.

and

$$f_2(r) = \frac{2}{\pi r^2}\frac{d}{dr}\left[\int_b^r \frac{t\,F_2(t)}{\sqrt{r^2-t^2}}\,dt\right], \qquad b < r < \infty. \tag{2.5.65}$$

Substituting Equation 2.5.64 and Equation 2.5.65 into Equation 2.5.42 and integrating by parts, we find that

$$\begin{aligned} A(k) = {} & \frac{4ka^3}{\pi}\int_0^1 \left[\int_t^1 \frac{X_1(a\tau)}{\sqrt{\tau^2-t^2}}\,d\tau\right] t\,J_0(akt)\,dt \\ & + \frac{4ka^3}{c^2\pi}\int_1^\infty \left[\int_1^t \frac{\tau^2 X_2(b\tau)}{\sqrt{t^2-\tau^2}}\,d\tau\right]\frac{J_2(akt/c)}{t}\,dt. \end{aligned} \tag{2.5.66}$$

Finally, Equation 2.5.37 gives the solution $u(r,z)$. Figure 2.5.3 illustrates this solution when $c = \frac{1}{2}$.

In the previous examples of the Reissner-Sagoci problem, we solved it in the half-space $z > 0$. Here we solve this problem[21] within a cylinder of radius b when the shear modulus of the material varies as $\mu_0 z^\alpha$, where $0 \le \alpha < 1$. Mathematically the problem is

$$\frac{\partial^2 u}{\partial r^2} + \frac{1}{r}\frac{\partial u}{\partial r} - \frac{u}{r^2} + \frac{\partial^2 u}{\partial z^2} + \frac{\alpha}{z}\frac{\partial u}{\partial z} = 0, \qquad 0 \le r < b, \quad 0 < z < \infty, \tag{2.5.67}$$

[21] Reprinted from *Int. J. Engng. Sci.*, **8**, M. K. Kassir, The Reissner-Sagoci problem for a non-homogeneous solid, 875–885, ©1970, with permission from Elsevier.

subject to the boundary conditions

$$\lim_{r\to 0} |u(r,z)| < \infty, \quad u(b,z) = 0, \qquad 0 < z < \infty, \tag{2.5.68}$$

$$\lim_{z\to\infty} u(r,z) \to 0, \qquad 0 \le r < b, \tag{2.5.69}$$

and

$$\begin{cases} u(r,0) = f(r), & 0 \le r < a, \\ z^{\alpha} u_z(r,z)\Big|_{z=0} = 0, & a < r \le b, \end{cases} \tag{2.5.70}$$

where $b > a$.

Using separation of variables, the solution to Equation 2.5.67 through Equation 2.5.69 is

$$u(r,z) = z^p \sum_{n=1}^{\infty} k_n^{-p} A_n(k_n) K_p(k_n z) J_1(k_n r), \tag{2.5.71}$$

where $2p = 1 - \alpha$, $0 < p \le \frac{1}{2}$. Here k_n denotes the nth root of $J_1(kb) = 0$ and $n = 1, 2, 3, \ldots$. Upon substituting Equation 2.5.71 into Equation 2.5.70, we obtain the dual series:

$$\sum_{n=1}^{\infty} k_n^{-2p} A_n J_1(k_n r) = \frac{2^{1-p}}{\Gamma(p)} f(r), \qquad 0 \le r \le a, \tag{2.5.72}$$

and

$$\sum_{n=1}^{\infty} A_n J_1(k_n r) = 0, \qquad a < r \le b. \tag{2.5.73}$$

Sneddon and Srivastav[22] studied dual Fourier-Bessel series of the form

$$\sum_{n=1}^{\infty} k_n^{-p} A_n J_\nu(k_n \rho) = f(\rho), \qquad 0 < \rho < 1, \tag{2.5.74}$$

and

$$\sum_{n=1}^{\infty} A_n J_\nu(k_n \rho) = f(\rho), \qquad 1 < \rho < a. \tag{2.5.75}$$

Applying here the results from their Section 4, we have that

$$A_n = \frac{2^{1-p}\Gamma(1-p) k_n^p}{b^2 J_2^2(k_n b)} \int_0^a t^{1-p} J_{1-p}(k_n t) g(t)\, dt. \tag{2.5.76}$$

---

[22] Sneddon, I. N., and R. P. Srivastav, 1966: Dual series relations. I. Dual relations involving Fourier-Bessel series. *Proc. R. Soc. Edinburgh, Ser. A*, **66**, 150–160.

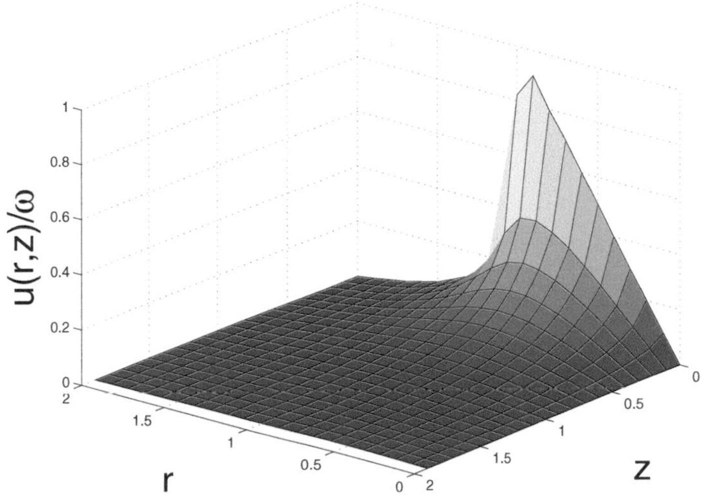

Figure 2.5.4: The solution $u(r, z)/\omega$ to the mixed boundary value problem governed by Equation 2.5.33 through Equation 2.5.36 when $a = 1$, $b = 2$ and $\alpha = \frac{1}{4}$.

The function $g(t)$ is determined via the Fredholm integral equation of the second kind

$$g(t) = h(t) + \frac{2}{\pi} \sin(\pi p) t^p \int_0^a \tau^{1-p} g(\tau) L(t, \tau) \, d\tau, \qquad \textbf{(2.5.77)}$$

where

$$h(t) = \frac{2^{1+p} \sin(\pi p)}{\pi \Gamma(1-p)} t^{2p-1} \int_0^1 \frac{d}{dr} \left[r f(r) \right] \frac{dr}{(t^2 - r^2)^p}, \qquad \textbf{(2.5.78)}$$

and

$$L(t, \tau) = \int_0^\infty \frac{K_1(by)}{I_1(by)} I_{1-p}(ty) I_{1-p}(\tau y) \, y \, dy. \qquad \textbf{(2.5.79)}$$

To illustrate our results, we choose $f(r) = \omega r$. Then

$$h(t) = \frac{2^{1+p} \omega \sin(\pi p)}{\pi \Gamma(2-p)} t. \qquad \textbf{(2.5.80)}$$

Figure 2.5.4 illustrates the case when $a = 1$, $b = 2$ and $\alpha = \frac{1}{4}$.

Problems

1. Solve Helmholtz's equation

$$\frac{\partial^2 u}{\partial r^2} + \frac{1}{r} \frac{\partial u}{\partial r} + \frac{\partial^2 u}{\partial z^2} - \left(\alpha^2 + \frac{1}{r^2} \right) u = 0, \qquad 0 \le r < \infty, \quad 0 < z < \infty,$$

subject to the boundary conditions

$$\lim_{r\to 0}|u(r,z)| < \infty, \qquad \lim_{r\to\infty} u(r,z) \to 0, \qquad 0 < z < \infty,$$

$$\lim_{z\to\infty} u(r,z) \to 0, \qquad 0 \le r < \infty,$$

and

$$\begin{cases} u(r,0) = q(r), & 0 \le r < 1, \\ u_z(r,0) = 0, & 1 < r < \infty. \end{cases}$$

Step 1: Show that the solution to the differential equation plus the first three boundary conditions is

$$u(r,z) = \int_0^\infty A(k) e^{-z\sqrt{k^2+\alpha^2}} J_1(kr)\, dk.$$

Step 2: Using the last boundary condition, show that we obtain the dual integral equations

$$\int_0^\infty A(k) J_1(kr)\, dk = q(r), \qquad 0 \le r < 1,$$

and

$$\int_0^\infty \sqrt{k^2+\alpha^2}\, A(k) J_1(kr)\, dk = 0, \qquad 1 < r < \infty.$$

Step 3: If $\sqrt{k^2+\alpha^2}\, A(k) = kB(k)$, then the dual integral equations become

$$\int_0^\infty \frac{k}{\sqrt{k^2+\alpha^2}} B(k) J_1(kr)\, dk = q(r), \qquad 0 \le r < 1,$$

and

$$\int_0^\infty kB(k) J_1(kr)\, dk = 0, \qquad 1 < r < \infty.$$

Step 4: Consider the first integral equation in Step 3. By multiplying both sides of this equation by $dr/\sqrt{t^2-r^2}$, integrating from 0 to t and using

$$\int_0^t \frac{J_1(kr)}{\sqrt{t^2-r^2}}\, dr = \frac{1-\cos(kt)}{kt},$$

show that

$$\int_0^\infty \frac{k}{\sqrt{k^2+\alpha^2}} B(k) \left[\frac{1-\cos(kt)}{kt}\right] dk = \int_0^t \frac{q(r)}{\sqrt{t^2-r^2}}\, dr, \qquad 0 \le t < 1,$$

or

$$\int_0^\infty \frac{k}{\sqrt{k^2+\alpha^2}} B(k)\sin(kt)\,dk = \frac{d}{dt}\left[\int_0^t \frac{t\,q(r)}{\sqrt{t^2-r^2}}\,dr\right], \qquad 0 \le t < 1.$$

Step 5: Consider next the second integral equation in Step 3. By multiplying both sides by $dr/\sqrt{r^2-t^2}$, integrating from t to ∞ and using

$$\int_t^\infty \frac{J_1(kr)}{\sqrt{r^2-t^2}}\,dr = \frac{\sin(kt)}{kt},$$

show that

$$\int_0^\infty B(k)\sin(kt)\,dk = 0, \qquad 1 \le t < \infty.$$

Step 6: If we introduce

$$g(t) = \int_0^\infty B(k)\sin(kt)\,dk,$$

show that the integral equations in Step 4 and Step 5 become

$$g(t) - \int_0^\infty \left[1 - \frac{k}{\sqrt{k^2+\alpha^2}}\right] B(k)\sin(kt)\,dk = \frac{d}{dt}\left[\int_0^t \frac{t\,q(r)}{\sqrt{t^2-r^2}}\,dr\right],$$

for $0 \le t < 1$, and

$$g(t) = 0, \qquad 1 < t < \infty.$$

Step 7: Because

$$B(k) = \frac{2}{\pi}\int_0^1 g(t)\sin(kt)\,dt,$$

show that the function $g(t)$ is governed by

$$g(t) - \frac{2}{\pi}\int_0^1 g(\tau)\left[\int_0^\infty \left(1 - \frac{k}{\sqrt{k^2+\alpha^2}}\right)\sin(kt)\sin(k\tau)\,dk\right] d\tau = \frac{d}{dt}\left[\int_0^t \frac{t\,q(r)}{\sqrt{t^2-r^2}}\,dr\right]$$

for $0 \le t < 1$. Once $g(t)$ is computed numerically, then $B(k)$ and $A(k)$ follow. Finally, the values of $A(k)$ are used in the integral solution given in Step 1.

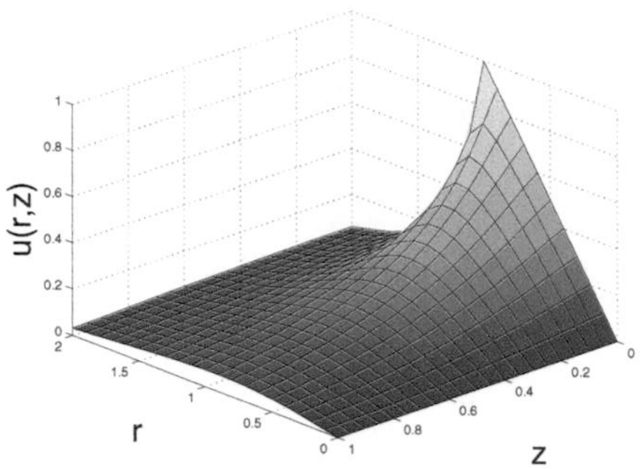

Problem 1

Step 8: Taking the limit as $\alpha \to 0$, show that you recover the solution given by Equation 2.5.11 and Equation 2.5.12 with $a = 1$ and $q(r) = r$. In the figure labeled Problem 1, we illustrate the solution when $\alpha = 1$ and $q(r) = r$.

2.6 STEADY ROTATION OF A CIRCULAR DISC

A problem that is similar to the Reissner-Sagoci problem involves finding the steady-state velocity field within a laminar, infinitely deep fluid that is driven by a slowly rotating disc of radius a in contact with the free surface. The disc rotates at the angular velocity ω. The Navier-Stokes equations for the angular component $u(r, z)$ of the fluid's velocity reduce to

$$\frac{\partial^2 u}{\partial r^2} + \frac{1}{r}\frac{\partial u}{\partial r} - \frac{u}{r^2} + \frac{\partial^2 u}{\partial z^2} = 0, \qquad 0 \le r < \infty, \quad 0 < z < \infty. \tag{2.6.1}$$

At infinity the velocity must tend to zero which yields the boundary conditions

$$\lim_{r\to 0} |u(r, z)| < \infty, \qquad \lim_{r\to\infty} u(r, z) \to 0, \qquad 0 < z < \infty, \tag{2.6.2}$$

and

$$\lim_{z\to\infty} u(r, z) \to 0, \qquad 0 \le r < \infty. \tag{2.6.3}$$

At the interface, the mixed boundary condition is

$$\begin{cases} u(r, 0) = \omega r, & 0 \le r < a, \\ \mu u_z(r, 0) + \eta u_{zz}(r, 0) = 0, & a < r < \infty. \end{cases} \tag{2.6.4}$$

Goodrich[23] was the first to attack this problem. Using Hankel transforms, the solution to Equation 2.6.1 through Equation 2.6.3 is

$$u(r, z) = \int_0^\infty A(k) J_1(kr) e^{-kz}\, dk. \tag{2.6.5}$$

---

[23] Taken with permission from Goodrich, F. C., 1969: The theory of absolute surface shear viscosity. I. *Proc. Roy. Soc. London, Ser. A*, **310**, 359–372.

Substituting Equation 2.6.5 into Equation 2.6.4, we have that

$$\int_0^\infty A(k)J_1(kr)\,dk = \omega r, \qquad 0 \le r < a, \tag{2.6.6}$$

and

$$\int_0^\infty (\eta k^2 - \mu k)A(k)J_1(kr)\,dk = 0, \qquad a < r < \infty. \tag{2.6.7}$$

Before Goodrich tackled the general problem, he considered the following special cases.

$\mu = 0$

In this special case, Equation 2.6.6 and Equation 2.6.7 simplify to

$$\int_0^\infty A(k)J_1(kr)\,dk = \omega r, \qquad 0 \le r < a, \tag{2.6.8}$$

and

$$\int_0^\infty k^2 A(k)J_1(kr)\,dk = 0, \qquad a < r < \infty. \tag{2.6.9}$$

Now, multiplying Equation 2.6.8 by r and differentiating with respect to r,

$$\int_0^\infty A(k)\frac{d}{dr}[rJ_1(kr)]\;dk = 2\omega r, \qquad 0 < r < a. \tag{2.6.10}$$

In the case of Equation 2.6.9, integrating both sides with respect to r, we obtain

$$\int_0^\infty k^2 A(k)\left[\int_r^\infty J_1(k\xi)\,d\xi\right]\,dk = 0, \qquad a < r < \infty. \tag{2.6.11}$$

From the theory of Bessel functions,[24]

$$\frac{d}{dr}[rJ_1(kr)] = krJ_0(kr), \tag{2.6.12}$$

and

$$\int_r^\infty J_1(k\xi)\,d\xi = \frac{J_0(kr)}{k}. \tag{2.6.13}$$

Substituting Equation 2.6.12 and Equation 2.6.13 into Equation 2.6.10 and Equation 2.6.11, respectively, they become

$$\int_0^\infty kA(k)J_0(kr)\,dk = 2\omega, \qquad 0 \le r < a, \tag{2.6.14}$$

---

[24] Gradshteyn and Ryzhik, op. cit., Formulas 8.472.3 and 8.473.4 with $z = kr$.

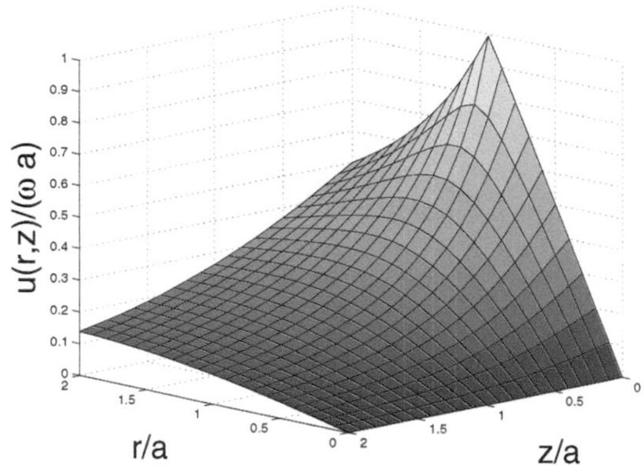

Figure 2.6.1: The solution to Laplace's equation, Equation 2.6.1, with the boundary conditions given by Equation 2.6.2 through Equation 2.6.4 when $\mu = 0$.

and

$$\int_0^\infty kA(k)J_0(kr)\,dk = 0, \qquad a < r < \infty. \tag{2.6.15}$$

Taking the inverse Hankel transform given by Equation 2.6.14 and Equation 2.6.15, we have that

$$A(k) = 2\omega \int_0^a r J_0(kr)\,dr = 2\omega a \frac{J_1(ak)}{k}. \tag{2.6.16}$$

Therefore,

$$u(r,z) = 2\omega a \int_0^\infty J_1(\xi) J_1(\xi r/a) e^{-\xi z/a} \frac{d\xi}{\xi}. \tag{2.6.17}$$

Figure 2.6.1 illustrates this solution.

$\eta = 0$

In this case Equation 2.6.6 and Equation 2.6.7 become

$$\int_0^\infty A(k)J_1(kr)\,dk = \omega r, \qquad 0 \le r < a, \tag{2.6.18}$$

and

$$\int_0^\infty kA(k)J_1(kr)\,dk = 0, \qquad a < r < \infty. \tag{2.6.19}$$

If we now introduce a function $g(r)$ such that

$$\int_0^\infty kA(k)J_1(kr)\,dk = g(r), \qquad 0 \le r < a, \tag{2.6.20}$$

then

$$A(k) = \int_0^a \xi\, g(\xi) J_1(k\xi)\, d\xi. \tag{2.6.21}$$

Upon substituting Equation 2.6.21 into Equation 2.6.18 and interchanging the order of integration,

$$\int_0^\infty \xi\, g(\xi) \left[\int_0^\infty J_1(k\xi) J_1(kr)\, dk \right] d\xi = \omega r. \tag{2.6.22}$$

Because[25]

$$\int_0^\infty J_\nu(k\xi) J_\nu(kx)\, dk = \frac{2(\xi x)^{-\nu}}{\pi} \int_0^{\min(\xi,x)} \frac{s^{2\nu}}{\sqrt{(\xi^2 - s^2)(x^2 - s^2)}}\, ds, \tag{2.6.23}$$

Equation 2.6.22 can be rewritten

$$\begin{aligned} \frac{2}{\pi} \int_0^r g(\xi) \left[\int_0^\xi \frac{s^2}{\sqrt{(r^2 - s^2)(\xi^2 - s^2)}}\, ds \right] d\xi \qquad \qquad \qquad \qquad (\mathbf{2.6.24}) \\ + \frac{2}{\pi} \int_r^\infty g(\xi) \left[\int_0^r \frac{s^2}{\sqrt{(r^2 - s^2)(\xi^2 - s^2)}}\, ds \right] d\xi = \omega r^2. \end{aligned}$$

Interchanging the order of integration,

$$\frac{2}{\pi} \int_0^r \frac{f(s)}{\sqrt{r^2 - s^2}}\, ds = \omega r^2, \qquad 0 \le r < a, \tag{2.6.25}$$

where we set

$$f(s) = s^2 \int_s^a \frac{g(\xi)}{\sqrt{\xi^2 - s^2}}\, d\xi. \tag{2.6.26}$$

From Equation 1.2.13 and Equation 1.2.14,

$$f(s) = \omega \frac{d}{ds} \left[\int_0^s \frac{r^3}{\sqrt{s^2 - r^2}}\, dr \right] = 2\omega s^2. \tag{2.6.27}$$

Substituting the results from Equation 2.6.27 into Equation 2.6.26 we find that

$$g(\xi) = -\frac{2}{\pi} \frac{d}{d\xi} \left[\int_\xi^a \frac{2\omega\tau}{\sqrt{\tau^2 - \xi^2}}\, d\tau \right] = \frac{4\omega\xi}{\pi\sqrt{a^2 - \xi^2}}, \qquad 0 \le \xi < a, \tag{2.6.28}$$

---

[25] Cooke, J. C., 1963: Triple integral equations. *Quart. J. Mech. Appl. Math.*, **16**, 193–203. See Appendix 1.

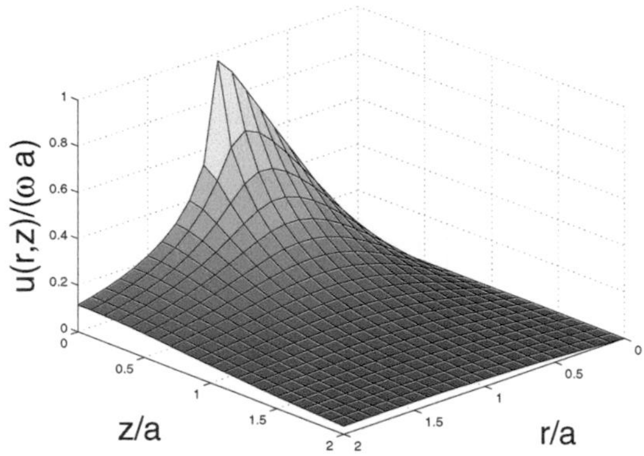

Figure 2.6.2: This is similar to Figure 2.6.1 except that $\mu \neq 0$ and $\eta = 0$.

after using Equation 1.2.15 and Equation 1.2.16. Therefore,[26]

$$A(k) = \frac{4\omega}{\pi} \int_0^a \frac{\xi^2}{\sqrt{a^2 - \xi^2}} J_1(k\xi)\, d\xi = 2\omega a^2 \sqrt{\frac{2}{\pi k a}}\, J_{\frac{3}{2}}(ka). \tag{2.6.29}$$

The solution $u(r, z)$ follows from Equation 2.6.5. Figure 2.6.2 illustrates this solution.

$\mu \neq 0$, $\eta \neq 0$

In this general case Equation 2.6.6 and Equation 2.6.7 become

$$\int_0^\infty A(k) J_1(kr)\, dk = \omega r, \qquad 0 \le r < a, \tag{2.6.30}$$

and

$$\int_0^\infty k(1 + a\lambda_0 k) A(k) J_1(kr)\, dk = 0, \qquad a < r < \infty, \tag{2.6.31}$$

where $\lambda_0 = -\eta/(a\mu)$. Let us turn to Equation 2.6.30 first. Multiplying Equation 2.6.30 by $(2/\pi)^{1/2} r^2 / \sqrt{x^2 - r^2}$ and integrating the resulting equation over r from 0 to $x \le a$, we find that

$$\int_0^\infty A(k) \left[\sqrt{\frac{2}{\pi}} \int_0^x \frac{r^2 J_1(kr)}{\sqrt{x^2 - r^2}}\, dr \right] dk = \omega \sqrt{\frac{2}{\pi}} \int_0^x \frac{r^3}{\sqrt{x^2 - r^2}}\, dr, \tag{2.6.32}$$

[26] Gradshteyn and Ryzhik, op. cit., Formula 6.567.1 with $\nu = 1$ and $\mu = -\frac{1}{2}$.

or

$$x^{3/2} \int_0^\infty A(k) \frac{J_{\frac{3}{2}}(kx)}{\sqrt{k}}\, dk = \frac{2\sqrt{2}\,\omega x^3}{3\sqrt{\pi}} \qquad \textbf{(2.6.33)}$$

after using integral tables. Differentiating both sides of Equation 2.6.33 with respect to x,

$$\int_0^\infty \sqrt{k}\, A(k) J_{\frac{1}{2}}(kx)\, dk = \frac{2\omega\sqrt{2x}}{\sqrt{\pi}}, \qquad 0 \le x < a. \qquad \textbf{(2.6.34)}$$

We now turn to Equation 2.6.31. Multiplying this equation by $(2x/\pi)^{1/2} / \sqrt{r^2 - x^2}$ and integrating the resulting equation over r from $x > a$ to ∞, we find that

$$\int_0^\infty k(1 + a\lambda_0 k) A(k) \left[\sqrt{\frac{2}{\pi}} \int_x^\infty \frac{\sqrt{x}\, J_1(kr)}{\sqrt{r^2 - x^2}}\, dr \right] dk = 0; \qquad \textbf{(2.6.35)}$$

or

$$\int_0^\infty \sqrt{k}(1 + a\lambda_0 k) A(k) J_{\frac{1}{2}}(kx)\, dk = 0, \qquad a < x < \infty. \qquad \textbf{(2.6.36)}$$

Let us now replace x with r in Equation 2.6.34 and Equation 2.6.36 and express $J_{\frac{1}{2}}(z) = \sqrt{2/(\pi z)}\, \sin(z)$. This yields

$$\int_0^\infty A(k) \sin(kr)\, dk = 2\omega r, \qquad 0 \le r < a, \qquad \textbf{(2.6.37)}$$

and

$$\int_0^\infty (1 + a\lambda_0 k) A(k) \sin(kr)\, dk = 0, \qquad a < r < \infty. \qquad \textbf{(2.6.38)}$$

We can rewrite Equation 2.6.38 as

$$\int_0^\infty A(k) \sin(kr)\, dk = -\lambda_0 \int_0^\infty akA(k) \sin(kr)\, dk, \qquad a < r < \infty. \quad \textbf{(2.6.39)}$$

From the theory of Fourier integrals,

$$\begin{aligned} A(k) &= \frac{4\omega}{\pi} \int_0^a r \sin(kr)\, dr \\ &\quad - \frac{2\lambda_0}{\pi} \int_a^\infty \left[\int_0^\infty akA(k) \sin(kr)\, dk \right] \sin(kr)\, dr \qquad \textbf{(2.6.40)} \\ &= 2\omega a^2 \sqrt{\frac{2}{\pi a k}}\, J_{\frac{3}{2}}(ak) \\ &\quad - \frac{2\lambda_0}{\pi} \int_0^\infty \left[\int_0^\infty a\xi A(\xi) \sin(\xi r)\, d\xi \right] \sin(kr)\, dr \\ &\quad + \frac{2\lambda_0}{\pi} \int_0^a \left[\int_0^\infty a\xi A(\xi) \sin(\xi r)\, d\xi \right] \sin(kr)\, dr. \qquad \textbf{(2.6.41)} \end{aligned}$$

The second term in Equation 2.6.41 equals $a\lambda_0 kA(k)$. Therefore,

$$(1 + a\lambda_0 k)A(k) = 2\omega a^2 \sqrt{\frac{2}{\pi ak}} J_{\frac{3}{2}}(ak) + \frac{2\lambda_0}{\pi} \int_0^a \left[\int_0^\infty a\xi A(\xi) \sin(\xi r)\, d\xi \right] \sin(kr)\, dr. \qquad \mathbf{(2.6.42)}$$

Interchanging the order of integration, we finally have that

$$(1 + a\lambda_0 k)A(k) = 2\omega a^2 \sqrt{\frac{2}{\pi ak}} J_{\frac{3}{2}}(ak) + \lambda_0 \int_0^a a\xi A(\xi) K(k,\xi)\, d\xi, \qquad \mathbf{(2.6.43)}$$

where

$$K(y,\xi) = \frac{2}{\pi} \int_0^a \sin(ry)\sin(r\xi)\, dr = \frac{1}{\pi} \left\{ \frac{\sin[a(y-\xi)]}{y-\xi} - \frac{\sin[a(y+\xi)]}{y+\xi} \right\}. \qquad \mathbf{(2.6.44)}$$

One of the intriguing aspects of Equation 2.6.43 is that the unknown is the Hankel transform $A(k)$. In general, the integral equations that we will see will involve an unknown which is related to $A(k)$ via an integral definition. See Equation 2.5.61, Equation 2.5.62 and Equation 2.5.65. In Goodrich's paper he solved the integral equation as a variational problem and found an approximate solution by the optimization of suitable solutions. However, we shall shortly outline an alternative method for any value of λ_0.

In 1978 Shail[27] reexamined Goodrich's paper for two reasons. First, the solution for the $\mu = 0$ case, Equation 2.6.16, creates a divergent integral when it is substituted back into dual equations, Equation 2.6.8 and Equation 2.6.9. Second, the governing Fredholm integral equation is over an infinite range and appears to be unsuitable for asymptotic solution as $\lambda_0 \to 0$ or $\lambda_0 \to \infty$.

Shail's analysis for the $\mu = 0$ case begins by noting that

$$\frac{\partial^2 u(r,0)}{\partial r^2} + \frac{1}{r}\frac{\partial u(r,0)}{\partial r} - \frac{u(r,0)}{r^2} = 0, \qquad a \le r < \infty \qquad \mathbf{(2.6.45)}$$

from Equation 2.6.1 and Equation 2.6.4. The general solution to Equation 2.6.45 is

$$u(r,0) = Cr + D/r, \qquad a < r < \infty. \qquad \mathbf{(2.6.46)}$$

The values of C and D follow from the limits that $u(r,0) \to 0$ as $r \to \infty$ and continuity at $u(r,0)$ at $r = a$. This yields

$$u(r,0) = \begin{cases} \omega r, & 0 \le r < a, \\ \omega a^2/r, & a < r < \infty. \end{cases} \qquad \mathbf{(2.6.47)}$$

---

[27] Taken from Shail, R., 1978: The torque on a rotating disk in the surface of a liquid with an adsorbed film. *J. Engng. Math.*, **12**, 59–76 with kind permission from Springer Science and Business Media.

If we use Equation 2.6.47 in place of Equation 2.6.4, then the solution Equation 2.6.17 follows directly.

Shail also examined the general case and solved it using the method of complementary representations for generalized axially symmetric potential functions. This method is very complicated and we introduce a greatly simplified version of an analysis first done by Chakrabarti.[28]

We begin by introducing the function $g(x)$ such that

$$(1 + a\lambda_0 k)A(k) = \frac{2}{\pi}\int_0^a g(x)\sin(kx)\,dx. \tag{2.6.48}$$

Turning to Equation 2.6.31 first, direct substitution yields

$$\int_0^\infty k(1 + a\lambda_0 k)A(k)J_1(kr)\,dk$$
$$= \int_0^a g(x)\left[\int_0^\infty k\sin(kx)J_1(kr)\,dk\right]dx \tag{2.6.49}$$
$$= -\int_0^a g(x)\frac{d}{dr}\left[\int_0^\infty \sin(kx)J_0(kr)\,dk\right]dx \tag{2.6.50}$$
$$= 0, \tag{2.6.51}$$

because $0 \le x \le a < r < \infty$. Thus, our choice of $A(k)$ satisfies Equation 2.6.31 identically.

Turning to Equation 2.6.30 next, direct substitution gives

$$\frac{2}{\pi}\int_0^a g(x)\left[\int_0^\infty \frac{\sin(kx)J_1(kr)}{1 + a\lambda_0 k}\,dk\right]dx = \omega r, \qquad 0 \le r < a; \tag{2.6.52}$$

or

$$\int_0^a g(x)\left[\int_0^\infty \sin(kx)J_1(kr)\,dk\right]dx \tag{2.6.53}$$
$$+ \int_0^a g(x)\left[\int_0^\infty \frac{1 - a\lambda_0 k}{1 + a\lambda_0 k}\sin(kx)J_1(kr)\,dk\right]dx = \omega\pi r, \qquad 0 \le r < a.$$

From tables,[29] the integral within the square brackets of the first integral in Equation 2.6.53 can be evaluated and Equation 2.6.53 simplifies to

$$\int_0^r \frac{x\,g(x)}{\sqrt{r^2 - x^2}}\,dx + \int_0^a g(\tau)\left[\int_0^\infty \frac{1 - a\lambda_0 k}{1 + a\lambda_0 k}\sin(k\tau)\,r\,J_1(kr)\,dk\right]d\tau = \omega\pi r^2 \tag{2.6.54}$$

---

[28] Chakrabarti, A., 1989: On some dual integral equations involving Bessel functions of order one. *Indian J. Pure Appl. Math.*, **20**, 483–492.

[29] Gradshteyn and Ryzhik, op. cit., Formula 6.671.1.

for $0 \le r < a$. Using Equation 1.2.13 and Equation 1.2.14, we can solve for $x\,g(x)$ and find that

$$x\,g(x) + \frac{2}{\pi}\int_0^a g(\tau)\left\{\int_0^\infty \frac{1-a\lambda_0 k}{1+a\lambda_0 k}\sin(k\tau)\frac{d}{dx}\left[\int_0^x \frac{\xi^2 J_1(k\xi)}{\sqrt{x^2-\xi^2}}\,d\xi\right]dk\right\}d\tau$$
$$= 2\omega\frac{d}{dx}\left[\int_0^x \frac{\xi^3}{\sqrt{x^2-\xi^2}}\,d\xi\right], \quad 0 \le x < a. \qquad \textbf{(2.6.55)}$$

From integral tables,[30]

$$\frac{d}{ds}\left[\int_0^s \frac{\xi^2 J_1(k\xi)}{\sqrt{s^2-\xi^2}}\,d\xi\right] = s\sin(ks), \qquad \textbf{(2.6.56)}$$

and Equation 2.6.55 becomes

$$g(x) + \frac{2}{\pi}\int_0^a g(\tau)\left[\int_0^\infty \frac{1-a\lambda_0 k}{1+a\lambda_0 k}\sin(k\tau)\sin(kx)\,dk\right]d\tau = 4\omega x, \quad 0 \le x < a. \qquad \textbf{(2.6.57)}$$

Equation 2.6.57 is identical to Chakrabarti's equations (50) and (52). Once we solve Equation 2.6.57 numerically, its values of $g(x)$ can be substituted into Equation 2.6.48. Finally the solution $u(r,z)$ follows from Equation 2.6.5.

The numerical solution of Equation 2.6.57 is nontrivial due to the nature of the integration over k. To solve it, we use a spectral method. If we take $g(\tau)$ to be an odd function over $(-a,a)$, we have that

$$g(\tau) = \sum_{n=1}^\infty A_n \sin\left(\frac{n\pi\tau}{a}\right), \qquad \textbf{(2.6.58)}$$

$$\sin(k\tau) = 2\sum_{n=1}^\infty \frac{(-1)^n n\pi}{k^2a^2 - n^2\pi^2}\sin(ka)\sin\left(\frac{n\pi\tau}{a}\right), \qquad \textbf{(2.6.59)}$$

and

$$x = -2a\sum_{n=1}^\infty \frac{(-1)^n}{n\pi}\sin\left(\frac{n\pi x}{a}\right). \qquad \textbf{(2.6.60)}$$

Substitution of Equation 2.6.58 through Equation 2.6.60 into Equation 2.6.57 gives the infinite set of equations

$$A_m + \sum_{n=1}^\infty H_{mn}A_n = C_m, \qquad m = 1, 2, 3, \ldots, \qquad \textbf{(2.6.61)}$$

[30] Ibid., Formula 6.567.1 with $\nu = 1$ and $\mu = -\frac{1}{2}$.

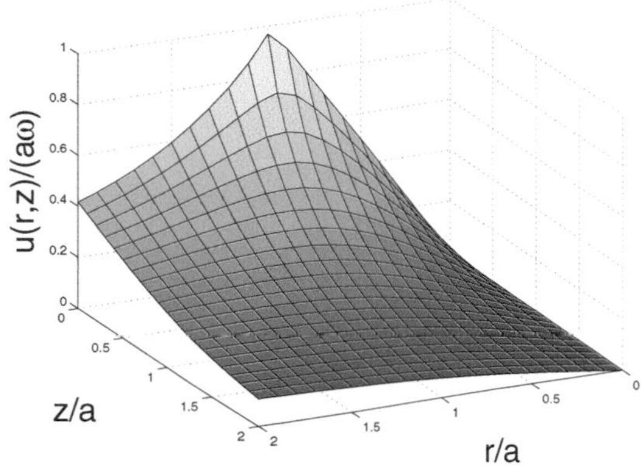

Figure 2.6.3: This is similar to Figure 2.6.1 except that $\mu, \eta \neq 0$ and $\lambda_0 = 5$.

where

$$H_{mn} = 4a(-1)^{n+m} nm\pi \int_0^\infty \frac{1 - a\lambda_0 k}{1 + a\lambda_0 k} \frac{\sin^2(ka)}{(k^2a^2 - n^2\pi^2)(k^2a^2 - m^2\pi^2)} \, dk, \tag{2.6.62}$$

and

$$C_m = \frac{8\omega a}{m\pi}(-1)^{m+1}. \tag{2.6.63}$$

The system of equations is then truncated to, say, N spectral components and the system is inverted to yield A_m for $m = 1, 2, \ldots, N$. Next, $A(k)$ can be found via

$$(1 + a\lambda_0 k)\frac{A(ak)}{a} = 2\sum_{n=1}^{N} \frac{n(-1)^n}{k^2a^2 - n^2\pi^2} \sin(ka) A_n. \tag{2.6.64}$$

The larger the value of N, the greater the accuracy. Finally,

$$u(r, z) = \int_0^\infty \frac{A(\xi)}{a} e^{-\xi z/a} J_1(\xi r/a) \, d\xi. \tag{2.6.65}$$

Figure 2.6.3 illustrates this solution when $\lambda_0 = 5$.

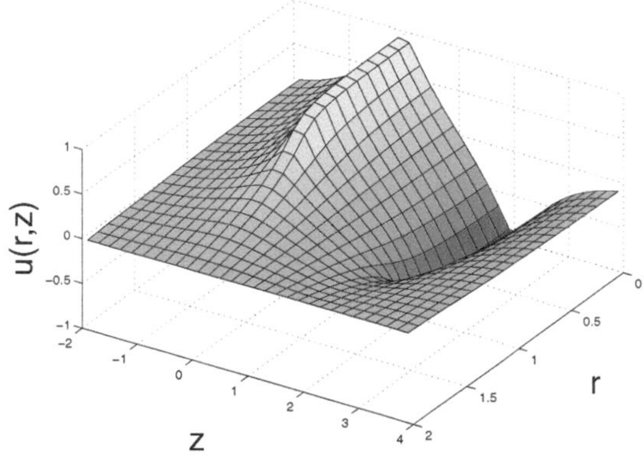

Chapter 3

Separation of Variables

Separation of variables is the most commonly used technique for solving boundary value problems. In the case of mixed boundary value problems they lead to dual or higher-numbered Fourier series which yield the Fourier coefficients. In this chapter we examine dual Fourier series in Section 3.1 and Section 3.2, while dual Fourier-Bessel series are treated in Section 3.3 and dual Fourier-Legendre series in Section 3.4. Finally Section 3.5 treats triple Fourier series.

In Example 1.1.1 we showed that the method of separation of variables led to the dual cosine series:

$$\sum_{n=1}^{\infty} \frac{a_n}{n-\frac{1}{2}} \cos\left[\left(n-\tfrac{1}{2}\right)x\right] = 1, \qquad 0 \le x < c, \tag{3.0.1}$$

and

$$\sum_{n=1}^{\infty} a_n \cos\left[\left(n-\tfrac{1}{2}\right)x\right] = 0, \qquad c < x \le \pi. \tag{3.0.2}$$

Equations 3.0.1 and 3.0.2 are examples of a larger class of dual trigonometric equations. The general form of these dual series can be written

$$\begin{cases} \sum_{n=1}^{\infty} n^p a_n \sin(nx) = f(x), & 0 \le x < c, \\ \sum_{n=1}^{\infty} a_n \sin(nx) = g(x), & c < x \le \pi, \end{cases} \tag{3.0.3}$$

$$\begin{cases} \sum_{n=1}^{\infty} \left(n - \frac{1}{2}\right)^p a_n \cos\left[\left(n - \frac{1}{2}\right) x\right] = f(x), & 0 \le x < c, \\ \sum_{n=1}^{\infty} a_n \cos\left[\left(n - \frac{1}{2}\right) x\right] = g(x), & c < x \le \pi, \end{cases} \tag{3.0.4}$$

$$\begin{cases} \sum_{n=1}^{\infty} \left(n - \frac{1}{2}\right)^p a_n \sin\left[\left(n - \frac{1}{2}\right) x\right] = f(x), & 0 \le x < c, \\ \sum_{n=1}^{\infty} a_n \sin\left[\left(n - \frac{1}{2}\right) x\right] = g(x), & c \le x \le \pi, \end{cases} \tag{3.0.5}$$

and

$$\begin{cases} \frac{1}{2}\alpha a_0 + \sum_{n=1}^{\infty} n^p a_n \cos(nx) = f(x), & 0 \le x < c, \\ \frac{1}{2}a_0 + \sum_{n=1}^{\infty} a_n \cos(nx) = g(x), & c < x \le \pi, \end{cases} \tag{3.0.6}$$

where $-1 \le p \le 1$. Comparing Equation 3.0.1 and Equation 3.0.2 with Equation 3.0.4, they are identical if we set $p = -1$. The purpose of this chapter is to focus on those mixed value problems that lead to these dual equations and solve them.

3.1 DUAL FOURIER COSINE SERIES

Tranter[1] examined dual trigonometric series of the form

$$\begin{cases} \sum_{n=1}^{\infty} \left(n - \frac{1}{2}\right)^p a_n \cos\left[\left(n - \frac{1}{2}\right) x\right] = F(x), & 0 \le x < c, \\ \sum_{n=1}^{\infty} a_n \cos\left[\left(n - \frac{1}{2}\right) x\right] = G(x), & c < x \le \pi, \end{cases} \tag{3.1.1}$$

where $p = \pm 1$. The most commonly encountered case is when $G(x) = 0$. When $p = 1$, he showed that

$$a_n = \frac{2}{\pi} \int_0^c h(x) \cos\left[\left(n - \tfrac{1}{2}\right) x\right] \, dx, \tag{3.1.2}$$

[1] Tranter, C. J., 1960: Dual trigonometrical series. *Proc. Glasgow Math. Assoc.*, **4**, 49–57.

where

$$h(x) = \int_\xi^1 \frac{\chi(\eta)}{\sqrt{\eta^2 - \xi^2}}\, d\eta, \tag{3.1.3}$$

$$\chi(\eta) = \frac{4\eta}{\pi} \sin(c/2) \int_0^\eta \frac{F\{2\arcsin[x\sin(c/2)]\}}{\sqrt{(\eta^2 - x^2)[1 - x^2\sin^2(c/2)]}}\, dx, \tag{3.1.4}$$

and $\xi = \sin(x/2)\csc(c/2)$.

When $p = -1$,

$$a_n = 2\,\chi(1)\sin(c/2) P_{n-1}[\cos(c)] \\ - 2\sin(c/2) \int_0^1 \chi'(\eta) P_{n-1}\left[1 - 2\eta^2 \sin^2(c/2)\right] d\eta, \tag{3.1.5}$$

where

$$\chi(\eta) = \frac{2}{\pi} \int_0^\eta \frac{x\, F'\{2\arcsin[x\sin(c/2)]\}}{\sqrt{\eta^2 - x^2}}\, dx + C. \tag{3.1.6}$$

Here, C is a constant whose value is determined by substituting Equation 3.1.6 into Equation 3.1.5 and then choosing C so that

$$\sum_{n=1}^\infty \left(n - \tfrac{1}{2}\right)^{-1} a_n = F(0). \tag{3.1.7}$$

• Example 3.1.1

To illustrate Tranter's solution, let us assume that $F(x) = 1$ if $0 \le x < c$. From Equation 3.1.6, we have that $\chi(\eta) = C$; from Equation 3.1.5, $a_n = 2C\sin(c/2)P_{n-1}[\cos(c)]$. To evaluate C, we substitute a_n into Equation 3.1.7 and find that

$$2C\sin(c/2) \sum_{n=1}^\infty \left(n - \tfrac{1}{2}\right)^{-1} P_{n-1}[\cos(c)] = 1. \tag{3.1.8}$$

From the generation formula for Legendre polynomials,

$$\sum_{n=1}^\infty x^{2n-2} P_{n-1}[\cos(c)] = [1 - 2x^2\cos(c) + x^4]^{-1/2}. \tag{3.1.9}$$

Integrating Equation 3.1.9 from 0 to 1, we have

$$\sum_{n=1}^\infty \left(n - \tfrac{1}{2}\right) P_{n-1}[\cos(c)] = 2\int_0^1 \frac{dx}{\sqrt{1 - 2x^2\cos(c) + x^4}} \tag{3.1.10}$$

$$= \int_0^{\pi/2} \frac{d\theta}{\sqrt{1 - \cos^2(c/2)\sin^2(\theta)}} \tag{3.1.11}$$

$$= K[\cos^2(c/2)], \tag{3.1.12}$$

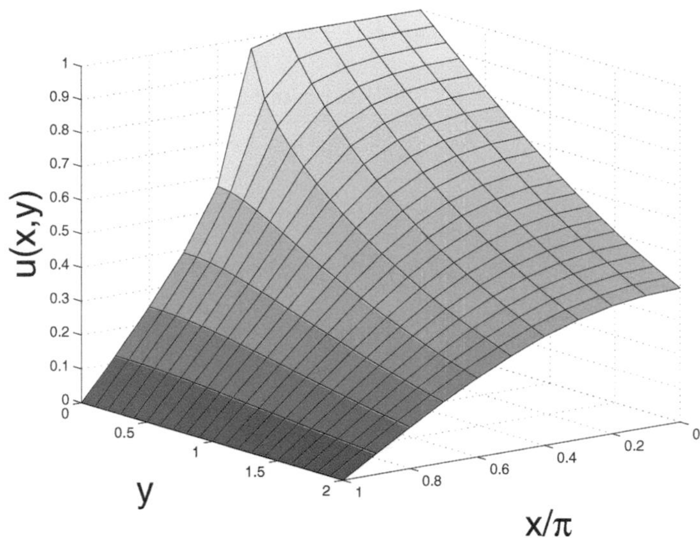

Figure 3.1.1: The solution $u(x, y)$ to the mixed boundary value problem posed in Example 1.1.1 when $c = \pi/2$.

where $K(\cdot)$ denotes the complete elliptic integral[2] and $\sin(\theta) = 2x/(1 + x^2)$. Therefore, $2C\sin(c/2)K[\cos^2(c/2)] = 1$, and $a_n = P_{n-1}[\cos(c)]/K[\cos^2(c/2)]$ is the solution to the dual Fourier cosine series Equation 3.0.1 and Equation 3.0.2.

Recall that Equation 3.0.1 and Equation 3.0.2 arose from the separation of variables solution of Equation 1.1.1 through Equation 1.1.4. Therefore, the solution to this particular mixed boundary value problem is

$$u(x,y) = \sum_{n=1}^{\infty} \frac{P_{n-1}[\cos(c)]}{K[\cos^2(c/2)]} \frac{\exp\left[-\left(n-\frac{1}{2}\right)y\right]}{n-\frac{1}{2}} \cos\left[\left(n-\frac{1}{2}\right)x\right]. \tag{3.1.13}$$

This solution is illustrated in Fig. 3.1.1 when $c = \pi/2$.

• Example 3.1.2

Let us solve[3]

$$\frac{\partial^2 u}{\partial x^2} + \frac{\partial^2 u}{\partial y^2} = 0, \qquad 0 < x < \pi, \quad 0 < y < \pi/2, \tag{3.1.14}$$

---

[2] See Milne-Thomson, L. M., 1965: Elliptic integrals. *Handbook of Mathematical Functions*, M. Abromowitz and I. A. Stegun, Eds., Dover, 587–626. See Section 17.3.

[3] Taken from Whiteman, J. R., 1968: Treatment of singularities in a harmonic mixed boundary value problem by dual series methods. *Quart. J. Mech. Appl. Math.*, **21**, 41–50 with permission of Oxford University Press.

subject to the boundary conditions

$$u_x(0,y) = 0, \quad u(\pi,y) = 1, \qquad 0 < y < \pi/2, \tag{3.1.15}$$

$$\begin{cases} u(x,0) = \frac{1}{2}, & 0 \le x < \pi/2, \\ u_y(x,0) = 0, & \pi/2 < x \le \pi, \end{cases} \tag{3.1.16}$$

and

$$u_y(x,\pi/2) = 0, \qquad 0 < x < \pi. \tag{3.1.17}$$

Using separation of variables, a solution to Equation 3.1.14 which also satisfies Equation 3.1.15 and Equation 3.1.17 is

$$u(x,y) = 1 - \sum_{n=0}^{\infty} \frac{B_n}{n+\frac{1}{2}} \frac{\cosh\left[\left(n+\frac{1}{2}\right)\left(y-\frac{\pi}{2}\right)\right]}{\sinh\left[\left(n+\frac{1}{2}\right)\pi/2\right]} \cos\left[\left(n+\frac{1}{2}\right)x\right]. \tag{3.1.18}$$

If we then substitute Equation 3.1.18 into the mixed boundary condition Equation 3.1.17, we obtain the dual series

$$\sum_{n=0}^{\infty} \frac{B_n}{2n+1} \coth\left[\left(n+\frac{1}{2}\right)\pi/2\right] \cos\left[\left(n+\frac{1}{2}\right)x\right] = \frac{1}{4}, \qquad 0 \le x < \pi/2, \tag{3.1.19}$$

and

$$\sum_{n=0}^{\infty} B_n \cos\left[\left(n+\frac{1}{2}\right)x\right] = 0, \qquad \pi/2 < x \le \pi. \tag{3.1.20}$$

The remaining challenge is to solve this dual series.

Fortunately, in the 1960s Tranter[4] showed that the dual trigonometrical series

$$\sum_{n=0}^{\infty} \frac{A_n}{2n+1} \cos\left[\left(n+\frac{1}{2}\right)x\right] = f(x), \qquad 0 < x < c, \tag{3.1.21}$$

and

$$\sum_{n=0}^{\infty} A_n \cos\left[\left(n+\frac{1}{2}\right)x\right] = 0, \qquad c < x < \pi, \tag{3.1.22}$$

has the solution

$$A_n = A_0 P_n[\cos(c)] - \int_0^c F(\theta) P_n'[\cos(\theta)] \sin(\theta)\, d\theta, \quad n = 1, 2, 3, \ldots, \tag{3.1.23}$$

where

$$F(\theta) = \frac{2\sqrt{2}}{\pi} \int_0^{\theta} \frac{f'(x)\sin(x)}{\sqrt{\cos(x) - \cos(\theta)}}\, dx, \tag{3.1.24}$$

---

[4] Tranter, C. J., 1964: An improved method for dual trigonometrical series. *Proc. Glasgow Math. Assoc.*, **6**, 136–140.

and A_0 is found by substituting Equation 3.1.23 into Equation 3.1.21.

Can we apply Tranter's results, Equation 3.1.21 through Equation 3.1.24, to solve Equation 3.1.19 and Equation 3.1.20? We begin by rewriting these equations as follows:

$$\sum_{n=0}^{\infty} \frac{B_n}{2n+1} \cos\left[\left(n+\tfrac{1}{2}\right)x\right] = \frac{1}{4} + \sum_{m=0}^{\infty} \frac{B_m}{2m+1} \left\{1 - \coth\left[\left(m+\tfrac{1}{2}\right)\pi/2\right]\right\} \times \cos\left[\left(m+\tfrac{1}{2}\right)x\right], \quad 0 \le x < \pi/2, \tag{3.1.25}$$

and

$$\sum_{n=0}^{\infty} B_n \cos\left[\left(n+\tfrac{1}{2}\right)x\right] = 0, \qquad \pi/2 < x \le \pi. \tag{3.1.26}$$

By inspection, we set

$$f(x) = \frac{1}{4} + \sum_{m=0}^{\infty} \frac{B_m}{2m+1} \left\{1 - \coth\left[\left(m+\tfrac{1}{2}\right)\pi/2\right]\right\} \cos\left[\left(m+\tfrac{1}{2}\right)x\right]. \tag{3.1.27}$$

Substituting Equation 3.1.27 into Equation 3.1.24,

$$F(\theta) = -\frac{\sqrt{2}}{\pi} \int_0^{\theta} \sum_{m=0}^{\infty} B_m \left\{1 - \coth\left[\left(m+\tfrac{1}{2}\right)\pi/2\right]\right\} \times \frac{\sin(x)\sin\left[\left(m+\tfrac{1}{2}\right)x\right]}{\sqrt{\cos(x)-\cos(\theta)}}\,dx \tag{3.1.28}$$

$$= -\frac{\sqrt{2}}{\pi} \sum_{m=0}^{\infty} B_m \left\{1 - \coth\left[\left(m+\tfrac{1}{2}\right)\pi/2\right]\right\} \times \int_0^{\theta} \frac{\sin(x)\sin\left[\left(m+\tfrac{1}{2}\right)x\right]}{\sqrt{\cos(x)-\cos(\theta)}}\,dx \tag{3.1.29}$$

$$= \frac{1}{\pi} \sum_{m=0}^{\infty} B_m \left\{1 - \coth\left[\left(m+\tfrac{1}{2}\right)\pi/2\right]\right\} \times \int_0^{\theta} \frac{\cos\left[\left(m+\tfrac{3}{2}\right)x\right] - \cos\left[\left(m-\tfrac{1}{2}\right)x\right]}{\sqrt{2[\cos(x)-\cos(\theta)]}}\,dx \tag{3.1.30}$$

$$= \frac{1}{2} \sum_{m=0}^{\infty} B_m \left\{1 - \coth\left[\left(m+\tfrac{1}{2}\right)\pi/2\right]\right\} \left\{P_{m+1}[\cos(\theta)] - P_{m-1}[\cos(\theta)]\right\}, \tag{3.1.31}$$

where we used Mehler formula, Equation 1.3.4. Substituting the results from Equation 3.1.31 into Equation 3.1.23, we have that

$$B_n = B_0 P_n(0) + \frac{1}{2} \sum_{m=0}^{\infty} B_m \left\{1 - \coth\left[\left(m+\tfrac{1}{2}\right)\pi/2\right]\right\} \times \left\{\left[P_{m+1}(0) - P_{m-1}(0)\right] P_n(0) + (2m+1) \int_0^1 P_n(t) P_m(t)\,dt\right\}, \tag{3.1.32}$$

because

$$\int_0^{\pi/2} \{P_{m+1}[\cos(\theta)] - P_{m-1}[\cos(\theta)]\} P_n'[\cos(\theta)] \sin(\theta)\, d\theta$$

$$= \int_0^1 [P_{m+1}(t) - P_{m-1}(t)]\, P_n'(t)\, dt \tag{3.1.33}$$

$$= P_n(t)P_{m+1}(t)\Big|_0^1 - P_n(t)P_{m-1}(t)\Big|_0^1 - \int_0^1 \left[P_{m+1}'(t) - P_{m-1}'(t)\right] P_n(t)\, dt \tag{3.1.34}$$

$$= P_n(0)P_{m-1}(0) - P_n(0)P_{m+1}(0) - (2m+1)\int_0^1 P_n(t)P_m(t)\, dt, \tag{3.1.35}$$

and $P_m(1) = 1$. Equation 3.1.32 can be expressed in the succinct form of

$$B_n = B_0 P_n(0) + \widetilde{C}(n,0)B_0 + \sum_{m=1}^{\infty} \widetilde{C}(n,m)B_m, \tag{3.1.36}$$

where

$$\widetilde{C}(n,0) = [\coth(\pi/4) - 1]\, P_n(0) + \int_0^1 P_n(t)\, dt, \tag{3.1.37}$$

and

$$\widetilde{C}(n,m) = \left\{1 - \coth\left[\left(m + \tfrac{1}{2}\right)\pi/2\right]\right\}\left\{[P_{m+1}(0) - P_{m-1}(0)]\, P_n(0) + (2m+1)\int_0^1 P_n(t)P_m(t)\, dt\right\}. \tag{3.1.38}$$

The coefficients B_m/B_0 for $m \geq 1$ are found by solving the linear equations

$$\sum_{m=1}^{\infty} C(n,m)\frac{B_m}{B_0} = D(n), \tag{3.1.39}$$

where

$$C(n,m) = \begin{cases} \widetilde{C}(n,m), & n \neq m, \\ \widetilde{C}(n,m) - 1, & n = m, \end{cases} \tag{3.1.40}$$

and

$$D(n) = -P_n(0) - \widetilde{C}(n,0). \tag{3.1.41}$$

Having found these B_m's, we use Equation 3.1.19 with $x = \pi/4$ to compute B_0 via

$$B_0 = \frac{1}{4}\Bigg/\left\{\sum_{n=0}^{\infty} \frac{B_n/B_0}{2n+1} \cosh\left[\left(n + \tfrac{1}{2}\right)\pi/2\right] \cos\left[\left(n + \tfrac{1}{2}\right)x\right]\right\}. \tag{3.1.42}$$

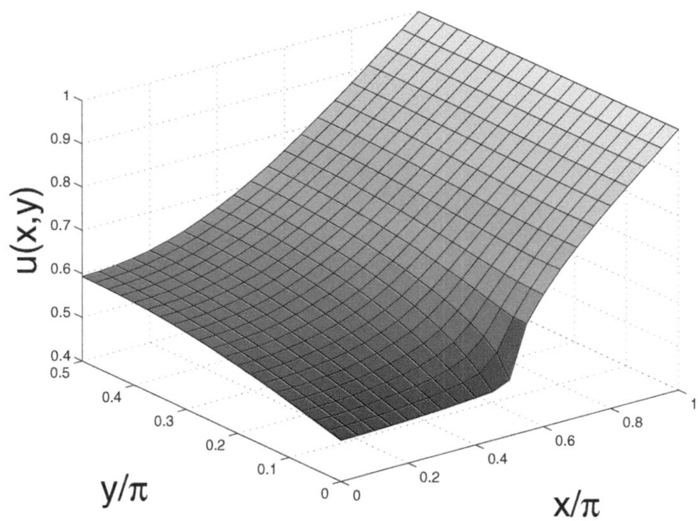

Figure 3.1.2: The solution $u(x, y)$ to the mixed boundary value problem given by Equation 3.1.14 through Equation 3.1.17.

Finally, $u(x, y)$ follows from Equation 3.1.18. We illustrate this solution in Figure 3.1.2.

• Example 3.1.3

Let us solve

$$\frac{\partial^2 u}{\partial x^2} + \frac{\partial^2 u}{\partial y^2} = 0, \qquad -b < x < b, \quad 0 < y < h, \tag{3.1.43}$$

subject to the boundary conditions

$$u(x, 0) = u(x, h) = 0, \quad u(x, c^-) = u(x, c^+), \qquad -b < x < b, \tag{3.1.44}$$

$$\begin{cases} u(x, c) = 1, & |x| < w, \\ \epsilon_1 u_y(x, c^-) = \epsilon_2 u_y(x, c^+), & w < |x| < b, \end{cases} \tag{3.1.45}$$

and

$$u(-b, y) = u(b, y) = 0, \qquad 0 < y < h. \tag{3.1.46}$$

Figure 3.1.3 illustrates the geometry for this problem.

If we use separation of variables, the solution to Equation 3.1.43 is

$$u(x, y) = \sum_{n=1}^{\infty} A_n \left[e^{k_n(y-c)} - e^{-k_n(y+c)} \right] \cos(k_n x), \qquad 0 < y < c, \tag{3.1.47}$$

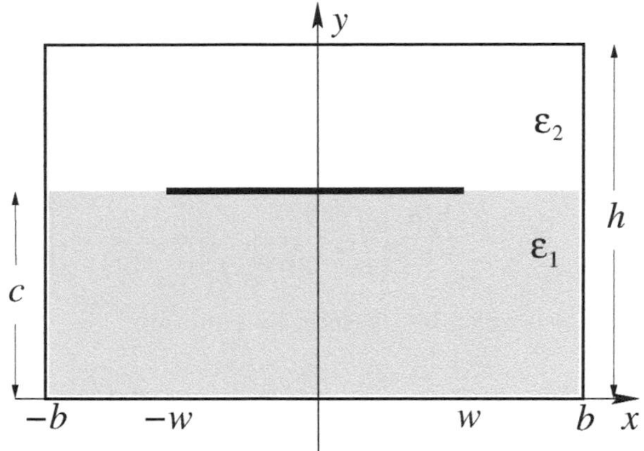

Figure 3.1.3: Schematic of the geometry of Example 3.1.3.

and

$$u(x,y) = \sum_{n=1}^{\infty} B_n \left[e^{k_n(y+c-2h)} - e^{k_n(c-y)} \right], \qquad c < y < h, \tag{3.1.48}$$

where $k_n = \left(n - \frac{1}{2}\right)\pi/b$. Equation 3.1.47 and Equation 3.1.48 satisfy not only Equation 3.1.43, but also Equation 3.1.46 and $u(x,0) = u(x,h) = 0$. Using the fact that $u(x,c^-) = u(x,c^+)$ for $|x| < b$ and Equation 3.1.47 and Equation 3.1.48, we find that

$$B_n = -A_n \frac{1 - e^{-2k_n c}}{1 - e^{-2k_n(h-c)}}. \tag{3.1.49}$$

Finally, we substitute Equation 3.1.47 and Equation 3.1.48 into Equation 3.1.45. The results can be written

$$\sum_{n=1}^{\infty} \frac{1 + H_n}{n - \frac{1}{2}} C_n \cos\left[\left(n - \tfrac{1}{2}\right)\theta\right] = f(\theta), \qquad 0 < \theta < d, \tag{3.1.50}$$

and

$$\sum_{n=1}^{\infty} C_n \cos\left[\left(n - \tfrac{1}{2}\right)\theta\right] = 0, \qquad d < \theta < \pi, \tag{3.1.51}$$

where

$$C_n = \left(n - \tfrac{1}{2}\right) A_n \frac{1 - e^{-2k_n(h-c) - 2k_n c} + \kappa\left[e^{-2k_n h} - e^{-2k_n(h-c)}\right]}{1 - e^{-2k_n(h-c)}}, \tag{3.1.52}$$

$$H_n = -\frac{e^{-2k_n c} + e^{-2k_n(h-c)} - 2e^{-2k_n h} + \kappa\left[e^{-2k_n c} - e^{-2k_n(h-c)}\right]}{1 - e^{-2k_n h} + \kappa\left[e^{-2k_n c} - e^{-2k_n(h-c)}\right]}, \tag{3.1.53}$$

$d = \pi w/b$, $\theta = \pi x/b$, and $\kappa = (\epsilon_1 - \epsilon_2)/(\epsilon_1 + \epsilon_2)$. Here $f(\theta) = 1$.

Kiyono and Shimasaki[5] developed a method for computing C_n. They showed that

$$C_n = \frac{2}{\pi} \int_0^d g(\varphi) \cos\left[\left(n - \tfrac{1}{2}\right) \varphi\right] d\varphi, \tag{3.1.54}$$

where

$$g(\varphi) = \frac{1}{\pi} \frac{d}{d\varphi} \left[\int_0^\varphi \frac{h(\eta)}{\sqrt{\cos(\eta) - \cos(\varphi)}} \, d\eta \right]. \tag{3.1.55}$$

The function $h(\eta)$ is given by the integral equation

$$h(\eta) + \int_0^d K(\eta, \xi) h(\xi) \, d\xi = -\frac{d}{d\eta} \left[\int_\eta^d \frac{\sin(\theta) f(\theta)}{\sqrt{\cos(\eta) - \cos(\theta)}} \, d\theta \right] \tag{3.1.56}$$

for $0 < \eta < d$, where

$$K(\eta, \xi) = K_0(\eta, \xi) + \sin(\eta) \sum_{n=1}^\infty H_n I(\eta, n-1) I(\xi, n-1), \tag{3.1.57}$$

$$\begin{aligned} K_0(\eta, \xi) = & \frac{2}{\pi^2} \frac{\sin(\eta)}{\cos(\eta) - \cos(\xi)} \left\{ \sqrt{\frac{\cos(\eta) - \cos(d)}{\cos(\xi) - \cos(d)}} \ln\left[\sqrt{\frac{1 + \cos(\eta)}{\cos(\eta) - \cos(d)}} \right] \right. \\ & \left. - \sqrt{\frac{\cos(\xi) - \cos(d)}{\cos(\eta) - \cos(d)}} \ln\left[\sqrt{\frac{1 + \cos(\xi)}{\cos(\xi) - \cos(d)}} \right] \right\}, \end{aligned} \tag{3.1.58}$$

$$I(\xi, n) = \frac{\sqrt{2} \cos\left[\left(n + \frac{1}{2}\right) d\right]}{\pi \sqrt{\left(n + \frac{1}{2}\right) [\cos(\xi) - \cos(d)]}} + \sqrt{n + \tfrac{1}{2}} \, R_n(\xi), \tag{3.1.59}$$

$$R_0(\xi) = 1 - \frac{2}{\pi} \arcsin\left[\frac{\cos(d/2)}{\cos(\xi/2)} \right], \tag{3.1.60}$$

$$R_1(\xi) = \cos(\xi) R_0(\xi) + \frac{2\sqrt{2}}{\pi} \cos(d/2) \sqrt{\cos(\xi) - \cos(d)}, \tag{3.1.61}$$

and

$$\begin{aligned} (n+1) R_{n+1}(\xi) - & (2n+1) \cos(\xi) R_n(\xi) + n R_{n-1}(\xi) \\ & = \frac{2\sqrt{2}}{\pi} \cos\left[\left(n + \tfrac{1}{2}\right) d\right] \sqrt{\cos(\xi) - \cos(d)}, \quad n \geq 1. \end{aligned} \tag{3.1.62}$$

---

[5] Kiyono, T., and M. Shimasaki, 1971: On the solution of Laplace's equation by certain dual series equations. *SIAM J. Appl. Math.*, **21**, 245–257.

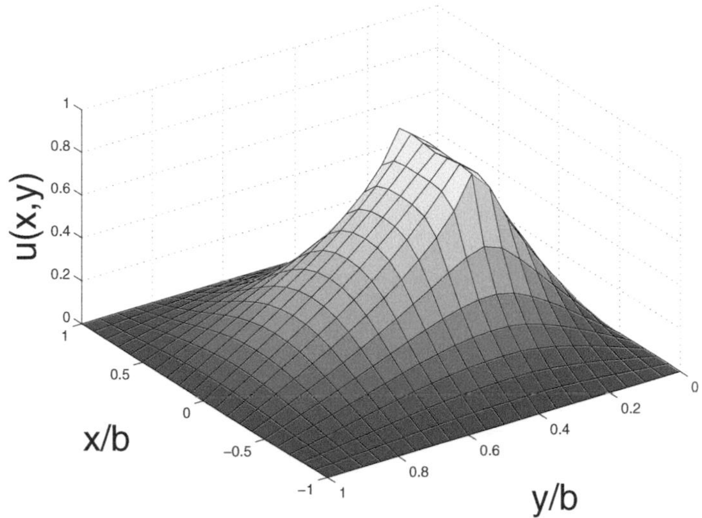

Figure 3.1.4: The solution $u(x, y)$ to the mixed boundary value problem given by Equation 3.1.43 through Equation 3.1.46.

Figure 3.1.4 illustrates the solution $u(x, y)$ when $b = h = 3$, $c = w = 1$, $\epsilon_1 = 1$ and $\epsilon_2 = 2$.

• Example 3.1.4

Let us solve[6]

$$\frac{\partial^2 u}{\partial x^2} + \frac{\partial^2 u}{\partial y^2} = 0, \qquad 0 \le x \le \pi, \quad -\infty < y < \infty, \tag{3.1.63}$$

subject to the boundary conditions

$$u_x(0, y) = u_x(\pi, y) = 0, \qquad -\infty < y < \infty, \tag{3.1.64}$$

$$\lim_{|y|\to\infty} u_y(x, y) \to 1, \qquad 0 \le x \le \pi, \tag{3.1.65}$$

$$u_y(x, 0^-) = u_y(x, 0^+), \qquad 0 \le x \le c, \tag{3.1.66}$$

and

$$\begin{cases} u(x, 0^-) = u(x, 0^+) = 0, & 0 \le x < c, \\ u_y(x, 0^-) = u_y(x, 0^+) = 0, & c < x \le \pi. \end{cases} \tag{3.1.67}$$

---

[6] Reprinted from *Int. J. Heat Mass Transfer*, **19**, J. Dundurs and C. Panek, Heat conduction between bodies with wavy surfaces, 731–736, ©1976, with permission of Elsevier.

If we use separation of variables, the solution to Equation 3.1.63 is

$$u(x,y) = y + A_0 + \sum_{n=1}^{\infty} A_n e^{-ny}\cos(nx), \qquad 0 < y < \infty, \tag{3.1.68}$$

and

$$u(x,y) = y - A_0 - \sum_{n=1}^{\infty} A_n e^{ny}\cos(nx), \qquad -\infty < y < 0. \tag{3.1.69}$$

Equation 3.1.68 and Equation 3.1.69 satisfy not only Equation 3.1.63, but also Equation 3.1.64 through Equation 3.1.66. Finally, we substitute Equation 3.1.68 and Equation 3.1.69 into Equation 3.1.67. This results in

$$A_0 + \sum_{n=1}^{\infty} A_n \cos(nx) = 0, \qquad 0 \le x < c, \tag{3.1.70}$$

and

$$\sum_{n=1}^{\infty} nA_n \cos(nx) = 1, \qquad c < x \le \pi. \tag{3.1.71}$$

To solve Equation 3.1.70 and Equation 3.1.71, we first substitute $x = \pi - \xi$, $c = \pi - \gamma$, $A_0 = a_0/2$ and $A_n = (-1)^n a_n$. We then obtain

$$\sum_{n=1}^{\infty} na_n \cos(n\xi) = 1, \qquad 0 \le \xi < \gamma, \tag{3.1.72}$$

and

$$\tfrac{1}{2}a_0 + \sum_{n=1}^{\infty} a_n \cos(n\xi) = 0, \qquad \gamma < \xi \le \pi. \tag{3.1.73}$$

Recently Sbragaglia and Prosperetti[7] also solved this dual series when it arose during their study of the effects of surface deformation on a type of superhydrophobic surface.

Equation 3.1.72 and Equation 3.1.73 are an example of Equation 3.0.6. For the special case $p = 1$, Sneddon[8] showed that the solution to Equation 3.0.6 is

$$a_0 = \frac{2}{\pi}\left[\frac{\pi}{\sqrt{2}}\int_0^c h(t)\,dt + \int_c^{\pi} g(t)\,dt\right], \tag{3.1.74}$$

---

[7] Sbragaglia, M., and A. Prosperetti, 2007: A note on the effective slip properties for microchannel flows with ultrahydrophobic surfaces. *Phys. Fluids*, **19**, Art. No. 043603.

[8] See Section 5.4.3 in Sneddon, I. N., 1966: *Mixed Boundary Value Problems in Potential Theory.* North Holland, 283 pp.

and

$$a_n = \frac{2}{\pi}\left[\frac{\pi}{2\sqrt{2}}\int_0^c h(t)\left\{P_n[\cos(t)] + P_{n-1}[\cos(t)]\right\} dt + \int_c^\pi g(t)\cos(nt)\, dt\right], \tag{3.1.75}$$

where $n = 1, 2, 3, \ldots$,

$$h(t) = \frac{2}{\pi}\frac{d}{dt}\left\{\int_0^t \frac{\sin(x/2)}{\sqrt{\cos(x) - \cos(t)}}\, dx \times \left[\int_0^x f(\xi)\, d\xi - \tfrac{1}{2}\alpha a_0 x + \sum_{n=1}^\infty b_n \sin(nx)\right]\right\}, \tag{3.1.76}$$

and

$$b_n = \frac{2}{\pi}\int_c^\pi g(\xi)\cos(n\xi)\, d\xi. \tag{3.1.77}$$

Applying these results to Equation 3.1.72 and Equation 3.1.73, we find that

$$h(t) = \frac{2}{\pi}\frac{d}{dt}\left[\int_0^t \frac{\xi\sin(\xi/2)}{\sqrt{\cos(\xi) - \cos(t)}}\, d\xi\right] = \frac{\sqrt{2}\sin(t)}{1 + \cos(t)} = \sqrt{2}\tan(t/2). \tag{3.1.78}$$

Therefore,

$$a_0 = \frac{2}{\sqrt{2}}\int_0^\gamma h(t)\, dt = -4\ln[\cos(\gamma/2)], \tag{3.1.79}$$

and

$$a_n = \frac{1}{\sqrt{2}}\int_0^\gamma h(t)\{P_n[\cos(\theta)] + P_{n-1}[\cos(t)]\}\, dt \tag{3.1.80}$$

$$= \frac{1}{n}\{P_{n-1}[\cos(\gamma)] - P_n[\cos(\gamma)]\}. \tag{3.1.81}$$

Returning to the original Fourier coefficient,

$$A_0 = -2\ln[\sin(c/2)], \tag{3.1.82}$$

and

$$A_n = -\frac{1}{n}\{P_n[\cos(c)] + P_{n-1}[\cos(c)]\}. \tag{3.1.83}$$

Figure 3.1.5 illustrates the solution when $c = \frac{1}{2}$.

• Example 3.1.5

Let us solve[9]

$$\frac{\partial^2 u}{\partial x^2} + \frac{\partial^2 u}{\partial y^2} = 0, \qquad 0 \le x \le L, \quad 0 < y < h, \tag{3.1.84}$$

[9] Adapted from Westmann, R. A., and W. H. Yang, 1967: Stress analysis of cracked rectangular beams. *J. Appl. Mech.*, **34**, 693–701.

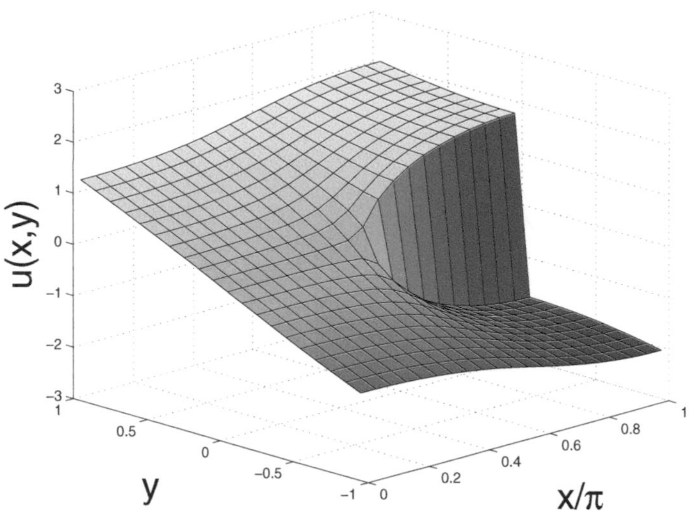

Figure 3.1.5: The solution $u(x, y)$ to the mixed boundary value problem given by Equation 3.1.63 through Equation 3.1.67.

subject to the boundary conditions

$$u_x(0, y) = u_x(L, y) = 0, \qquad 0 < y < h, \tag{3.1.85}$$

$$u_y(x, h) = S, \qquad 0 < x < L, \tag{3.1.86}$$

and

$$\begin{cases} u_y(x, 0) = 0, & 0 < x < a, \\ u(x, 0) = 0, & a < x < L, \end{cases} \tag{3.1.87}$$

where $L > a$.

If we use separation of variables, the solution to Equation 3.1.84 is

$$u(x, y) = Sy + \sum_{n=0}^{\infty} A_n \frac{\cosh[n\pi(h - y)/L]}{\cosh(n\pi h/L)} \cos\left(\frac{n\pi x}{L}\right). \tag{3.1.88}$$

Equation 3.1.88 satisfies not only Equation 3.1.84, but also Equation 3.1.85 and Equation 3.1.86. Upon substituting Equation 3.1.88 into Equation 3.1.87, we find the dual series

$$\sum_{n=0}^{\infty} \left(\frac{n\pi}{L}\right) A_n \tanh\left(\frac{n\pi h}{L}\right) \cos\left(\frac{n\pi x}{L}\right) = S, \qquad 0 < x < a, \tag{3.1.89}$$

and

$$\sum_{n=0}^{\infty} A_n \cos\left(\frac{n\pi x}{L}\right) = 0, \qquad a < x < L. \tag{3.1.90}$$

To solve the dual equations, Equation 3.1.89 and Equation 3.1.90, let us assume that $u(x,0)$ is known for $0 < x < L$. Then

$$u(x,0) = \sum_{n=0}^{\infty} A_n \cos\left(\frac{n\pi x}{L}\right). \tag{3.1.91}$$

From the theory of Fourier series we have that

$$A_0 = \frac{1}{L}\int_0^L u(x,0)\,dx, \tag{3.1.92}$$

and

$$A_n = \frac{2}{L}\int_0^L u(x,0)\cos\left(\frac{n\pi x}{L}\right)\,dx. \tag{3.1.93}$$

Let us assume that we can express $u(x,0)$ as

$$u(x,0) = \int_x^a \frac{h(t)}{\sqrt{t^2 - x^2}}\,dt, \qquad 0 < x < a. \tag{3.1.94}$$

Upon substituting Equation 3.1.94 into Equation 3.1.92 and Equation 3.1.93,

$$A_0 = \frac{1}{L}\int_0^a \left[\int_x^a \frac{h(t)}{\sqrt{t^2 - x^2}}\,dt\right]dx = \frac{\pi}{2L}\int_0^a h(t)\,dt, \tag{3.1.95}$$

and

$$A_n = \frac{2}{L}\int_0^a \left[\int_x^a \frac{h(t)}{\sqrt{t^2 - x^2}}\,dt\right]\cos\left(\frac{n\pi x}{L}\right)\,dx \tag{3.1.96}$$

$$= \frac{\pi}{L}\int_0^a h(t) J_0\left(\frac{n\pi t}{L}\right)\,dt \tag{3.1.97}$$

for $n = 1, 2, 3, \ldots$. Next, we substitute Equation 3.1.95 and Equation 3.1.97 into Equation 3.1.89,

$$\sum_{n=1}^{\infty} \frac{n\pi}{L}\left[\int_0^a h(t) J_0\left(\frac{n\pi t}{L}\right)\,dt\right]\tanh\left(\frac{n\pi h}{L}\right)\cos\left(\frac{n\pi x}{L}\right) = \frac{LS}{\pi}. \tag{3.1.98}$$

Integrating both sides of Equation 3.1.98 from 0 to x and interchanging the order of integration and summation,

$$\int_0^a h(t)\left[\sum_{n=1}^{\infty}\tanh\left(\frac{n\pi h}{L}\right) J_0\left(\frac{n\pi t}{L}\right)\sin\left(\frac{n\pi x}{L}\right)\right]dt = \frac{LSx}{\pi}. \tag{3.1.99}$$

If we integrate the function $\csc(\pi z)e^{i\pi z}J_0(\pi t z/L)\sin(\pi x z/L)$ around a contour which consists of 1) the positive real axis, 2) the positive imaginary axis, and 3) the arc in the first quadrant of the circle $|z| = N + \frac{1}{2}$, Sneddon and Srivastav[10] showed that

$$\sum_{n=1}^{\infty} J_0\left(\frac{n\pi t}{L}\right) \sin\left(\frac{n\pi x}{L}\right) = \frac{L\,H(x-t)}{\pi\sqrt{x^2-t^2}} - \frac{1}{\pi}\int_0^{\infty} \frac{e^{-\xi}\sinh(\xi x/L) I_0(\xi t/L)}{\sinh(\xi)}\, d\xi. \tag{3.1.100}$$

Therefore, Equation 3.1.99 can be rewritten and made nondimensional with respect to a so that it becomes

$$\int_0^x \frac{h(t)}{\sqrt{x^2-t^2}}\, dt = Sax - \int_0^1 K(x,\eta)h(\eta)\, d\eta, \qquad 0 < x < 1, \tag{3.1.101}$$

where

$$K(x,\eta) = \frac{a\pi}{L}\sum_{n=1}^{\infty}\left[\tanh\left(\frac{n\pi h}{L}\right) - 1\right] J_0\left(\frac{n\pi a\eta}{L}\right) \sin\left(\frac{n\pi a x}{L}\right) - \frac{a}{L}\int_0^{\infty} \frac{e^{-\xi}\sinh(\xi a x/L) I_0(\xi a t/L)}{\sinh(\xi)}\, d\xi. \tag{3.1.102}$$

Equation 3.1.101 is an integral equation of the Abel type. Applying Equation 1.2.14 and Equation 1.2.15, we find that

$$h(t) = Sat + \int_0^1 L(t,\eta)h(\eta)\, d\eta, \qquad 0 \le t \le 1, \tag{3.1.103}$$

where

$$L(t,\eta) = -\frac{2}{\pi}\frac{d}{dt}\left[\int_0^t \frac{x\,K(x,\eta)}{\sqrt{t^2-x^2}}\, dx\right] \tag{3.1.104}$$

$$= \left(\frac{a\pi}{L}\right)^2 \sum_{n=1}^{\infty} nt\left[1 - \tanh\left(\frac{n\pi h}{L}\right)\right] J_0\left(\frac{n\pi a\eta}{L}\right) J_0\left(\frac{n\pi a t}{L}\right) + \frac{a^2}{L^2}\int_0^{\infty} \frac{\xi t e^{-\xi} I_0(\xi a t/L) I_0(\xi a\eta/L)}{\sinh(\xi)}\, d\xi. \tag{3.1.105}$$

Figure 3.1.6 illustrates the solution when $a/L = 0.5$ and $h/L = 1$. Keer and Sve[11] applied this technique to the biharmonic equation. Later, Sezgin[12] used this method to solve a coupled set of Laplace-like equations.

---

[10] Sneddon, I. N., and R. P. Srivastav, 1964: Dual series relationships. *Proc. Roy. Soc. Edinburgh, Ser. A*, **66**, 150–191.

[11] Keer, L. M., and C. Sve, 1970: On the bending of cracked plates. *Int. J. Solids Struct.*, **6**, 1545–1559.

[12] Sezgin, M., 1987: Magnetohydrodynamic flow in a rectangular duct. *Int. J. Numer. Meth. Fluids*, **7**, 697–718.

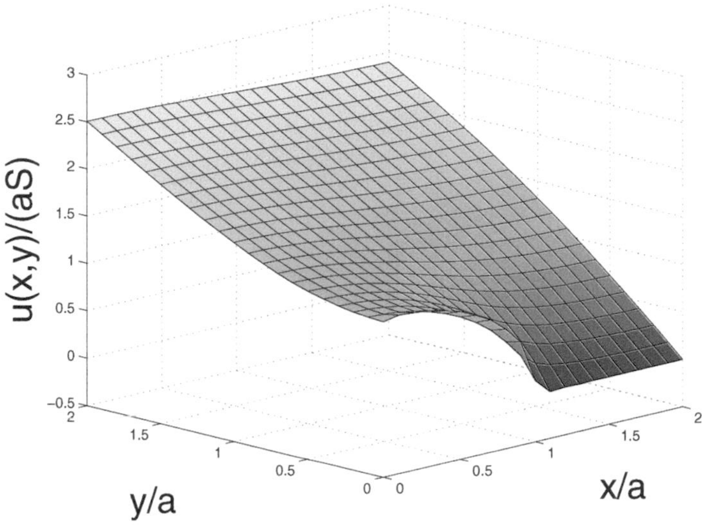

Figure 3.1.6: The solution $u(x, y)$ to the mixed boundary value problem given by Equation 3.1.84 through Equation 3.1.87.

• Example 3.1.6

So far we have only encountered Fourier cosine series in a rectilinear problem. Consider now the following problem in spherical coordinates:[13]

$$\frac{1}{r^2}\frac{\partial}{\partial r}\left(r^2\frac{\partial u}{\partial r}\right) + \frac{\partial^2 u}{\partial \varphi^2} = 0, \qquad a < r < \infty, \quad 0 \le \varphi \le \pi, \tag{3.1.106}$$

subject to the boundary conditions

$$\lim_{\varphi\to 0} |u(r,\varphi)| < \infty, \quad \lim_{\varphi\to\pi} |u(r,\varphi)| < \infty, \qquad a < r < \infty, \tag{3.1.107}$$

$$\lim_{r\to\infty} u(r,\varphi) = u_0, \qquad 0 \le \varphi \le \pi, \tag{3.1.108}$$

and

$$\begin{cases} u(a,\varphi) = 0, & 0 < \varphi < \varphi_0, \\ u_r(a,\varphi) = 0, & \varphi_0 < \varphi < \pi. \end{cases} \tag{3.1.109}$$

If we use separation of variables, the solution to Equation 3.1.106 is

$$u(r,\varphi) = u_0 + \sum_{n=0}^{\infty} A_n \frac{e^{n(a-r)}}{r}\cos(n\varphi). \tag{3.1.110}$$

---

[13] See Baldo, M., A. Grassi, and A. Raudino, 1989: Modeling the mechanisms of enzyme reactivity by the rototranslational diffusion equation. *Phys. Review, Ser. A*, **39**, 3700–3702.

Equation 3.1.110 satisfies not only Equation 3.1.106, but also Equation 3.1.107 and Equation 3.1.108. Upon substituting Equation 3.1.110 into Equation 3.1.109, we find the dual series

$$\sum_{n=0}^{\infty} A_n \cos(n\varphi) = -u_0 a, \qquad 0 < \varphi < \varphi_0, \tag{3.1.111}$$

and

$$\sum_{n=0}^{\infty} A_n (1 + na) \cos(n\varphi) = 0, \qquad \varphi_0 < \varphi < \pi. \tag{3.1.112}$$

Let us rewrite Equation 3.1.112 as

$$\sum_{n=0}^{\infty} A_n \cos(n\varphi) = -\sum_{m=0}^{\infty} A_m ma \cos(m\varphi), \qquad \varphi_0 < \varphi < \pi. \tag{3.1.113}$$

Because the left side of both Equation 3.1.111 and Equation 3.1.113 are the same, the Fourier cosine series expresses the function

$$f(\varphi) = \begin{cases} -u_0 a, & 0 < \varphi < \varphi_0, \\ -\sum_{m=0}^{\infty} A_m ma \cos(m\varphi), & \varphi_0 < \varphi < \pi, \end{cases} \tag{3.1.114}$$

which is given by the right side of Equation 3.1.111 and Equation 3.1.113. From the theory of Fourier series,

$$A_0 = -\frac{a u_0 \varphi_0}{\pi} + \frac{1}{\pi} \sum_{m=1}^{\infty} A_m \sin(m\varphi_0), \tag{3.1.115}$$

and

$$\begin{aligned} A_n = &-\frac{2au_0 \sin(n\varphi_0)}{n\pi} - \frac{n}{\pi} A_n \left[\pi - \varphi_0 - \frac{\sin(2n\varphi_0)}{2n}\right] \\ &+ \frac{1}{\pi} \sum_{\substack{m=1 \\ m\neq n}}^{\infty} m A_m \left\{ \frac{\sin[(m-n)\varphi_0]}{m-n} + \frac{\sin[(m+n)\varphi_0]}{m+n} \right\}. \end{aligned} \tag{3.1.116}$$

Equation 3.1.115 and Equation 3.1.116 yield an infinite set of equations. If we only retain the first N terms, we can invert these equations and find approximate values for the A_n's. The potential then follows from Equation 3.1.110. Figure 3.1.7 illustrates this solution when $\varphi = \pi/3$ and $N = 100$.

Problems

1. Solve Laplace's equation

$$\frac{1}{r}\frac{\partial}{\partial r}\left(r\frac{\partial u}{\partial r}\right) + \frac{1}{r^2}\frac{\partial^2 u}{\partial \theta^2} = 0, \qquad 0 \le r < \infty, \quad 0 < \theta < 2\pi,$$

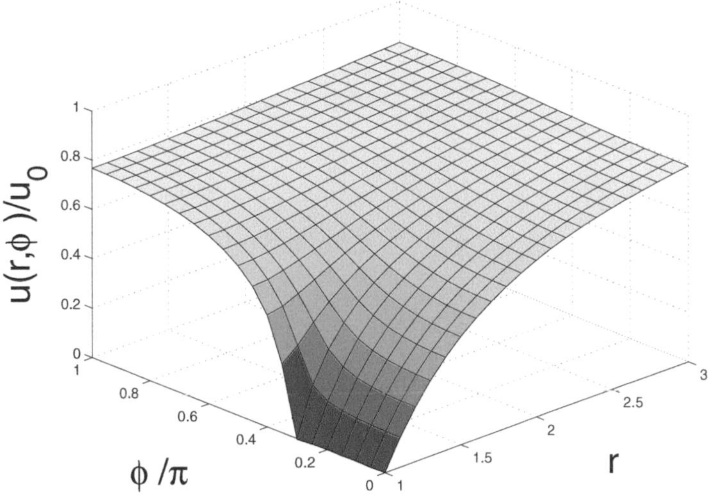

Figure 3.1.7: The solution $u(r, \varphi)$ to the mixed boundary value problem given by Equation 3.1.106 through Equation 3.1.109 when $a = 1$ and $\varphi_0 = \pi/3$.

subject to the boundary conditions

$$\lim_{r\to 0} |u(r,\theta)| < \infty, \qquad \lim_{r\to\infty} u(r,\theta) \to Vr\cos(\theta), \qquad 0 < \theta < 2\pi,$$

$$u(1^-,\theta) = u(1^+,\theta), \qquad 0 < \theta < 2\pi,$$

and

$$\begin{cases} \kappa u_r(1^-,\theta) = u_r(1^+,\theta), & -\pi/2 < \theta < \pi/2, \\ u(1,\theta) = 0, & \pi/2 < \theta < 3\pi/2, \end{cases}$$

where 1^- and 1^+ denote points slightly inside and outside of the circle $r = 1$, respectively.

Step 1: Use separation of variables and show that the general solution to the problem is

$$u(r,\theta) = Vr\cos(\theta) + \sum_{n=0}^{\infty} D_n r^n \cos(n\theta), \qquad 0 < r < 1,$$

and

$$u(r,\theta) = Vr\cos(\theta) + \sum_{n=0}^{\infty} D_n r^{-n} \cos(n\theta), \qquad 1 < r < \infty.$$

Note that this solution satisfies not only the differential equation, but also the first three boundary conditions.

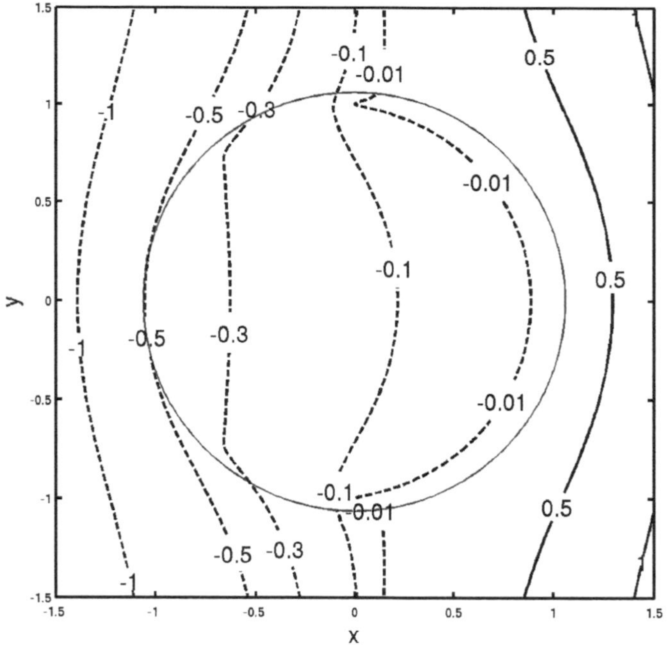

Problem 1

Step 2: Using the final boundary condition and the results given by Equation 2.3.10 through Equation 2.3.16, show that

$$D_0 = -\frac{V}{\kappa + 1}, \qquad D_1 = \frac{1 - 2\kappa}{2(\kappa + 1)} V, \qquad D_2 = -\frac{V}{2(\kappa + 1)},$$

and

$$D_{2n-1} = -D_{2n} = (-1)^{n+1} \frac{1 \cdot 3 \cdot \cdots \cdot (2n-3)V}{2 \cdot 4 \cdots 2n},$$

where $n = 2, 3, 4, \ldots$. The figure entitled Problem 1 illustrates the solution $u(r, \theta)/V$ when $\kappa = 6$.

3.2 DUAL FOURIER SINE SERIES

Dual Fourier sine series arise in the same manner as dual Fourier cosine series in mixed boundary value problems in rectangular domains. However, for some reason, they do not appear as often and the following example is the only one that we present.

Consider Laplace's equation on a semi-infinite strip

$$\frac{\partial^2 u}{\partial x^2} + \frac{\partial^2 u}{\partial y^2} = 0, \qquad 0 < x < L, \quad 0 < y < \infty, \tag{3.2.1}$$

subject to the boundary conditions

$$u(0,y)=u(L,y)=0, \qquad 0<y<\infty, \tag{3.2.2}$$

$$\lim_{y\to\infty} u(x,y)\to 0, \qquad 0<x<L, \tag{3.2.3}$$

and

$$\begin{cases} u(x,0)=U_0(x), & 0<x<\ell, \\ u_y(x,0)=0, & \ell<x<L, \end{cases} \tag{3.2.4}$$

where $L>\ell$ and $U_0(x)$ is a known odd function.

The solution to Equation 3.2.1 and the boundary conditions Equation 3.2.2 and Equation 3.2.3 is

$$u(x,y)=\sum_{n=1}^{\infty} A_n e^{-\lambda_n y}\sin(\lambda_n x), \tag{3.2.5}$$

where $\lambda_n=n\pi/L$. Substituting Equation 3.2.5 into Equation 3.2.4 yields the dual series of

$$\sum_{n=1}^{\infty} A_n \sin(\lambda_n x)=U_0(x), \qquad 0<x<\ell, \tag{3.2.6}$$

and

$$\sum_{n=1}^{\infty} \lambda_n A_n \sin(\lambda_n x)=0, \qquad \ell<x<L. \tag{3.2.7}$$

To solve these dual equations, let us follow Williams[14] and introduce $p(x)$ such that

$$\sum_{n=1}^{\infty} \lambda_n A_n \sin(\lambda_n x)=\begin{cases} p(x), & 0<x<\ell, \\ 0, & \ell<x<L. \end{cases} \tag{3.2.8}$$

Therefore, the Fourier coefficient is given by

$$\lambda_n A_n=\frac{2}{L}\int_0^{\ell} p(x)\sin(\lambda_n x)\,dx, \tag{3.2.9}$$

or

$$n\pi A_n=\int_{-\ell}^{\ell} p(x)\sin(\lambda_n x)\,dx \tag{3.2.10}$$

if we assume that $p(x)$ is an odd function. Upon substituting Equation 3.2.10 into Equation 3.2.6,

$$\frac{1}{\pi}\sum_{n=1}^{\infty}\frac{1}{n}\sin\left(\frac{n\pi x}{L}\right)\left[\int_{-\ell}^{\ell} p(t)\sin\left(\frac{n\pi t}{L}\right)dt\right]=U_0(x), \qquad |x|<\ell. \tag{3.2.11}$$

---

[14] Williams, W. E., 1964: The solution of dual series and dual integral equations. *Proc. Glasgow Math. Assoc.*, **6**, 123–129.

Let us assume that we can Chebyshev express $p(t)$ by the expansion

$$p(t) = \frac{1}{\sqrt{\ell^2 - t^2}} \sum_{m=0}^{\infty} B_{2m+1} T_{2m+1}(t/\ell), \tag{3.2.12}$$

where $T_n(\cdot)$ is the nth Chebyshev polynomial. Substituting Equation 3.2.12 into Equation 3.2.11 and noting[15] that

$$\int_{-t}^{t} e^{ip\tau} T_n(\tau/t) \frac{d\tau}{\sqrt{t^2 - \tau^2}} = i^n \pi J_n(pt), \tag{3.2.13}$$

then

$$\sum_{n=1}^{\infty} \frac{1}{n} \left[\sum_{m=0}^{\infty} (-1)^m B_{2m+1} J_{2m+1}\left(\frac{n\pi\ell}{L}\right) \right] \sin\left(\frac{n\pi x}{L}\right) = U_0(x). \tag{3.2.14}$$

Because the Chebyshev polynomial expansion for $\sin(ax)$ is

$$\sin(ax) = 2 \sum_{k=0}^{\infty} (-1)^k J_{2k+1}(a) T_{2k+1}(x), \qquad a > 0, \tag{3.2.15}$$

and reexpressing $U_0(x)$ as

$$U_0(x) = \sum_{k=0}^{\infty} b_{2k+1} T_{2k+1}(x/\ell), \tag{3.2.16}$$

we obtain from Equation 3.2.14 the following set of simultaneous equations

$$\sum_{m=0}^{\infty} (-1)^{m+k} B_{2m+1} C_{2k+1,2m+1} = b_{2k+1}, \qquad k = 0, 1, 2, \ldots, \tag{3.2.17}$$

where

$$C_{2k+1,2m+1} = 2 \sum_{n=1}^{\infty} \frac{1}{n} J_{2m+1}\left(\frac{n\pi\ell}{L}\right) J_{2k+1}\left(\frac{n\pi\ell}{L}\right). \tag{3.2.18}$$

Finally, we note that Equation 3.2.10 and Equation 3.2.12 yield

$$nA_n = \sum_{m=0}^{\infty} (-1)^m B_{2m+1} J_{2m+1}\left(\frac{n\pi\ell}{L}\right). \tag{3.2.19}$$

---

[15] Gradshteyn, I. S., and I. M. Ryzhik, 1965: *Table of Integrals, Series, and Products*. Academic Press, Formula 7.355.1 and 7.355.2.

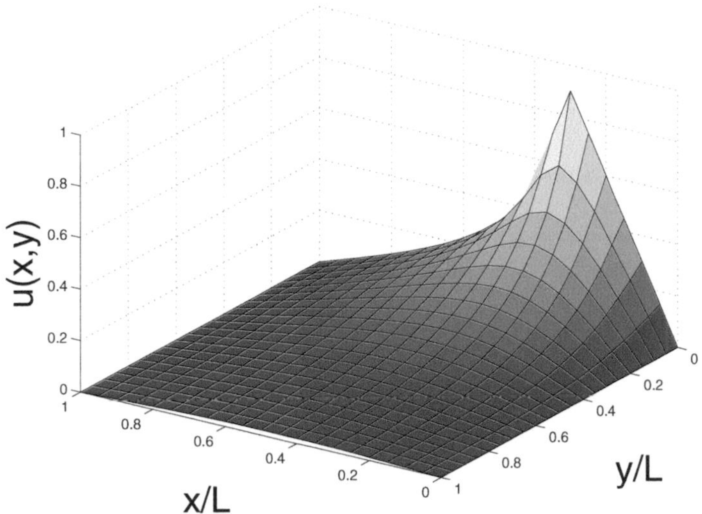

Figure 3.2.1: The solution $u(x, y)$ to the mixed boundary value problem given by Equation 3.2.1 through Equation 3.2.4 when $U_0(x) = x/\ell$ and $L/\ell = 3$.

For a given $U_0(x)$, we compute b_{2n+1}. Equation 3.2.17 then yields B_{2m+1} while Equation 3.2.19 yields A_n. Finally, $u(x, y)$ follows from Equation 3.2.5. Figure 3.2.1 illustrates the solution when $U_0(x) = x/\ell$ and $L/\ell = 3$.

An alternative method of solving the dual equations Equation 3.2.6 and Equation 3.2.7 involves introducing $B_n = \lambda_n A_n$, $C_n = (-1)^{n-1}B_n$, $\xi = \pi x/L$, $\eta = \pi - \xi$, and $\gamma = \pi - \pi\ell/L$. Then, these dual equations become

$$\sum_{n=1}^{\infty} nC_n \sin(n\eta) = \pi U_0(\pi - \eta)/\pi, \qquad 0 < \eta < \gamma, \tag{3.2.20}$$

and

$$\sum_{n=1}^{\infty} C_n \sin(n\eta) = 0, \qquad \gamma < \eta < \pi. \tag{3.2.21}$$

Why did we derive Equation 3.2.20 and Equation 3.2.21? Consider the dual series:

$$\begin{cases} \sum_{n=1}^{\infty} na_n \sin(nx) = f(x), & 0 \le x < c, \\ \sum_{n=1}^{\infty} a_n \sin(nx) = 0, & c < x \le \pi. \end{cases} \tag{3.2.22}$$

To find a_n, we introduce the Fourier sine series:

$$y(x) = \sum_{n=1}^{\infty} a_n \sin(nx). \tag{3.2.23}$$

Clearly, to satisfy Equation 3.2.22, $y(x) = 0$ if $c < x \leq \pi$. On the interval $[0, c)$, we assume that

$$y(x) = \sin\left(\frac{x}{2}\right) \int_x^c \frac{h(t)}{\sqrt{\cos(x) - \cos(t)}}\, dt, \qquad 0 \leq x < c, \tag{3.2.24}$$

where $h(x)$ is unknown. Why have we chosen such an unusual definition for $y(x)$?

Recall that we seek an a_n that satisfies Equation 3.2.22. From the theory of Fourier series, we know that

$$a_n = \frac{2}{\pi} \int_0^c y(x) \sin(nx)\, dx \tag{3.2.25}$$

$$= \frac{1}{\pi} \int_0^c h(t) \left\{ \int_0^t \frac{\cos\left[\left(n - \frac{1}{2}\right) x\right] - \cos\left[\left(n + \frac{1}{2}\right) x\right]}{\sqrt{\cos(x) - \cos(t)}}\, dx \right\} dt. \tag{3.2.26}$$

We now simplify Equation 3.2.26 by applying Mehler integral, Equation 1.3.4:

$$P_n[\cos(x)] = \frac{\sqrt{2}}{\pi} \int_0^x \frac{\cos\left[\left(n + \frac{1}{2}\right) \xi\right]}{\sqrt{\cos(\xi) - \cos(x)}}\, d\xi, \tag{3.2.27}$$

and find that

$$a_n = \frac{1}{\sqrt{2}} \int_0^c h(t) \left\{ P_{n-1}[\cos(t)] - P_n[\cos(t)] \right\} dt. \tag{3.2.28}$$

Turning now to the first equation in Equation 3.2.22, we eliminate n inside of the summation by integrating both sides:

$$\sum_{n=1}^{\infty} a_n \left[1 - \cos(nx)\right] = \int_0^x f(\xi)\, d\xi, \tag{3.2.29}$$

or

$$\int_0^c h(t) \left(\frac{1}{\sqrt{2}} \sum_{n=1}^{\infty} \left[1 - \cos(nx)\right] \left\{ P_{n-1}[\cos(t)] - P_n[\cos(t)] \right\} \right) dt = \int_0^x f(\xi)\, d\xi. \tag{3.2.30}$$

Using the results from Example 1.3.1, we can simplify Equation 3.2.30 to

$$\sin\left(\frac{x}{2}\right) \int_0^c h(t) \frac{H(x-t)}{\sqrt{\cos(t) - \cos(x)}}\, dt = \int_0^x f(\xi)\, d\xi, \tag{3.2.31}$$

or

$$\int_0^x \frac{h(t)}{\sqrt{\cos(t) - \cos(x)}}\, dt = \csc\left(\frac{x}{2}\right) \int_0^x f(\xi)\, d\xi. \tag{3.2.32}$$

Finally, using the results given in Equation 1.2.9 and Equation 1.2.10, the solution to the integral equation Equation 3.2.32 is

$$h(t) = \frac{2}{\pi}\frac{d}{dt}\left\{\int_0^t \frac{\cos(x/2)}{\sqrt{\cos(x)-\cos(t)}}\left[\int_0^x f(\xi)\,d\xi\right]dx\right\}, \tag{3.2.33}$$

or

$$h(t) = \frac{2}{\pi}\cot\left(\frac{t}{2}\right)\int_0^t \frac{\sin(x/2)f(x)}{\sqrt{\cos(x)-\cos(t)}}\,dx. \tag{3.2.34}$$

Consider now the set of dual equation similar to Equation 3.2.22 is

$$\begin{cases} \sum_{n=1}^{\infty} n a_n \sin(nx) = 0, & 0 \le x < c, \\ \sum_{n=1}^{\infty} a_n \sin(nx) = f(x), & c < x \le \pi. \end{cases} \tag{3.2.35}$$

To find a_n, we begin by introducing a function $g(x)$, defined by

$$g(x) = \sum_{n=1}^{\infty} n a_n \sin(nx), \qquad c < x \le \pi. \tag{3.2.36}$$

Next, we assume that $g(x)$ can be represented by

$$g(x) = -\frac{d}{dx}\left[\sin\left(\frac{x}{2}\right)\int_c^x \frac{h(t)}{\sqrt{\cos(t)-\cos(x)}}\,dt\right] \tag{3.2.37}$$

over the range $c < x \le \pi$, where $h(t)$ is unknown. From the properties of Fourier sine series,

$$n a_n = \frac{2}{\pi}\int_c^{\pi} g(x)\sin(nx)\,dx \tag{3.2.38}$$

$$= -\frac{2}{\pi}\int_c^{\pi} \frac{d}{dx}\left[\sin\left(\frac{x}{2}\right)\int_c^x \frac{h(t)}{\sqrt{\cos(t)-\cos(x)}}\,dt\right]\sin(nx)\,dx \tag{3.2.39}$$

$$= \frac{2n}{\pi}\int_c^{\pi} \sin\left(\frac{x}{2}\right)\left[\int_c^x \frac{h(t)}{\sqrt{\cos(t)-\cos(x)}}\,dt\right]\cos(nx)\,dx \tag{3.2.40}$$

by integration by parts. Interchanging the order of integration and using Equation 1.3.5, we have

$$a_n = \frac{1}{\sqrt{2}}\int_c^{\pi} h(t)\left\{P_n[\cos(t)] - P_{n-1}[\cos(t)]\right\}dt. \tag{3.2.41}$$

Substituting Equation 3.2.41 into Equation 3.2.35, we obtain the integral equation

$$\frac{1}{\sqrt{2}}\int_c^\pi h(t)\left[\sum_{n=1}^\infty \{P_n[\cos(t)] - P_{n-1}[\cos(t)]\}\sin(nx)\right] dt = f(x). \quad \textbf{(3.2.42)}$$

Using the results from Problem 3 in Section 1.3, Equation 3.2.42 simplifies to

$$\int_x^\pi \frac{h(t)}{\sqrt{\cos(x) - \cos(t)}}\, dt = -\csc\left(\frac{x}{2}\right) f(x), \qquad c < x \le \pi. \quad \textbf{(3.2.43)}$$

From Equation 1.2.11 and Equation 1.2.12, we obtain

$$h(t) = \frac{2}{\pi}\frac{d}{dt}\left[\int_t^\pi \frac{f(x)\,\cos(x/2)}{\sqrt{\cos(t) - \cos(x)}}\, dx\right]. \quad \textbf{(3.2.44)}$$

Using the results from Equations 3.2.22, 3.2.28, 3.2.34, 3.2.35, 3.2.41, and 3.2.44, the solution to the dual equations

$$\begin{cases} \sum_{n=1}^\infty nc_n \sin(ny) = g(\pi - y), & 0 \le y < \gamma, \\ \sum_{n=1}^\infty c_n \sin(ny) = f(\pi - y), & \gamma < y \le \pi, \end{cases} \quad \textbf{(3.2.45)}$$

is

$$c_n = \frac{1}{\sqrt{2}}\int_0^\pi h(t)\,\{P_{n-1}[\cos(t)] - P_n[\cos(t)]\}\, dt, \quad \textbf{(3.2.46)}$$

where

$$h(t) = \frac{2}{\pi}\cot\left(\frac{t}{2}\right)\int_0^t \frac{g(\pi - \xi)\,\sin(\xi/2)}{\sqrt{\cos(\xi) - \cos(t)}}\, d\xi, \qquad 0 \le t < \gamma, \quad \textbf{(3.2.47)}$$

and

$$h(t) = -\frac{2}{\pi}\frac{d}{dt}\left[\int_t^\pi \frac{f(\pi - \xi)\,\cos(\xi/2)}{\sqrt{\cos(t) - \cos(\xi)}}\, d\xi\right], \qquad \gamma < t \le \pi. \quad \textbf{(3.2.48)}$$

Therefore, the solution to Equation 3.2.20 and Equation 3.2.21 is

$$C_n = \frac{1}{\sqrt{2}}\int_0^\gamma h(t)\,\{P_{n-1}[\cos(t)] - P_n[\cos(t)]\}\, dt, \quad \textbf{(3.2.49)}$$

where

$$h(t) = \frac{2}{L}\cot\left(\frac{t}{2}\right)\int_0^t \frac{U_0[L(\pi - \xi)/\pi]\,\sin(\xi/2)}{\sqrt{\cos(\xi) - \cos(t)}}\, d\xi, \qquad 0 \le t < \gamma. \quad \textbf{(3.2.50)}$$

Consequently, making the back substitution, the dual series

$$\begin{cases} \displaystyle\sum_{n=1}^{\infty} \frac{b_n}{n} \sin(nx) = f(x) & 0 \le x < c, \\ \displaystyle\sum_{n=1}^{\infty} b_n \sin(nx) = g(x). & c < x \le \pi, \end{cases} \quad (\mathbf{3.2.51})$$

has the solution

$$b_n = \frac{n}{\sqrt{2}} \int_0^{\pi} k(t) \{P_{n-1}[\cos(t)] + P_n[\cos(t)]\}\, dt, \quad (\mathbf{3.2.52})$$

where

$$k(t) = \frac{2}{\pi} \frac{d}{dt} \left[\int_0^c \frac{f(\xi) \sin(\xi/2)}{\sqrt{\cos(\xi) - \cos(t)}}\, d\xi \right], \qquad 0 \le t < c, \quad (\mathbf{3.2.53})$$

and

$$k(t) = \frac{2}{\pi} \tan\left(\frac{t}{2}\right) \int_t^{\pi} \frac{g(\xi) \cos(\xi/2)}{\sqrt{\cos(t) - \cos(\xi)}}\, d\xi, \qquad c < t \le \pi. \quad (\mathbf{3.2.54})$$

Using Equation 3.2.51 through Equation 3.2.54, we finally have that

$$A_n = \frac{1}{\sqrt{2}} \int_0^{\pi\ell/L} k(t) \{P_{n-1}[\cos(t)] + P_n[\cos(t)]\}\, dt, \quad (\mathbf{3.2.55})$$

where

$$k(t) = \frac{2}{\pi} \frac{d}{dt} \left[\int_0^t \frac{U_0(L\xi/\pi) \sin(\xi/2)}{\sqrt{\cos(\xi) - \cos(t)}}\, d\xi \right], \qquad 0 \le t < \pi\ell/L. \quad (\mathbf{3.2.56})$$

3.3 DUAL FOURIER-BESSEL SERIES

Dual Fourier-Bessel series arise during mixed boundary value problems in cylindrical coordinates where the radial dimension is of finite extent. Here we show a few examples.

• Example 3.3.1

Let us find[16] the potential for Laplace's equation in cylindrical coordinates:

$$\frac{\partial^2 u}{\partial r^2} + \frac{1}{r} \frac{\partial u}{\partial r} + \frac{\partial^2 u}{\partial z^2} = 0, \qquad 0 \le r < 1, \quad 0 < z < \infty, \quad (\mathbf{3.3.1})$$

---

[16] Originally solved by Borodachev, N. M., and F. N. Borodacheva, 1967: Considering the effect of the walls for an impact of a circular disk on liquid. *Mech. Solids*, **2(1)**, 118.

subject to the boundary conditions

$$\lim_{r\to 0}|u(r,z)| < \infty, \quad u_r(1,z) = 0, \qquad 0 < z < \infty, \tag{3.3.2}$$

$$\begin{cases} u_z(r,0) = 1, & 0 \le r < a, \\ u(r,0) = 0, & a < r < 1, \end{cases} \tag{3.3.3}$$

and

$$\lim_{z\to\infty} u(r,z) \to 0, \qquad 0 \le r < 1, \tag{3.3.4}$$

where $a < 1$.

Separation of variables yields the potential, namely

$$u(r,z) = A_0 + \sum_{n=1}^{\infty} A_n e^{-k_n z} J_0(k_n r), \tag{3.3.5}$$

where k_n is the nth positive root of $J_0'(k) = -J_1(k) = 0$. Equation 3.3.5 satisfies Equation 3.3.1, Equation 3.3.2, and Equation 3.3.4. Substituting Equation 3.3.5 into Equation 3.3.3, we obtain the dual series:

$$\sum_{n=1}^{\infty} k_n A_n J_0(k_n r) = -1, \qquad 0 \le r < a, \tag{3.3.6}$$

and

$$A_0 + \sum_{n=1}^{\infty} A_n J_0(k_n r) = 0, \qquad a < r < 1. \tag{3.3.7}$$

Srivastav[17] showed that this dual Dini series has the solution

$$A_0 = -2\int_0^a t\, h(t)\, dt, \tag{3.3.8}$$

and

$$A_n = -\frac{2}{k_n J_0^2(k_n)} \int_0^a h(t)\sin(k_n t)\, dt, \tag{3.3.9}$$

where the unknown function $h(t)$ is given by the regular Fredholm integral equation of the second kind:

$$h(t) + \int_0^a L(t,\eta) h(\eta)\, d\eta = t, \qquad 0 \le t < a, \tag{3.3.10}$$

and

$$L(t,\eta) = \frac{4}{\pi^2}\int_0^\infty \frac{K_1(x)}{I_1(x)} \sinh(tx)\sinh(\eta x)\, dx. \tag{3.3.11}$$

---

[17] Srivastav, R. P., 1961/1962: Dual series relations. II. Dual relations involving Dini series. *Proc. R. Soc. Edinburgh, Ser. A*, **66**, 161–172.

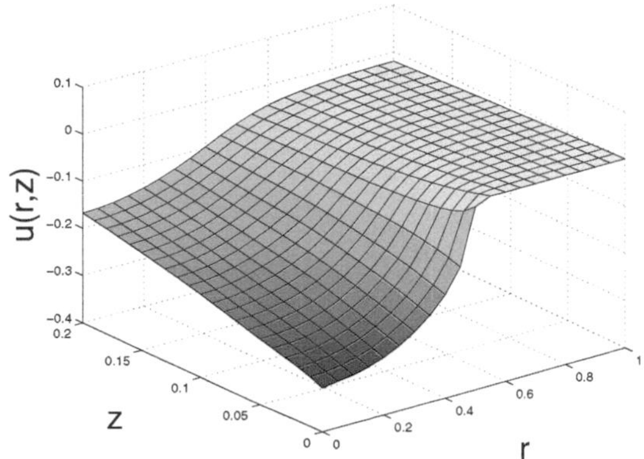

Figure 3.3.1: The solution to Laplace's equation subject to the boundary conditions given by Equation 3.3.2, Equation 3.3.3, and Equation 3.3.4 when $a = 0.5$.

Figure 3.3.1 illustrates this solution when $a = 0.5$.

• Example 3.3.2

A similar problem[18] to the previous one arises during the solution of Laplace's equation in cylindrical coordinates:

$$\frac{\partial^2 u}{\partial r^2} + \frac{1}{r}\frac{\partial u}{\partial r} + \frac{\partial^2 u}{\partial z^2} = 0, \qquad 0 \le r < 1, \quad 0 < z < \infty, \tag{3.3.12}$$

subject to the boundary conditions

$$\lim_{r \to 0} |u(r,z)| < \infty, \quad u_r(1,z) = 0, \qquad 0 < z < \infty, \tag{3.3.13}$$

$$\begin{cases} u(r,0) = 1, & 0 \le r < a, \\ u_z(r,0) = 0, & a < r < 1, \end{cases} \tag{3.3.14}$$

and

$$\lim_{z \to \infty} |u_z(r,z)| < \infty, \qquad 0 \le r < 1, \tag{3.3.15}$$

where $a < 1$.

Separation of variables gives

$$u(r,z) = A_0 z + \sum_{n=1}^{\infty} A_n e^{-k_n z} \frac{J_0(k_n r)}{k_n}, \tag{3.3.16}$$

[18] See Hunter, A., and A. Williams, 1969: Heat flow across metallic joints – The constriction alleviation factor. *Int. J. Heat Mass Transfer*, **12**, 524–526.

where k_n is the nth positive root of $J_0'(k) = -J_1(k) = 0$. Equation 3.3.16 satisfies Equation 3.3.12, Equation 3.3.13, and Equation 3.3.15. Substituting Equation 3.3.16 into Equation 3.3.14, we obtain the dual series:

$$\sum_{n=1}^{\infty} A_n \frac{J_0(k_n r)}{k_n} = 1, \qquad 0 \le r < a, \tag{3.3.17}$$

and

$$A_0 - \sum_{n=1}^{\infty} A_n J_0(k_n r) = 0, \qquad a < r < 1. \tag{3.3.18}$$

Srivastav[19] has given the solution to the dual Fourier-Bessel series

$$\alpha a_0 + \sum_{n=1}^{\infty} a_n \frac{J_0(k_n r)}{k_n} = f(r), \qquad 0 \le r < a, \tag{3.3.19}$$

and

$$a_0 + \sum_{n=1}^{\infty} a_n J_0(k_n r) = 0, \qquad a < r < 1. \tag{3.3.20}$$

Then,

$$a_0 = 2 \int_0^a h(t)\, dt, \tag{3.3.21}$$

and

$$a_n = \frac{2}{J_0^2(k_n)} \int_0^a h(t) \cos(k_n t)\, dt, \tag{3.3.22}$$

where the function $h(t)$ is given by the integral equation

$$h(t) - \int_0^a K(t,\tau) h(\tau)\, d\tau = x(t), \quad 0 < t < a, \tag{3.3.23}$$

$$x(t) = \frac{2}{\pi} \frac{d}{dt} \left[\int_0^t \frac{r f(r)}{\sqrt{t^2 - r^2}}\, dr \right], \tag{3.3.24}$$

and

$$K(t,\tau) = \frac{4}{\pi}(1-\alpha) + \frac{4}{\pi^2} \int_0^\infty \frac{K_1(\xi)}{\xi I_1(\xi)} \left[2I_1(\xi) - \xi \cosh(\tau\xi) \cosh(t\xi) \right] d\xi. \tag{3.3.25}$$

---

[19] Ibid. See also Sneddon, op. cit., Equation 5.3.27 through Equation 5.3.35.

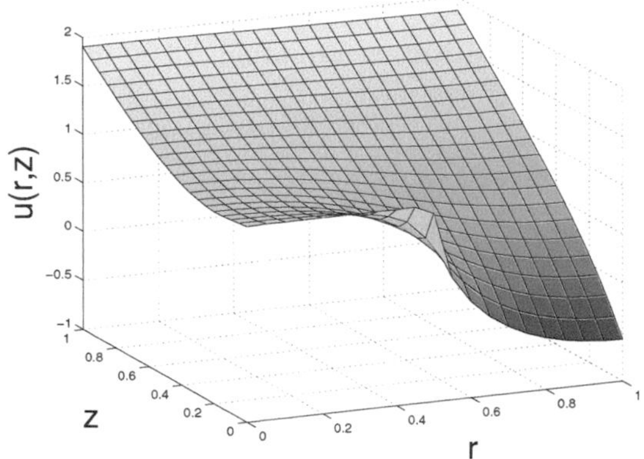

Figure 3.3.2: The solution to Laplace's equation subject to the boundary conditions given by Equation 3.3.13 through Equation 3.3.15 when $a = \frac{1}{2}$.

Equation 3.3.19 through Equation 3.3.25 provide the answer to our problem if we set $\alpha = 0$ and $f(r) = 1$. Figure 3.3.2 illustrates the solution when $a = \frac{1}{2}$.

• Example 3.3.3

Let us solve Laplace equation[20]

$$\frac{\partial^2 u}{\partial r^2} + \frac{1}{r}\frac{\partial u}{\partial r} + \frac{\partial^2 u}{\partial z^2} = 0, \qquad 0 \le r < a, \quad 0 < z < \infty, \tag{3.3.26}$$

subject to the boundary conditions

$$\lim_{r\to 0} |u(r,z)| < \infty, \quad u(a,z) = 0, \qquad 0 < z < \infty, \tag{3.3.27}$$

$$\begin{cases} u_z(r,0) = 1, & 0 \le r < 1, \\ u_r(r,0) = 0, & 1 < r < a, \end{cases} \tag{3.3.28}$$

and

$$\lim_{z\to\infty} u(r,z) \to 0, \qquad 0 \le r < a, \tag{3.3.29}$$

where $a > 1$.

The method of separation of variables yields the product solution

$$u(r,z) = \sum_{n=1}^{\infty} A_n J_0(k_n r) e^{-k_n z}, \tag{3.3.30}$$

[20] See Sherwood, J. D., and H. A. Stone, 1997: Added mass of a disc accelerating within a pipe. *Phys. Fluids*, **9**, 3141–3148.

where k_n is the nth root of $J_1(ka) = 0$. Equation 3.3.30 satisfies not only Laplace's equation, but also the boundary conditions given by Equation 3.3.27 and Equation 3.3.29. Substituting Equation 3.3.30 into Equation 3.3.28, we obtain the dual series

$$\sum_{n=1}^{\infty} A_n k_n J_0(k_n r) = -1, \qquad 0 \le r < 1, \tag{3.3.31}$$

and

$$\sum_{n=1}^{\infty} A_n k_n J_1(k_n r) = 0, \qquad 1 \le r < a. \tag{3.3.32}$$

We begin our solution of these dual equations by applying the identity[21]

$$\sum_{n=1}^{\infty} \frac{J_{\nu+2m+1-p}(\zeta_n) J_\nu(\zeta_n r)}{\zeta_n^{1-p} J_{\nu+1}^2(\zeta_n a)} = 0, \qquad 1 < r < a, \tag{3.3.33}$$

where $|p| \le \frac{1}{2}$, $\nu > p - 1$, $m = 0, 1, 2, \ldots$, and ζ_n denotes the nth root of $J_\nu(\zeta a) = 0$. By direction substitution it is easily seen that Equation 3.3.32 is satisfied if $\nu = 1$ and

$$k_n^{2-p} J_2^2(k_n a) A_n = \sum_{m=0}^{\infty} C_m J_{2m+2-p}(k_n). \tag{3.3.34}$$

Here p is still a free parameter. Substituting Equation 3.3.34 into Equation 3.3.31,

$$\sum_{n=1}^{\infty} \sum_{m=0}^{\infty} C_m \frac{J_{2m+2-p}(k_n) J_0(k_n r)}{k_n^{1-p} J_2^2(k_n a)} = -1, \qquad 0 \le r < 1. \tag{3.3.35}$$

Our remaining task is to compute C_m. Although Equation 3.3.35 holds for any r between 0 and 1, it would be better if we did not have to deal with its presence. It can be eliminated as follows: From Sneddon's book,[22]

$$\int_0^{\infty} \eta^{1-k} J_{\nu+2m+k}(\eta) J_\nu(r\eta)\, d\eta = \frac{\Gamma(\nu+m+1) r^\nu (1-r^2)^{k-1}}{2^{k-1}\Gamma(\nu+1)\Gamma(m+k)} P_m^{(k+\nu,\nu+1)}(r^2) \tag{3.3.36}$$

---

[21] Tranter, C. J., 1959: On the analogies between some series containing Bessel functions and certain special cases of the Weber-Schafheitlin integral. *Quart. J. Math., Ser. 2*, **10**, 110–114.

[22] Sneddon, op. cit., Equation 2.1.33 and Equation 2.1.34.

if $0 \le r < 1$; this integral equals 0 if $1 < r < \infty$. Here $P_m^{(a,b)}(x) = {}_2F_1(-m, a+m; b; x)$ is the Jacobi polynomial. If we view Equation 3.3.36 as a Hankel transform of $\eta^{-k} J_{\nu+2m+k}(\eta)$, then its inverse is

$$\eta^{-k} J_{\nu+2m+k}(\eta) = \int_0^1 \frac{\Gamma(\nu+m+1) r^{1+\nu} \left(1-r^2\right)^{k-1}}{2^{k-1}\Gamma(\nu+1)\Gamma(m+k)} J_\nu(r\eta) P_m^{(k+\nu,\nu+1)}\left(r^2\right) dr. \quad \textbf{(3.3.37)}$$

We also have from the orthogonality condition[23] of Jacobi polynomials that

$$\int_0^1 r^{2\nu+1} \left(1-r^2\right)^{k-1} P_m^{(k+\nu,\nu+1)}\left(r^2\right) dr = \frac{\Gamma(\nu+1)\Gamma(k)}{2\Gamma(\nu+k+1)} \delta_{0m}, \quad \textbf{(3.3.38)}$$

where δ_{nm} is the Kronecker delta. Multiplying both sides of Equation 3.3.35 by $r\left(1-r^2\right)^{-p} P_m^{(1-p,1)}(r^2)$ and applying Equation 3.3.38, we obtain

$$-\frac{\Gamma(1-p)}{2\Gamma(2-p)} \delta_{0j} = \sum_{n=1}^{\infty} \sum_{m=0}^{\infty} C_m \frac{J_{2m+2-p}(k_n) J_{2j+1-p}(k_n) \Gamma(j+1-p)}{2^p \Gamma(j+1) k_n^{2-2p} J_2^2(k_n a)}; \quad \textbf{(3.3.39)}$$

or

$$\sum_{m=0}^{\infty} A_{jm} C_m = B_j, \qquad j = 0, 1, 2, \ldots, \quad \textbf{(3.3.40)}$$

where

$$A_{jm} = \sum_{n=1}^{\infty} \frac{J_{2m+2-p}(k_n) J_{2j+1-p}(k_n)}{k_n^{2-2p} J_2^2(k_n a)}, \quad \textbf{(3.3.41)}$$

and

$$B_j = \begin{cases} -2^{p-1}/\Gamma(2-p), & j = 0, \\ 0, & \text{otherwise}. \end{cases} \quad \textbf{(3.3.42)}$$

For a given p, we can solve Equation 3.3.40 after we truncate the infinite number of equations to just M. For a given k_n we solve the truncated Equation 3.3.40, which yields C_m for $m = 0, 1, 2, \ldots, M$. Then Equation 3.3.34 gives A_n. Finally, the potential $u(r, z)$ follows from Equation 3.3.30. Figure 3.3.3 illustrates this solution when $a = 2$ and $p = 0.5$.

• Example 3.3.4

In the previous example we solved Laplace's equation over a semi-infinite right cylinder. Here, let us solve Laplace's equation[24] when the cylinder has

---

[23] See page 83 in Magnus, W., and F. Oberhettinger, 1954: *Formulas and Theorems for the Functions of Mathematical Physics*. Chelsea Publ. Co., 172 pp.

[24] See Galceran, J., J. Cecilia, E. Companys, J. Salvador, and J. Puy, 2000: Analytical expressions for feedback currents at the scanning electrochemical microscope. *J. Phys. Chem., Ser. B*, **104**, 7993-8000.

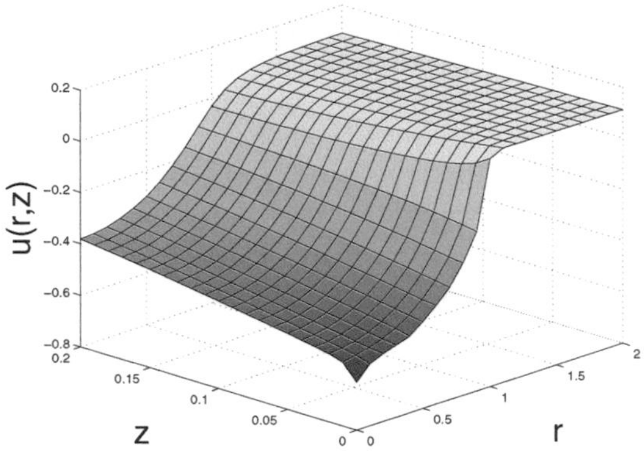

Figure 3.3.3: The solution to Equation 3.3.26 subject to the boundary conditions given by Equation 3.3.27 through Equation 3.3.29 when $a = 2$ and $p = 0.5$.

a height b. Our problem now reads

$$\frac{\partial^2 u}{\partial r^2} + \frac{1}{r}\frac{\partial u}{\partial r} + \frac{\partial^2 u}{\partial z^2} = 0, \qquad 0 \le r < a, \quad 0 < z < b, \tag{3.3.43}$$

subject to the boundary conditions

$$\lim_{r\to 0} |u(r,z)| < \infty, \quad u(a,z) = 0, \qquad 0 < z < b, \tag{3.3.44}$$

$$\begin{cases} u(r,0) = 1, & 0 \le r < 1, \\ u_z(r,0) = 0, & 1 < r < a, \end{cases} \tag{3.3.45}$$

and

$$\lim_{z\to\infty} u(r,z) \to 0, \qquad 0 \le r < a, \tag{3.3.46}$$

where $a > 1$.

The method of separation of variables yields the product solution

$$u(r,z) = \sum_{n=1}^{\infty} A_n \coth(k_n b) \left[\cosh(k_n z) - \tanh(k_n b)\sinh(k_n z)\right] \frac{J_0(k_n r)}{k_n}, \tag{3.3.47}$$

where k_n is the nth root of $J_0(ka) = 0$. Equation 3.3.47 satisfies not only Laplace's equation, but also the boundary conditions given by Equation 3.3.44 and Equation 3.3.46. Substituting Equation 3.3.47 into Equation 3.3.45, we obtain the dual series

$$\sum_{n=1}^{\infty} A_n \frac{\coth(k_n b)}{k_n} J_0(k_n r) = 1, \qquad 0 \le r < 1, \tag{3.3.48}$$

and

$$\sum_{n=1}^{\infty} A_n J_0(k_n r) = 0, \qquad 1 \le r < a. \tag{3.3.49}$$

Let us reexpress A_n as follows:

$$A_n = \frac{1}{\sqrt{k_n}\, J_1^2(k_n a)} \sum_{m=0}^{\infty} B_m J_{2m+\frac{1}{2}}(k_m). \tag{3.3.50}$$

Substituting Equation 3.3.50 into Equation 3.3.49, we have[25] that

$$\sum_{n=1}^{\infty} A_n J_0(k_n r) = \sum_{m=o}^{\infty} B_m \left[\sum_{n=1}^{\infty} \frac{J_{2m+\frac{1}{2}}(k_m) J_0(k_n r)}{\sqrt{k_n}\, J_1^2(k_n a)} \right] = 0 \tag{3.3.51}$$

for $1 < r \le a$. Therefore, Equation 3.3.49 is satisfied identically with this definition of A_n.

Next, we substitute Equation 3.3.50 into Equation 3.3.48, multiply both sides of the resulting equation by $r\ {}_2F_1\left(-s, s+\frac{1}{2}, 1, r^2\right) dr/\sqrt{1-r^2}$ and integrate between $r = 0$ and $r = 1$. We find that

$$\sum_{m=0}^{\infty} C_{m,s} B_m = \frac{\sqrt{2}\,\Gamma(s+1)}{\Gamma\left(s+\frac{1}{2}\right)} \int_0^1 \frac{r}{\sqrt{1-r^2}}\, {}_2F_1\left(-s, s+\frac{1}{2}, 1, r^2\right) dr \tag{3.3.52}$$

$$= \begin{cases} \sqrt{2/\pi}, & s = 0, \\ 0, & s > 0, \end{cases} \tag{3.3.53}$$

where

$$C_{m,s} = \sum_{n=1}^{\infty} \frac{\coth(k_n b) J_{2m+\frac{1}{2}}(k_n) J_{2s+\frac{1}{2}}(k_n)}{k_n^2 J_1^2(k_n a)} \tag{3.3.54}$$

and $s = 0, 1, 2, \ldots$. We used

$$\frac{J_{2s+\frac{1}{2}}(k_n)}{\sqrt{k_n}} = \frac{\sqrt{2}\,\Gamma(s+1)}{\Gamma\left(s+\frac{1}{2}\right)} \int_0^1 \frac{r}{\sqrt{1-r^2}}\, {}_2F_1\left(-s, s+\frac{1}{2}, 1, r^2\right) J_0(k_n r)\, dr. \tag{3.3.55}$$

Equation 3.3.55 is now solved to yield B_m. Next, we compute A_n from Equation 3.3.50. Finally $u(r, z)$ follows from Equation 3.3.47.

Figure 3.3.4 illustrates the solution to Equation 3.3.43 through Equation 3.3.46 when $a = 2$ and $b = 1$. As suggested by Galceran et al.,[26] the

---

[25] Tranter, op. cit.

[26] Galceran, Cecilia, Companys, Salvador, and Puy, op. cit.

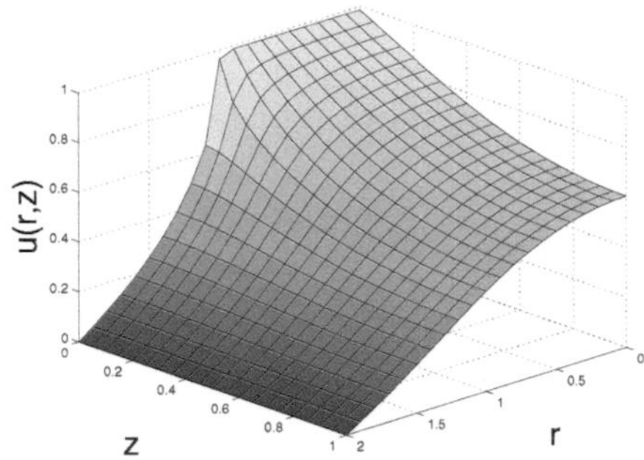

Figure 3.3.4: The solution to Equation 3.3.43 through Equation 3.3.46 when $a = 2$ and $b = 1$.

computation of B_m is assisted by noting that

$$C_{m,s} = \sum_{n=1}^{\infty} \frac{[\coth(k_n b) - 1] J_{2m+\frac{1}{2}}(k_n) J_{2s+\frac{1}{2}}(k_n)}{k_n^2 J_1^2(k_n a)} - \sum_{n=1}^{\infty} \frac{J_{2m+\frac{1}{2}}(k_n) J_{2s+\frac{1}{2}}(k_n)}{k_n^2 J_1^2(k_n a)} \tag{3.3.56}$$

$$= \sum_{n=1}^{\infty} \frac{[\coth(k_n b) - 1] J_{2m+\frac{1}{2}}(k_n) J_{2s+\frac{1}{2}}(k_n)}{k_n^2 J_1^2(k_n a)} - \frac{a^2}{2}\left[\frac{\delta_{ms}}{4s+1} - \frac{2(-1)^{m+s}}{\pi} \int_0^{\infty} \frac{K_0(t)}{t\, I_0(t)} I_{2m+\frac{1}{2}}\left(\frac{t}{a}\right) I_{2s+\frac{1}{2}}\left(\frac{t}{a}\right) dt\right], \tag{3.3.57}$$

where δ_{ms} is the Kronecker delta and we used results from a paper[27] by Tranter to replace the second summation in Equation 3.3.56 with an integral.

• Example 3.3.5

Let us find[28] the electrostatic potential due to a parallel plate condenser that lies within a hollow cylinder of radius $a > 1$ that is grounded and infinitely long. See Figure 3.3.5. The governing equations are

$$\frac{\partial^2 u}{\partial r^2} + \frac{1}{r}\frac{\partial u}{\partial r} + \frac{\partial^2 u}{\partial z^2} = 0, \qquad 0 \le r < a, \quad -\infty < z < \infty, \tag{3.3.58}$$

subject to the boundary conditions

$$\lim_{r \to 0} |u(r, z)| < \infty, \quad u(a, z) = 0, \qquad -\infty < z < \infty, \tag{3.3.59}$$

[27] Tranter, op. cit.

[28] See Singh, B. M., 1973: On mixed boundary value problem in electrostatics. *Indian J. Pure Appl. Math.*, **6**, 166–176.

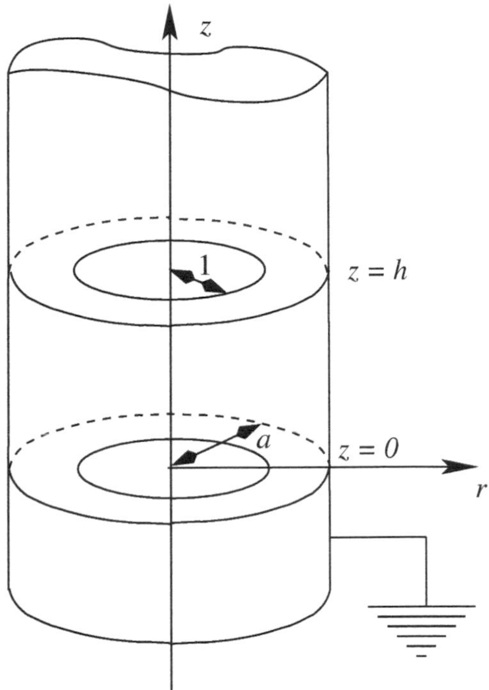

Figure 3.3.5: Schematic of a hollow cylinder containing discs at $z = 0$ and $z = h$.

$$u(r, 0^-) = u(r, 0^+), \quad u(r, h^-) = u(r, h^+), \qquad 0 \le r < a, \tag{3.3.60}$$

$$\begin{cases} u(r, 0) = G(r), \quad u(r, h) = F(r), & 0 \le r < 1, \\ u_z(r, 0^-) = u_z(r, 0^+), \quad u_z(r, h^-) = u_z(r, h^+), & 1 < r < a, \end{cases} \tag{3.3.61}$$

and

$$\lim_{|z|\to\infty} u(r, z) \to 0, \qquad 0 \le r < a. \tag{3.3.62}$$

Here, $G(r)$ and $F(r)$ denote the prescribed potential on the discs at $z = 0$ and $z = h$, respectively. The parameters h^+ and h^- denote points that are slightly above or below h, respectively.

Separation of variables yields the potential, namely

$$u(r, z) = \sum_{n=0}^{\infty} A_n e^{-k_n(z-h)} \frac{J_0(k_n r)}{k_n}, \qquad h \le z < \infty, \tag{3.3.63}$$

$$u(r, z) = \sum_{n=0}^{\infty} \left\{ A_n \frac{\sinh(k_n z)}{\sinh(k_n h)} + B_n \frac{\sinh[k_n(h - z)]}{\sinh(k_n h)} \right\} \frac{J_0(k_n r)}{k_n}, \qquad 0 \le z \le h, \tag{3.3.64}$$

and

$$u(r, z) = \sum_{n=0}^{\infty} B_n e^{k_n z} \frac{J_0(k_n r)}{k_n}, \qquad -\infty < z \le 0, \tag{3.3.65}$$

where k_n is the nth positive root of $J_0(ka) = 0$. Equation 3.3.63 through Equation 3.3.65 satisfy Equation 3.3.58, Equation 3.3.59, Equation 3.3.60, and Equation 3.3.62. Substituting Equation 3.3.63 through Equation 3.3.65 into Equation 3.3.61, we obtain the following system of simultaneous dual series equations:

$$\sum_{n=0}^{\infty} A_n \frac{J_0(k_n r)}{k_n} = F(r), \qquad 0 \le r < 1, \tag{3.3.66}$$

$$\sum_{n=0}^{\infty} B_n \frac{J_0(k_n r)}{k_n} = G(r), \qquad 0 \le r < 1, \tag{3.3.67}$$

$$\sum_{n=0}^{\infty} \left\{ [1+\coth(k_n h)]A_n - \frac{B_n}{\sinh(k_n h)} \right\} J_0(k_n r) = 0, \quad 1 < r < a, \tag{3.3.68}$$

and

$$\sum_{n=0}^{\infty} \left\{ [1+\coth(k_n h)]B_n - \frac{A_n}{\sinh(k_n h)} \right\} J_0(k_n r) = 0, \quad 1 < r < a. \tag{3.3.69}$$

For $0 \le r < 1$, let us augment Equation 3.3.68 and Equation 3.3.69 with

$$\sum_{n=0}^{\infty} \left\{ [1+\coth(k_n h)]A_n - \frac{B_n}{\sinh(k_n h)} \right\} J_0(k_n r) = -\frac{1}{r}\frac{d}{dr}\left[\int_r^1 \frac{t\, g(t)}{\sqrt{t^2-r^2}}\, dt\right], \tag{3.3.70}$$

and

$$\sum_{n=0}^{\infty} \left\{ [1+\coth(k_n h)]B_n - \frac{A_n}{\sinh(k_n h)} \right\} J_0(k_n r) = -\frac{1}{r}\frac{d}{dr}\left[\int_r^1 \frac{t\, h(t)}{\sqrt{t^2-r^2}}\, dt\right], \tag{3.3.71}$$

where $g(t)$ and $h(t)$ are *unknown* functions.

Taken together, Equation 3.3.68 through Equation 3.3.71 are a Fourier-Bessel series over the interval $0 \le r < a$. From Equation 1.4.16 and Equation 1.4.17, it follows that

$$[1+\coth(k_n h)]A_n - \frac{B_n}{\sinh(k_n h)}$$

$$= -\frac{2}{a^2 J_1^2(k_n a)} \int_0^1 \frac{d}{dr}\left[\int_r^1 \frac{t\, g(t)}{\sqrt{t^2-r^2}}\, dt\right] J_0(k_n r)\, dr \tag{3.3.72}$$

$$= -\frac{2}{a^2 J_1^2(k_n a)} \left[\int_r^1 \frac{t\, g(t)}{\sqrt{t^2-r^2}}\, dt\right] J_0(k_n r)\bigg|_0^1$$

$$- \frac{2}{a^2 J_1^2(k_n a)} \int_0^1 k_n J_1(k_n r) \left[\int_r^1 \frac{t\, g(t)}{\sqrt{t^2-r^2}}\, dt\right] dr \tag{3.3.73}$$

$$= \frac{2}{a^2 J_1^2(k_n a)} \int_0^1 g(t)\, dt$$

$$- \frac{2k_n}{a^2 J_1^2(k_n a)} \int_0^1 t\, g(t) \left[\int_0^t \frac{J_1(k_n r)}{\sqrt{t^2-r^2}}\, dr\right] dt. \tag{3.3.74}$$

Now,

$$\int_0^t \frac{J_1(k_n r)}{\sqrt{t^2 - r^2}}\, dr = \int_0^1 \frac{J_1(k_n t\eta)}{\sqrt{1-\eta^2}}\, d\eta = \frac{\pi}{2} J_{\frac{1}{2}}^2(k_n t/2) = \frac{1 - \cos(k_n t)}{k_n t}, \quad \textbf{(3.3.75)}$$

where we used tables[29] to evaluate the integral. Therefore,

$$[1 + \coth(k_n h)]A_n - \frac{B_n}{\sinh(k_n h)} = \frac{2}{a^2 J_1^2(k_n a)} \int_0^1 g(t) \cos(k_n t)\, dt. \quad \textbf{(3.3.76)}$$

In a similar manner,

$$[1 + \coth(k_n h)]B_n - \frac{A_n}{\sinh(k_n h)} = \frac{2}{a^2 J_1^2(k_n a)} \int_0^1 h(t) \cos(k_n t)\, dt. \quad \textbf{(3.3.77)}$$

Solving for A_n and B_n, we find that

$$A_n = \frac{1}{a^2 J_1^2(k_n a)} \left[\int_0^1 g(t) \cos(k_n t)\, dt + e^{-k_n h} \int_0^1 h(t) \cos(k_n t)\, dt \right], \quad \textbf{(3.3.78)}$$

and

$$B_n = \frac{1}{a^2 J_1^2(k_n a)} \left[\int_0^1 h(t) \cos(k_n t)\, dt + e^{-k_n h} \int_0^1 g(t) \cos(k_n t)\, dt \right]. \quad \textbf{(3.3.79)}$$

Substituting A_n and B_n into Equation 3.3.66 and 3.3.67 and interchanging the order of integration and summation,

$$\int_0^1 g(t) \left[\sum_{n=0}^{\infty} \frac{J_0(k_n r) \cos(k_n t)}{a^2 k_n J_1^2(k_n a)} \right] dt + \int_0^1 h(t) \left[\sum_{n=0}^{\infty} \frac{e^{-k_n h} J_0(k_n r) \cos(k_n t)}{a^2 k_n J_1^2(k_n a)} \right] dt = F(r), \quad \textbf{(3.3.80)}$$

and

$$\int_0^1 h(t) \left[\sum_{n=0}^{\infty} \frac{J_0(k_n r) \cos(k_n t)}{a^2 k_n J_1^2(k_n a)} \right] dt + \int_0^1 g(t) \left[\sum_{n=0}^{\infty} \frac{e^{-k_n h} J_0(k_n r) \cos(k_n t)}{a^2 k_n J_1^2(k_n a)} \right] dt = G(r). \quad \textbf{(3.3.81)}$$

---

[29] Gradshteyn and Ryzhik, op. cit., Formula 6.552.4.

It is readily shown[30] that

$$\frac{2}{a^2}\sum_{n=0}^{\infty}\frac{J_0(k_nr)\cos(k_nt)}{k_nJ_1^2(k_na)} = \int_0^\infty J_0(r\eta)\cos(t\eta)\,d\eta - \frac{2}{\pi}\int_0^\infty \frac{K_0(a\eta)}{I_0(a\eta)}I_0(r\eta)\cosh(t\eta)\,d\eta. \quad \textbf{(3.3.82)}$$

In a similar manner,

$$\frac{2}{a^2}\sum_{n=0}^{\infty}\frac{e^{-k_nh}J_0(k_nr)\cos(k_nt)}{k_nJ_1^2(k_na)} = \int_0^\infty J_0(r\eta)\cos(t\eta)e^{-h\eta}\,d\eta - \frac{2}{\pi}\int_0^\infty \frac{K_0(a\eta)}{I_0(a\eta)}I_0(r\eta)\cosh(t\eta)\cos(h\eta)\,d\eta. \quad \textbf{(3.3.83)}$$

Substituting Equation 1.4.14, Equation 3.3.82, and Equation 3.3.83 into Equation 3.3.80 and Equation 3.3.81, we have that

$$\begin{aligned}\int_0^r \frac{g(t)}{\sqrt{r^2-t^2}}\,dt = {} & \frac{2}{\pi}\int_0^1 g(t)\left[\int_0^\infty \frac{K_0(a\eta)}{I_0(a\eta)}I_0(r\eta)\cosh(t\eta)\,d\eta\right]dt \\ & - \frac{2}{\pi}\int_0^1 h(t)\left[\int_0^\infty \frac{K_0(a\eta)}{I_0(a\eta)}I_0(r\eta)\cosh(t\eta)\cos(h\eta)\,d\eta\right]dt \\ & - \int_0^1 h(t)\left[\int_0^\infty J_0(r\eta)\cos(t\eta)e^{-h\eta}\,d\eta\right]dt + 2F(r), \quad \textbf{(3.3.84)}\end{aligned}$$

and

$$\begin{aligned}\int_0^r \frac{h(t)}{\sqrt{r^2-t^2}}\,dt = {} & \frac{2}{\pi}\int_0^1 h(t)\left[\int_0^\infty \frac{K_0(a\eta)}{I_0(a\eta)}I_0(r\eta)\cosh(t\eta)\,d\eta\right]dt \\ & - \frac{2}{\pi}\int_0^1 g(t)\left[\int_0^\infty \frac{K_0(a\eta)}{I_0(a\eta)}I_0(r\eta)\cosh(t\eta)\cos(h\eta)\,d\eta\right]dt \\ & - \int_0^1 g(t)\left[\int_0^\infty J_0(r\eta)\cos(t\eta)e^{-h\eta}\,d\eta\right]dt + 2G(r). \quad \textbf{(3.3.85)}\end{aligned}$$

Equation 3.3.84 and Equation 3.3.85 are integral equations of the Abel type. From Equation 1.2.13 and Equation 1.2.14, we find

$$\begin{aligned}g(t) = {} & \frac{4}{\pi}\frac{d}{dt}\left[\int_0^t \frac{r\,F(r)}{\sqrt{t^2-r^2}}\,dr\right] \\ & + \frac{4}{\pi^2}\int_0^1 g(\xi)\left[\int_0^\infty \frac{K_0(a\eta)}{I_0(a\eta)}\cosh(t\eta)\cosh(\xi\eta)\,d\eta\right]d\xi \\ & + \frac{4}{\pi^2}\int_0^1 h(\xi)\left[\int_0^\infty \frac{K_0(a\eta)}{I_0(a\eta)}\cosh(t\eta)\cosh(\xi\eta)\cos(h\eta)\,d\eta\right]d\xi \\ & - \frac{2}{\pi}\int_0^1 h(\xi)\left[\int_0^\infty e^{-h\eta}\cos(\xi\eta)\cos(t\eta)\,d\eta\right]d\xi, \quad \textbf{(3.3.86)}\end{aligned}$$

30 See Section 2.2 in Sneddon, op. cit.

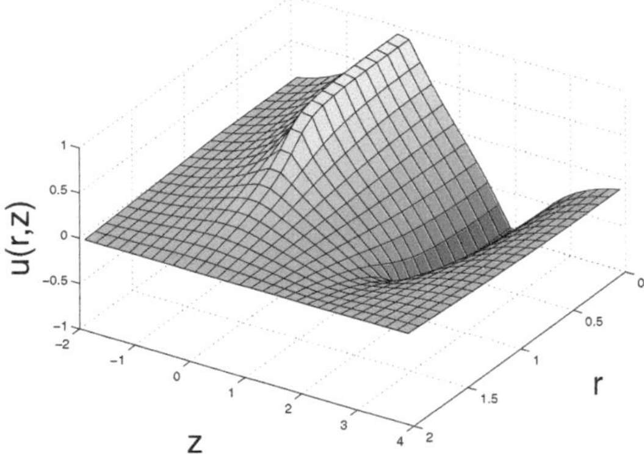

Figure 3.3.6: The electrostatic potential within an infinitely long, grounded, and hollow cylinder of radius 2 when two discs with potential -1 and 1 are placed at $z = 0$ and $z = 2$, respectively.

and

$$
\begin{aligned}
h(t) = {} & \frac{4}{\pi}\frac{d}{dt}\left[\int_0^t \frac{r\,G(r)}{\sqrt{t^2-r^2}}\,dr\right] \\
& + \frac{4}{\pi^2}\int_0^1 g(\xi)\left[\int_0^\infty \frac{K_0(a\eta)}{I_0(a\eta)}\cosh(t\eta)\cosh(\xi\eta)\cos(h\eta)\,d\eta\right] d\xi \\
& + \frac{4}{\pi^2}\int_0^1 h(\xi)\left[\int_0^\infty \frac{K_0(a\eta)}{I_0(a\eta)}\cosh(t\eta)\cosh(\xi\eta)\,d\eta\right] d\xi \\
& - \frac{2}{\pi}\int_0^1 g(\xi)\left[\int_0^\infty e^{-h\eta}\cos(\xi\eta)\cos(t\eta)\,d\eta\right] d\xi. \qquad (3.3.87)
\end{aligned}
$$

For a given $F(r)$ and $G(r)$, we solve the integral equations Equation 3.3.86 and Equation 3.3.87 for $g(t)$ and $h(t)$. Equation 3.3.78 and Equation 3.3.79 give A_n and B_n. Finally, we can use Equation 3.3.63 through Equation 3.3.65 to evaluate the potential for any given r and z. Figure 3.3.6 illustrates the electrostatic potential when $F(r) = 1$, $G(r) = -1$, and $a = h = 2$.

• Example 3.3.6: Electrostatic problem

In electrostatics the potential due to a point charge located at $r = 0$ and $z = h$ in the upper half-plane $z > 0$ above a grounded plane $z = 0$ is

$$
u(r, z) = \frac{1}{\sqrt{r^2 + (z-h)^2}} - \frac{1}{\sqrt{r^2 + (z+h)^2}}. \qquad (3.3.88)
$$

Let us introduce a unit circular hole at $z = 0$ and attach an infinite pipe to

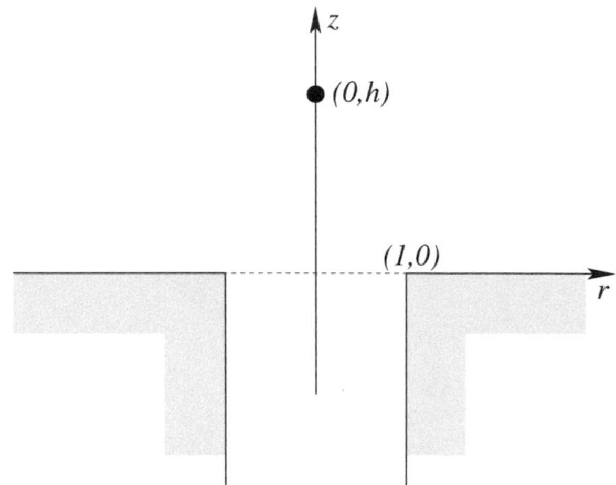

Figure 3.3.7: Schematic of the spatial domain for which we are finding the potential in Example 3.3.6.

this hole. See Figure 3.3.7. Let us find the potential[31] in this case.

The potential in this new configuration is

$$u(r,z) = \frac{1}{\sqrt{r^2+(z-h)^2}} - \frac{1}{\sqrt{r^2+(z+h)^2}} + \frac{1}{2i}\int_{-1}^{1} \frac{g(t)}{\sqrt{r^2+(z+it)^2}}\,dt \tag{3.3.89}$$

when $0 \le r < \infty$ and $0 \le z < \infty$ and

$$u(r,z) = \sum_{n=1}^{\infty} A_n J_0(k_n r) e^{k_n z} \tag{3.3.90}$$

when $0 \le r \le 1$ and $-\infty < z \le 0$. The integral[32] in Equation 3.3.89 vanishes when $z = 0$ and $1 \le r < \infty$. Here $g(t)$ is an odd real-valued function and k_n denotes the nth root of $J_0(k) = 0$. To determine $g(t)$ and the Fourier coefficients A_n, the potential and its normal derivative must be continuous across the aperture $z = 0$ and $0 \le r \le 1$. Mathematically these conditions are

$$u(r,0^-) = u(r,0^+), \qquad 0 \le r \le 1, \tag{3.3.91}$$

and

$$u_z(r,0^-) = u_z(r,0^+), \qquad 0 \le r \le 1. \tag{3.3.92}$$

---

[31] Taken from Shail, R., and B. A. Packham, 1986: Some potential problems associated with the sedimentation of a small particle into a semi-infinite fluid-filled pore. *IMA J. Appl. Math.*, **37**, 37–66 with permission of Oxford University Press.

[32] See Section 5.10 in Green, A. E., and W. Zerna, 1992: *Theoretical Elasticity*. New York: Dover, 457 pp.

Equation 3.3.91 yields

$$\sum_{n=1}^{\infty} A_n J_0(k_n r) = -\int_r^1 \frac{g(t)}{\sqrt{t^2 - r^2}}\, dt, \qquad 0 \le r \le 1. \tag{3.3.93}$$

In deriving Equation 3.3.93, we used

$$\sqrt{r^2 + (z + it)^2} = \xi e^{i\eta/2}, \quad \sqrt{r^2 + (z - it)^2} = \xi e^{-i\eta/2}, \tag{3.3.94}$$

with $\xi^2 \cos(\eta) = r^2 + z^2 - t^2$, $\xi^2 \sin(\eta) = 2zt$, $\xi \ge 0$, and $0 \le \eta \le \pi$. On the other hand, Equation 3.3.92 gives

$$\frac{1}{r}\frac{\partial}{\partial r}\left[\int_0^r \frac{t\, h(t)}{\sqrt{r^2 - t^2}}\, dt\right] = \sum_{n=1}^{\infty} k_n A_n J_0(k_n r) - \frac{2h}{(r^2 + h^2)^{3/2}}, \qquad 0 \le r \le 1. \tag{3.3.95}$$

Using Equation 1.2.13 and Equation 1.2.14, we can solve for $g(t)$ in Equation 3.3.93 and find that

$$g(t) = \frac{2}{\pi} \sum_{n=1}^{\infty} k_n A_n \int_0^t \frac{r J_0(k_n r)}{\sqrt{t^2 - r^2}}\, dr - \frac{4h}{\pi} \int_0^t \frac{r}{\sqrt{(t^2 - r^2)(r^2 + h^2)^3}}\, dr \tag{3.3.96}$$

$$= \frac{2}{\pi} \sum_{n=1}^{\infty} A_n \sin(k_n t) - \frac{4t}{\pi(t^2 + h^2)}, \qquad 0 \le t \le 1. \tag{3.3.97}$$

We used Equation 1.4.9 and tables[33] to evaluate the integrals in Equation 3.3.96.

Substituting the results from Equation 3.3.97 into Equation 3.3.95, we obtain

$$\sum_{n=1}^{\infty} A_n J_0(k_n r) = -\frac{2}{\pi} \sum_{n=1}^{\infty} A_n \int_r^1 \frac{\sin(k_n t)}{\sqrt{t^2 - r^2}}\, dt + \frac{4}{\pi} \int_r^1 \frac{t}{(t^2 + h^2)\sqrt{t^2 - r^2}}\, dt. \tag{3.3.98}$$

The left side of Equation 3.3.98 is a Fourier-Bessel expansion. Multiplying both sides of this equation by $r J_0(k_m r)$ and integrating with respect to r from 0 to 1, we find that

$$\begin{aligned} \tfrac{1}{2} J_1^2(k_m) A_m = &-\frac{2}{\pi} \sum_{n=1}^{\infty} A_n \int_0^1 r J_0(k_m r) \left[\int_r^1 \frac{\sin(k_n t)}{\sqrt{t^2 - r^2}}\, dt\right] dr \\ &+ \frac{4}{\pi} \int_0^1 r J_0(k_m r) \left[\int_r^1 \frac{dt}{(t^2 + h^2)\sqrt{t^2 - r^2}}\right] dr. \end{aligned} \tag{3.3.99}$$

[33] Gradshteyn and Ryzhik, op. cit., Formula 2.252, Point II.

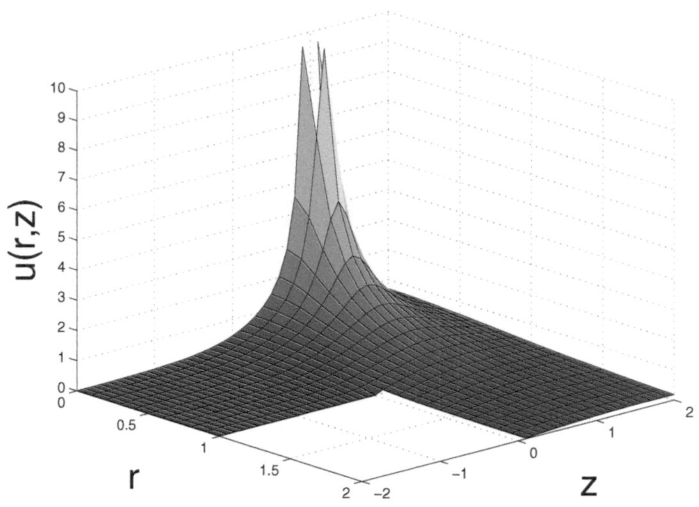

Figure 3.3.8: The electrostatic potential when a unit point charge is placed at $r = 0$ and $z = h$ for the structure illustrated in Figure 3.3.7.

Interchanging the order of integration in Equation 3.3.99 and evaluating the inner r integrals, we finally achieve

$$\tfrac{1}{2}\pi J_1^2(k_m)A_m + \sum_{n=1}^{\infty} A_n \left[\frac{\sin(k_n - k_m)}{k_n - k_m} - \frac{\sin(k_n + k_m)}{k_n + k_m}\right] = 4\int_0^1 \frac{t}{t^2+h^2}\sin(k_m t)\,dt, \qquad \textbf{(3.3.100)}$$

where $m = 1, 2, 3, \ldots$. To find A_m, we truncate the infinite number of equations given by Equation 3.3.100 to a finite number, say M. As M increases the A_m's with a smaller m increase in accuracy. The procedure is stopped when the leading A_m's are sufficiently accurate for the evaluation of Equation 3.3.89, Equation 3.3.90 and Equation 3.3.97. Figure 3.3.8 illustrates this solution when $h = 1$ and the first 300 terms have been retained when $M = 600$.

3.4 DUAL FOURIER-LEGENDRE SERIES

Here we turn to problems in spherical coordinates which lead to dual Fourier-Legendre series. We now present several examples of their solution.

• Example 3.4.1

Let us solve[34] the axisymmetric Laplace equation within the unit sphere:

$$\frac{\partial}{\partial r}\left(r^2\frac{\partial u}{\partial r}\right)+\frac{1}{\sin(\theta)}\frac{\partial}{\partial\theta}\left[\sin(\theta)\frac{\partial u}{\partial\theta}\right]=0,\qquad 0\le r<1,\quad 0<\theta<\pi,\ \textbf{(3.4.1)}$$

subject to the boundary conditions

$$\lim_{\theta\to0}|u(r,\theta)|<\infty,\quad \lim_{\theta\to\pi}|u(r,\theta)|<\infty,\qquad 0\le r\le 1,\qquad \textbf{(3.4.2)}$$

$$\lim_{r\to0}|u(r,\theta)|<\infty,\qquad 0\le\theta\le\pi,\qquad \textbf{(3.4.3)}$$

and

$$\begin{cases} u(1,\theta)=1, & 0\le\theta<\alpha,\\ u_r(1,\theta)=-bu(1,\theta), & \alpha<\theta\le\pi.\end{cases}\qquad \textbf{(3.4.4)}$$

Separation of variables yields the solution

$$u(r,\theta)=\sum_{n=0}^{\infty}A_n r^n P_n[\cos(\theta)].\qquad \textbf{(3.4.5)}$$

Equation 3.4.5 satisfies not only Equation 3.4.1, but also Equation 3.4.2 and Equation 3.4.3. Substituting Equation 3.4.5 into Equation 3.4.4 yields the dual Fourier-Legendre series

$$\sum_{n=0}^{\infty}A_nP_n[\cos(\theta)]=1,\qquad 0\le\theta<\alpha,\qquad \textbf{(3.4.6)}$$

and

$$\sum_{n=0}^{\infty}(b+n)A_nP_n[\cos(\theta)]=0,\qquad \alpha\le\theta<\pi.\qquad \textbf{(3.4.7)}$$

Consider Equation 3.4.6. Using Equation 1.3.4 to eliminate $P_n[\cos(\theta)]$ and then interchanging the order of integration and summation, we have

$$\int_0^\theta \frac{1}{\sqrt{\cos(\varphi)-\cos(\theta)}}\left\{\sum_{n=0}^{\infty}A_n\cos\left[\left(n+\tfrac12\right)\varphi\right]\right\}d\varphi=\frac{\pi}{\sqrt2},\qquad 0\le\varphi<\alpha.\qquad \textbf{(3.4.8)}$$

Applying the results from Equation 1.2.9 and Equation 1.2.10,

$$\sum_{n=0}^{\infty}A_n\cos\left[\left(n+\tfrac12\right)\varphi\right]=\frac{1}{\sqrt2}\frac{d}{d\varphi}\left[\int_0^\varphi\frac{\sin(\tau)}{\sqrt{\cos(\tau)-\cos(\varphi)}}\,d\tau\right]\qquad \textbf{(3.4.9)}$$

$$=\sqrt2\,\frac{d}{d\varphi}\left[\sqrt{1-\cos(\varphi)}\right]\qquad \textbf{(3.4.10)}$$

$$=\cos(\varphi/2).\qquad \textbf{(3.4.11)}$$

[34] See Ramachandran, M. P., 1993: A note on the integral equation method to a diffusion-reaction problem. *Appl. Math. Lett.*, **6**, 27–30.

In a similar manner, Equation 3.4.7 can be rewritten as

$$\int_\theta^\pi \frac{1}{\sqrt{\cos(\theta)-\cos(\varphi)}} \left\{\sum_{n=0}^\infty (b+n)A_n \sin\left[\left(n+\tfrac{1}{2}\right)\varphi\right]\right\} d\varphi = 0 \tag{3.4.12}$$

for $\alpha < \theta \le \pi$. Therefore,

$$\sum_{n=0}^\infty (b+n)A_n \sin\left[\left(n+\tfrac{1}{2}\right)\varphi\right] = \sum_{n=0}^\infty \left(n+\tfrac{1}{2}-\gamma\right) A_n \sin\left[\left(n+\tfrac{1}{2}\right)\varphi\right] = 0 \tag{3.4.13}$$

with $\alpha < \varphi \le \pi$ and $\gamma = \frac{1}{2} - b$. Integrating Equation 3.4.13 with respect to φ, we have

$$\sum_{n=0}^\infty \left(1 - \frac{\gamma}{n+\frac{1}{2}}\right) A_n \cos\left[\left(n+\tfrac{1}{2}\right)\varphi\right] = 0. \tag{3.4.14}$$

The constant of integration equals zero because the left side of Equation 3.4.14 vanishes when $\varphi = \pi$.

At this point, let us supplement Equation 3.4.11 with

$$\psi(\varphi) = \sum_{n=0}^\infty A_n \cos\left[\left(n+\tfrac{1}{2}\right)\varphi\right], \qquad \alpha < \varphi \le \pi. \tag{3.4.15}$$

Therefore, Equation 3.4.14 can be rewritten

$$\psi(\varphi) - \gamma \sum_{n=0}^\infty A_n \frac{\cos\left[\left(n+\frac{1}{2}\right)\varphi\right]}{n+\frac{1}{2}} = 0. \tag{3.4.16}$$

From the definition of half-range Fourier series,

$$A_n = \frac{2}{\pi}\int_0^\alpha \cos(\tilde{\varphi}/2)\cos\left[\left(n+\tfrac{1}{2}\right)\tilde{\varphi}\right]\,d\tilde{\varphi} + \frac{2}{\pi}\int_\alpha^\pi \psi(\tilde{\varphi})\cos\left[\left(n+\tfrac{1}{2}\right)\tilde{\varphi}\right]\,d\tilde{\varphi} \tag{3.4.17}$$

$$= \frac{\sin(n\alpha)}{n\pi} + \frac{\sin[(n+1)\alpha]}{(n+1)\pi} + \frac{2}{\pi}\int_\alpha^\pi \psi(\tilde{\varphi})\cos\left[\left(n+\tfrac{1}{2}\right)\tilde{\varphi}\right]\,d\tilde{\varphi}. \tag{3.4.18}$$

Substituting Equation 3.4.18 into Equation 3.4.16 and interchanging the order of integration and summation,

$$\begin{aligned}\psi(\varphi) = \frac{2\gamma}{\pi}\Bigg[&\int_0^\alpha \cos(\tilde{\varphi}/2)\left\{\sum_{n=0}^\infty \frac{\cos\left[\left(n+\frac{1}{2}\right)\varphi\right]\cos\left[\left(n+\frac{1}{2}\right)\tilde{\varphi}\right]}{n+\frac{1}{2}}\right\}d\tilde{\varphi} \\ &+ \int_\alpha^\pi \psi(\tilde{\varphi})\left\{\sum_{n=0}^\infty \frac{\cos\left[\left(n+\frac{1}{2}\right)\varphi\right]\cos\left[\left(n+\frac{1}{2}\right)\tilde{\varphi}\right]}{n+\frac{1}{2}}\right\}d\tilde{\varphi}\Bigg].\end{aligned} \tag{3.4.19}$$

Because[35]

$$\sum_{n=0}^{\infty} \frac{\cos\left[\left(n+\frac{1}{2}\right)\varphi\right]\cos\left[\left(n+\frac{1}{2}\right)\tilde{\varphi}\right]}{n+\frac{1}{2}} = \frac{1}{2}\ln\left[\frac{\cos(\tilde{\varphi}/2)+\cos(\varphi/2)}{\cos(\tilde{\varphi}/2)-\cos(\varphi/2)}\right], \quad (3.4.20)$$

Equation 3.4.19 becomes

$$\psi(\varphi) = \frac{\gamma}{\pi}\int_0^{\alpha} \cos(\tilde{\varphi}/2)L(\tilde{\varphi},\varphi)\,d\tilde{\varphi} + \frac{\gamma}{\pi}\int_{\alpha}^{\pi} \psi(\tilde{\varphi})L(\tilde{\varphi},\varphi)\,d\tilde{\varphi}, \quad (3.4.21)$$

where

$$L(\tilde{\varphi},\varphi) = \ln\left|\frac{\cos(\tilde{\varphi}/2)+\cos(\varphi/2)}{\cos(\tilde{\varphi}/2)-\cos(\varphi/2)}\right|. \quad (3.4.22)$$

To simplify Equation 3.4.21, we set $\psi(\varphi) = \cos(\varphi/2) + r(\varphi)$ and it becomes

$$r(\varphi) - \frac{\gamma}{\pi}\int_{\alpha}^{\pi} r(\tilde{\varphi})L(\tilde{\varphi},\varphi)\,d\tilde{\varphi} = (2\gamma-1)\cos(\varphi/2), \qquad \alpha < \varphi \le \pi. \quad (3.4.23)$$

To numerically solve Equation 3.4.23, we introduce n nodal points at $\varphi_i = \alpha + ih$, $i = 0, 1, \ldots, n-1$, where $h = (\pi - \alpha)/n$. We do not have to compute $r(\pi)$ because it equals zero since $\psi(\pi) = 0$ from Equation 3.4.15. Then, Equation 3.4.23 becomes

$$r(\varphi_i) - \frac{\gamma}{\pi}\int_{\alpha}^{\pi} r(t)L(t,\varphi_i)\,dt = (2\gamma-1)\cos(\varphi_i/2), \quad (3.4.24)$$

where

$$\begin{aligned} L(t,\varphi) = {} & \ln\bigl|2\cos[(t+\varphi)/4]\bigr| + \ln\bigl|\cos(t-\varphi)/4\bigr| \\ & - \ln\bigl|2\sin[(t+\varphi)/4]\bigr| - \ln\left|\frac{\sin[(t-\varphi)/4]}{t-\varphi}\right| - \ln\bigl|t-\varphi\bigr|. \end{aligned} \quad (3.4.25)$$

Why have we written $L(t,\varphi)$ in the form given in Equation 3.4.25? It clearly shows that the integral equation, Equation 3.4.23, contains a weakly singular kernel. Consequently, the finite difference representation of the integral in Equation 3.4.24 consists of two parts. A simple trapezoidal rule is used for the first four terms given in Equation 3.4.25. For the fifth term, we employ a numerical method devised by Atkinson[36] for kernels with singularities. Therefore, for a particular `alpha` and `b`, we compute `dt = (pi-alpha)/N`,

---

[35] Parihar, K. S., 1971: Some triple trigonometrical series equations and their application. *Proc. R. Soc. Edinburgh, Ser. A*, **69**, 255–265.

[36] Atkinson, K. E., 1967: The numerical solution of Fredholm integral equations of the second kind. *SIAM J. Numer. Anal.*, **4**, 337–348. See Section 5 in particular.

where N is the number of nodal points. The MATLAB code for computing $\psi(\varphi)$ is

```
for j = 0:N
tt(j+1) = alpha + j*dt;
pphi(j+1) = alpha + j*dt;
end

% Solve the integral equation Equation 3.4.24 to find r(φ_i).
% Note that r(π) = 0.

for n = 0:N-1 % rows loop (top to bottom in the matrix)

phi = pphi(n+1); bb(n+1) = (2*gamma-1)*cos(phi/2);

for m = 0:N-1 % columns loop (left to right in the matrix)

t = tt(m+1);

% Introduce the first terms from Equation 3.4.24.

if (n==m) AA(n+1,m+1) = 1;
else AA(n+1,m+1) = 0; end

% Compute integral in Equation 3.4.24.  Add in the contribution
% from the first four terms of Equation 3.4.25.  Use the
% trapezoidal rule.  Recall that r(π) = 0.

if (m < N)
arg_1 = (t+phi)/4; arg_2 = (t-phi)/4;
AA(n+1,m+1) = AA(n+1,m+1) ...
            - 0.5*dt*gamma*log(abs(2*cos(arg_1)))/pi ...
            - 0.5*dt*gamma*log(abs(cos(arg_2)))/pi...
            + 0.5*dt*gamma*log(abs(2*sin(arg_1)))/pi;
if (arg_2 == 0)
AA(n+1,m+1) = AA(n+1,m+1) + 0.5*dt*gamma*log(0.25)/pi;
else
AA(n+1,m+1) = AA(n+1,m+1) ...
            + 0.5*dt*gamma*log(abs(sin(arg_2)/(t-phi)))/pi;
end
end

if (m > 0)
arg_1 = (t+phi)/4; arg_2 = (t-phi)/4;
AA(n+1,m+1) = AA(n+1,m+1) ...
```

```
            - 0.5*dt*gamma*log(abs(2*cos(arg_1)))/pi ...
            - 0.5*dt*gamma*log(abs(cos(arg_2)))/pi...
            + 0.5*dt*gamma*log(abs(2*sin(arg_1)))/pi;
if (arg_2 == 0)
AA(n+1,m+1) = AA(n+1,m+1) + 0.5*dt*gamma*log(0.25)/pi;
else
AA(n+1,m+1) = AA(n+1,m+1) ...
            + 0.5*dt*gamma*log(abs(sin(arg_2)/(t-phi)))/pi;
end
end

% Use Atkinson's technique to treat the fifth term in
% Equation 3.4.25.  See Section 5 of his paper.

if (m > 0)
kk = n+1-m; Psi_0 = -1;
Psi_1 = 0.25*(kk*kk-(kk-1)*(kk-1));
if ( kk ~= 0 )
Psi_0 = Psi_0 + kk*log(abs(kk));
Psi_1 = Psi_1 - 0.50*kk*kk*log(abs(kk));
end
if ( kk ~= 1 )
Psi_0 = Psi_0 - (kk-1)*log(abs(kk-1));
Psi_1 = Psi_1 + 0.50*(kk-1)*(kk-1)*log(abs(kk-1));
end
Psi_1 = Psi_1 + kk*Psi_0;
W_0 = Psi_0 - Psi_1; W_1 = Psi_1;
aalpha = 0.5*dt*log(dt) + dt*W_0;
bbeta = 0.5*dt*log(dt) + dt*W_1;
AA(n+1,m ) = AA(n+1,m ) + gamma*aalpha/pi;
AA(n+1,m+1) = AA(n+1,m+1) + gamma*bbeta/pi;
end

end % end of column loop
end % end of rows loop

% compute r(φ_i)
r = AA\bb'

% compute ψ(φ)
for n = 0:N-1
theta = tt(n+1); psi(n+1) = cos(theta/2) + r(n+1);
end
```

Once $\psi(\varphi_i)$ is found with $\psi(\pi) = 0$, we compute A_n via Equation 3.4.18. The MATLAB code to compute A_n is

```
for m = 0:M

if ( m == 0 )
A(m+1) = alpha/pi + sin(alpha)/pi;
else
A(m+1) = sin(m*alpha)/(m*pi) + sin((m+1)*alpha)/((m+1)*pi);
end

% This is the n = 0 term for Simpson's rule.
A(m+1) = A(m+1) + 2*psi( 1 )*cos((m+0.5)*tt( 1 ))*dt/(3*pi);

% Recall that ψ(π) = 0.  Therefore, we do not need the n = N
% term in the numerical integration.
for n = 1:N-1
if ( mod(n+1,2) == 0 )
A(m+1) = A(m+1) + 8*psi(n+1)*cos((m+0.5)*tt(n+1))*dt/(3*pi);
else
A(m+1) = A(m+1) + 4*psi(n+1)*cos((m+0.5)*tt(n+1))*dt/(3*pi);
end; end; end
```

The final solution follows from Equation 3.4.5. The MATLAB code to realize this solution is

```
for j = 1:41
y = 0.05*(j-21);
for i = 1:41
x = 0.05*(i-21);
u(i,j) = NaN; r = sqrt(x*x + y*y); theta = abs(atan2(y,x));

if (r <= 1)
mu = cos(theta);
% Compute the Legendre polynomials for a given theta
% via the recurrence formula
Legendre(1) = 1; Legendre(2) = mu;
for m = 2:M
Legendre(m+1) = (2-1/m)*mu*Legendre(m) - (1-1/m)*Legendre(m-1);
end

% For a point within the sphere, find the solution.
u(i,j) = 0; power = 1;
for m = 0:M
u(i,j) = u(i,j) + A(m+1)*Legendre(m+1)*power;
```

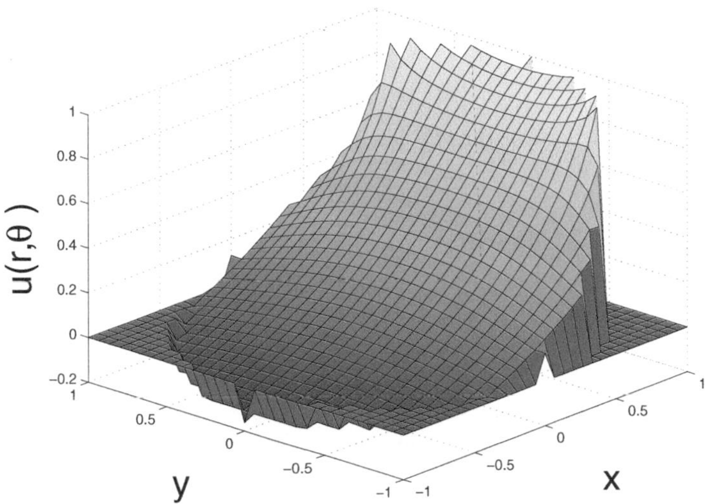

Figure 3.4.1: The solution $u(r,\theta)$ to the mixed boundary value problem posed in Example 3.4.1 with $\alpha = \pi/4$ and $b = 0.25$.

```
power = power*r;
end; end; end; end
```

Figure 3.4.1 illustrates the solution when $\alpha = \pi/4$ and $b = 0.25$.

An alternative method for solving Equation 3.4.6 and Equation 3.4.7 is to introduce the following integral definition for A_n:

$$A_n = \frac{2n+1}{n+b}\int_0^\alpha h(t)\cos\left[\left(n+\frac{1}{2}\right)t\right]\,dt. \tag{3.4.26}$$

Integration by parts gives

$$(n+b)A_n = 2h(\alpha)\sin\left[\left(n+\frac{1}{2}\right)\alpha\right] - 2\int_0^\alpha h'(t)\sin\left[\left(n+\frac{1}{2}\right)t\right]\,dt. \tag{3.4.27}$$

Then,

$$\sum_{n=0}^\infty (n+b)A_n P_n[\cos(\theta)] = 2h(\alpha)\sum_{n=0}^\infty \sin\left[\left(n+\frac{1}{2}\right)\alpha\right] P_n[\cos(\theta)] \tag{3.4.28}$$

$$- 2\int_0^\alpha h'(t)\left\{\sum_{n=0}^\infty \sin\left[\left(n+\frac{1}{2}\right)t\right] P_n[\cos(\theta)]\right\}\,dt$$

$$= 0. \tag{3.4.29}$$

Therefore, our choice for the A_n satisfies Equation 3.4.7 identically. Here we used results from Problem 1 at the end of Section 1.3.

Turning to Equation 3.4.6, we have for A_n that

$$\sum_{n=0}^{\infty} \frac{n+\frac{1}{2}}{n+b} P_n[\cos(\theta)] \int_0^{\alpha} h(t) \cos\left[\left(n+\frac{1}{2}\right)t\right] dt = \frac{1}{2}. \tag{3.4.30}$$

Breaking the summation on the left side of Equation 3.4.30 into two parts, we can rewrite it as

$$\begin{aligned}\sum_{n=0}^{\infty} P_n[\cos(\theta)] \int_0^{\alpha} h(t) \cos\left[\left(n+\frac{1}{2}\right)t\right] dt \qquad & \qquad (3.4.31)\\ = \frac{1}{2} - \sum_{n=0}^{\infty} \frac{\gamma}{n+b} P_n[\cos(\theta)] \int_0^{\alpha} h(\tau) \cos\left[\left(n+\frac{1}{2}\right)\tau\right] d\tau. &\end{aligned}$$

On the left side of Equation 3.4.31, we interchange the order of integration and summation and employ again the results from Problem 1. Equation 3.4.31 simplifies to

$$\begin{aligned}\int_0^{\theta} \frac{h(t)}{\sqrt{2[\cos(t)-\cos(\theta)]}}\, dt \qquad & \qquad (3.4.32)\\ = \frac{1}{2} - \sum_{n=0}^{\infty} \frac{\gamma}{n+b} P_n[\cos(\theta)] \int_0^{\alpha} h(\tau) \cos\left[\left(n+\frac{1}{2}\right)\tau\right] d\tau. &\end{aligned}$$

Upon employing Equation 1.2.9 and Equation 1.2.10, we can solve for $h(t)$ and find that

$$\begin{aligned}h(t) &= \frac{1}{\pi}\frac{d}{dt}\left\{\int_0^{t} \frac{\sin(\theta)}{\sqrt{2[\cos(\theta)-\cos(t)]}}\, d\theta\right\}\\ &\quad - \frac{2}{\pi}\int_0^{\alpha} h(\tau)\left(\sum_{n=0}^{\infty}\frac{\gamma}{n+b}\cos\left[\left(n+\frac{1}{2}\right)\tau\right]\right. \qquad (3.4.33)\\ &\qquad \left.\times \frac{d}{dt}\left\{\int_0^{t} \frac{\sin(\theta)P_n[\cos(\theta)]}{\sqrt{2[\cos(\theta)-\cos(t)]}}\, d\theta\right\}\right) d\tau\\ &= \frac{1}{\pi}\frac{d}{dt}\left\{\sqrt{2[1-\cos(t)]}\right\} \qquad (3.4.34)\\ &\quad - \frac{2}{\pi}\int_0^{\alpha} h(\tau)\left\{\sum_{n=0}^{\infty}\frac{\gamma}{n+b}\cos\left[\left(n+\frac{1}{2}\right)\tau\right]\cos\left[\left(n+\frac{1}{2}\right)t\right]\right\} dt.\end{aligned}$$

We used results from Problem 2 at the end of Section 1.3 to evaluate the integral within the wavy brackets in Equation 3.4.33. Therefore, we now have the Fredholm integral equation

$$h(t) + \int_0^{\alpha} [K(t-\tau) + K(t+\tau)] h(\tau)\, d\tau = \frac{\sin(t)}{\pi\sqrt{2[1-\cos(t)]}}, \tag{3.4.35}$$

where

$$K(z) = \frac{1}{\pi}\sum_{n=0}^{\infty}\frac{\gamma}{n+b}\cos\left[\left(n+\frac{1}{2}\right)z\right]. \tag{3.4.36}$$

Solving Equation 3.4.35 is straightforward. Having determined $h(t)$, A_n follows from Equation 3.4.26. Finally Equation 3.4.5 yields $u(r,z)$.

• **Example 3.4.2**

Let us solve the axisymmetric Laplace equation[37]

$$\frac{\partial}{\partial r}\left(r^2\frac{\partial u}{\partial r}\right) + \frac{1}{\sin(\theta)}\frac{\partial}{\partial\theta}\left[\sin(\theta)\frac{\partial u}{\partial\theta}\right] = 0, \qquad 0 \le r < \infty, \quad 0 < \theta < \pi, \tag{3.4.37}$$

subject to the boundary conditions

$$\lim_{\theta\to 0}|u(r,\theta)| < \infty, \quad \lim_{\theta\to\pi}|u(r,\theta)| < \infty, \qquad 0 \le r < \infty, \tag{3.4.38}$$

$$\lim_{r\to 0}|u(r,\theta)| < \infty, \quad \lim_{r\to\infty}u(r,\theta) \to 0, \qquad 0 \le \theta \le \pi, \tag{3.4.39}$$

and

$$\begin{cases} u_r(a^-,\theta) = u_r(a^+,\theta) = -U\cos(\theta), & 0 \le \theta < \alpha, \\ u_r(a^-,\theta) = u_r(a^+,\theta), & \alpha < \theta \le \pi, \end{cases} \tag{3.4.40}$$

where a^- and a^+ denote points located just inside and outside of $r = a$.

Separation of variables yields the solution

$$u(r,\theta) = Ua\sum_{n=0}^{\infty}A_n\left(\frac{r}{a}\right)^n P_n[\cos(\theta)], \qquad 0 \le r < a, \tag{3.4.41}$$

and

$$u(r,\theta) = -Ua\sum_{n=0}^{\infty}\frac{n}{n+1}A_n\left(\frac{a}{r}\right)^{n+1}P_n[\cos(\theta)], \qquad a < r < \infty. \tag{3.4.42}$$

Equation 3.4.41 and Equation 3.4.42 satisfy not only Equation 3.4.37, but also Equation 3.4.38 and Equation 3.4.39. They also yield a continuous value of $u_r(a,\theta)$ for $0 \le \theta \le \pi$.

Substituting Equation 3.4.41 and Equation 3.4.42 into Equation 3.4.40 gives the dual Fourier-Legendre series

$$\sum_{n=0}^{\infty}nA_nP_n[\cos(\theta)] = -\cos(\theta), \qquad 0 \le \theta < \alpha, \tag{3.4.43}$$

---

[37] Reprinted from *Int. J. Solids Struct.*, **38**, P. A. Martin, The spherical-cap crack revisited, 4759–4776, ©2001, with permission of Elsevier.

and

$$\sum_{n=0}^{\infty}(2n+1)\frac{A_n}{n+1}P_n[\cos(\theta)] = 0, \qquad \alpha \le \theta < \pi. \tag{3.4.44}$$

Turning to Equation 3.4.44 first, let us introduce the function $h(t)$ such that

$$\frac{A_n}{n+1} = \frac{1}{2n+1}\int_0^{\alpha} h(t)\sin\left[\left(n+\tfrac{1}{2}\right)t\right]\,dt, \quad h(0)=0. \tag{3.4.45}$$

Therefore,

$$\sum_{n=0}^{\infty}(2n+1)\frac{A_n}{n+1}P_n[\cos(\theta)] = \int_0^{\alpha} h(t)\left\{\sum_{n=0}^{\infty}\sin\left[\left(n+\tfrac{1}{2}\right)t\right]P_n[\cos(\theta)]\right\}dt = 0. \tag{3.4.46}$$

This follows from Problem 1 in Section 1.3 since $0 \le t \le \alpha < \theta < \pi$. Consequently, our choice for A_n automatically satisfies Equation 3.4.44.

How do we find $h(t)$? We begin by integrating Equation 3.4.45 by parts. This yields

$$2\left(n+\tfrac{1}{2}\right)^2\frac{A_n}{n+1} = \int_0^{\alpha} h'(t)\cos\left[\left(n+\tfrac{1}{2}\right)t\right]\,dt - h(\alpha)\cos\left[\left(n+\tfrac{1}{2}\right)\alpha\right]. \tag{3.4.47}$$

Next, we rewrite Equation 3.4.43 as

$$\sum_{n=0}^{\infty}2\left(n+\tfrac{1}{2}\right)^2\frac{A_n}{n+1}P_n[\cos(\theta)] - \frac{1}{2}\sum_{n=0}^{\infty}\frac{A_n}{n+1}P_n[\cos(\theta)] = -2\cos(\theta). \tag{3.4.48}$$

Then using Equation 3.4.45, Equation 3.4.47, the results from Problem 1 in Section 1.3 and the fact that

$$\sum_{n=0}^{\infty}P_n[\cos(\theta)]\frac{\sin\left[\left(n+\frac{1}{2}\right)\right]}{n+\frac{1}{2}} = \int_0^{t}\frac{H(\theta-\tau)}{\sqrt{2\cos(\tau)-2\cos(\theta)}}\,d\tau, \tag{3.4.49}$$

we find that

$$\int_0^{\theta}\frac{h'(t)}{\sqrt{2\cos(t)-2\cos(\theta)}}\,dt - \frac{1}{4}\int_0^{\theta}\frac{1}{\sqrt{2\cos(t)-\cos(\theta)}}\left[\int_t^{\alpha}h(\tau)\,d\tau\right]dt = -2\cos(\theta). \tag{3.4.50}$$

Finally, we employ Equation 1.2.11 and Equation 1.2.12 and find the inhomogeneous Fredholm equation of the second kind:

$$h'(t) - \frac{1}{4}\int_t^{\alpha}h(\tau)\,d\tau = -\frac{4}{\pi}\cos(3t/2), \qquad h(0)=0. \tag{3.4.51}$$

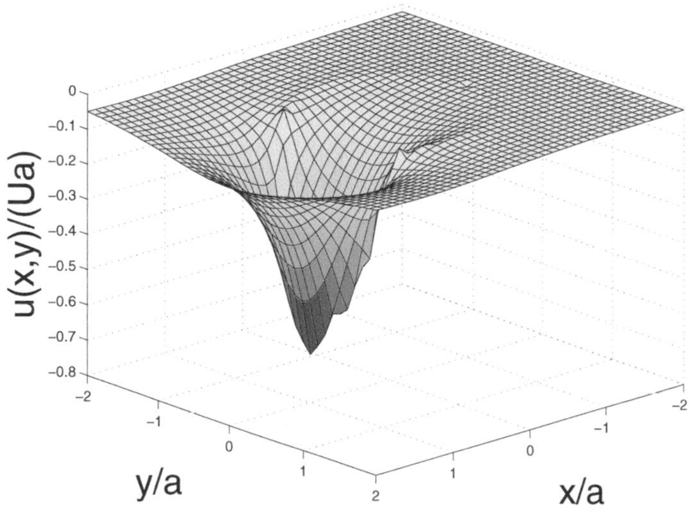

Figure 3.4.2: The solution $u(r,\theta)$ to the mixed boundary value problem posed by Equation 3.4.37 through Equation 3.4.40 when $\alpha = \pi/4$.

To solve Equation 3.4.51, we rewrite it as

$$h'(t) + \frac{1}{4}\int_0^t h(\tau)\,d\tau = \frac{M_1}{4} - \frac{4}{\pi}\cos(3t/2), \qquad \textbf{(3.4.52)}$$

where $M_1 = \int_0^\alpha h(t)\,dt$, an unknown constant. Taking the Laplace transform of Equation 3.4.52,

$$\left(s^2 + \tfrac{1}{4}\right)H(s) = \frac{M_1}{4} - \frac{4}{\pi}\frac{s^2}{s^2 + \frac{9}{4}}, \qquad \textbf{(3.4.53)}$$

or

$$h(t) = \left(\frac{M_1}{2} + \frac{1}{\pi}\right)\sin(t/2) - \frac{3}{\pi}\sin(3t/2). \qquad \textbf{(3.4.54)}$$

From the definition of M_1, $M_1 = -(4/\pi)\sin(\alpha)\sin(\alpha/2)$. Therefore,

$$h(t) = \sec(\alpha/2)\cos(3\alpha/2)\sin(t/2)/\pi - 3\sin(3t/2)/\pi. \qquad \textbf{(3.4.55)}$$

Figure 3.4.2 illustrates the solution when $\alpha = \pi/4$.

- **Example 3.4.3**

Let us solve[38]

$$\frac{\partial}{\partial r}\left(r^2\frac{\partial u}{\partial r}\right)+\frac{1}{\sin(\theta)}\frac{\partial}{\partial \theta}\left[\sin(\theta)\frac{\partial u}{\partial \theta}\right]=0, \qquad 0\le r<\infty, \quad 0<\theta<\pi, \tag{3.4.56}$$

subject to the boundary conditions

$$\lim_{r\to 0}|u(r,\theta)|<\infty, \quad \lim_{r\to\infty}u(r,\theta)\to 0, \qquad 0<\theta<\pi, \tag{3.4.57}$$

and

$$\begin{cases} u(a^-,\theta)=u(a^+,\theta)=1, & 0\le\theta<\alpha, \\ u(a^-,\theta)=u(a^+,\theta), u_r(a^-,\theta)=u_r(a^+,\theta), & \alpha<\theta<\pi-\alpha, \\ u(a^-,\theta)=u(a^+,\theta)=(-1)^m, & \pi-\alpha<\theta<\pi, \end{cases} \tag{3.4.58}$$

where $m = 0$ or 1 and $0<\alpha<\pi/2$. The solution must be finite at the $\theta=0,\pi$.

Separation of variables yields the solution

$$u(r,\theta)=\sum_{n=0}^{\infty}A_{2n+m}\left(\frac{r}{a}\right)^{2n+m}P_{2n+m}[\cos(\theta)], \qquad 0\le r<a, \tag{3.4.59}$$

and

$$u(r,\theta)=\sum_{n=0}^{\infty}A_{2n+m}\left(\frac{a}{r}\right)^{2n+m+1}P_{2n+m}[\cos(\theta)], \qquad a<r<\infty. \tag{3.4.60}$$

We have written the solution in this form so that we can take advantage of symmetry and limit θ between 0 and $\pi/2$ rather than $0<\theta<\pi$. Equation 3.4.59 and Equation 3.4.60 satisfy not only Equation 3.4.56, but also Equation 3.4.57. Substituting Equation 3.4.59 and Equation 3.4.60 into Equation 3.4.58 yields the dual Fourier-Legendre series

$$\sum_{n=0}^{\infty}A_{2n+m}P_{2n+m}[\cos(\theta)]=1, \qquad 0<\theta<\alpha, \tag{3.4.61}$$

and

$$\sum_{n=0}^{\infty}\left(2n+m+\tfrac{1}{2}\right)A_{2n+m}P_{2n+m}[\cos(\theta)]=0, \qquad \alpha<\theta<\pi/2. \tag{3.4.62}$$

---

[38] Minkov, I. M., 1963: Electrostatic field of a sectional spherical capacitor. *Sov. Tech. Phys.*, **7**, 1041–1043.

At this point, we introduce

$$A_{2n+m} = \int_0^\alpha g(t) \cos\left[\left(2n + m + \tfrac{1}{2}\right) t\right]\, dt \tag{3.4.63}$$

$$= g(\alpha)\frac{\sin\left[\left(2n + m + \frac{1}{2}\right)\alpha\right]}{2n + m + \frac{1}{2}} - \int_0^\alpha g'(t)\frac{\sin\left[\left(2n + m + \frac{1}{2}\right) t\right]}{2n + m + \frac{1}{2}}\, dt. \tag{3.4.64}$$

Now,

$$\sum_{n=0}^{\infty} \left(2n + m + \tfrac{1}{2}\right) A_{2n+m} P_{2n+m}[\cos(\theta)]$$

$$= g(\alpha) \sum_{n=0}^{\infty} P_{2n+m}[\cos(\theta)] \sin\left[\left(2n + m + \tfrac{1}{2}\right)\alpha\right] \tag{3.4.65}$$

$$- \int_0^\alpha g'(t) \left\{ \sum_{n=0}^{\infty} P_{2n+m}[\cos(\theta)] \sin\left[\left(2n + m + \tfrac{1}{2}\right) t\right] \right\} dt = 0.$$

Equation 3.4.65 follows from Problem 4 at the end of Section 1.3 as well as the facts that $0 \le t \le \alpha < \theta < \pi/2$. Therefore, our choice for A_{2n+m} satisfies Equation 3.4.62 identically.

Turning to Equation 3.4.61,

$$\int_0^\alpha g(t) \left\{ \sum_{n=0}^{\infty} P_{2n+m}[\cos(\theta)] \cos\left[\left(2n + m + \tfrac{1}{2}\right) t\right] \right\} dt = 1. \tag{3.4.66}$$

Again, using the results from Problem 4 at the end of Section 1.3, we have

$$\int_0^\theta \frac{g(t)}{\sqrt{2[\cos(t) - \cos(\theta)]}}\, dt = 2 - (-1)^m \int_0^\alpha \frac{g(\tau)}{\sqrt{2[\cos(\tau) - \cos(\theta)]}}\, d\tau, \tag{3.4.67}$$

where $0 \le \theta \le \alpha$. Applying Equation 1.2.9 and Equation 1.2.10,

$$g(t) = \frac{4}{\pi}\frac{d}{dt}\left\{ \int_0^t \frac{\sin(\theta)}{\sqrt{2[\cos(\theta) - \cos(t)]}}\, d\theta \right\} - \frac{(-1)^m}{2\pi} \int_0^\alpha K(t,\tau) g(\tau)\, d\tau, \tag{3.4.68}$$

where

$$K(t,\tau) = 2\frac{d}{dt}\left[\int_0^t \frac{\sin(\theta)}{\sqrt{\cos(\tau) + \cos(\theta)}\, \sqrt{\cos(\theta) - \cos(t)}}\, d\theta \right] \tag{3.4.69}$$

$$= \sec[(t + \tau)/2] + \sec[(t - \tau)/2]. \tag{3.4.70}$$

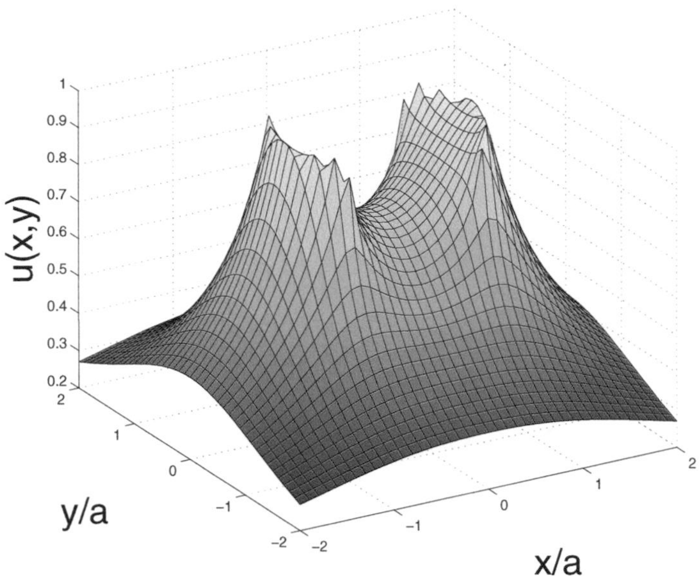

Figure 3.4.3: The solution $u(x, y)$ to the mixed boundary value problem governed by Equation 3.4.56 through Equation 3.4.58 when $\alpha = \pi/4$ and $m = 0$.

Evaluating the first integral in Equation 3.4.68, the integral equation that governs $g(t)$ is

$$g(t) + \frac{(-1)^m}{2\pi} \int_0^\alpha K(t,\tau) g(\tau)\, d\tau = \frac{4}{\pi} \cos(t/2), \qquad 0 \le t \le \alpha. \tag{3.4.71}$$

Figure 3.4.3 illustrates our solution when $\alpha = \pi/4$ and $m = 0$.

• Example 3.4.4

Let us solve

$$\frac{\partial}{\partial r}\left(r^2 \frac{\partial u}{\partial r}\right) + \frac{1}{\sin(\theta)} \frac{\partial}{\partial \theta}\left[\sin(\theta) \frac{\partial u}{\partial \theta}\right] = 0, \qquad 0 \le r < \infty, \quad 0 < \theta < \pi, \tag{3.4.72}$$

subject to the boundary conditions

$$\lim_{\theta \to 0} |u(r,\theta)| < \infty, \quad \lim_{\theta \to \pi} |u(r,\theta)| < \infty, \qquad 0 < r < \infty, \tag{3.4.73}$$

$$\lim_{r \to 0} |u(r,\theta)| < \infty, \quad \lim_{r \to \infty} u(r,\theta) \to 0, \qquad 0 < \theta < \pi, \tag{3.4.74}$$

and

$$\begin{cases} u(1^-,\theta) = u(1^+,\theta) = 1, & 0 \le \theta < \alpha, \\ u_r(1^-,\theta) = u_r(1^+,\theta), & \alpha < \theta < \beta, \\ u(1^-,\theta) = u(1^+,\theta) = 0, & \beta < \theta < \pi. \end{cases} \tag{3.4.75}$$

Separation of variables yields the solution

$$u(r, \theta) = \sum_{n=0}^{\infty} A_n r^n P_n[\cos(\theta)], \qquad 0 \le r < 1, \tag{3.4.76}$$

and

$$u(r, \theta) = \sum_{n=0}^{\infty} A_n r^{-n-1} P_n[\cos(\theta)], \qquad 1 < r < \infty. \tag{3.4.77}$$

Equation 3.4.76 and Equation 3.4.77 satisfy not only Equation 3.4.72, but also Equation 3.4.73 and Equation 3.4.74. Substitution of Equation 3.4.76 and Equation 3.4.77 into Equation 3.4.75 yields the triple Fourier-Legendre series

$$\sum_{n=0}^{\infty} A_n P_n[\cos(\theta)] = 1, \qquad 0 \le \theta < \alpha, \tag{3.4.78}$$

$$\sum_{n=0}^{\infty} (2n+1) A_n P_n[\cos(\theta)] = 0, \qquad \alpha < \theta < \beta, \tag{3.4.79}$$

and

$$\sum_{n=0}^{\infty} A_n P_n[\cos(\theta)] = 0, \qquad \beta < \theta \le \pi. \tag{3.4.80}$$

How do we determine A_n? Recently Singh et al.[39] solved the triple series equation

$$\sum_{n=0}^{\infty} A_n P_n[\cos(\theta)] = f_1(\theta), \qquad 0 \le \theta < \alpha, \tag{3.4.81}$$

$$\sum_{n=0}^{\infty} (2n+1) A_n P_n[\cos(\theta)] = f_2(\theta), \qquad \alpha < \theta < \beta, \tag{3.4.82}$$

and

$$\sum_{n=0}^{\infty} A_n P_n[\cos(\theta)] = f_3(\theta), \qquad \beta < \theta \le \pi. \tag{3.4.83}$$

They showed that A_n is given by

$$A_n = \tfrac{1}{2} \left\{ \int_0^{\alpha} g_1(\eta) \sin(\eta) P_n[\cos(\eta)]\, d\eta + \int_{\alpha}^{\beta} f_2(\eta) \sin(\eta) P_n[\cos(\eta)]\, d\eta \right. \\ \left. + \int_{\beta}^{\pi} g_3(\eta) \sin(\eta) P_n[\cos(\eta)]\, d\eta \right\}, \tag{3.4.84}$$

---

[39] Results quoted with permission from Singh, B. M., R. S. Dhaliwal, and J. Rokne, 2002: The elementary solution of triple series equations involving series of Legendre polynomials and their application to an electrostatic problem. *Z. Angew. Math. Mech.*, **82**, 497–503.

where

$$\sin(x)g_1(x) = -\frac{1}{\pi}\frac{d}{dx}\left[\int_x^\alpha \frac{G_1(\eta)\sin(\eta)}{\sqrt{\cos(x)-\cos(\eta)}}\,d\eta\right], \qquad 0 < x < \alpha, \quad \textbf{(3.4.85)}$$

$$\sin(x)g_3(x) = \frac{1}{\pi}\frac{d}{dx}\left[\int_\beta^x \frac{G_3(\eta)\sin(\eta)}{\sqrt{\cos(\eta)-\cos(x)}}\,d\eta\right], \qquad \beta < x < \pi, \quad \textbf{(3.4.86)}$$

$$G_1(x) + \frac{2}{\pi}\cos(x/2)\int_\beta^\pi \frac{\sin(\eta/2)G_3(\eta)}{\cos(x)-\cos(\eta)}\,d\eta = \frac{d}{dx}\left[\int_0^x \frac{F_1(\theta)\sin(\theta)}{\sqrt{\cos(\theta)-\cos(x)}}\,d\theta\right] \quad \textbf{(3.4.87)}$$

for $0 < x < \alpha$,

$$G_3(x) = -\frac{2}{\pi}\sin(x/2)\int_0^\alpha \frac{\cos(\eta/2)G_1(\eta)}{\cos(\eta)-\cos(x)}\,d\eta - \frac{d}{dx}\left[\int_x^\pi \frac{F_3(\theta)\sin(\theta)}{\sqrt{\cos(x)-\cos(\theta)}}\,d\theta\right] \quad \textbf{(3.4.88)}$$

for $\beta < x < \pi$,

$$F_1(\theta) = 2f_1(\theta) - \int_\alpha^\beta f_2(\eta)\sin(\eta)K(\eta,\theta)\,d\eta, \qquad 0 < \theta < \alpha, \quad \textbf{(3.4.89)}$$

$$F_3(\theta) = 2f_3(\theta) - \int_\alpha^\beta f_2(\eta)\sin(\eta)K(\eta,\theta)\,d\eta, \qquad \beta < \theta < \pi, \quad \textbf{(3.4.90)}$$

and

$$K(\eta,\theta) = \sum_{n=0}^\infty P_n[\cos(\theta)]P_n[\cos(\eta)]. \quad \textbf{(3.4.91)}$$

Let us apply these results to our problem. Because $f_1(\theta) = 1$ and $f_2(\theta) = f_3(\theta) = 0$, $F_1(\theta) = 2$ and $F_3(\theta) = 0$. Therefore,

$$A_n = \tfrac{1}{2}\left\{\int_0^\alpha g_1(\eta)\sin(\eta)P_n[\cos(\eta)]\,d\eta + \int_\beta^\pi g_3(\eta)\sin(\eta)P_n[\cos(\eta)]\,d\eta\right\}, \quad \textbf{(3.4.92)}$$

where

$$G_1(x) + \frac{2}{\pi}\cos(x/2)\int_\beta^\pi \frac{\sin(\eta/2)G_3(\eta)}{\cos(x)-\cos(\eta)}\,d\eta = \frac{2\,\sin(x)}{\sqrt{1-\cos(x)}}, \quad \textbf{(3.4.93)}$$

for $0 < x < \alpha$, and

$$G_3(x) = -\frac{2}{\pi}\sin(x/2)\int_0^\alpha \frac{\cos(\eta/2)G_1(\eta)}{\cos(\eta)-\cos(x)}\,d\eta \quad \textbf{(3.4.94)}$$

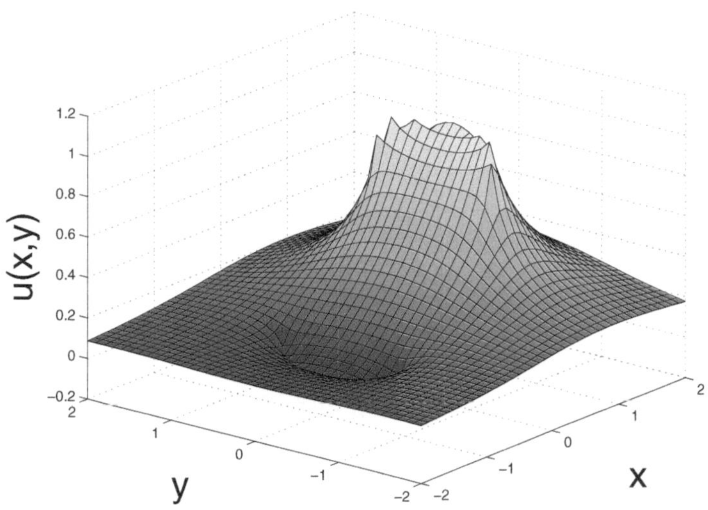

Figure 3.4.4: The solution $u(x,y)$ to the mixed boundary value problem governed by Equation 3.4.72 and Equation 3.4.75 when $\alpha = \pi/4$ and $\beta = 3\pi/4$.

with Equation 3.4.85 and Equation 3.4.86. Figure 3.4.4 illustrates our solution when $\alpha = \pi/4$ and $\beta = 3\pi/4$.

A special case of particular interest occurs when $\beta \to \pi$. Here, Equation 3.4.81 through Equation 3.4.83 reduce to

$$\sum_{n=0}^{\infty} A_n P_n[\cos(\theta)] = f_1(\theta), \qquad 0 \le \theta < \alpha, \tag{3.4.95}$$

and

$$\sum_{n=0}^{\infty} (2n+1) A_n P_n[\cos(\theta)] = 0, \qquad \alpha < \theta \le \pi. \tag{3.4.96}$$

From Equation 3.4.92, we have that

$$A_n = \tfrac{1}{2} \int_0^{\alpha} g_1(x) \sin(x) P_n[\cos(x)]\, dx \tag{3.4.97}$$

$$= \frac{1}{\sqrt{2}\,\pi} \int_0^{\alpha} g_1(x) \sin(x) \left\{ \int_0^{x} \frac{\cos\left[\left(n+\frac{1}{2}\right)\eta\right]}{\sqrt{\cos(\eta)-\cos(x)}}\, d\eta \right\} dx \tag{3.4.98}$$

$$= \frac{1}{\sqrt{2}\,\pi} \int_0^{\alpha} \left[\int_{\eta}^{\alpha} \frac{g_1(x)\sin(x)}{\sqrt{\cos(\eta)-\cos(x)}}\, dx \right] \cos\left[\left(n+\tfrac{1}{2}\right)\eta\right] d\eta \tag{3.4.99}$$

$$= \frac{1}{\sqrt{2}\,\pi} \int_0^{\alpha} G_1(\eta) \cos\left[\left(n+\tfrac{1}{2}\right)\eta\right] d\eta, \tag{3.4.100}$$

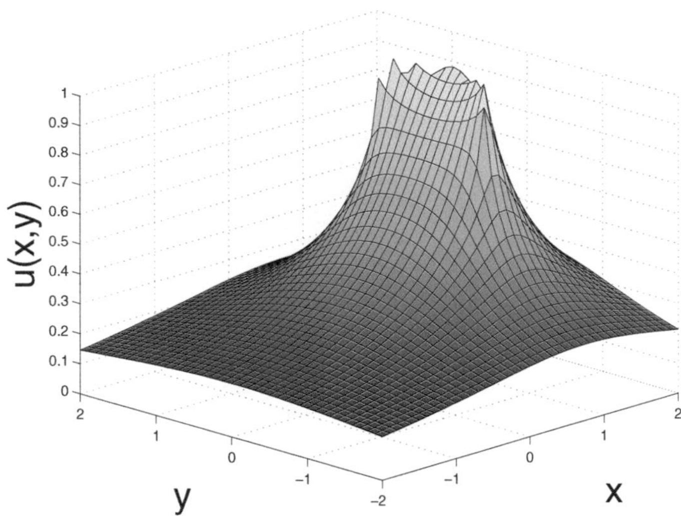

Figure 3.4.5: The solution $u(x, y)$ to the mixed boundary value problem governed by Equation 3.4.72 through Equation 3.4.74 and Equation 3.4.104 when $\alpha = \pi/4$.

where we used the Mehler integral representation of $P_n[\cos(x)]$ and interchanged the order of integration in Equation 3.4.98. We also used the fact that

$$G_1(\eta) = \int_\eta^\alpha \frac{g_1(x)\sin(x)}{\sqrt{\cos(\eta) - \cos(x)}}\,dx \tag{3.4.101}$$

which follows from Equation 3.4.85, Equation 1.2.11, and Equation 1.2.12.

Sneddon[40] was the first to derive the solution to the dual series of Equation 3.4.95 and Equation 3.4.96; Table 3.4.1 summarizes the results. Subsequently Boridy[41] derived several additional solutions and they have also been included in the table.

To illustrate these results, we apply them to a case examined by Collins.[42] For $f_1(\theta) = 1$,

$$G_1(\eta) = \frac{2\sin(\eta)}{\sqrt{1 - \cos(\eta)}} = 2\sqrt{2}\cos(\eta/2). \tag{3.4.102}$$

---

40 Sneddon, op. cit., Section 5.6.

41 Boridy, E., 1987: Solution of some electrostatic potential problems involving spherical conductors: A dual series approach. *IEEE Trans. Electromagn. Compat.*, **EMC-29**, 132–140. ©1987 IEEE.

42 Collins, W. D., 1961: On some dual series equations and their application to electrostatic problems for spheroidal caps. *Proc. Cambridge Philosoph. Soc.*, **57**, 367–384.

Table 3.4.1: The Solution (given in the right column) of Dual Series Involving Legendre Polynomials (given in the left column). Taken from Boridy, E., 1987: Solution of some electrostatic potential problems involving spherical conductors: A dual series approach. *IEEE Trans. Electromagn. Compat.*, **EMC-29**, 132–140. ©1987 IEEE.

Dual series	Solution
$\sum_{n=0}^{\infty} A_n P_n[\cos(\theta)] = F(\theta), \quad 0 \le \theta < \theta_0$ $\sum_{n=0}^{\infty} (2n+1) A_n P_n[\cos(\theta)] = 0, \quad \theta_0 < \theta \le \pi$	$A_n = \frac{\sqrt{2}}{\pi} \int_0^{\theta_0} f(\xi) \cos\left[\left(n+\frac{1}{2}\right)\xi\right] d\xi$ $f(\xi) = \frac{d}{d\xi}\left[\int_0^{\xi} \frac{F(\theta)\sin(\theta)}{\sqrt{\cos(\theta)-\cos(\xi)}}\, d\theta\right]$
$\sum_{n=0}^{\infty} A_n P_n[\cos(\theta)] = 0, \quad 0 \le \theta < \theta_0$ $\sum_{n=0}^{\infty} (2n+1) A_n P_n[\cos(\theta)] = F(\theta), \quad \theta_0 < \theta \le \pi$	$A_n = -\frac{1}{\sqrt{2}\,\pi} \int_{\theta_0}^{\pi} f(\xi) \cos\left[\left(n+\frac{1}{2}\right)\xi\right] d\xi$ $f(\xi) = \int_{\xi}^{\pi} \frac{F(\theta)\sin(\theta)}{\sqrt{\cos(\xi)-\cos(\theta)}}\, d\theta$
$\sum_{n=0}^{\infty} (2n+1) A_n P_n[\cos(\theta)] = F(\theta), \quad 0 \le \theta < \theta_0$ $\sum_{n=0}^{\infty} A_n P_n[\cos(\theta)] = 0, \quad \theta_0 < \theta \le \pi$	$A_n = \frac{(-1)^{n+1}}{\sqrt{2}\,\pi} \int_{\pi-\theta_0}^{\pi} f(\xi) \cos\left[\left(n+\frac{1}{2}\right)\xi\right] d\xi$ $f(\xi) = \int_{\xi}^{\pi} \frac{F(\pi-\theta)\sin(\theta)}{\sqrt{\cos(\xi)-\cos(\theta)}}\, d\theta$
$\sum_{n=0}^{\infty} (2n+1) A_n P_n[\cos(\theta)] = 0, \quad 0 \le \theta < \theta_0$ $\sum_{n=0}^{\infty} A_n P_n[\cos(\theta)] = F(\theta), \quad \theta_0 < \theta \le \pi$	$A_n = \frac{(-1)^n\sqrt{2}}{\pi} \int_0^{\pi-\theta_0} f(\xi) \cos\left[\left(n+\frac{1}{2}\right)\xi\right] d\xi$ $f(\xi) = \frac{d}{d\xi}\left[\int_0^{\xi} \frac{F(\pi-\theta)\sin(\theta)}{\sqrt{\cos(\theta)-\cos(\xi)}}\, d\theta\right]$

Substituting Equation 3.4.102 into Equation 3.4.100 and carrying out the integration, we find that

$$A_n = \frac{\sin(n\alpha)}{n\pi} + \frac{\sin[(n+1)\alpha]}{(n+1)\pi}. \tag{3.4.103}$$

Figure 3.4.5 illustrates the solution to Equation 3.4.72 through Equation 3.4.74 and

$$\begin{cases} u(1^-,\theta) = u(1^+,\theta) = 1, & 0 \le \theta < \pi/4, \\ u_r(1^-,\theta) = u_r(1^+,\theta), & \pi/4 < \theta < \pi. \end{cases} \tag{3.4.104}$$

- **Example 3.4.5**

Let us solve[43]

$$\frac{\partial}{\partial r}\left(r^2\frac{\partial u}{\partial r}\right) + \frac{1}{\sin(\theta)}\frac{\partial}{\partial\theta}\left[\sin(\theta)\frac{\partial u}{\partial\theta}\right] = 0, \qquad 0 \le r < b, \quad 0 \le \theta \le \pi, \tag{3.4.105}$$

subject to the boundary conditions that

$$\lim_{\theta\to 0}|u(r,\theta)| < \infty, \quad \lim_{\theta\to\pi}|u(r,\theta)| < \infty, \qquad 0 \le r < b, \tag{3.4.106}$$

$$\lim_{r\to 0}|u(r,\theta)| < \infty, \quad u(b,\theta) = 0, \qquad 0 \le \theta \le \pi, \tag{3.4.107}$$

and

$$\begin{cases} u_r(a^-,\theta) = u_r(a^+,\theta), & 0 \le \theta < \theta_0, \\ u(a,\theta) = V_0, & \theta_0 < \theta \le \pi. \end{cases} \tag{3.4.108}$$

Before we solve our original problem, let us find the solution to a simpler one when we replace Equation 3.4.108 with

$$u(a,\theta) = V_0, \qquad 0 \le \theta \le \pi. \tag{3.4.109}$$

The solution to this new problem is

$$\begin{cases} u(r,\theta) = V_0, & 0 \le r \le a, \\ u(r,\theta) = \dfrac{aV_0}{b-a}\left(\dfrac{b}{r} - 1\right), & a \le r \le b. \end{cases} \tag{3.4.110}$$

Let us return to our original problem. We can view the introduction of the aperture between $\theta_0 < \theta \le \pi$ as a perturbation on the solution given by Equation 3.4.110. Therefore, we write the solution as

$$u(r,\theta) = V_0 + \sum_{n=0}^{\infty}\left[1 - \left(\frac{a}{b}\right)^{2n+1}\right] A_n \left(\frac{r}{a}\right)^n P_n[\cos(\theta)], \qquad 0 \le r \le a, \tag{3.4.111}$$

[43] Boridy, op. cit.

and

$$u(r,\theta) = \frac{aV_0}{b-a}\left(\frac{b}{r}-1\right) + \sum_{n=0}^{\infty} A_n \left[\left(\frac{a}{r}\right)^{n+1} - \left(\frac{a}{b}\right)^{2n+1}\left(\frac{r}{a}\right)^n\right] P_n[\cos(\theta)] \tag{3.4.112}$$

for $a \le r \le b$. The coefficients in Equation 3.4.111 and Equation 3.4.112 were chosen so that Equation 3.4.107 is satisfied and $u(r,\theta)$ is continuous at $r = a$.

Turning to the mixed boundary condition, direct substitution yields

$$\sum_{n=0}^{\infty}(2n+1)A_n P_n[\cos(\theta)] = -\frac{bV_0}{b-a}, \qquad 0 \le \theta < \theta_0, \tag{3.4.113}$$

and

$$\sum_{n=0}^{\infty}\left[1 - \left(\frac{a}{b}\right)^{2n+1}\right] A_n P_n[\cos(\theta)] = 0, \qquad \theta_0 < \theta \le \pi. \tag{3.4.114}$$

At this point, we would like to use the results given in Table 3.4.1 but Equation 3.4.114 is not in the correct form. To circumvent this difficulty, let us set $x = a/b < 1$. Then we can rewrite Equation 3.4.114 as

$$\sum_{n=0}^{\infty} A_n P_n[\cos(\theta)] = \sum_{n=0}^{\infty} A_n x^{2n+1} P_n[\cos(\theta)], \qquad \theta_0 < \theta \le \pi. \tag{3.4.115}$$

Setting $\xi = \cos(\theta)$, let us integrate Equation 3.4.115 from -1 to ξ. We then have

$$\sum_{n=0}^{\infty} A_n \int_{-1}^{\xi} P_n(\xi)\,d\xi = \sum_{n=0}^{\infty} A_n x^{2n+1} \int_{-1}^{\xi} P_n(\xi)\,d\xi. \tag{3.4.116}$$

However, because

$$\int_{-1}^{1} P_n(\xi)\,d\xi = 2\delta_{n0}, \tag{3.4.117}$$

where δ_{ij} is the Kronecker delta,

$$2A_0 - \sum_{n=0}^{\infty} A_n \int_{\xi}^{1} P_n(\xi)\,d\xi = 2A_0 x - \sum_{n=0}^{\infty} A_n x^{2n+1} \int_{\xi}^{1} P_n(\xi)\,d\xi. \tag{3.4.118}$$

If we differentiate Equation 3.4.118 with respect of x, then differentiate it with respect of ξ, and finally multiply by x, we obtain

$$\sum_{n=0}^{\infty}(2n+1)A_n x^{2n+1} P_n[\cos(\theta)] = 0, \qquad 0 \le \theta < \theta_0. \tag{3.4.119}$$

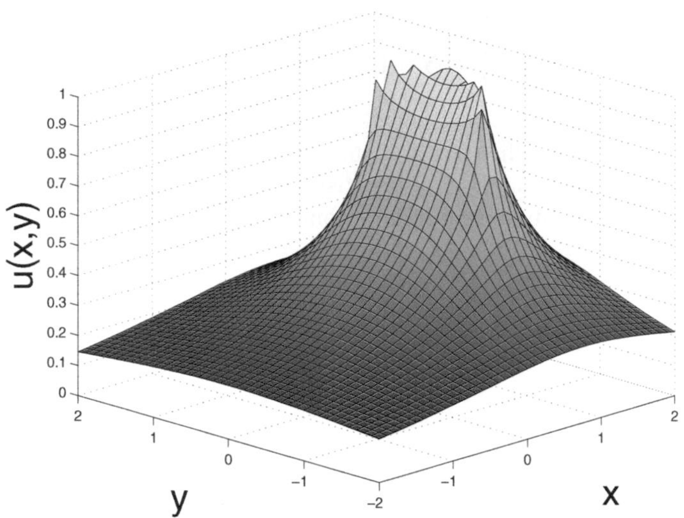

Figure 3.4.6: The solution $u(x, y)$ to the mixed boundary value problem governed by Equation 3.4.105 through Equation 3.4.108 when $a = 1$, $b = 2$ and $\theta_0 = \pi/2$.

Subtracting Equation 3.4.119 from Equation 3.4.113, we obtain the following dual equations:

$$\sum_{n=0}^{\infty}(2n+1)A_n\left[1-\left(\frac{a}{b}\right)^{2n+1}\right]P_n[\cos(\theta)] = -\frac{bV_0}{b-a}, \qquad 0 \le \theta < \theta_0, \tag{3.4.120}$$

and

$$\sum_{n=0}^{\infty}A_n\left[1-\left(\frac{a}{b}\right)^{2n+1}\right]P_n[\cos(\theta)] = 0, \qquad \theta_0 < \theta \le \pi. \tag{3.4.121}$$

If we set $C_n = A_n\left[1-\left(\frac{a}{b}\right)^{2n+1}\right]$, then we can immediately use the results from Table 3.4.1 and find that

$$A_n = -\frac{bV_0}{\pi(b-a)\left[1-\left(\frac{a}{b}\right)^{2n+1}\right]}\left\{\frac{\sin(n\theta_0)}{n} - \frac{\sin[(n+1)\theta_0]}{n+1}\right\}. \tag{3.4.122}$$

Figure 3.4.6 illustrates our solution when $a = 1$, $b = 2$ and $\theta_0 = \pi/2$.

• **Example 3.4.6**

A problem[44] that is similar to Example 3.4.3 consists of solving

$$\frac{\partial}{\partial r}\left(r^2\frac{\partial u}{\partial r}\right)+\frac{1}{\sin(\theta)}\frac{\partial}{\partial\theta}\left[\sin(\theta)\frac{\partial u}{\partial\theta}\right]=0, \qquad 0\le r<1, \quad 0<\theta<\pi, \tag{3.4.123}$$

subject to the boundary conditions

$$\lim_{\theta\to 0}|u(r,\theta)|<\infty, \quad \lim_{\theta\to\pi}|u(r,\theta)|<\infty, \qquad 0\le r<1, \tag{3.4.124}$$

$$\lim_{r\to 0}|u(r,\theta)|<\infty, \qquad 0<\theta<\pi, \tag{3.4.125}$$

and

$$\begin{cases} u(1,\theta)=\cos(\theta), & 0\le\theta<\alpha,\\ u_r(1,\theta)=0, & \alpha<\theta<\beta,\\ u(1,\theta)=\cos(\theta), & \beta<\theta\le\pi. \end{cases} \tag{3.4.126}$$

Separation of variables yields the solution

$$u(r,\theta)=rP_1[\cos(\theta)]-\sum_{n=1}^{\infty}\frac{2n+1}{n}C_n r^n P_n[\cos(\theta)]. \tag{3.4.127}$$

From the nature of the boundary conditions, we anticipate that $C_0=C_2=C_4=\ldots=0$. Upon substituting Equation 3.4.127 into Equation 3.4.126,

$$\sum_{n=1}^{\infty}C_n(1+H_n)P_n[\cos(\theta)]=0, \qquad 0\le\theta<\alpha, \quad \beta<\theta\le\pi, \tag{3.4.128}$$

and

$$\sum_{n=1}^{\infty}C_n(2n+1)P_n[\cos(\theta)]=P_1[\cos(\theta)], \qquad \alpha<\theta<\beta, \tag{3.4.129}$$

where $H_n=1/(2n)$.

To find $C_1, C_3, C_5, \ldots$, let us set $C_n=A_n+B_n$ with $B_n=(-1)^{n+1}A_n$. Therefore, Equation 3.4.128 and Equation 3.4.129 can be rewritten

$$\sum_{n=0}^{\infty}(A_n+B_n)(1+H_n)P_n[\cos(\theta)]=0, \qquad 0\le\theta<\alpha, \tag{3.4.130}$$

$$\sum_{n=0}^{\infty}A_n(2n+1)P_n[\cos(\theta)]-\tfrac{1}{2}P_1[\cos(\theta)]=0, \qquad \alpha<\theta\le\pi, \tag{3.4.131}$$

---

[44] See Dryden, J. R., and F. W. Zok, 2004: Effective conductivity of partially sintered solids. *J. Appl. Phys.*, **95**, 156–160.

$$\sum_{n=0}^{\infty} B_n(2n+1)P_n[\cos(\theta)] - \tfrac{1}{2}P_1[\cos(\theta)] = 0, \qquad 0 \le \theta < \beta, \tag{3.4.132}$$

and

$$\sum_{n=0}^{\infty}(A_n + B_n)(1 + H_n)P_n[\cos(\theta)] = 0, \qquad \beta < \theta \le \pi. \tag{3.4.133}$$

In this formulation, A_0 and B_0 are nonzero although $A_0 + B_0 = 0$. For convenience we introduce $H_0 \equiv 1$ so that no difficulty arises in solving this system of equations.

Due to symmetry, we must only solve Equation 3.4.130 and Equation 3.4.131. Using the integral representation of Legendre polynomials, Equation 1.3.4, we can rewrite Equation 3.4.130 as

$$\int_0^{\theta} \left\{ \sum_{n=0}^{\infty}(A_n + B_n)(1 + H_n)\cos\left[\left(n + \tfrac{1}{2}\right)t\right] \right\} \frac{dt}{\sqrt{\cos(t) - \cos(\theta)}} = 0 \tag{3.4.134}$$

after interchanging the order of integration and summation. In a similar manner, we can use Equation 1.3.5 to write Equation 3.4.131 as

$$\int_{\theta}^{\pi} \frac{d}{dt} \left\{ \sum_{n=0}^{\infty} A_n \cos\left[\left(n + \tfrac{1}{2}\right)t\right] - \tfrac{1}{6}\cos\left(\frac{3t}{2}\right) \right\} \frac{dt}{\sqrt{\cos(\theta) - \cos(t)}} = 0. \tag{3.4.135}$$

Equation 3.4.134 and Equation 3.4.135 are integral equations of the Abel type; see Equation 1.2.9 and Equation 1.2.12. In the case of Equation 3.4.134 the quantity within the wavy brackets must vanish. In the case of Equation 3.4.135 the t-derivative of the quantity within the wavy brackets must vanish. Actually the quantity within the wavy brackets also vanishes because this quantity equals zero since it vanishes when $t = \pi$. Consequently,

$$\sum_{n=0}^{\infty}(A_n + B_n)(1 + H_n)\cos\left[\left(n + \tfrac{1}{2}\right)t\right] = 0, \quad 0 \le t < \alpha, \tag{3.4.136}$$

and

$$\sum_{n=0}^{\infty} A_n \cos\left[\left(n + \tfrac{1}{2}\right)t\right] - \tfrac{1}{6}\cos\left(\frac{3t}{2}\right) = 0, \quad \alpha < t \le \pi. \tag{3.4.137}$$

To find A_n, let us introduce an unknown function $h(t)$ such that

$$\sum_{n=0}^{\infty} A_n \cos\left[\left(n + \tfrac{1}{2}\right)t\right] = \begin{cases} \cos(3t/2)/6 - h(t)/2, & 0 \le t < \alpha, \\ \cos(3t/2)/6, & \alpha < t \le \pi. \end{cases} \tag{3.4.138}$$

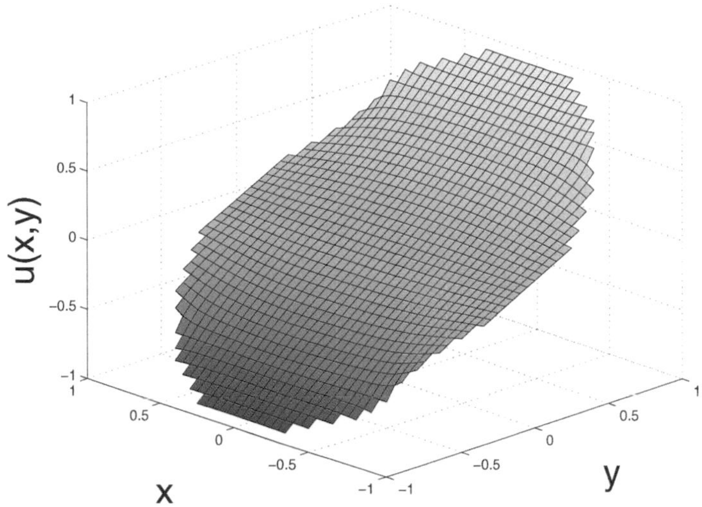

Figure 3.4.4: The solution $u(x, y)$ to the mixed boundary value problem governed by Equation 3.4.123 through Equation 3.4.126 when $\alpha = \pi/4$ and $\beta = \pi/2$.

Applying the orthogonality properties of $\cos\left[\left(n + \frac{1}{2}\right)t\right]$ over $[0, \pi]$,

$$A_n = \tfrac{1}{6}\delta_{1n} - \frac{1}{\pi}\int_0^\alpha h(\xi)\cos\left[\left(n + \tfrac{1}{2}\right)\xi\right]\,d\xi, \tag{3.4.139}$$

where δ_{nm} is the Kronecker delta. Upon setting $B_n = (-1)^{n+1}A_n$ and substituting Equation 3.4.138 into Equation 3.4.136, we obtain

$$h(t) + 2\sum_{n=0}^{\infty} A_n\left\{(-1)^n - H_n\left[1 - (-1)^n\right]\right\}\cos\left[\left(n + \tfrac{1}{2}\right)t\right] = \tfrac{1}{3}\cos(3t/2). \tag{3.4.140}$$

Finally, substituting for A_n, we obtain an integral equation that governs $h(t)$:

$$h(t) = \cos(3t/2) + \frac{1}{\pi}\int_0^\alpha K(\tau, t)h(\tau)\,d\tau, \qquad 0 \le t < \alpha, \tag{3.4.141}$$

where $K(\tau, t) = G(t - \tau) + G(t + \tau)$ and

$$G(\xi) = \sum_{n=0}^{\infty}\left\{(-1)^n - H_n\left[1 - (-1)^n\right]\right\}\cos\left[\left(n + \tfrac{1}{2}\right)\xi\right] \tag{3.4.142}$$

$$= \frac{1}{4}\left\{2\sec\left(\frac{\xi}{2}\right) + \pi\sin\left|\frac{\xi}{2}\right| + \cos\left(\frac{\xi}{2}\right)\ln\left[\tan^2\left(\frac{\xi}{2}\right)\right]\right\}. \tag{3.4.143}$$

Once $h(t)$ is computed numerically from Equation 3.4.141, we can find A_n from Equation 3.4.139. Finally, $C_{2n+1} = 2A_{2n+1}$ and $u(r, \theta)$ follows from Equation 3.4.127. Figure 3.4.4 illustrates the solution when $\alpha = \pi/4$ and $\beta = \pi/2$.

Problems

1. Solve Laplace's equation

$$\frac{\partial}{\partial r}\left(r^2\frac{\partial u}{\partial r}\right)+\frac{1}{\sin(\theta)}\frac{\partial}{\partial \theta}\left[\sin(\theta)\frac{\partial u}{\partial \theta}\right]=0, \qquad a<r<\infty, \quad 0\le\theta\le\pi,$$

subject to the boundary conditions that

$$\lim_{\theta\to 0}|u(r,\theta)|<\infty, \quad \lim_{\theta\to\pi}|u(r,\theta)|<\infty, \qquad a<r<\infty, \tag{1}$$

$$u(a,\theta)=V_0, \quad \lim_{r\to\infty}u(r,\theta)\to 0, \qquad 0\le\theta\le\pi, \tag{2}$$

and

$$\begin{cases} u_r(b^-,\theta)=u_r(b^+,\theta), & 0\le\theta<\theta_0, \\ u(b,\theta)=0, & \theta_0<\theta\le\pi, \end{cases} \tag{3}$$

where b^- and b^+ denote points slightly inside and outside of $r=b$, respectively, and $0<\theta_0<\pi$.

Step 1: First solve the simpler problem when we replace Equation (3) with $u(b,\theta)=0$ for $0\le\theta\le\pi$ and show that

$$\begin{cases} u(r,\theta)=\dfrac{aV_0}{b-a}\left(\dfrac{b}{r}-1\right), & a<r<b, \\ u(r,\theta)=0, & b<r<\infty. \end{cases}$$

Step 2: Returning to the original problem, show that the solution to the partial differential equation plus the first two boundary conditions is

$$u(r,\theta)=\frac{aV_0}{b-a}\left(\frac{b}{r}-1\right)+\sum_{n=0}^{\infty}A_n\left[\left(\frac{r}{b}\right)^n-\left(\frac{a}{b}\right)^{2n+1}\left(\frac{b}{r}\right)^{n+1}\right]P_n[\cos(\theta)]$$

for $a\le r\le b$, and

$$u(r,\theta)=\sum_{n=0}^{\infty}\left[1-\left(\frac{a}{b}\right)^{2n+1}\right]A_n\left(\frac{b}{r}\right)^{n+1}P_n[\cos(\theta)], \qquad b\le r<\infty.$$

Step 3: Using Equation (3), show that A_n is given by the dual series:

$$\sum_{n=0}^{\infty}(2n+1)A_nP_n[\cos(\theta)]=\frac{aV_0}{b-a}, \qquad 0\le\theta<\theta_0,$$

and

$$\sum_{n=0}^{\infty}\left[1-\left(\frac{a}{b}\right)^{2n+1}\right]A_nP_n[\cos(\theta)]=0, \qquad \theta_0<\theta\le\pi.$$

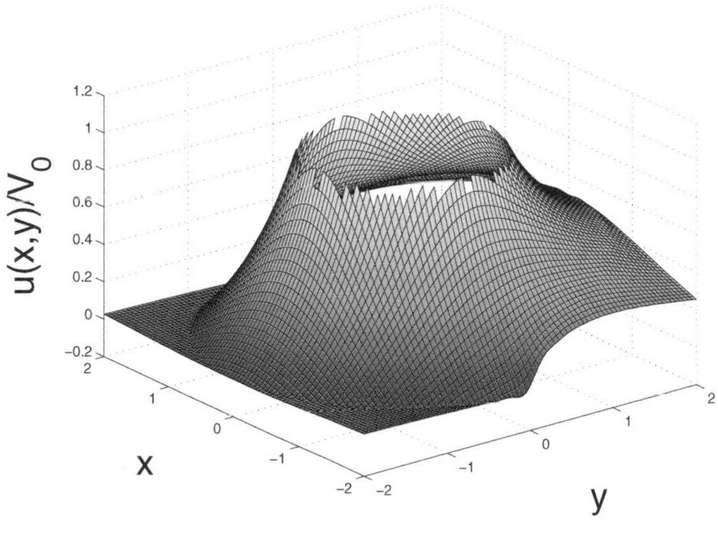

Problem 1

Step 4: Following the analysis given by Equation 3.4.115 through Equation 3.4.121, show that

$$A_n = \frac{aV_0}{\pi(b-a)\left[1-\left(\frac{a}{b}\right)^{2n+1}\right]} \left\{ \frac{\sin(n\theta_0)}{n} - \frac{\sin[(n+1)\theta_0]}{n+1} \right\}.$$

The figure labeled Problem 1 illustrates this solution when $a = 1$, $b = 2$ and $\theta_0 = \pi/2$.

2. A problem similar to the previous one involves finding the electrostatic potential when an uniform external electric field is applied along the z-axis. In this case,

$$\frac{\partial}{\partial r}\left(r^2 \frac{\partial u}{\partial r}\right) + \frac{1}{\sin(\theta)} \frac{\partial}{\partial \theta}\left[\sin(\theta) \frac{\partial u}{\partial \theta}\right] = 0, \qquad a < r < \infty, \quad 0 \le \theta \le \pi,$$

subject to the boundary conditions that

$$\lim_{\theta \to 0} |u(r,\theta)| < \infty, \quad \lim_{\theta \to \pi} |u(r,\theta)| < \infty, \qquad a < r < \infty, \tag{1}$$

$$u(a,\theta) = V_0, \quad \lim_{r \to \infty} u(r,\theta) \to E_0 r \cos(\theta), \qquad 0 \le \theta \le \pi, \tag{2}$$

and

$$\begin{cases} u_r(b^-,\theta) = u_r(b^+,\theta), & 0 \le \theta < \theta_0, \\ u(b,\theta) = 0, & \theta_0 < \theta \le \pi, \end{cases} \tag{3}$$

where b^- and b^+ denote points slightly inside and outside of $r = b$, respectively, and $0 < \theta_0 < \pi$.

Step 1: First solve the simpler problem when we replace Equation (3) with $u(b,\theta) = 0$ for $0 \le \theta \le \pi$ and show that

$$\begin{cases} u(r,\theta) = \dfrac{aV_0}{b-a}\left(\dfrac{b}{r} - 1\right), & a < r < b, \\ u(r,\theta) = E_0 r\cos(\theta) - E_0 b^3 \cos(\theta)/r^2, & b < r < \infty. \end{cases}$$

Step 2: Returning to the original problem, show that the solution to the partial differential equation plus the first two boundary conditions is

$$u(r,\theta) = \frac{aV_0}{b-a}\left(\frac{b}{r} - 1\right) + \sum_{n=0}^{\infty} A_n \left[\left(\frac{r}{b}\right)^n - \left(\frac{a}{b}\right)^{2n+1}\left(\frac{b}{r}\right)^{n+1}\right] P_n[\cos(\theta)]$$

for $a \le r \le b$, and

$$u(r,\theta) = E_0 r\cos(\theta) - E_0\frac{b^3}{r^2}\cos(\theta) + \sum_{n=0}^{\infty}\left[1 - \left(\frac{a}{b}\right)^{2n+1}\right] A_n \left(\frac{b}{r}\right)^{n+1} P_n[\cos(\theta)]$$

for $b \le r < \infty$.

Step 3: Using the third boundary condition, show that A_n is given by the dual series:

$$\sum_{n=0}^{\infty}(2n+1)A_n P_n[\cos(\theta)] = \frac{aV_0}{b-a} + 3E_0 b\cos(\theta), \qquad 0 \le \theta < \theta_0,$$

and

$$\sum_{n=0}^{\infty}\left[1 - \left(\frac{a}{b}\right)^{2n+1}\right] A_n P_n[\cos(\theta)] = 0, \qquad \theta_0 < \theta \le \pi.$$

Step 4: Following the analysis given by Equation 3.4.115 through Equation 3.4.121, show that

$$\begin{aligned} A_n = {} & \frac{aV_0}{\pi(b-a)\left[1 - \left(\frac{a}{b}\right)^{2n+1}\right]}\left\{\frac{\sin(n\theta_0)}{n} - \frac{\sin[(n+1)\theta_0]}{n+1}\right\} \\ & - \frac{E_0 b}{\pi\left[1 - \left(\frac{a}{b}\right)^{2n+1}\right]}\left\{\frac{\sin[(n-1)\theta_0]}{n-1} - \frac{\sin[(n+2)\theta_0]}{n+2}\right\}. \end{aligned}$$

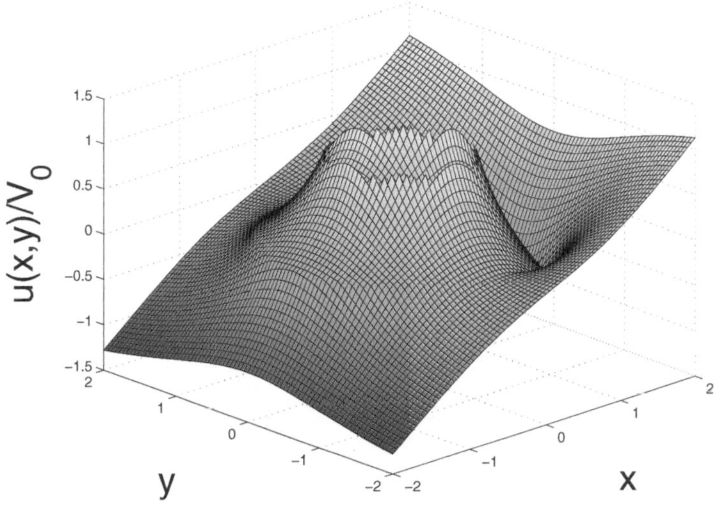

Problem 2

The figure labeled Problem 2 illustrates this solution when $a = 0.7$, $b = 1.4$, $\theta_0 = \pi/2$ and $V_0 = bE_0$.

3. Solve[45] Laplace equation

$$\frac{\partial}{\partial r}\left(r^2\frac{\partial u}{\partial r}\right) + \frac{1}{\sin(\theta)}\frac{\partial}{\partial \theta}\left[\sin(\theta)\frac{\partial u}{\partial \theta}\right] = 0, \qquad 0 \le r < \infty, \quad 0 \le \theta \le \pi,$$

subject to the boundary conditions that

$$\lim_{\theta\to 0}|u(r,\theta)| < \infty, \quad \lim_{\theta\to \pi}|u(r,\theta)| < \infty, \qquad 0 \le r < \infty, \tag{1}$$

$$\lim_{r\to 0}|u(r,\theta)| < \infty, \quad \lim_{r\to \infty}|u(r,\theta)| < \infty, \qquad 0 \le \theta \le \pi, \tag{2}$$

and

$$\begin{cases} u(a^-,\theta) = u(a^+,\theta) = C_1 + C_2\cos(\theta), & 0 \le \theta < \alpha, \\ u_r(a^-,\theta) = u_r(a^+,\theta), & \alpha < \theta \le \pi, \end{cases} \tag{3}$$

where a^- and a^+ denote points slightly inside and outside of $r = a$, respectively, and $0 < \alpha < \pi$. The parameter C_2 is nonzero.

---

[45] Taken with permission from Casey, K. F., 1985: Quasi-static electric- and magnetic-field penetration of a spherical shield through a circular aperture. *IEEE Trans. Electromag. Compat.*, **EMC-27**, 13–17. ©1985 IEEE.

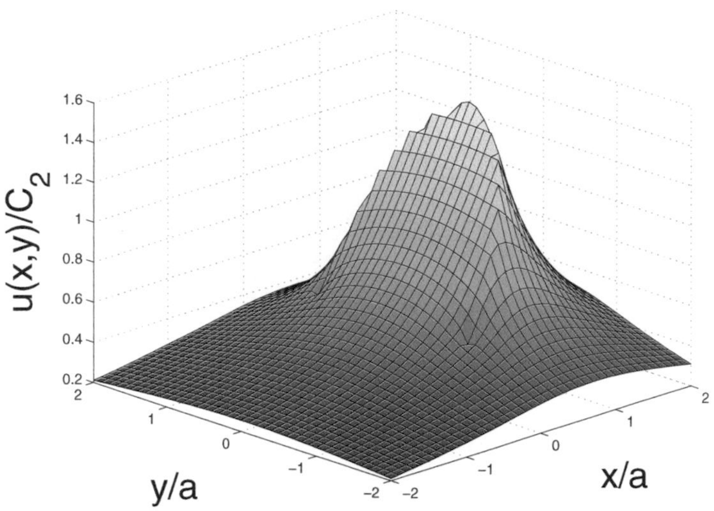

Problem 3

Step 1: Show that the solution to the differential equation and first two boundary conditions are

$$u(r,\theta) = C_2 \sum_{n=0}^{\infty} A_n \left(\frac{r}{a}\right)^n P_n[\cos(\theta)], \qquad 0 \le r \le a,$$

and

$$u(r,\theta) = C_2 \sum_{n=0}^{\infty} A_n \left(\frac{a}{r}\right)^{n+1} P_n[\cos(\theta)], \qquad a \le r < \infty.$$

Step 2: Using the third boundary condition, show that A_n is given by the dual series:

$$\sum_{n=0}^{\infty} A_n P_n[\cos(\theta)] = C_1/C_2 + \cos(\theta), \qquad 0 \le \theta < \alpha,$$

and

$$\sum_{n=0}^{\infty} (2n+1) A_n P_n[\cos(\theta)] = 0, \qquad \alpha < \theta \le \pi.$$

Step 3: Using Table 3.4.1, show that

$$A_n = \frac{C_1}{\pi C_2}\left\{\frac{\sin(n\alpha)}{n} + \frac{\sin[(n+1)\alpha]}{n+1}\right\} + \frac{1}{\pi}\left\{\frac{\sin[(n-1)\alpha]}{n-1} + \frac{\sin[(n+2)\alpha]}{n+2}\right\}.$$

The figure labeled Problem 3 illustrates this solution when $C_1/C_2 = 0.5$ and $\alpha = \pi/2$.

3.5 TRIPLE FOURIER SINE SERIES

In this closing section we illustrate a mixed boundary value problem that yields a triple Fourier sine series.

Let us find[46] the potential for Laplace's equation in cylindrical coordinates:

$$\frac{\partial^2 u}{\partial r^2} + \frac{1}{r}\frac{\partial u}{\partial r} + \frac{\partial^2 u}{\partial z^2} = 0, \qquad 0 \le r < a, \quad 0 < z < \pi, \tag{3.5.1}$$

subject to the boundary conditions

$$\lim_{r\to 0} |u(r,z)| < \infty, \qquad 0 < z < \pi, \tag{3.5.2}$$

$$\begin{cases} u_r(a,z) = 0, & 0 \le z < \alpha, \\ u(a,z) = 1, & \alpha < z < \beta, \\ u_r(a,z) = 0, & \beta < z < \pi, \end{cases} \tag{3.5.3}$$

and

$$u(r,0) = u_z(r,\pi) = 0, \qquad 0 \le r < a. \tag{3.5.4}$$

Separation of variables yields the potential, namely

$$u(r,z) = \sum_{n=0}^{\infty} A_n \sin\left[\left(n+\tfrac{1}{2}\right)z\right] \frac{I_0\left[\left(n+\frac{1}{2}\right)r\right]}{I_1\left[\left(n+\frac{1}{2}\right)a\right]}. \tag{3.5.5}$$

Equation 3.5.5 satisfies Equation 3.5.1, Equation 3.5.2, and Equation 3.5.4. Substituting Equation 3.5.5 into Equation 3.5.3, we obtain the triple Fourier sine series:

$$\sum_{n=0}^{\infty} \left(n+\tfrac{1}{2}\right) A_n \sin\left[\left(n+\tfrac{1}{2}\right)z\right] = 0, \qquad 0 < z < \alpha, \tag{3.5.6}$$

$$\sum_{n=0}^{\infty} (1+M_n) A_n \sin\left[\left(n+\tfrac{1}{2}\right)z\right] = 1, \qquad \alpha < z < \beta, \tag{3.5.7}$$

and

$$\sum_{n=0}^{\infty} \left(n+\tfrac{1}{2}\right) A_n \sin\left[\left(n+\tfrac{1}{2}\right)z\right] = 0, \qquad \beta < z < \pi, \tag{3.5.8}$$

where

$$M_n = \frac{I_0\left[\left(n+\frac{1}{2}\right)a\right]}{I_1\left[\left(n+\frac{1}{2}\right)a\right]} - 1. \tag{3.5.9}$$

---

[46] See Zanadvorov, N. P., V. A. Malinov, and A. V. Charukhchev, 1983: Radial transmission distribution in a cylindrical electrooptical shutter with a large aperture. *Opt. Spectrosc. (USSR)*, **54**, 212–215.

To solve Equation 3.5.6 through Equation 3.5.8, we first note that

$$\sum_{n=0}^{\infty} \left(n+\tfrac{1}{2}\right) A_n \sin\left[\left(n+\tfrac{1}{2}\right) z\right] = -\frac{d}{dz}\left\{\sum_{n=0}^{\infty} A_n \cos\left[\left(n+\tfrac{1}{2}\right) z\right]\right\}. \tag{3.5.10}$$

Following Tranter and Cooke,[47] we introduce

$$A_n = \sum_{k=0}^{\infty} B_k \int_0^{\infty} J_{2k+1}[x\sin(\beta/2)] J_{2n+1}(x) \frac{dx}{x}. \tag{3.5.11}$$

The integral in Equation 3.5.11 can be evaluated[48] in terms of hypergeometric functions. Then,

$$\sum_{n=0}^{\infty} A_n \cos\left[\left(n+\tfrac{1}{2}\right) z\right] = \frac{1}{4\sqrt{2}} \sum_{k=0}^{\infty} B_k \int_z^{\pi} \frac{\sin(\eta)}{\sqrt{\cos(z)-\cos(\eta)}} \times \left\{\int_0^{\infty} J_{2k+1}[x\sin(\beta/2)] J_0[x\sin(\eta/2)]\, dx\right\} d\eta. \tag{3.5.12}$$

Because[49]

$$\int_0^{\infty} J_{2k+1}[x\sin(\beta/2)] J_0[x\sin(\eta/2)]\, dx = \begin{cases} 0, & \eta > \beta, \\ \csc(\beta/2)\, {}_2F_1[k+1, -k; 1; \sin^2(\eta/2)/\sin^2(\beta/2)], & \eta < \beta, \end{cases} \tag{3.5.13}$$

$$\sum_{n=0}^{\infty} A_n \cos\left[\left(n+\tfrac{1}{2}\right) z\right] = 0 \tag{3.5.14}$$

if $z > \beta$. Therefore, it follows from Equation 3.5.10 that Equation 3.5.8 is also satisfied. On the other hand, if $0 < z < \beta$,

$$\sum_{n=0}^{\infty} A_n \cos\left[\left(n+\tfrac{1}{2}\right) z\right] = \frac{\csc(\beta/2)}{4\sqrt{2}} \sum_{k=0}^{\infty} B_k \int_z^{\beta} \frac{\sin(\eta)\, {}_2F_1[k+1, -k; 1; \sin^2(\eta/2)/\sin^2(\beta/2)]}{\sqrt{\cos(z)-\cos(\eta)}}\, d\eta \tag{3.5.15}$$

$$= \frac{1}{4\sqrt{2}} \sum_{k=0}^{\infty} B_k \int_y^{\pi} \frac{\sin(\theta)\, P_k[\cos(\theta)]}{\sqrt{\cos(y)-\cos(\theta)}}\, d\theta \tag{3.5.16}$$

$$= \frac{1}{4} \sum_{k=0}^{\infty} B_k \frac{\cos\left[\left(k+\tfrac{1}{2}\right) y\right]}{k+\tfrac{1}{2}}, \qquad 0 < y < \pi, \tag{3.5.17}$$

---

[47] Tranter, C. J., and J. C. Cooke, 1973: A Fourier-Neumann series and its application to the reduction of triple cosine series. *Glasgow Math. J.*, **14**, 198–201.

[48] Gradshteyn and Ryzhik, op. cit., Formula 6.574.1

[49] Ibid., Formula 6.512.2 with $\nu = n + 1$.

where we substituted $\sin(\theta/2) = \sin(\eta/2)/\sin(\beta/2)$ and $\sin(y/2) = \sin(z/2)$ $/\sin(\beta/2)$. We also used Equation 1.3.4 to simplify Equation 3.5.16. Upon substituting Equation 3.5.17 into Equation 3.5.10 and carrying out the differentiation, we find that Equation 3.5.6 becomes

$$\sum_{k=0}^{\infty} B_k \sin\left\{ (2k+1) \arcsin\left[\frac{\sin(z/2)}{\sin(\beta/2)} \right] \right\} = 0, \qquad 0 < z < \alpha. \tag{3.5.18}$$

Finally, consider Equation 3.5.7. We can rewrite it

$$\sum_{n=0}^{\infty} A_n \sin\left[\left(n+\tfrac{1}{2}\right) z \right] = 1 - \sum_{n=0}^{\infty} M_n A_n \sin\left[\left(n+\tfrac{1}{2}\right) z \right]. \tag{3.5.19}$$

Substituting Equation 3.5.11, we find that

$$\begin{aligned} &\sum_{k=0}^{\infty} B_k \int_0^{\infty} J_{2k+1}[x\sin(\beta/2)] \left\{ \sum_{n=0}^{\infty} \sin\left[\left(n+\tfrac{1}{2}\right) z \right] J_{2n+1}(x) \right\} \frac{dx}{x} \\ &\quad = 1 - \sum_{k=0}^{\infty} B_k \int_0^{\infty} J_{2k+1}[x\sin(\beta/2)] \left\{ \sum_{n=0}^{\infty} M_n \sin\left[\left(n+\tfrac{1}{2}\right) z \right] J_{2n+1}(x) \right\} \frac{dx}{x}. \end{aligned} \tag{3.5.20}$$

The summation over n on the left side of Equation 3.5.20 can be replaced[50] by $\sin[x\sin(z/2)]/2$ so that we now have

$$\begin{aligned} &\sum_{k=0}^{\infty} B_k \int_0^{\infty} J_{2k+1}[x\sin(\beta/2)] \sin[x\sin(z/2)] \frac{dx}{x} \\ &\quad = 2 - 2\sum_{k=0}^{\infty} B_k \int_0^{\infty} J_{2k+1}[x\sin(\beta/2)] \left\{ \sum_{n=0}^{\infty} M_n \sin\left[\left(n+\tfrac{1}{2}\right) z \right] J_{2n+1}(x) \right\} \frac{dx}{x}. \end{aligned} \tag{3.5.21}$$

Evaluating[51] the integral on the left side of Equation 3.5.21, we finally obtain

$$\begin{aligned} &\sum_{k=0}^{\infty} \frac{B_k}{2k+1} \sin\left\{ (2k+1) \arcsin\left[\frac{\sin(z/2)}{\sin(\beta/2)} \right] \right\} \\ &\quad = 2 - 2\sum_{k=0}^{\infty} B_k \int_0^{\infty} J_{2k+1}[x\sin(\beta/2)] \left\{ \sum_{n=0}^{\infty} M_n \sin\left[\left(n+\tfrac{1}{2}\right) z \right] J_{2n+1}(x) \right\} \frac{dx}{x}. \end{aligned} \tag{3.5.22}$$

In summary, by introducing Equation 3.5.11, we reduced the triple Fourier sine equations, Equation 3.5.6 through Equation 3.5.8, to the dual Fourier sine series

$$\sum_{k=0}^{\infty} B_k \sin\left[\left(k+\tfrac{1}{2}\right) \varphi \right] = 0, \tag{3.5.23}$$

---

50 Ibid., Formula 8.514.6.

51 Ibid., Formula 6.693.1.

and

$$\sum_{k=0}^{\infty} \frac{B_k}{k+\frac{1}{2}} \sin\left[\left(k+\tfrac{1}{2}\right)\varphi\right] = 4 - 4\sum_{k=0}^{\infty} B_k F_k(\varphi), \tag{3.5.24}$$

where

$$F_k(\varphi) = \sum_{n=0}^{\infty} M_n \sin\{(2n+1)\arcsin[\sin(\varphi/2)\sin(\beta/2)]\} \\ \times \int_0^{\infty} J_{2k+1}[x\sin(\beta/2)] J_{2n+1}(x)\, \frac{dx}{x}, \tag{3.5.25}$$

and $\varphi = 2\arcsin[\sin(z/2)/\sin(\beta/2)]$.

Our final task is to compute B_k. To this end, let us introduce

$$B_k = AP_k[\cos(\varphi_0)] + \int_{\varphi_0}^{\pi} f(\tau)\frac{d}{d\tau}\{P_k[\cos(\tau)]\}\, d\tau, \tag{3.5.26}$$

where A is a free parameter and $\varphi_0 = 2\arcsin[\sin(\alpha/2)/\sin(\beta/2)]$. Substituting Equation 3.5.26 into Equation 3.5.23, we obtain

$$\sum_{k=0}^{\infty} B_k \sin\left[\left(k+\tfrac{1}{2}\right)\varphi\right] = A\sum_{k=0}^{\infty} \sin\left[\left(k+\tfrac{1}{2}\right)\varphi\right] P_k[\cos(\varphi_0)] \tag{3.5.27}$$

$$+ \int_{\varphi_0}^{\pi} f(\tau)\frac{d}{d\tau}\left\{\sum_{k=0}^{\infty} \sin\left[\left(k+\tfrac{1}{2}\right)\varphi\right] P_k[\cos(\tau)]\right\} d\tau,$$

$$= A\frac{H(\varphi-\varphi_0)}{\sqrt{2\cos(\varphi_0) - 2\cos(\varphi)}} \tag{3.5.28}$$

$$+ \int_{\varphi_0}^{\pi} f(\tau)\frac{d}{d\tau}\left[\frac{H(\varphi-\tau)}{\sqrt{2\cos(\tau)-2\cos(\varphi)}}\right] d\tau,$$

where we used results from Problem 1 at the end of Section 1.3. Because $\varphi < \varphi_0$, both Heaviside functions in Equation 3.5.28 equal zero and our choice for B_k satisfies Equation 3.5.23.

Turning to Equation 3.5.24, we take its derivative with respect to φ and obtain

$$\sum_{k=0}^{\infty} B_k \cos\left[\left(k+\tfrac{1}{2}\right)\varphi\right] = -4\sum_{k=0}^{\infty} B_k F_k'(\varphi). \tag{3.5.29}$$

Next, we substitute for B_k and find that

$$A\sum_{k=0}^{\infty} \cos\left[\left(k+\tfrac{1}{2}\right)\varphi\right] P_k[\cos(\varphi_0)]$$

$$+ \int_{\varphi_0}^{\pi} f(t)\frac{d}{dt}\left\{\sum_{k=0}^{\infty} \cos\left[\left(k+\tfrac{1}{2}\right)\varphi\right] P_k[\cos(t)]\right\} dt \tag{3.5.30}$$

$$= -4A\sum_{k=0}^{\infty} F_k'(\varphi)P_k[\cos(\varphi_0)] - 4\int_{\varphi_0}^{\pi} f(t)\frac{d}{dt}\left\{\sum_{k=0}^{\infty} F_k'(\varphi)P_k[\cos(t)]\right\} dt.$$

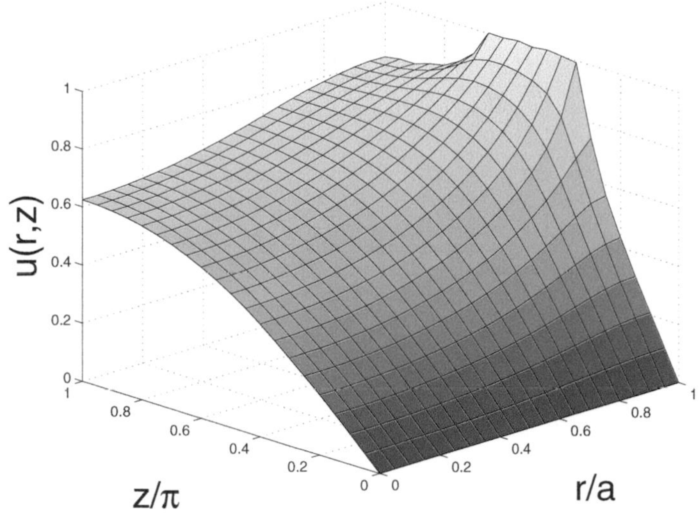

Figure 3.5.1: The solution $u(r,z)$ to the mixed boundary value problem governed by Equation 3.5.1 through Equation 3.5.3 when $a = \pi$, $\alpha = \pi/3$, and $\beta = 2\pi/3$.

If we now integrate the second term in Equation 3.5.30 by parts and again introduce the results from Problem 1 from Section 1.3, we derive

$$\frac{A\,H(\varphi_0-\varphi)}{\sqrt{2\cos(\varphi)-2\cos(\varphi_0)}} - f(\varphi_0)\sum_{k=0}^{\infty}\cos\left[\left(k+\tfrac{1}{2}\right)\varphi\right]P_k[\cos(\varphi_0)]$$
$$-\int_{\varphi_0}^{\pi}\frac{f'(t)H(t-\varphi)}{\sqrt{2\cos(\varphi)-2\cos(t)}}\,dt + \int_{\varphi_0}^{\pi} f(t)\frac{d}{dt}\left\{\sum_{k=0}^{\infty}4F_k'(\varphi)P_k[\cos(t)]\right\}dt$$
$$= -A\sum_{k=0}^{\infty}4F_k'(\varphi)P_k[\cos(\varphi_0)]. \qquad \textbf{(3.5.31)}$$

The first two terms in Equation 3.5.31 vanish while the limits of integration for the integral in the third term run from φ to π. Finally, let us multiply Equation 3.5.31 by $\sin(\varphi)\,d\varphi/\sqrt{2\cos(\tau)-2\cos(\varphi)}$ and then integrate from τ to π. We find then that

$$-\int_{\tau}^{\pi}\frac{\sin(\varphi)}{\sqrt{2\cos(\tau)-2\cos(\varphi)}}\left\{\int_{\varphi}^{\pi}\frac{f'(t)}{\sqrt{2\cos(\varphi)-2\cos(t)}}\,dt\right\}d\varphi$$
$$+\int_{\varphi_0}^{\pi} f(t)\frac{d}{dt}\left\{\int_{\tau}^{\pi}\frac{\sin(\varphi)}{\sqrt{2\cos(\tau)-2\cos(\varphi)}}\sum_{k=0}^{\infty}4F_k'(\varphi)P_k[\cos(t)]\,d\varphi\right\}dt$$
$$= -A\int_{\tau}^{\pi}\frac{\sin(\varphi)}{\sqrt{2\cos(\tau)-2\cos(\varphi)}}\sum_{k=0}^{\infty}4F_k'(\varphi)P_k[\cos(\varphi_0)]\,d\varphi. \qquad \textbf{(3.5.32)}$$

Using results given by Equation 1.2.11 and Equation 1.2.12, the first term in Equation 3.5.32 equals $f(\tau)$ and Equation 3.5.32 becomes

$$f(\tau) + \frac{2}{\pi} \int_{\varphi_0}^{\pi} f(t) \frac{dL(\tau, t)}{dt} \, dt = -\frac{2A}{\pi} L(\tau, \varphi_0), \qquad \varphi_0 < \tau < \pi, \qquad \mathbf{(3.5.33)}$$

where

$$L(\tau, t) = \int_{\tau}^{\pi} \frac{\sin(\varphi)}{\sqrt{2\cos(\tau) - 2\cos(\varphi)}} \sum_{k=0}^{\infty} 4F_k'(\varphi) P_k[\cos(t)] \, d\varphi. \qquad \mathbf{(3.5.34)}$$

It is clear from Equation 3.5.33 that $f(\tau)$ is proportional to A. Consequently, both B_k and A_n also are proportional to A. Therefore, A must be chosen to that $u(r, z) = 1$ for $\alpha < z < \beta$. Figure 3.5.1 illustrates this solution when $a = \pi$, $\alpha = \pi/3$ and $\beta = 2\pi/3$. It is better to use

$$B_k = [A - f(\varphi_0)] P_k[\cos(\varphi_0)] + f(\pi)(-1)^k - \int_{\varphi_0}^{\pi} f'(\tau) P_k[\cos(\tau)] \, d\tau \quad \mathbf{(3.5.35)}$$

rather than Equation 3.5.26 so that we avoid large values of the derivative of the Legendre polynomials for large k near $\tau = \pi$.

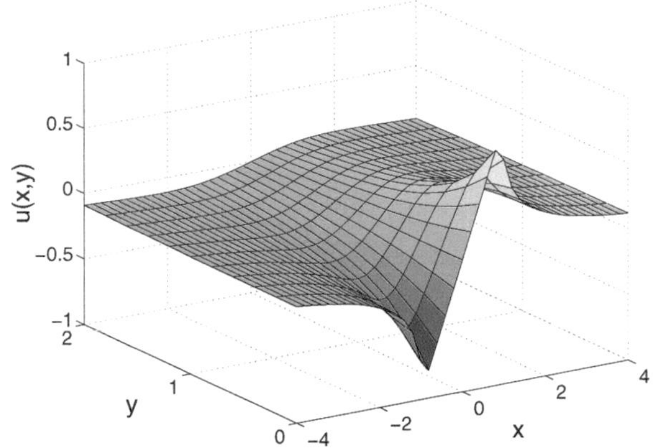

Chapter 4
Transform Methods

In Example 1.1.2 we showed that applying a Fourier cosine transform leads to the dual integral equations:

$$-\frac{2}{\pi}\int_0^\infty k\coth(kh)A(k)\cos(kx)\,dk = 1/h, \qquad 0 \le x < 1, \tag{4.0.1}$$

and

$$\frac{2}{\pi}\int_0^\infty A(k)\cos(kx)\,dk = 0, \qquad 1 < x < \infty. \tag{4.0.2}$$

The purpose of this chapter is to illustrate how these dual integral equations are solved. In Sections 4.1 and 4.2 we focus on Fourier-type of integrals while Sections 4.3 and 4.4 treat Fourier-Bessel integrals. Finally Section 4.5 deals with situations where we have a mixture of Fourier series and transforms, Fourier and Fourier-Bessel transforms and Fourier series and Laplace transforms.

Before we proceed to our study of dual and triple integral equations, let us finish Example 1.1.2. We begin by introducing

$$u(x,0) = \frac{2}{\pi}\int_0^\infty A(k)\cos(kx)\,dk. \tag{4.0.3}$$

Referring back to Equation 1.1.15, we see that $u(x,0)$ is the solution to Equation 1.1.11 along the boundary $y = 0$. Next, for convenience, let us define

$$g(x) = -\frac{du(x,0)}{dx} = \frac{2}{\pi}\int_0^\infty kA(k)\sin(kx)\,dk. \tag{4.0.4}$$

From Equation 4.0.2, $u(x,0)$ is nonzero only if $0 < x < 1$. Consequently, $g(x)$ is nonzero only between $0 < x < 1$. Taking the Fourier sine transform of $g(x)$,

$$kA(k) = \int_0^1 g(x)\sin(kx)\,dx. \tag{4.0.5}$$

If we integrate Equation 4.0.1 with respect to x, we have that

$$-\frac{2}{\pi}\int_0^\infty \coth(kh)A(k)\sin(kx)\,dk = \frac{x}{h}, \qquad 0 \le x < 1. \tag{4.0.6}$$

Substituting Equation 4.0.5 into Equation 4.0.6, we have the integral equation

$$-\frac{2}{\pi}\int_0^1 g(\xi)\left[\int_0^\infty \coth(kh)\sin(k\xi)\sin(kx)\,\frac{dk}{k}\right]d\xi = \frac{x}{h}. \qquad 0 \le x < 1. \tag{4.0.7}$$

The integral within the square brackets in Equation 4.0.7 can be evaluated[1] exactly and the integral equation simplifies to

$$-\frac{1}{\pi}\int_0^1 g(\xi)\ln\left|\frac{\tanh(\beta x)+\tanh(\beta\xi)}{\tanh(\beta x)-\tanh(\beta\xi)}\right|d\xi = \frac{x}{h}, \qquad 0 \le x < 1, \tag{4.0.8}$$

where $\beta = \pi/(2h)$. The results from Example 1.2.3 can be employed to solve Equation 4.0.8 after substituting $x' = \tanh(\beta x)/\tanh(\beta)$. This yields

$$g(\xi) = \frac{1}{h^2}\frac{d}{d\xi}\left\{\int_\xi^1 \frac{\tanh(\beta x)}{\cosh^2(\beta x)\sqrt{\tanh^2(\beta x)-\tanh^2(\beta\xi)}} \times\left[\int_0^x \frac{d\tau}{\sqrt{\tanh^2(\beta x)-\tanh^2(\beta\tau)}}\right]dx\right\} \tag{4.0.9}$$

$$= \frac{\tanh(\beta\xi)}{h^2\cosh^2(\beta\xi)\sqrt{\tanh^2(\beta)-\tanh^2(\beta\xi)}} \times\int_0^1 \frac{\sqrt{\tanh^2(\beta)-\tanh^2(\beta x)}}{\tanh^2(\beta x)-\tanh^2(\beta\xi)}\,dx \tag{4.0.10}$$

$$= -\frac{\pi\tanh(\beta\xi)}{2h^2\beta\cosh(\beta)\sqrt{\tanh^2(\beta)-\tanh^2(\beta\xi)}}. \tag{4.0.11}$$

[1] Gradshteyn, I. S., and I. M. Ryzhik, 1965: *Table of Integrals, Series, and Products.* Academic Press, Formula 4.116.3.

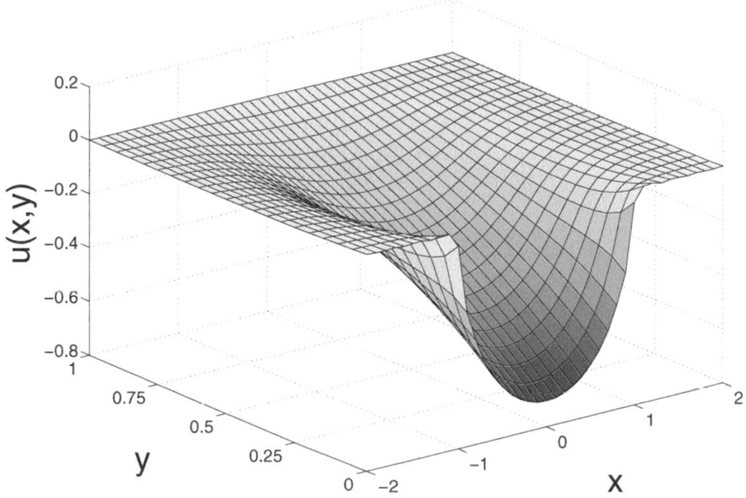

Figure 4.0.1: The solution to Equation 1.1.11 subject to the mixed boundary conditions given by Equation 1.1.12, Equation 1.1.13, and Equation 1.1.14 when $h = 1$.

Substituting Equation 4.0.11 into Equation 4.0.5, $A(k)$ follows via numerical integration. Finally, we can use this $A(k)$ to find the solution to Equation 1.1.11 subject to the boundary conditions given by Equation 1.1.12, Equation 1.1.13, and Equation 1.1.14 by numerically integrating Equation 1.1.17. Figure 4.0.1 illustrates this solution.

4.1 DUAL FOURIER INTEGRALS

A common technique in solving boundary value problems in rectangular coordinates involves Fourier transforms. In the case of mixed boundary value problems, this leads to sets of integral equations. In this section we focus on commonly occuring cases of dual integral equations.

• Example 4.1.1

Let us solve Laplace's equation:[2]

$$\frac{\partial^2 u}{\partial x^2} + \frac{\partial^2 u}{\partial y^2} = 0, \qquad -\infty < x < \infty, \quad 0 < y < \infty, \tag{4.1.1}$$

[2] See Iossel, Yu. Ya., and R. A. Pavlovskii, 1966: A plane steady heat conduction problem. *J. Engng. Phys.*, **10**, 163–166.

subject to the boundary conditions

$$\lim_{|x|\to\infty} u(x,y) \to 0, \qquad 0 < y < \infty, \tag{4.1.2}$$

$$\begin{cases} u(x,0) - hu_y(x,0) = C, & |x| < 1, \\ u(x,0) = 0, & |x| > 1, \end{cases} \tag{4.1.3}$$

and

$$\lim_{y\to\infty} u(x,y) \to 0, \qquad -\infty < x < \infty, \tag{4.1.4}$$

with $h > 0$.

By using separation of variables or transform methods, the general solution to Equation 4.1.1, Equation 4.1.2 and Equation 4.1.4 is

$$u(x,y) = \int_0^\infty A(k)e^{-ky}\cos(kx)\,dk. \tag{4.1.5}$$

Direct substitution of Equation 4.1.5 into Equation 4.1.3 yields the dual integral equations:

$$\int_0^\infty (1+kh)A(k)\cos(kx)\,dk = C, \qquad |x| < 1, \tag{4.1.6}$$

and

$$\int_0^\infty A(k)\cos(kx)\,dk = 0, \qquad |x| > 1. \tag{4.1.7}$$

We begin our solution of these dual equations by introducing

$$A(k) = \int_0^1 g'(t)J_0(kt)\,dt = g(1)J_0(k) + k\int_0^1 g(t)J_1(kt)\,dt, \tag{4.1.8}$$

if we assume that $g(0) = 0$. Turning first to Equation 4.1.7, if we substitute Equation 4.1.8 into Equation 4.1.7 and interchange the order of integration, we find that

$$\int_0^\infty A(k)\cos(kx)\,dk = \int_0^1 g'(t)\left[\int_0^\infty J_0(kt)\cos(kx)\,dk\right]dt. \tag{4.1.9}$$

From Equation 1.4.14 and noting that $x > t$ here, the integral vanishes within the square brackets and we see that our choice of $A(k)$ satisfies Equation 4.1.7 identically.

If we now integrate Equation 4.1.6 with respect to x,

$$\int_0^\infty A(k)\sin(kx)\,dk + \frac{1}{h}\int_0^\infty A(k)\sin(kx)\,\frac{dk}{k} = \frac{Cx}{h}. \tag{4.1.10}$$

Substituting for $A(k)$ from Equation 4.1.8 and interchanging the order of integration,

$$\int_0^1 g'(t)\left[\int_0^\infty J_0(kt)\sin(kx)\,dk\right]\,dt + \frac{g(1)}{h}\int_0^\infty J_0(k)\sin(kx)\,\frac{dk}{k} + \frac{1}{h}\int_0^1 g(t)\left[\int_0^\infty J_1(kt)\sin(kx)\,dk\right]\,dt = \frac{Cx}{h}. \tag{4.1.11}$$

Evaluating the integrals,[3]

$$\int_0^x \frac{g'(t)}{\sqrt{x^2-t^2}}\,dt + \frac{g(1)}{h}\arcsin(x) + \frac{1}{h}\int_0^1 \psi(x,t)g(t)\,dt = \frac{Cx}{h} \tag{4.1.12}$$

with $0 < x < 1$, where

$$\psi(x,t) = \begin{cases} x/\left[t\sqrt{t^2-x^2}\right], & 0 < x < t, \\ 0, & 0 < t < x. \end{cases} \tag{4.1.13}$$

If we introduce

$$f(x) = \int_0^x \frac{g'(t)}{\sqrt{x^2-t^2}}\,dt, \tag{4.1.14}$$

then by Equation 1.2.13 and Equation 1.2.14, we find that

$$g(t) = \frac{2}{\pi}\int_0^t \frac{\xi f(\xi)}{\sqrt{t^2-\xi^2}}\,d\xi. \tag{4.1.15}$$

Substituting Equation 4.1.14 and Equation 4.1.15 into Equation 4.1.12,

$$hf(x) + \frac{2}{\pi}\int_0^1 N(x,\xi)f(\xi)\,d\xi = Cx - \frac{2}{\pi}\arcsin(x)\int_0^1 \frac{\xi f(\xi)}{\sqrt{1-\xi^2}}\,d\xi, \tag{4.1.16}$$

where

$$N(x,\eta) = \eta\int_\eta^1 \frac{\psi(x,t)}{\sqrt{t^2-\eta^2}}\,dt = \frac{1}{2}\ln\left[\frac{(x+\eta)|x-\eta|}{\left(x\sqrt{1-\eta^2}-\eta\sqrt{1-x^2}\right)^2}\right]. \tag{4.1.17}$$

Equation 4.1.17 shows that the integral on the left side of Equation 4.1.16 is weakly singular. Therefore, this integral is divided into two parts. We use a simple trapezoidal rule for the nonsingular term. For the singular term, we employ a numerical method devised by Atkinson.[4] Defining `dx = 1/N` so that $x_n = \left(n-\frac{1}{2}\right)\Delta x$, $n = 1, 2, \cdots, N$, the MATLAB® code to find $f(x)$ is

[3] Gradshteyn and Ryzhik, op. cit., Formula 6.671.1 and Formula 6.693.7.

[4] Atkinson, K. E., 1967: The numerical solution of Fredholm integral equations of the second kind. *SIAM J. Numer. Anal.*, **4**, 337–348. See Section 5 in particular.

```
for j = 1:N
xx(j) = (j-0.5)*dx; xxi(j) = (j-0.5)*dx;
end
% the value of x at the interfaces
for j = 0:N
xx_e(j+1) = j*dx; xxi_e(j+1) = j*dx;
end
% ************************************************************
% Solve the integral equation, Equation 4.1.16
% ************************************************************
for n = 1:N
x = xx(n); b(n) = x; % the right side of Equation 4.1.16
for m = 1:N
xi = xxi(m);
if (n == m)
% the first term on the left side of Equation 4.1.16
AA(n,m) = h;
else AA(n,m) = 0; end
if (n == m)
NN(n,m) = log(sqrt(1-x*x));
else
NN(n,m) = log((x-xi) / (x*sqrt(1-xi*xi)-xi*sqrt(1-x*x))) ;
end
NN(n,m) = NN(n,m) + 0.5*log(x+xi);
% ************************************************************
% Find the non-singular contribution from the integrals
% in Equation 4.1.16
% ************************************************************
AA(n,m) = AA(n,m) + 2*NN(n,m)*dx/pi ...
        + 2*asin(x)*xi*dx/(pi*sqrt(1-xi*xi));
end; end
% ************************************************************
% Add in the contribution from the singular term
% ************************************************************
for n = 1:N
for m = 2:N
k = n-m+1;
psi_0 = -1;
psi_1 = 0.25*(k*k-(k-1)*(k-1));
if (k ~= 0)
psi_0 = psi_0 + k*log(abs(k));
psi_1 = psi_1 - 0.5*k*k*log(abs(k));
end
if (k ~= 1)
psi_0 = psi_0 - (k-1)*log(abs(k-1));
```

```
psi_1 = psi_1 + 0.5*(k-1)*(k-1)*log(abs(k-1));
end
psi_1 = psi_1 + k*psi_0;
alpha = 0.5*dx*log(dx) + dx*(psi_0-psi_1);
beta = 0.5*dx*log(dx) + dx*psi_1;
AA(n,m-1) = AA(n,m-1) - alpha/pi;
AA(n, m ) = AA(n, m ) - beta/pi;
end; end
f = AA\b' % Compute f(x) from Equation 4.1.16
```

Having found $f(x)$ from Equation 4.1.16, $g(t)$ follows from Equation 4.1.15.

```
g(1) = 0;
for n = 2:N+1
t = xx_e(n); g(n) = 0;
for m = 1:n-1
xi = xxi(m);
g(n) = g(n) + 2*xi*f(m)*dx/(pi*sqrt(t*t-xi*xi));
end; end
```

With $g(t)$, we can compute $A(k)$ and $u(x, y)$. The maximum number of wavenumbers included in the computations is `K_max*dk`. The MATLAB code is

```
% ***********************************************************
% Compute A(k) from Equation 4.1.8.
% Use Simpson's rule.
% ***********************************************************
for n = 1:N
derivative(n) = (g(n+1)-g(n))/dx;
end
for k = 0:K_max
ak = k*dk; t = xx(1);
A(k+1) = derivative(1)*besselj(0,ak*t); % k = 0 term
for n = 2:N-1
t = xx(n);
if ( mod(n,2) == 0)
A(k+1) = A(k+1) + 4*derivative(n)*besselj(0,ak*t);
else
A(k+1) = A(k+1) + 2*derivative(n)*besselj(0,ak*t);
end; end
t = xx(N);
% k = k_max term
A(k+1) = A(k+1) +   derivative(N)*besselj(0,ak*t);
A(k+1) = A(k+1)*dx/3;
```

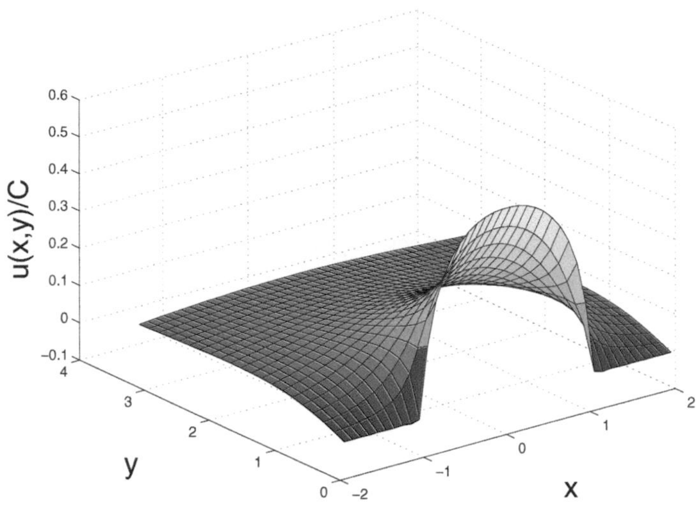

Figure 4.1.1: The solution to Equation 4.1.1 subject to the mixed boundary conditions given by Equation 4.1.2 through Equation 4.1.4 when $h = 1$.

```
end
% ***********************************************************
% Compute the solution u(x,y) for a given x and y.
% Use Simpson's rule.
% ***********************************************************
u(i,j) = 0;
% integral contribution from wavenumber
% between k = 0 and k = K_max
% Use Equation 4.1.5.
for k = 0:K_max
ak = k*dk; factor = A(k+1)*exp(-ak*y)*cos(ak*x);
if ( (k>0) & (k<K_max) )
if (mod(k+1,2) == 0)
u(i,j) = u(i,j) + 4*factor;
else
u(i,j) = u(i,j) + 2*factor;
end
else
u(i,j) = u(i,j) +   factor;
end; end
u(i,j) = dk*u(i,j)/3;
```

Figure 4.1.1 illustates the solution when $h = 1$.

• **Example 4.1.2**

Let us solve Laplace's equation:[5]

$$\frac{\partial^2 u}{\partial x^2} + \frac{\partial^2 u}{\partial y^2} = 0, \qquad -\infty < x < \infty, \quad 0 < y < \infty, \tag{4.1.18}$$

subject to the boundary conditions

$$\lim_{|x|\to\infty} u(x,y) \to 0, \qquad 0 < y < \infty, \tag{4.1.19}$$

$$\lim_{y\to\infty} u(x,y) \to 0, \qquad -\infty < x < \infty, \tag{4.1.20}$$

and

$$\begin{cases} -u_y(x,0) + hu(x,0) = f(x), & |x| < 1, \\ u_y(x,0) = 0, & |x| > 1. \end{cases} \tag{4.1.21}$$

Using separation of variables or transform methods, the general solution to Equation 4.1.18, Equation 4.1.19 and Equation 4.1.20 is $u(x,y) = u_+(x,y) + u_-(x,y)$, where

$$u_+(x,y) = \int_0^\infty A(k)e^{-ky}\cos(kx)\,\frac{dk}{k}, \tag{4.1.22}$$

and

$$u_-(x,y) = \int_0^\infty B(k)e^{-ky}\sin(kx)\,\frac{dk}{k}. \tag{4.1.23}$$

The idea here is that the solution consists of two parts: an even portion denoted by $u_+(x,y)$ and an odd portion denoted by $u_-(x,y)$. In a similar manner, $f(x) = f_+(x) + f_-(x)$.

Direct substitution of Equation 4.1.22 into Equation 4.1.21 yields the dual integral equations:

$$\int_0^\infty (k+h)A(k)\cos(kx)\,\frac{dk}{k} = f_+(x), \qquad |x| < 1, \tag{4.1.24}$$

and

$$\int_0^\infty A(k)\cos(kx)\,dk = 0, \qquad |x| > 1. \tag{4.1.25}$$

To solve these dual integral equations, we introduce

$$A(k) = \int_0^1 g_+'(t)\,[J_0(k) - J_0(kt)]\,dt, \tag{4.1.26}$$

[5] See Kuz'min, Yu. N., 1967: Plane-layer problem in the theory of heat conductivity for mixed boundary conditions. *Sov. Tech. Phys.*, **11**, 996–999.

because

$$\int_0^\infty A(k) J_0(kr)\, dk = \int_0^1 g'_+(t) \left[\int_0^\infty \cos(kx) J_0(k)\, dk\right] dt - \int_0^1 g'_+(t) \left[\int_0^\infty \cos(kx) J_0(kt)\, dk\right] dt = 0. \quad \textbf{(4.1.27)}$$

Thus, our choice for $A(k)$ identically satisfies Equation 4.1.25. This follows from Equation 1.4.14 since $|x| > 1$ and $0 \le t \le 1$.

Next, we integrate Equation 4.1.26 by parts and find that

$$A(k) = -k \int_0^1 g_+(t) J_1(kt)\, dt \quad \textbf{(4.1.28)}$$

provided that we require that $g_+(0) = 0$. Substituting Equation 4.1.26 and Equation 4.1.28 into Equation 4.1.24, we obtain

$$\int_0^1 g'_+(t) \left\{\int_0^\infty [J_0(k) - J_0(kt)] \cos(kx)\, dk\right\} dt = f_+(x) + h \int_0^1 g_+(\tau) \left[\int_0^\infty \cos(kx) J_1(k\tau)\, dk\right] d\tau. \quad \textbf{(4.1.29)}$$

Applying Equation 1.4.14, Equation 4.1.29 simplifies to

$$\int_x^1 \frac{g'_+(t)}{\sqrt{t^2 - x^2}}\, dt = -f_+(x) - h \int_0^1 g_+(\tau) \left[\int_0^\infty \cos(kx) J_1(k\tau)\, dk\right] d\tau. \quad \textbf{(4.1.30)}$$

Using Equation 1.2.15 and Equation 1.2.16, we solve for $g'_+(t)$ and find that

$$g'_+(t) = \frac{2}{\pi} \frac{d}{dt} \left[\int_t^1 \frac{x f_+(x)}{\sqrt{x^2 - t^2}}\, dx\right] + \frac{2h}{\pi} \frac{d}{dt} \left(\int_t^1 \frac{x}{\sqrt{x^2 - t^2}} \left\{\int_0^1 g_+(\tau) \left[\int_0^\infty \cos(kx) J_1(k\tau)\, dk\right] d\tau\right\} dx\right). \quad \textbf{(4.1.31)}$$

Integrating both sides of Equation 4.1.31,

$$g_+(t) = \frac{2}{\pi} \left[\int_t^1 \frac{x f_+(x)}{\sqrt{x^2 - t^2}}\, dx\right] + \frac{2h}{\pi} \int_t^1 \frac{x}{\sqrt{x^2 - t^2}} \left\{\int_0^1 g_+(\tau) \left[\int_0^\infty \cos(kx) J_1(k\tau)\, dk\right] d\tau\right\} dx + C, \quad \textbf{(4.1.32)}$$

where C denotes the arbitrary constant of integration. We must choose C so that $g_+(0) = 0$. Thus, Equation 4.1.32 becomes

$$g_+(t) = \frac{2}{\pi} \left[\int_t^1 \frac{x f_+(x)}{\sqrt{x^2 - t^2}}\, dx - \int_0^1 f_+(x)\, dx\right] + \frac{2}{\pi} \int_0^1 K(t, \tau) g_+(\tau)\, d\tau, \quad \textbf{(4.1.33)}$$

where

$$
\begin{aligned}
K(t,\tau) &= h \int_t^1 \frac{x}{\sqrt{x^2-t^2}} \left[\int_0^\infty \cos(kx) J_1(k\tau)\, dk \right] dx \\
&\quad - h \int_0^\infty \sin(k) J_1(k\tau)\, \frac{dk}{k} && \textbf{(4.1.34)} \\
&= \frac{h}{\tau} \left(\sqrt{1-\tau^2} + \sqrt{1-t^2} - 1 - \int_s^1 \frac{x^2}{\sqrt{x^2-t^2}\,\sqrt{x^2-\tau^2}}\, dx \right) && \textbf{(4.1.35)} \\
&= \frac{h}{\tau} \Big\{ \sqrt{1-\tau^2} + \sqrt{1+t^2} - 1 - \sqrt{1-t^2}\,\sqrt{1-\tau^2} \\
&\qquad - s\,[K(\kappa) - E(\kappa) - F(\theta,\kappa) + E(\theta,\kappa)] \Big\}, && \textbf{(4.1.36)}
\end{aligned}
$$

where $s = \max(t,\tau)$, $p = \min(t,\tau)$, $\kappa = p/s$, $\theta = \arcsin(s)$, $F(\cdot,\cdot)$ and $E(\cdot,\cdot)$ are elliptic integrals of the first and second kind, respectively, $K(\cdot) = F(\pi/2,\cdot)$, and $E(\cdot) = E(\pi/2,\cdot)$.

Turning to $B(k)$, we substitute Equation 4.1.23 into Equation 4.1.21 and obtain

$$
\int_0^\infty (k+h) B(k) \sin(kx)\, \frac{dk}{k} = f_-(x), \qquad |x| < 1, \tag{4.1.37}
$$

and

$$
\int_0^\infty B(k) \sin(kx)\, dk = 0, \qquad |x| > 1. \tag{4.1.38}
$$

We now set

$$
B(k) = -\int_0^1 t\, g_-'(t) J_1(kt)\, dt = k \int_0^1 t\, g_-(t) J_0(kt)\, dt \tag{4.1.39}
$$

if $g_-(1) = 0$. Because

$$
\int_0^\infty B(k) \sin(kx)\, dk = -\int_0^1 t g_-'(t) \left[\int_0^\infty \sin(kx) J_1(kt)\, dk \right] dt = 0 \tag{4.1.40}
$$

from Equation 1.4.13 if $|x| > 1$ and $0 \le t \le 1$, Equation 4.1.38 is identically satisfied by our choice for $B(k)$.

Substituting Equation 4.1.39 into Equation 4.1.37,

$$
\begin{aligned}
-\int_0^1 t\, g_-'(t) &\left[\int_0^\infty \sin(kx) J_1(kt)\, dk \right] dt \\
&= f_-(x) - h \int_0^1 \tau\, g_-(\tau) \left[\int_0^\infty \sin(kx) J_0(k\tau)\, dk \right] d\tau. && \textbf{(4.1.41)}
\end{aligned}
$$

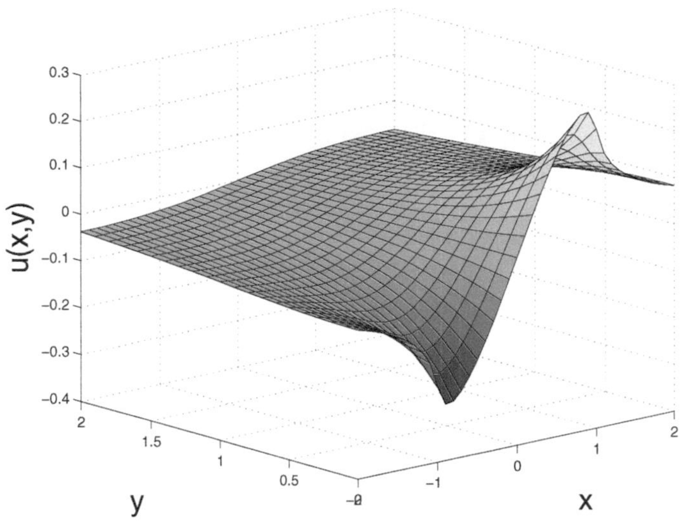

Figure 4.1.2: The solution to Equation 4.1.18 subject to the mixed boundary conditions given by Equation 4.1.19, Equation 4.1.20, and Equation 4.1.21 when $f_+(x) = 0$, $f_-(x) = x$ and $h = 1$.

From Equation 1.4.13,

$$x \int_x^1 \frac{g'_-(t)}{\sqrt{t^2 - x^2}}\, dt = -f_-(x) + h \int_0^1 \tau\, g_-(\tau) \left[\int_0^\infty \sin(kx) J_0(k\tau)\, dk \right] d\tau. \tag{4.1.42}$$

Using Equation 1.2.15 and Equation 1.2.16 and integrating,

$$g_-(t) = \frac{2}{\pi} \int_t^1 \frac{f_-(x)}{\sqrt{x^2 - t^2}}\, dx - \frac{2}{\pi} \int_0^1 L(t, \tau) g_-(\tau)\, d\tau, \tag{4.1.43}$$

where

$$L(t, \tau) = h\tau \int_t^1 \left[\int_0^\infty \sin(kx) J_0(k\tau)\, dk \right] \frac{dx}{\sqrt{x^2 - t^2}} \tag{4.1.44}$$

$$= h\tau \int_0^1 \frac{dx}{\sqrt{x^2 - t^2}\,\sqrt{x^2 - \tau^2}} = \frac{h\tau}{s} F(\vartheta, \kappa), \tag{4.1.45}$$

where $\vartheta = \arcsin\left[\sqrt{(1 - s^2)/(1 - p^2)}\right]$. Figure 4.1.2 illustrates this solution when $f_+(x) = 0$, $f_-(x) = x$ and $h = 1$.

• **Example 4.1.3**

Let us solve Laplace's equation:[6]

$$\frac{\partial^2 u}{\partial x^2} + \frac{\partial^2 u}{\partial y^2} = 0, \qquad -\infty < x < \infty, \quad 0 < y < h, \tag{4.1.46}$$

subject to the boundary conditions

$$\lim_{|x|\to\infty} u(x,y) \to 0, \qquad 0 < y < h, \tag{4.1.47}$$

$$\begin{cases} u_y(x,0) = \mathcal{A}, & |x| < 1, \\ u(x,0) = 0, & |x| > 1, \end{cases} \tag{4.1.48}$$

and

$$u_y(x,h) = 0, \qquad -\infty < x < \infty. \tag{4.1.49}$$

Using separation of variables or transform methods, the general solution to Equation 4.1.46, Equation 4.1.47 and Equation 4.1.49 is

$$u(x,y) = \frac{2}{\pi}\int_0^\infty A(k)\cosh[k(y-h)]\cos(kx)\,dk. \tag{4.1.50}$$

Direct substitution of Equation 4.1.50 into Equation 4.1.48 yields the dual integral equations:

$$\frac{2}{\pi}\int_0^\infty kA(k)\sinh(kh)\cos(kx)\,dk = -\mathcal{A}, \qquad |x| < 1, \tag{4.1.51}$$

and

$$\frac{2}{\pi}\int_0^\infty A(k)\cosh(kh)\cos(kx)\,dk = 0, \qquad |x| > 1. \tag{4.1.52}$$

We begin our solution of these dual integral equations by noting that for $|x| < 1$,

$$u(x,0) = \frac{2}{\pi}\int_0^\infty A(k)\cosh(kh)\cos(kx)\,dk \tag{4.1.53}$$

with $u(1,0) = 0$. Because Equation 4.1.53 is the Fourier cosine representation of $u(x,0)$,

$$A(k)\cosh(kh) = \int_0^1 u(\xi,0)\cos(k\xi)\,d\xi = -\frac{1}{k}\int_0^1 h(\xi)\sin(k\xi)\,d\xi, \tag{4.1.54}$$

[6] See Yang, F., and J. C. M. Li, 1993: Impression creep of a thin film by vacancy diffusion. I. Straight punch. *J. Appl. Phys.*, **74**, 4382–4389.

since $u(1,0) = 0$ and $h(\xi) = u_\xi(\xi,0)$. Substituting Equation 4.1.54 into Equation 4.1.51,

$$\frac{2}{\pi}\int_0^\infty \left[\int_0^1 h(\xi)\sin(k\xi)\,d\xi\right]\tanh(kh)\cos(kx)\,dk = \mathcal{A}, \qquad |x|<1; \tag{4.1.55}$$

or

$$\frac{d}{dx}\left\{\int_0^\infty \left[\int_0^1 h(\xi)\sin(k\xi)\,d\xi\right]\tanh(kh)\sin(kx)\,\frac{dk}{k}\right\} = \frac{\pi\mathcal{A}}{2}, \qquad |x|<1, \tag{4.1.56}$$

and

$$\frac{d}{dx}\left\{\int_0^1 h(\xi)\left[\int_0^\infty \tanh(kh)\sin(k\xi)\sin(kx)\,\frac{dk}{k}\right]d\xi\right\} = \frac{\pi\mathcal{A}}{2}, \qquad |x|<1. \tag{4.1.57}$$

Now,

$$\begin{aligned}\int_0^\infty \tanh(kh)\sin(k\xi)\sin(kx)\,\frac{dk}{k}&\\ &= \tfrac{1}{2}\int_0^\infty \tanh(kh)\left\{\cos[k(x-\xi)]-\cos[k(x+\xi)]\right\}\frac{dk}{k} \qquad (4.1.58)\\ &= \tfrac{1}{2}\ln\left|\frac{\sinh(\beta x)+\sinh(\beta\xi)}{\sinh(\beta x)-\sinh(\beta\xi)}\right|, \qquad (4.1.59)\end{aligned}$$

where $\beta = \pi/(2h)$, since[7]

$$\int_0^\infty \cos(\alpha x)\tanh(\beta x)\frac{dx}{x} = \ln\left[\coth\left(\frac{\alpha\pi}{4\beta}\right)\right], \qquad \alpha,\beta>0. \tag{4.1.60}$$

Substituting Equation 4.1.59 into Equation 4.1.57 and integrating, we obtain the integral equation

$$\int_0^1 h(\xi)\ln\left|\frac{\sinh(\beta x)+\sinh(\beta\xi)}{\sinh(\beta x)-\sinh(\beta\xi)}\right|d\xi = \pi\mathcal{A}x. \tag{4.1.61}$$

If we define

$$\gamma = \frac{\sinh(\beta\xi)}{\sinh(\beta)} \qquad \text{and} \qquad \gamma_0 = \frac{\sinh(\beta x)}{\sinh(\beta)}, \tag{4.1.62}$$

we find that

$$\int_0^1 g(\gamma)\ln\left|\frac{\gamma+\gamma_0}{\gamma-\gamma_0}\right|d\gamma = \frac{\pi^2\mathcal{A}\,\mathrm{arcsinh}[\gamma_0\sinh(\beta)]}{2h\beta}, \tag{4.1.63}$$

---

[7] Gradshteyn and Ryzhik, op. cit., Formula 4.116.2.

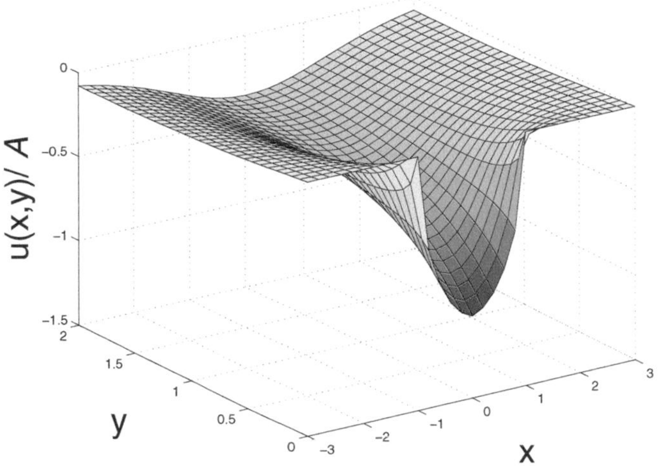

Figure 4.1.3: The solution to Equation 4.1.46 subject to the mixed boundary conditions given by Equation 4.1.47, Equation 4.1.48, and Equation 4.1.49 when $h = 2$.

where

$$g(\gamma) = \frac{\sinh(\beta)}{\cosh(\beta\xi)}\, h\left\{\frac{\operatorname{arcsinh}[\gamma\sinh(\beta)]}{\beta}\right\}. \tag{4.1.64}$$

Therefore, from Example 1.2.3,

$$\frac{\partial u(\xi,0)}{\partial \xi} = -\frac{\mathcal{A}}{2h}\frac{d}{d\xi}\left\{\int_\xi^1 \frac{\sinh(2\beta\tau)}{\sqrt{\sinh^2(\beta\tau)-\sinh^2(\beta\xi)}} \times\left[\int_0^\tau \frac{d\eta}{\sqrt{\sinh^2(\beta\tau)-\sinh^2(\beta\eta)}}\right] d\tau\right\}, \tag{4.1.65}$$

or, upon integrating,

$$u(x,0) = -\frac{\mathcal{A}}{2h}\int_x^1 \frac{\sinh(2\beta\tau)}{\sqrt{\sinh^2(\beta\tau)-\sinh^2(\beta x)}} \times\left[\int_0^\tau \frac{d\eta}{\sqrt{\sinh^2(\beta\tau)-\sinh^2(\beta\eta)}}\right] d\tau. \tag{4.1.66}$$

To compute $u(x,y)$, we first evaluate $u(x,0)$. Next, we numerically integrate Equation 4.1.54 to find $A(k)$. Finally, we employ Equation 4.1.50. Figure 4.1.3 illustrates this solution when $h = 2$.

At this point,[8] we can also show how to solve Equation 4.1.46, Equation

---

[8] See also Problem 1 in Singh, B. M., T. B. Moodie, and J. B. Haddow, 1981: Closed-form solutions for finite length crack moving in a strip under anti-plane shear stress. *Acta Mech.*, **38**, 99–109.

4.1.47, and Equation 4.1.48, while modifying Equation 4.1.49 to read

$$u(x,h) = 0, \qquad -\infty < x < \infty. \tag{4.1.67}$$

We begin once again using separation of variables or transform methods to find the general solution to Equation 4.1.46, Equation 4.1.47 and Equation 4.1.67. This now gives

$$u(x,y) = \frac{2}{\pi}\int_0^\infty A(k)\sinh[k(y-h)]\cos(kx)\,dk. \tag{4.1.68}$$

Direct substitution of Equation 4.1.68 into Equation 4.1.48 yields the dual integral equations:

$$\frac{2}{\pi}\int_0^\infty kA(k)\cosh(kh)\cos(kx)\,dk = \mathcal{A}, \qquad |x| < 1, \tag{4.1.69}$$

and

$$\frac{2}{\pi}\int_0^\infty A(k)\sinh(kh)\cos(kx)\,dk = 0, \qquad |x| > 1. \tag{4.1.70}$$

To solve Equation 4.1.69 and Equation 4.1.70, we introduce

$$A(k)\sinh(kh) = \int_0^1 g(\xi)\sin(k\xi)\,d\xi. \tag{4.1.71}$$

Equation 4.1.71 identically satisfies Equation 4.1.70. Substituting Equation 4.1.71 into Equation 4.1.69,

$$\int_0^\infty \left[\int_0^1 g(\xi)\sin(k\xi)\,d\xi\right]\coth(kh)\cos(kx)\,dk = \frac{\pi\mathcal{A}}{2}, \qquad |x| < 1; \tag{4.1.72}$$

or

$$\frac{d}{dx}\left\{\int_0^\infty \left[\int_0^1 g(\xi)\sin(k\xi)\,d\xi\right]\coth(kh)\sin(kx)\,\frac{dk}{k}\right\} = \frac{\pi\mathcal{A}}{2}, \qquad |x| < 1, \tag{4.1.73}$$

and

$$\frac{d}{dx}\left\{\int_0^1 g(\xi)\left[\int_0^\infty \coth(kh)\sin(k\xi)\sin(kx)\,\frac{dk}{k}\right]d\xi\right\} = \frac{\pi\mathcal{A}}{2}, \qquad |x| < 1. \tag{4.1.74}$$

Now,

$$\int_0^\infty \coth(kh)\sin(k\xi)\sin(kx)\,\frac{dk}{k}$$

$$= \tfrac{1}{2}\int_0^\infty \coth(kh)\left\{\cos[k(x-\xi)] - \cos[k(x+\xi)]\right\}\frac{dk}{k} \tag{4.1.75}$$

$$= \tfrac{1}{2}\ln\left|\frac{\tanh(\beta x) + \tanh(\beta\xi)}{\tanh(\beta x) - \tanh(\beta\xi)}\right|, \tag{4.1.76}$$

since[9]

$$\int_0^\infty \cos(\alpha x)\coth(\beta x)\frac{dx}{x} = -\ln\left[2\sinh\left(\frac{\alpha\pi}{2\beta}\right)\right], \qquad \alpha, \Re(\beta) > 0. \quad \textbf{(4.1.77)}$$

Substituting Equation 4.1.76 into Equation 4.1.74 and integrating, we obtain the integral equation

$$\int_0^1 g(\xi)\ln\left|\frac{\tanh(\beta x)+\tanh(\beta\xi)}{\tanh(\beta x)-\tanh(\beta\xi)}\right| d\xi = \pi\mathcal{A}x. \quad \textbf{(4.1.78)}$$

If we define

$$\gamma = \frac{\tanh(\beta\xi)}{\tanh(\beta)} \quad \text{and} \quad \gamma_0 = \frac{\tanh(\beta x)}{\tanh(\beta)}, \quad \textbf{(4.1.79)}$$

we find that Equation 4.1.78 becomes

$$\int_0^1 h(\gamma)\ln\left|\frac{\gamma+\gamma_0}{\gamma-\gamma_0}\right| d\gamma = \frac{\pi\mathcal{A}}{\beta}\operatorname{arctanh}[\gamma_0\tanh(\beta)], \quad \textbf{(4.1.80)}$$

where

$$h(\gamma) = \frac{\tanh(\beta)}{d[\tanh(\beta\xi)]/d\xi}\, g\left\{\frac{\operatorname{arctanh}[\gamma\tanh(\beta)]}{\beta}\right\}. \quad \textbf{(4.1.81)}$$

Again, using Cooke's results from Example 1.2.3,

$$g(\xi) = -\frac{\mathcal{A}}{h}\frac{d}{d\xi}\left\{\int_\xi^1 \frac{\tanh(2\beta\tau)}{\cosh^2(\beta\tau)\sqrt{\tanh^2(\beta\tau)-\tanh^2(\beta\xi)}} \times\left[\int_0^\tau \frac{d\eta}{\sqrt{\tanh^2(\beta\tau)-\tanh^2(\beta\eta)}}\right] d\tau\right\}. \quad \textbf{(4.1.82)}$$

To compute $u(x,y)$, we first evaluate $g(\xi)$. Next, we numerically integrate Equation 4.1.71 to find $A(k)$. Finally, we employ Equation 4.1.68. Figure 4.1.4 illustrates this solution when $h = 2$.

Konishi and Atsumi[10] have given an alternative method of attacking the problem given by Equation 4.1.46 through Equation 4.1.48 and Equation 4.1.67. For clarity let us restate the problem:

$$\frac{\partial^2 u}{\partial x^2} + \frac{\partial^2 u}{\partial y^2} = 0, \qquad -\infty < x < \infty, \quad 0 < y < h, \quad \textbf{(4.1.83)}$$

---

[9] Ibid., Formula 4.116.3.

[10] Konishi, Y., and A. Atsumi, 1973: The linear thermoelastic problem of uniform heat flow disturbed by a two-dimensional crack in a strip. *Int. J. Engng. Sci.*, **11**, 1–7.

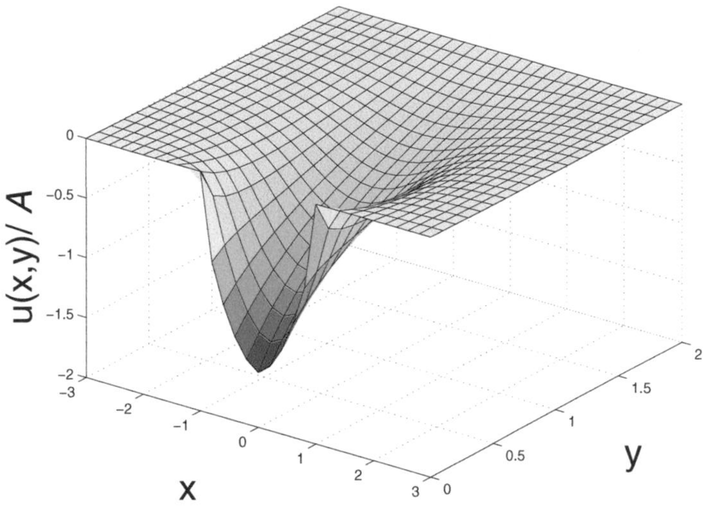

Figure 4.1.4: The solution to Equation 4.1.46 subject to the mixed boundary conditions given by Equation 4.1.47, Equation 4.1.48, and Equation 4.1.67 when $h = 2$.

subject to the boundary conditions

$$\lim_{|x|\to\infty} u(x,y) \to 0, \qquad 0 < y < \infty, \tag{4.1.84}$$

$$\begin{cases} u_y(x,0) = 1, & |x| < a, \\ u(x,0) = 0, & |x| > a, \end{cases} \tag{4.1.85}$$

and

$$u(x,h) = 0, \qquad -\infty < x < \infty. \tag{4.1.86}$$

We begin by applying Fourier cosine transforms and find that

$$u(x,y) = \frac{2}{\pi}\int_0^\infty A(k)\sinh[k(y-h)]\cos(kx)\,dk. \tag{4.1.87}$$

Equation 4.1.87 satisfies the differential equation plus the boundary conditions Equation 4.1.84 and Equation 4.1.86. Substituting Equation 4.1.87 into Equation 4.1.85, we obtain the dual integral equations

$$\frac{2}{\pi}\int_0^\infty kA(k)[1+M(kh)]\sinh(kh)\cos(kx)\,dk = 1, \qquad 0 \le |x| \le a, \tag{4.1.88}$$

where $M(kh) = e^{-kh}/\sinh(kh)$; and

$$\frac{2}{\pi}\int_0^\infty A(k)\sinh(kh)\cos(kx)\,dk = 0, \qquad a \le |x| \le \infty. \tag{4.1.89}$$

To solve these dual equations, we set

$$\sinh(kh)A(k) = \frac{\pi}{2}\int_0^a h(t)J_0(kt)\,dt. \quad \textbf{(4.1.90)}$$

We have chosen this definition for $A(k)$ because

$$\frac{2}{\pi}\int_0^\infty A(k)\sinh(kh)\cos(kx)\,dk = \int_0^a h(t)\left[\int_0^\infty \cos(kx)J_0(kt)\,dk\right]dt = 0, \quad \textbf{(4.1.91)}$$

where we used integral tables[11] with $0 \le t \le a < |x| < \infty$.

Turning to Equation 4.1.80, we have

$$\int_0^a h(t)\left\{\int_0^\infty k[1+M(kh)]\cos(kx)J_0(kt)\,dk\right\}dt = 1, \quad \textbf{(4.1.92)}$$

or

$$\int_0^a h(t)\left\{\int_0^\infty k\cos(kx)J_0(kt)\,dk\right\}dt + \int_0^a h(t)\left\{\int_0^\infty kM(kh)\cos(kx)J_0(kt)\,dk\right\}dt = 1. \quad \textbf{(4.1.93)}$$

Integrating Equation 4.1.93 with respect to x,

$$\int_0^a h(t)\left\{\int_0^\infty \sin(kx)J_0(kt)\,dk\right\}dt + \int_0^a h(t)\left\{\int_0^\infty M(kh)\sin(kx)J_0(kt)\,dk\right\}dt = x. \quad \textbf{(4.1.94)}$$

Applying Equation 1.4.13,

$$\int_0^x \frac{h(t)}{\sqrt{x^2-t^2}}\,dt + \int_0^a h(\tau)\left\{\int_0^\infty M(kh)\sin(kx)J_0(k\tau)\,dk\right\}d\tau = x. \quad \textbf{(4.1.95)}$$

From Equation 1.2.13 and Equation 1.2.14, we have that

$$h(t) = \frac{2}{\pi}\frac{d}{dt}\left[\int_0^t \frac{x^2}{\sqrt{t^2-x^2}}\,dx\right] - \frac{2}{\pi}\int_0^a h(\tau)\left\{\int_0^\infty M(kh)J_0(k\tau)\frac{d}{dt}\left[\int_0^t \frac{x\sin(kx)}{\sqrt{t^2-x^2}}\,dx\right]dk\right\}d\tau. \quad \textbf{(4.1.96)}$$

---

[11] Gradshteyn and Ryzhik, op. cit., Formula 6.671.8.

Now, from integral tables,[12]

$$\int_0^t \frac{x\sin(kx)}{\sqrt{t^2-x^2}}\,dx = \frac{\pi t}{2}J_1(kt) \tag{4.1.97}$$

and

$$\int_0^t \frac{x^2}{\sqrt{t^2-x^2}}\,dx = \frac{\pi t^2}{4}. \tag{4.1.98}$$

Substituting Equation 4.1.97 and Equation 4.1.98 into Equation 4.1.96 and taking the derivatives, we finally have

$$h(t) + t\int_0^a h(\tau)\left\{\int_0^\infty kM(kh)J_0(k\tau)J_0(kt)\,dk\right\}d\tau = t. \tag{4.1.99}$$

Therefore, the numerical solution of Equation 4.1.99 yields $h(t)$. This gives $A(k)$ from Equation 4.1.90. Finally, the solution $u(x,y)$ follows from Equation 4.1.87.

• **Example 4.1.4**

For our fourth example, let us solve Laplace's equation[13]

$$\frac{\partial^2 u}{\partial x^2} + \frac{\partial^2 u}{\partial y^2} = 0, \qquad -\infty < x < \infty, \quad 0 < y < h, \tag{4.1.100}$$

with the boundary conditions

$$\lim_{|x|\to\infty} u(x,y) \to 0, \qquad 0 < y < h, \tag{4.1.101}$$

$$\begin{cases} u_y(x,0) = -p(x), & |x| < a, \\ u(x,0) = 0, & |x| > a, \end{cases} \tag{4.1.102}$$

and

$$u_y(x,h) = 0, \qquad -\infty < x < \infty, \tag{4.1.103}$$

where $p(x)$ is an even function.

We begin by applying Fourier cosine transforms to solve Equation 4.1.100. This yields the solution

$$u(x,y) = \frac{2}{\pi}\int_0^\infty A(k)\frac{e^{-ky}+e^{ky-2kh}}{1+e^{-2kh}}\cos(kx)\,dk. \tag{4.1.104}$$

---

[12] Ibid., Formula 3.753.5.

[13] See Singh, Moodie, and Haddow, op. cit.

Equation 4.1.104 satisfies not only Equation 4.1.100, but also Equation 4.1.101 and Equation 4.1.103. Substituting Equation 4.1.104 into Equation 4.1.102, we obtain

$$\frac{2}{\pi}\int_0^\infty A(k)\cos(kx)\,dk = 0, \qquad |x| > a, \tag{4.1.105}$$

and

$$\frac{2}{\pi}\int_0^\infty k\tanh(kh)A(k)\cos(kx)\,dk = p(x), \qquad |x| < a. \tag{4.1.106}$$

To solve this set of dual integral equations, we introduce

$$kA(k) = \frac{\pi}{2}\int_0^a g(\tau)\sin(k\tau)\,d\tau. \tag{4.1.107}$$

Substituting Equation 4.1.107 into Equation 4.1.105, we find that

$$\frac{2}{\pi}\int_0^\infty A(k)\cos(kx)\,dk = \int_0^\infty \left[\int_0^a g(\tau)\sin(k\tau)\,d\tau\right]\cos(kx)\,\frac{dk}{k} \tag{4.1.108}$$

$$= \int_0^a g(\tau)\left[\int_0^\infty \sin(k\tau)\cos(kx)\frac{dk}{k}\right]d\tau = 0, \tag{4.1.109}$$

where the integral[14] within the square brackets vanishes since $|x| > a$ and $0 \le \tau \le a$. Thus, Equation 4.1.107 satisfies Equation 4.1.105 identically.

We now turn our attention to Equation 4.1.106. Substituting Equation 4.1.107 into Equation 4.1.106, we have that

$$\int_0^\infty \tanh(kh)\left[\int_0^a g(\tau)\sin(k\tau)\,d\tau\right]\cos(kx)\,dk = p(x), \tag{4.1.110}$$

or

$$\frac{d}{dx}\left\{\int_0^a g(\tau)\left[\int_0^\infty \tanh(kh)\sin(k\tau)\sin(kx)\,\frac{dk}{k}\right]d\tau\right\} = p(x), \qquad 0 < x < a, \tag{4.1.111}$$

after we interchange the order of integration in Equation 4.1.110. Following Equation 4.1.59 through Equation 4.1.61, we can show that

$$\int_0^\infty \tanh(kh)\sin(k\tau)\sin(kx)\,\frac{dk}{k} = \frac{1}{2}\ln\left|\frac{\sinh(cx)+\sinh(c\tau)}{\sinh(cx)-\sinh(c\tau)}\right|, \tag{4.1.112}$$

where $c = \pi/(2h)$. Therefore, substituting Equation 4.1.112 into Equation 4.1.111 and integrating,

$$\int_0^a g(\tau)\ln\left|\frac{\sinh(cx)+\sinh(c\tau)}{\sinh(cx)-\sinh(c\tau)}\right|\,d\tau = 2\int_0^x p(\xi)\,d\xi = F(x), \qquad 0 < x < a. \tag{4.1.113}$$

[14] Gradshteyn and Ryzhik, op. cit., Formula 3.741.2.

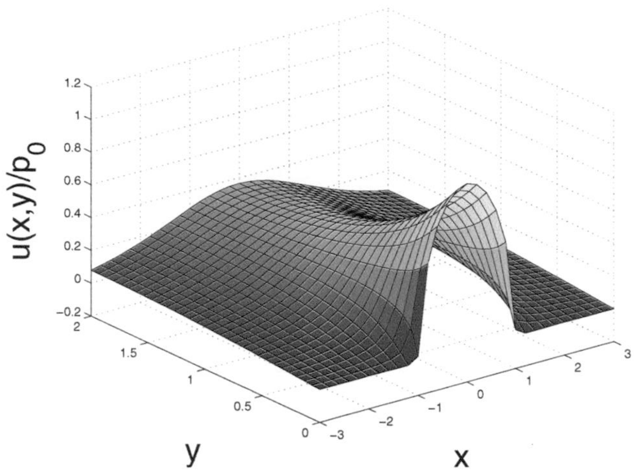

Figure 4.1.5: The solution to Equation 4.1.100 subject to the mixed boundary conditions given by Equation 4.1.101, Equation 4.1.102, and Equation 4.1.103 when $a = 1$ and $h = 2$.

Using the results from Example 1.2.3, the solution to the integral equation Equation 4.1.113 is

$$g(\tau) = -\frac{c\sinh(2c\tau)}{\pi^2\sqrt{\sinh^2(ca)-\sinh^2(c\tau)}}\int_0^a \frac{F'(x)\sqrt{\sinh^2(ca)-\sinh^2(cx)}}{\sinh^2(cx)-\sinh^2(c\tau)}\,dx + \frac{2cF(0)\cosh(c\tau)}{\pi^2\sinh(c\tau)\sqrt{\sinh^2(ca)-\sinh^2(c\tau)}}, \qquad 0 < \tau < a. \tag{4.1.114}$$

Because $F'(x) = 2p(x)$ with $F(0) = 0$, Equation 4.1.114 simplifies to

$$g(\tau) = -\frac{2c\sinh(2c\tau)}{\pi^2\sqrt{\sinh^2(ca)-\sinh^2(c\tau)}}\int_0^a \frac{p(x)\sqrt{\sinh^2(ca)-\sinh^2(cx)}}{\sinh^2(cx)-\sinh^2(c\tau)}\,dx, \tag{4.1.115}$$

if $0 < \tau < a$. Figure 4.1.5 illustrates the special solution when $a = 1$, $h = 2$, and $p(x) = p_0$, a constant.

• Example 4.1.5

Let us solve Laplace's equation[15]

$$\frac{\partial^2 u}{\partial x^2} + \frac{\partial^2 u}{\partial y^2} = 0, \qquad -\infty < x < \infty, \quad 0 < y < h, \tag{4.1.116}$$

[15] See Yang, F., 1997: Solution of a dual integral equation for crack and indentation problems. *Theoret. Appl. Fract. Mech.*, **26**, 211–217.

with the boundary conditions

$$\lim_{|x|\to\infty} u(x,y) \to 0, \qquad 0 < y < h, \tag{4.1.117}$$

$$\begin{cases} u(x,0) = \operatorname{sgn}(x), & |x| < 1, \\ u_y(x,0) = 0, & |x| > 1, \end{cases} \tag{4.1.118}$$

and

$$u(x,h) = 0, \qquad -\infty < x < \infty. \tag{4.1.119}$$

We begin by applying Fourier sine transforms to solve Equation 4.1.116. This yields the solution

$$u(x,y) = \frac{2}{\pi}\int_0^\infty A(k)\frac{\sinh[k(y-h)]}{\cosh(kh)}\sin(kx)\,dk. \tag{4.1.120}$$

Equation 4.1.120 satisfies not only Equation 4.1.116, but also Equation 4.1.117 and Equation 4.1.118. Substituting Equation 4.1.120 into Equation 4.1.118, we obtain

$$\frac{2}{\pi}\int_0^\infty A(k)\tanh(kh)\sin(kx)\,dk = -1, \qquad |x| < 1, \tag{4.1.121}$$

and

$$\int_0^\infty kA(k)\sin(kx)\,dk = 0, \qquad |x| > 1. \tag{4.1.122}$$

To solve this set of dual integral equations, we introduce

$$kA(k) = \int_0^1 g(t)\sin(kt)\,dt. \tag{4.1.123}$$

Substituting Equation 4.1.123 into Equation 4.1.122, we find that

$$\int_0^\infty kA(k)\sin(kx)\,dk = -\frac{d}{dx}\left[\int_0^\infty A(k)\cos(kx)\,dk\right] \tag{4.1.124}$$

$$= -\int_0^1 g(t)\frac{d}{dx}\left[\int_0^\infty \sin(kt)\cos(kt)\,\frac{dk}{k}\right]\,dt \tag{4.1.125}$$

$$= 0, \tag{4.1.126}$$

where the integral[16] within the square brackets vanishes since $|x| > 1$ and $0 \le t \le 1$. Thus, Equation 4.1.123 satisfies Equation 4.1.122 identically.

---

[16] Gradshteyn and Ryzhik, op. cit., Formula 3.741.2.

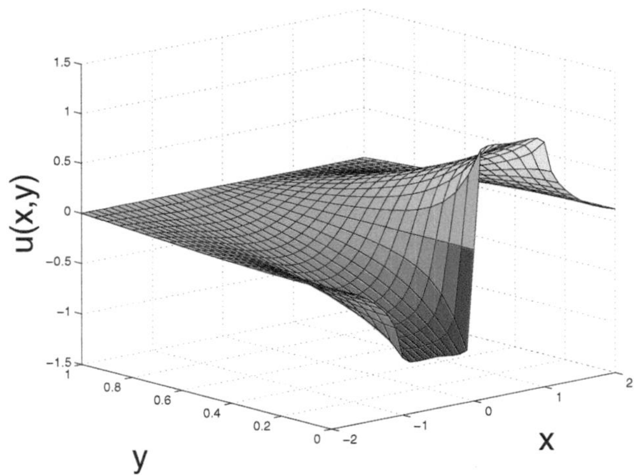

Figure 4.1.6: The solution to Equation 4.1.116 subject to the mixed boundary conditions given by Equation 4.1.117 through Equation 4.1.119 when $h = 1$.

We now turn our attention to Equation 4.1.121. Substituting Equation 4.1.123 into Equation 4.1.121, we have that

$$\int_0^1 g(t)\left[\int_0^\infty \tanh(kh)\sin(kx)\sin(kt)\,\frac{dk}{k}\right]dt = -\frac{\pi}{2}, \qquad (\mathbf{4.1.127})$$

or

$$\int_0^1 g(t)\ln\left|\frac{\sinh[\pi t/(2h)]+\sinh[\pi x/(2h)]}{\sinh[\pi t/(2h)]-\sinh[\pi x/(2h)]}\right|dt = -\pi. \qquad (\mathbf{4.1.128})$$

Upon introducing

$$\xi = \sinh[\pi t/(2h)]/\sinh[\pi/(2h)], \qquad \eta = \sinh[\pi x/(2h)]/\sinh[\pi/(2h)], \qquad (\mathbf{4.1.129})$$

and

$$F(\xi) = g\left\{\frac{2h}{\pi}\operatorname{arcsinh}\left[\xi\sinh\left(\frac{\pi}{2h}\right)\right]\right\}\frac{\sinh[\pi/(2h)]}{\sqrt{1+\xi^2\sinh^2[\pi/(2h)]}}, \qquad (\mathbf{4.1.130})$$

Equation 4.1.128 becomes

$$\int_0^1 F(\xi)\ln\left|\frac{\xi+\eta}{\xi-\eta}\right|d\xi = -\frac{\pi}{2h}. \qquad (\mathbf{4.1.131})$$

Applying the results from Equation 1.2.31 through Equation 1.2.33,

$$g(t) = -\frac{2\sinh[\pi/(2h)]\cosh[\pi t/(2h)]}{h\sqrt{\cosh(\pi t/h)-1}\sqrt{\cosh(\pi/h)-\cosh(\pi t/h)}} \qquad (\mathbf{4.1.132})$$

after back substitution. Figure 4.1.6 illustrates the special case when $h = 1$.

• **Example 4.1.6**

Our next example involves solving Laplace's equation[17]

$$\frac{\partial^2 u}{\partial x^2} + \frac{\partial^2 u}{\partial y^2} = 0, \qquad -\infty < x < \infty, \quad -h < y < \infty, \tag{4.1.133}$$

with the boundary conditions

$$\lim_{|x|\to\infty} u(x,y) \to 0, \qquad -h < y < \infty, \tag{4.1.134}$$

$$u(x,-h) = 0, \qquad -\infty < x < \infty, \tag{4.1.135}$$

$$u_y(x,0^-) = u_y(x,0^+), \qquad -\infty < x < \infty, \tag{4.1.136}$$

$$\begin{cases} u_y(x,0^-) = u_y(x,0^+) = -q(x), & |x| < 1, \\ u(x,0^-) = u(x,0^+), & |x| > 1, \end{cases} \tag{4.1.137}$$

and

$$\lim_{y\to\infty} u(x,y) \to 0, \qquad -\infty < x < \infty, \tag{4.1.138}$$

where $q(x)$ is an even function.

Because $q(x)$ is an even function, this suggests that we should apply Fourier cosine transforms. Therefore, the solution to Equation 4.1.133 through Equation 4.1.136 and Equation 4.1.138 is

$$u(x,y) = \frac{2}{\pi}\int_0^\infty A(k)\frac{\sinh[k(y+h)]}{\cosh(kh)}\cos(kx)\,dk, \qquad -h < y < 0, \tag{4.1.139}$$

and

$$u(x,y) = -\frac{2}{\pi}\int_0^\infty A(k)e^{-ky}\cos(kx)\,dk, \qquad 0 < y < \infty. \tag{4.1.140}$$

Substituting Equation 4.1.139 and Equation 4.1.140 into Equation 4.1.137, we obtain

$$\int_0^\infty C(k)\cos(kx)\,dk = 0, \qquad |x| > 1, \tag{4.1.141}$$

and

$$\int_0^\infty kC(k)\left(1 + e^{-2kh}\right)\cos(kx)\,dk = -\pi q(x), \qquad |x| < 1, \tag{4.1.142}$$

where $C(k) = A(k)[1 + \tanh(kh)]$.

---

[17] Suggested by a problem solved by Kit, G. S., and M. V. Khai, 1973: Thermoelastic state of a half-plane weakened by a rectilinear slit. *Mech. Solids*, **8(5)**, 36–41.

To solve this set of dual integral equations, we introduce

$$C(k) = \int_0^1 h'(\tau) J_0(k\tau)\, d\tau, \tag{4.1.143}$$

where the prime denotes differentiation with respect to the argument of $h(\tau)$. Substituting Equation 4.1.143 into Equation 4.1.141, we find that

$$\int_0^\infty C(k)\cos(kx)\, dk = \int_0^\infty \left[\int_0^1 h'(\tau)J_0(k\tau)\, d\tau\right]\cos(kx)\, dk \tag{4.1.144}$$

$$= \int_0^1 h'(\tau)\left[\int_0^\infty \cos(kx)J_0(k\tau)\, dk\right] d\tau = 0, \tag{4.1.145}$$

where the integral inside the square brackets vanishes by Equation 1.4.14 since $|x| > 1$ and $0 \le \tau \le 1$. Thus, Equation 4.1.143 satisfies Equation 4.1.141 identically.

We now turn our attention to Equation 4.1.142. Substituting Equation 4.1.143 into Equation 4.1.142, we have, after integrating with respect to x, that

$$\int_0^\infty \left[\int_0^1 h'(\tau)J_0(k\tau)\, d\tau\right]\sin(kx)\, dk + \int_0^\infty \left[\int_0^1 h'(\tau)J_0(k\tau)\, d\tau\right] e^{-2kh}\sin(kx)\, dk = -\pi p(x), \tag{4.1.146}$$

where $p(x) = \int_0^x q(\eta)\, d\eta$. Now,

$$\int_0^\infty \left[\int_0^1 h'(\tau)J_0(k\tau)\, d\tau\right]\sin(kx)\, dk$$

$$= \int_0^1 h'(\tau)\left[\int_0^\infty \sin(kx)J_0(k\tau)\, d\tau\right]\sin(kx)\, d\tau \tag{4.1.147}$$

$$= \int_0^x \frac{h'(\tau)}{\sqrt{x^2-\tau^2}}\, d\tau \tag{4.1.148}$$

after using Equation 1.4.13. Therefore, Equation 4.1.146 becomes

$$\int_0^x \frac{h'(\tau)}{\sqrt{x^2-\tau^2}}\, d\tau = -\pi p(x) - \int_0^1 h'(\tau)\left[\int_0^\infty e^{-2kh}\sin(kx)J_0(k\tau)\, dk\right] d\tau. \tag{4.1.149}$$

Comparing Equation 4.1.149 with Equation 1.2.13, we have from Equation 1.2.14 that

$$h(x) = -2\int_0^x \frac{\tau\, p(\tau)}{\sqrt{x^2-\tau^2}}\, d\tau - \frac{2}{\pi}\int_0^1 h'(t)\left\{\int_0^\infty e^{-2kh}J_0(kt)\left[\int_0^x \frac{\tau\sin(k\tau)}{\sqrt{x^2-\tau^2}}\, d\tau\right] dk\right\} dt \tag{4.1.150}$$

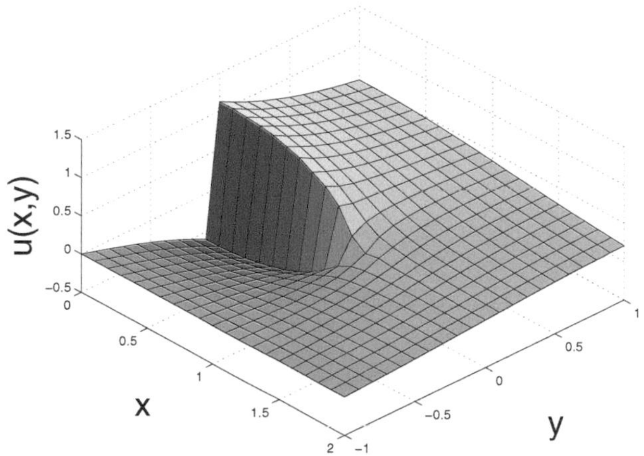

Figure 4.1.7: The solution to Equation 4.1.133 subject to the mixed boundary conditions given by Equation 4.1.134 through Equation 4.1.138 when $q(x) = h = l$.

after integrating with respect to x. Using integral tables,[18] Equation 4.1.150 simplifies to

$$h(x) = -2\int_0^x \frac{\tau\, p(\tau)}{\sqrt{x^2-\tau^2}}\,d\tau - x\int_0^1 h'(\tau)\left[\int_0^\infty e^{-2kh} J_0(k\tau)J_1(kx)\,dk\right]d\tau. \tag{4.1.151}$$

To compute the potential, we first find $h'(t)$ via Equation 4.1.151. Then $C(k)$ or $A(k)$ follows from 4.1.143. Finally, we employ Equation 4.1.139 and Equation 4.1.140. Figure 4.1.7 illustrates this solution when $q(x) = h = 1$.

• Example 4.1.7

In this problem we find the solution to Laplace's equation[19] in the upper half-plane $z > 0$ into which we insert a semi-circular cylinder of radius a that has a potential of 1. See Figure 4.1.8. Mathematically this problem is

$$\frac{\partial^2 u}{\partial r^2} + \frac{1}{r}\frac{\partial u}{\partial r} + \frac{\partial^2 u}{\partial z^2} = 0, \qquad \begin{cases} 0 \le r < a, & 0 < z < b, \\ a < r < \infty, & b < z < \infty, \end{cases} \tag{4.1.152}$$

with the boundary conditions

$$\lim_{r\to 0} |u(r,z)| < \infty, \qquad \lim_{r\to\infty} u(r,z) \to 0, \qquad 0 < z < \infty, \tag{4.1.153}$$

---

[18] Gradshteyn and Ryzhik, op. cit., Formula 3.771.10.

[19] Adapted from Shapiro, Yu. A., 1962: Electrostatic fields of an immersion electron lens consisting of two semi-infinite cylinders. *Sov. Tech. Phys.*, **7**, 501–506.

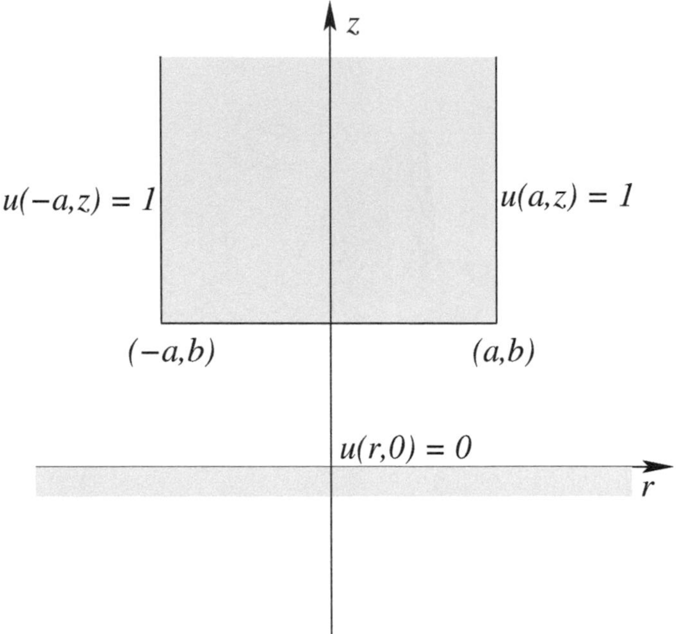

Figure 4.1.8: The geometry for Example 4.1.7. It consists of the upper half-plane $z > 0$ into which we insert a semi-circular cylinder of radius a that has a potential of 1.

$$\lim_{z\to\infty} |u(r,z)| < \infty, \qquad a \le r < \infty, \tag{4.1.154}$$

$$u(r,0) = 0, \qquad 0 \le r < \infty, \tag{4.1.155}$$

and

$$\begin{cases} u(a^+,z) = u(a^-,z), & 0 < z < b, \\ u(a,z) = 1, & b < z < \infty. \end{cases} \tag{4.1.156}$$

The solution to Equation 4.1.152 through Equation 4.1.155 is

$$\begin{aligned} u(r,z) = {} & \frac{\pi}{2}\int_0^\infty \frac{\cos(kb) I_0(kr)}{(\pi^2/4 - k^2b^2) I_0(ka)} \sin(kz)\,\frac{dk}{k} \\ & - \int_0^\infty A(k) \frac{I_0(kr)}{I_0(ka)} \sin(kz)\,dk \end{aligned} \tag{4.1.157}$$

for $0 < r < a$; and

$$\begin{aligned} u(r,z) = {} & \frac{\pi}{2}\int_0^\infty \frac{\cos(kb) K_0(kr)}{(\pi^2/4 - k^2b^2) K_0(ka)} \sin(kz)\,\frac{dk}{k} \\ & - \int_0^\infty A(k) \frac{K_0(kr)}{K_0(ka)} \sin(kz)\,dk \end{aligned} \tag{4.1.158}$$

for $a < r < \infty$. The first integrals in Equation 4.1.157 and Equation 4.1.158 satisfy the boundary condition $u(a,z) = 1$ if $z > b$, $u(a,z) = \sin[\pi z/(2b)]$

if $z < b$, $\lim_{r\to\infty} u(r,z) \to 0$, and $|u(r,z)| < \infty$ as $z \to \infty$. Thus the first integrals are particular solutions to our problem while the second integrals are homogeneous solutions.

Upon substituting Equation 4.1.157 and Equation 4.1.158 into Equation 4.1.156, we obtain the dual integral equations

$$\int_0^\infty \frac{A(k)}{g(k)}[1-\cos(kz)]\,\frac{dk}{k} = \Psi(z), \qquad 0 \le z < b, \tag{4.1.159}$$

and

$$\int_0^\infty A(k)\sin(kz)\,dk = 0, \qquad b < z < \infty, \tag{4.1.160}$$

where

$$\Psi(z) = \frac{\pi}{2}\int_0^\infty \frac{[1-\cos(kz)]\cos(kb)}{g(k)\,(\pi^2/4 - k^2b^2)}\,\frac{dk}{k^2}, \tag{4.1.161}$$

and $g(k) = 2aI_0(ka)K_0(ka)$. To solve these dual integral equations, we introduce a function $\psi(t)$ such that

$$A(k) = \int_0^b t\,\psi(t)J_1(kt)\,dt. \tag{4.1.162}$$

If we substitute Equation 4.1.162 into Equation 4.1.160, interchange the order of integration, and then use integral tables,[20] we can show that this choice for $A(k)$ satisfies Equation 4.1.160 identically.

Upon substituting Equation 4.1.162 into Equation 4.1.159,

$$\int_0^\infty \frac{1-\cos(kz)}{g(k)}\left[\int_0^\infty \psi(t)J_1(kt)\,t\,dt\right]\frac{dk}{k} = \Psi(z). \tag{4.1.163}$$

Setting

$$h(k) = 1 - \frac{1}{k\,g(k)}, \tag{4.1.164}$$

Equation 4.1.163 can be rewritten as

$$\begin{aligned}\int_0^\infty [1-\cos(kz)]\left[\int_0^b \psi(t)J_1(kt)\,t\,dt\right]dk \\ - \int_0^\infty h(k)[1-\cos(kz)]\left[\int_0^b \psi(t)J_1(kt)\,t\,dt\right]dk \\ = \frac{\pi}{2}\int_0^\infty \frac{[1-\cos(kz)]\cos(kb)}{g(k)\,(\pi^2/4-k^2b^2)}\,\frac{dk}{k^2}.\end{aligned} \tag{4.1.165}$$

[20] Gradshteyn and Ryzhik, op. cit., Formula 6.671.1.

If we now interchange the order of integration in Equation 4.1.165 and use integral tables,[21]

$$\begin{aligned}\frac{1}{z}\int_0^z &\left\{\frac{t^2}{\sqrt{z^2-t^2}\,\left[z+\sqrt{z^2-t^2}\,\right]}+1\right\}\psi(t)\,dt \\ &-\int_0^b t\,\psi(t)\left[\int_0^\infty \frac{1-\cos(kz)}{kz}h(k)\,k\,J_1(kt)\,dk\right]\,dt \\ &= \frac{\pi}{2}\int_0^\infty \frac{[1-\cos(kz)]\cos(kb)}{k^2 z\,g(k)(\pi^2/4-k^2b^2)}\,dk. \quad (\mathbf{4.1.166})\end{aligned}$$

Because[22]

$$\frac{1-\cos(x)}{x} = \int_0^{\pi/2} J_1[x\sin(\theta)]\,d\theta, \qquad (\mathbf{4.1.167})$$

$$\begin{aligned}\int_0^{\pi/2} &\psi[z\sin(\theta)]\,d\theta \qquad (\mathbf{4.1.168}) \\ &-\int_0^{\pi/2}\int_0^b t\,\psi(t)\left[\int_0^\infty h(k)J_1[kz\sin(\theta)]J_1(kt)\,k\,dk\right]\,dt\,d\theta \\ &= \frac{\pi}{2}\int_0^{\pi/2}\int_0^\infty J_1[kz\sin(\theta)]\frac{\cos(kb)}{g(k)(\pi^2/4-k^2b^2)}\,\frac{dk}{k}\,d\theta.\end{aligned}$$

Equation 4.1.168 is satisfied identically if $\psi(t)$ satisfies the integral equation

$$\psi(x)-\int_0^b \psi(t)K(x,t)\,dt = G(x), \qquad (\mathbf{4.1.169})$$

where

$$K(x,t) = t\int_0^\infty \left[1-\frac{1}{k\,g(k)}\right] J_1(kx)J_1(kt)\,k\,dk, \qquad (\mathbf{4.1.170})$$

and

$$G(x) = \frac{\pi}{2}\int_0^\infty \frac{\cos(kb)J_1(kx)}{g(k)(\pi^2/4-k^2b^2)}\,\frac{dk}{k}. \qquad (\mathbf{4.1.171})$$

Figure 4.1.9 illustrates the solution when $a=b=1$.

---

[21] Ibid., Formula 6.671.2.

[22] Ibid., Formula 6.519.2 with $\nu=\frac{1}{2}$ and $z=x/2$.

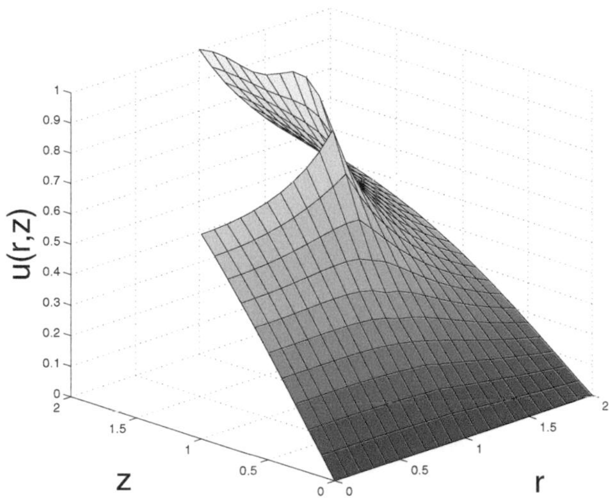

Figure 4.1.9: The solution to the mixed boundary value problem governed by Equation 4.1.152 through Equation 4.1.156.

• Example 4.1.8

Consider the problem of solving Laplace's equation[23] over the quarter plane

$$\frac{\partial^2 u}{\partial x^2} + \frac{\partial^2 u}{\partial y^2} = 0, \qquad 0 < x, y < \infty, \tag{4.1.172}$$

with the boundary conditions

$$\lim_{x\to\infty} u(x,y) \to 0, \qquad 0 < y < \infty, \tag{4.1.173}$$

$$\lim_{y\to\infty} u(x,y) \to 0, \qquad 0 < x < \infty, \tag{4.1.174}$$

$$\begin{cases} u(x,0) = f(x), & 0 \le x < 1, \\ u_y(x,0) = 0, & 1 < x < \infty, \end{cases} \tag{4.1.175}$$

and

$$\begin{cases} u(0,y) = f(y), & 0 \le y < 1, \\ u_x(0,y) = 0, & 1 < y < \infty. \end{cases} \tag{4.1.176}$$

[23] Taken with permission from Gupta, O. P., and S. K. Gupta, 1975: Mixed boundary value problems in electrostatics. *Z. Angew. Math. Mech.*, **55**, 715–720.

From the form of the boundary conditions along $x = 0$ and $y = 0$, we anticipate that we should use Fourier cosine transforms. Therefore, the solution to Equation 4.1.172 through Equation 4.1.174 is

$$u(x, y) = \int_0^\infty A(k)e^{-ky} \cos(kx)\, dk + \int_0^\infty A(k)e^{-kx} \cos(ky)\, dk. \quad \textbf{(4.1.177)}$$

Substituting Equation 4.1.177 into Equation 4.1.175, we obtain the dual integral equations

$$\int_0^\infty A(k) \cos(kx)\, dk + \int_0^\infty A(k)e^{-kx}\, dk = f(x), \qquad 0 \le x < 1, \quad \textbf{(4.1.178)}$$

and

$$\int_0^\infty k\, A(k) \cos(kx)\, dk = 0, \qquad 1 < x < \infty. \quad \textbf{(4.1.179)}$$

To solve Equation 4.1.178 and Equation 4.1.179, we introduce

$$A(k) = k \int_0^1 t\, h(t) J_0(kt)\, dt = h(1)J_1(k) - \int_0^1 t\, h'(t) J_1(kt)\, dt. \quad \textbf{(4.1.180)}$$

Turning to Equation 4.1.179 first, we have that

$$\begin{aligned}\int_0^\infty k\, A(k) \cos(kx)\, dk &= h(1) \int_0^\infty k\, J_1(k) \cos(kx)\, dk \\ &\quad - \int_0^1 t\, h'(t) \left[\int_0^\infty k\, J_1(kt) \cos(kx)\, dk\right] dt \quad \textbf{(4.1.181)} \\ &= 0. \quad \textbf{(4.1.182)}\end{aligned}$$

This follows by noting[24]

$$\int_0^\infty k J_1(kt) \cos(kx)\, dk = \frac{d}{dx}\left[\int_0^\infty J_1(kt) \sin(kx)\, dk\right] = 0 \quad \textbf{(4.1.183)}$$

if $0 \le t \le 1 < x < \infty$. On the other hand, from Equation 4.1.178,

$$\begin{aligned}&\int_0^1 t\, h(t) \left[\int_0^\infty k \cos(kx) J_0(kt)\, dk\right] dt \\ &\quad + \int_0^1 t\, h(t) \left[\int_0^\infty k\, e^{-kx} J_0(kt)\, dk\right] dt = f(x), \quad \textbf{(4.1.184)}\end{aligned}$$

[24] Gradshteyn and Ryzhik, op. cit., Formula 6.693.1.

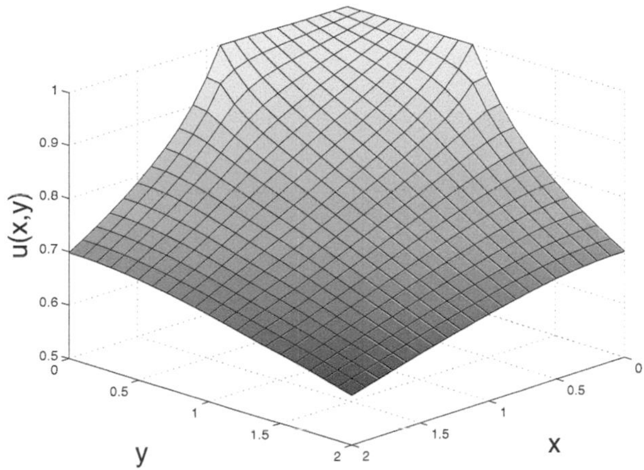

Figure 4.1.10: The solution to Equation 4.1.172 subject to the mixed boundary conditions given by Equation 4.1.173 through Equation 4.1.176.

or

$$\frac{\partial}{\partial x}\left\{\int_0^1 t\,h(t)\left[\int_0^\infty \sin(kx)J_0(kt)\,dk\right]dt\right\} + \int_0^1 t\,h(t)\left[\int_0^\infty k\,e^{-kx}J_0(kt)\,dk\right]dt = f(x). \tag{4.1.185}$$

Using integral tables,[25] Equation 4.1.185 simplifies to

$$\frac{\partial}{\partial x}\left[\int_0^x \frac{t\,h(t)}{\sqrt{x^2-t^2}}\,dt\right] + x\int_0^1 \frac{t\,h(t)}{(x^2+t^2)^{3/2}}\,dt = f(x), \quad 0 \le x < 1. \tag{4.1.186}$$

Using Equation 1.2.13 and Equation 1.2.14, we can solve for $h(t)$ and find that

$$h(t) + \int_0^1 K(t,\tau)h(\tau)\,d\tau = \frac{2}{\pi}\int_0^t \frac{f(x)}{\sqrt{t^2-x^2}}\,dx, \tag{4.1.187}$$

where

$$K(t,\tau) = \frac{2\tau}{\pi}\int_0^t \frac{x}{\sqrt{t^2-x^2}\,(x^2+\tau^2)^{3/2}}\,dx = \frac{2t}{\pi(t^2+\tau^2)}. \tag{4.1.188}$$

In the case when $f(x) = 1$, Equation 4.1.187 simplifies to

$$h(t) + \int_0^1 K(t,\tau)h(\tau)\,d\tau = 1. \tag{4.1.189}$$

Figure 4.1.10 illustrates the solution.

---

[25] Ibid., Formula 6.671.7 and Formula 6.621.4.

Problems

1. Solve Laplace's equation[26]

$$\frac{\partial^2 u}{\partial x^2} + \frac{\partial^2 u}{\partial y^2} = 0, \qquad -\infty < x < \infty, \quad 0 < y < \infty,$$

with the boundary conditions

$$\lim_{|x|\to\infty} u(x,y) \to 0, \qquad 0 < y < \infty,$$

$$\lim_{y\to\infty} u(x,y) \to 0, \qquad -\infty < x < \infty,$$

and

$$\begin{cases} u_y(x,0) = 1, & 0 \le |x| < 1, \\ u(x,0) = 0, & 1 < |x| < \infty. \end{cases} \tag{1}$$

Step 1: Using separation of variables or transform methods, show that the general solution to the problem is

$$u(x,y) = \int_0^\infty A(k) e^{-ky} \cos(kx)\, dk.$$

Step 2: Using boundary condition (1), show that $A(k)$ satisfies the dual integral equations

$$\int_0^\infty kA(k) \cos(kx)\, dk = -\frac{\pi}{2}, \qquad 0 \le |x| < 1,$$

and

$$\int_0^\infty A(k) \cos(kx)\, dk = 0, \qquad 1 < |x| < \infty.$$

Step 3: Using Equation 1.4.14, show that

$$A(k) = -\frac{\pi}{2} \int_0^1 t J_0(kt)\, dt = -\frac{\pi J_1(k)}{2k}$$

satisfies both integral equations given in Step 2.

---

[26] See Yang, F.-Q., and J. C. M. Li, 1995: Impression and diffusion creep of anisotropic media. *J. Appl. Phys.*, **77**, 110–117. See also Shindo, Y., H. Tamura, and Y. Atobe, 1990: Transient singular stresses of a finite crack in an elastic conductor under electromagnetic force (in Japanese). *Nihon Kikai Gakkai Rombunshu (Trans. Japan Soc. Mech. Engrs.), Ser. A*, **56**, 278–282.

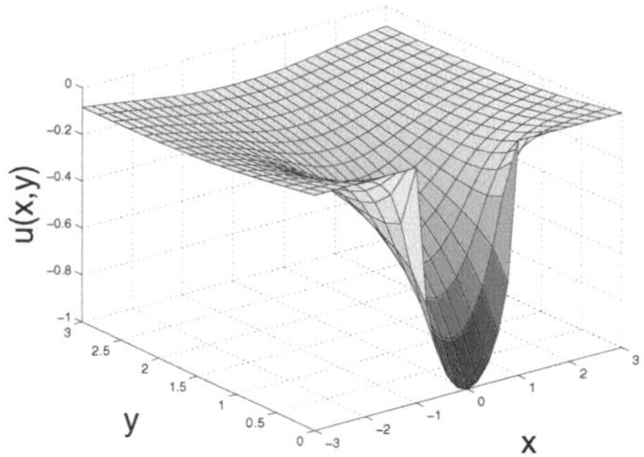

Problem 1

Step 4: Show that the solution to this problem is

$$u(x,y) = -\int_0^\infty J_1(k)e^{-ky}\cos(kx)\frac{dk}{k}.$$

In particular, verify that $u(x,0) = -\sqrt{1-x^2}$ if $|x| < 1$. The figure labeled Problem 1 illustrates the solution $u(x,y)$.

2. Solve Laplace's equation

$$\frac{\partial^2 u}{\partial x^2} + \frac{\partial^2 u}{\partial y^2} = 0, \qquad -\infty < x < \infty, \quad 0 < y < \infty,$$

with the boundary conditions

$$\lim_{|x|\to\infty} u(x,y) \to 0, \qquad 0 < y < \infty,$$

$$\lim_{y\to\infty} u(x,y) \to 0, \qquad -\infty < x < \infty,$$

and

$$\begin{cases} u(x,0) = x, & 0 \le |x| < 1, \\ u_y(x,0) = 0, & 1 < |x| < \infty. \end{cases} \tag{1}$$

Step 1: Using separation of variables, show that the general solution to the problem is

$$u(x,y) = \int_0^\infty A(k)e^{-ky}\frac{\sin(kx)}{k}\,dk.$$

Step 2: Using boundary condition (1), show that $A(k)$ satisfies the dual integral equations

$$\int_0^\infty \frac{A(k)}{k}\sin(kx)\,dk = x, \qquad 0 \le |x| < 1,$$

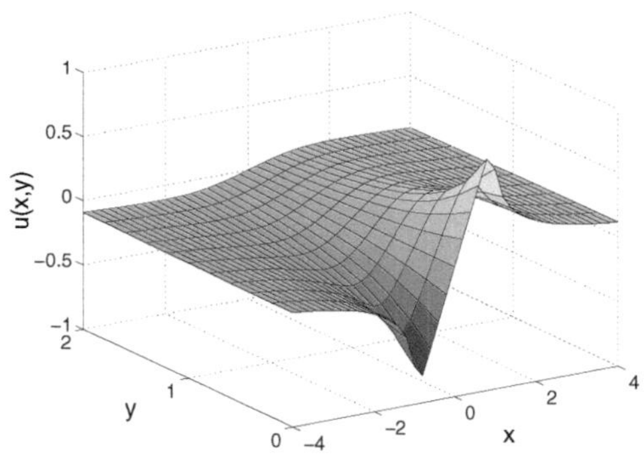

Problem 2

and

$$\int_0^\infty A(k)\sin(kx)\,dk = 0, \qquad 1 < |x| < \infty.$$

Step 3: Fredricks[27] showed that dual integral equations of the form

$$\int_0^\infty \frac{A(k)}{k}\sin(kx)\,dk = f(x), \qquad 0 \le x < a,$$

and

$$\int_0^\infty A(k)\sin(kx)\,dk = 0, \qquad a < x < \infty,$$

have the solution

$$A(k) = 2\sum_{n=1}^\infty \sum_{m=0}^\infty (2m+1)B_n J_{2m+1}(n\pi)J_{2m+1}(ka),$$

where B_n is given by the Fourier sine series

$$f(x) = \sum_{n=1}^\infty B_n \sin(n\pi x/a), \qquad 0 \le x < a.$$

Clearly $f(0) = 0$. Use this result to show that

$$A(k) = \frac{4}{\pi}\sum_{n=1}^\infty \sum_{m=0}^\infty \frac{(-1)^{n+1}}{n}(2m+1)B_n J_{2m+1}(n\pi)J_{2m+1}(k).$$

The figure labeled Problem 2 illustrates this solution $u(x,y)$.

---

27 Fredricks, R. W., 1958: Solution of a pair of integral equations from elastostatics. *Proc. Natl. Acad. Sci.*, **44**, 309–312. See also Sneddon, I. N., 1962: Dual integral equations with trigonometrical kernels. *Proc. Glasgow Math. Assoc.*, **5**, 147–152.

3. Following Example 4.1.3, solve Laplace's equation

$$\frac{\partial^2 u}{\partial x^2} + \frac{\partial^2 u}{\partial y^2} = 0, \qquad 0 < x < \infty, \quad 0 < y < h,$$

with the boundary conditions

$$u_x(0, y) = 0, \quad \lim_{x \to \infty} u(x, y) \to 0, \qquad 0 < y < h,$$

$$\begin{cases} u_y(x, 0) = -g(x), & 0 < x < a, \\ u(x, 0) = 0, & a < x < \infty, \end{cases} \tag{1}$$

and

$$u(x, h) = 0, \qquad 0 < x < \infty.$$

Step 1: Using separation of variables or transform methods, show that the general solution to the problem is

$$u(x, y) = \frac{2}{\pi} \int_0^\infty A(k) \sinh[k(h - y)] \cos(kx)\, dk.$$

Step 2: Using boundary condition (1), show that $A(k)$ satisfies the dual integral equations

$$\frac{2}{\pi} \int_0^\infty kA(k)\, [1 + M(kh)] \sinh(kh) \cos(kx)\, dk = g(x), \qquad 0 < x < a,$$

and

$$\frac{2}{\pi} \int_0^\infty A(k) \sinh(kh) \cos(kx)\, dk = 0, \qquad a < x < \infty,$$

where $M(kh) = e^{-kh} / \sinh(kh)$.

Step 3: Setting

$$\sinh(kh) A(k) = \int_0^a \tau\, h(\tau) J_0(k\tau)\, d\tau,$$

show that the second integral equation in Step 2 is identically satisfied.

Step 4: Show that the first integral equation in Step 2 leads to the integral equation

$$h(t) + \int_0^a \tau\, h(\tau) \left[\int_0^\infty k\, M(kh) J_0(k\tau) J_0(kt)\, dk \right] d\tau = \int_0^t \frac{g(x)}{\sqrt{t^2 - x^2}}\, dx$$

with $0 < t < a$.

Step 5: Simplify your results in the case $g(x) = 1$ and show that they are identical with the results given in Example 4.1.3 by Equation 4.1.83 through Equation 4.1.86.

4. Following Example 4.1.2, solve Laplace's equation[28]

$$\frac{\partial^2 u}{\partial x^2} + \frac{\partial^2 u}{\partial y^2} = 0, \qquad -\infty < x < \infty, \quad 0 < y < h,$$

with the boundary conditions

$$\lim_{|x|\to\infty} u(x,y) \to 0, \qquad 0 < y < h,$$

$$\begin{cases} u_y(x,0) = -p(x)/h, & |x| < a, \\ u(x,0) = 0, & |x| > a, \end{cases} \tag{1}$$

and

$$u(x,h) = 0, \qquad -\infty < x < \infty.$$

Step 1: Using separation of variables or transform methods, show that the general solution to the problem is

$$u(x,y) = \frac{2}{\pi}\int_0^\infty A(k)\frac{e^{-ky} - e^{ky-2kh}}{1 - e^{-2kh}}\cos(kx)\,dk.$$

Step 2: Using boundary condition (1), show that $A(k)$ satisfies the dual integral equations

$$\frac{2}{\pi}\int_0^\infty k\coth(kh)A(k)\cos(kx)\,dk = \frac{p(x)}{h}, \qquad |x| < a,$$

and

$$\frac{2}{\pi}\int_0^\infty A(k)\cos(kx)\,dk = 0, \qquad |x| > a.$$

Step 3: Setting

$$kA(k) = \frac{\pi}{2}\int_0^a g(\tau)\sin(k\tau)\,d\tau,$$

show that the second integral equation in Step 2 is identically satisfied.

Step 4: Show that the first integral equation in Step 2 can be rewritten

$$\frac{d}{dx}\left\{\int_0^a g(\tau)\left[\int_0^\infty \coth(kh)\sin(k\tau)\sin(kx)\,\frac{dk}{k}\right]d\tau\right\} = \frac{p(x)}{h}, \qquad 0 < x < a.$$

---

[28] Suggested from Singh, B. M., T. B. Moodie, and J. B. Haddow, 1981: Closed-form solutions for finite length crack moving in a strip under anti-plane shear stress. *Acta Mech.*, **38**, 99–109.

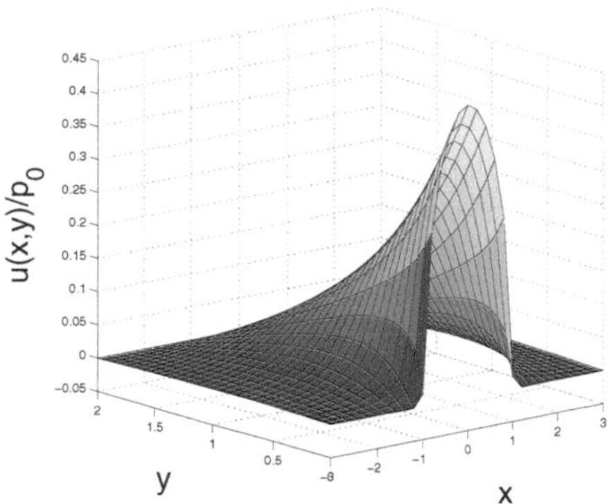

Problem 4

Step 5: Using the relationship that

$$\int_0^\infty \coth(kh)\sin(k\tau)\sin(kx)\,\frac{dk}{k} = \frac{1}{2}\ln\left|\frac{\tanh(cx)+\tanh(c\tau)}{\tanh(cx)-\tanh(c\tau)}\right|,$$

where $c = \pi/(2h)$, show that Step 4 can be written

$$\int_0^a g(\tau)\ln\left|\frac{\tanh(cx)+\tanh(c\tau)}{\tanh(cx)-\tanh(c\tau)}\right|\,d\tau = \frac{2}{h}\int_0^x p(\xi)\,d\xi = F(x), \qquad 0 < x < a.$$

Step 6: By using the results from Example 1.2.3, show that the solution to the integral equation in Step 5 is

$$g(\tau) = -\frac{2c\tanh(c\tau)\operatorname{sech}^2(c\tau)}{\pi^2\sqrt{\tanh^2(ca)-\tanh^2(c\tau)}}\int_0^a \frac{F'(x)\sqrt{\tanh^2(ca)-\tanh^2(cx)}}{\tanh^2(cx)-\tanh^2(c\tau)}\,dx$$
$$+\frac{4cF(0)\tanh(ca)}{\pi^2\sinh(2c\tau)\sqrt{\tanh^2(ca)-\tanh^2(c\tau)}}, \qquad 0 < \tau < a.$$

Step 7: Because $F'(x) = 2p(x)/h$ with $F(0) = 0$, show that Step 6 simplifies to

$$g(\tau) = -\frac{4c\tanh(c\tau)\operatorname{sech}^2(c\tau)}{\pi^2 h\sqrt{\tanh^2(ca)-\tanh^2(c\tau)}}\int_0^a \frac{p(x)\sqrt{\tanh^2(ca)-\tanh^2(cx)}}{\tanh^2(cx)-\tanh^2(c\tau)}\,dx,$$

if $0 < \tau < a$.

Step 8: For the special case $p(x) = p_0$, a constant, show that

$$g(\tau) = \frac{2p_0\sinh(c\tau)}{\pi h\sqrt{\sinh^2(ca)-\sinh^2(c\tau)}}, \qquad 0 < \tau < a.$$

The figure entitled Problem 4 illustrates this special solution when $a = 1$ and $h = 2$.

4.2 TRIPLE FOURIER INTEGRALS

In the previous section we considered the case where the mixed boundary conditions led to two integral equations that are in the form of a Fourier integral. In the present section we take the next step and examine the situation where the mixed boundary condition contains different boundary conditions along three segments.

• Example 4.2.1

For our first example, let us solve Laplace's equation

$$\frac{\partial^2 u}{\partial x^2} + \frac{\partial^2 u}{\partial y^2} = 0, \qquad -\infty < x < \infty, \quad 0 < y < \infty, \tag{4.2.1}$$

subject to the boundary conditions

$$\lim_{|x|\to\infty} u(x,y) \to 0, \qquad 0 < y, \tag{4.2.2}$$

$$\lim_{y\to\infty} u(x,y) \to 0, \qquad -\infty < x < \infty, \tag{4.2.3}$$

$$u(x,0) = \begin{cases} -1, & -b < x < -a, \\ 1, & a < x < b, \end{cases} \tag{4.2.4}$$

and

$$u_y(x,0) = 0, \qquad 0 < |x| < a, \quad b < |x| < \infty. \tag{4.2.5}$$

The interesting aspect of this problem is the boundary condition along $y = 0$. For a portion of the boundary ($-b < x < -a$ and $a < x < b$), it consists of a Dirichlet condition; otherwise, it is a Neumann condition.

If we employ separation of variables or transform methods, the most general solution is

$$u(x,y) = \int_0^\infty \frac{A(k)}{k} e^{-ky} \sin(kx)\, dk. \tag{4.2.6}$$

Substituting Equation 4.2.6 into the boundary conditions given by Equation 4.2.4 and Equation 4.2.5, we obtain the following set of integral equations:

$$\int_0^\infty A(k) \sin(kx)\, dk = 0, \qquad 0 \le x < a, \tag{4.2.7}$$

$$\int_0^\infty \frac{A(k)}{k} \sin(kx)\, dk = 1, \qquad a < x < b, \tag{4.2.8}$$

and

$$\int_0^\infty A(k) \sin(kx)\, dk = 0, \qquad b < x < \infty. \tag{4.2.9}$$

We must now solve for $A(k)$ which appears in a set of integral equations. Tranter[29] showed that triple integral equations of the form given by Equation 4.2.7 through Equation 4.2.9 have the solution

$$A(k) = 2\sum_{n=1}^{\infty}(-1)^{n-1}A_n J_{2n-1}(bk), \tag{4.2.10}$$

where the constants A_n are the solution of the dual series relationship

$$\sum_{n=1}^{\infty}(-1)^{n-1}A_n \sin\left[\left(n-\tfrac{1}{2}\right)\varphi\right] = 0, \qquad 0 \le \varphi < \gamma, \tag{4.2.11}$$

$$\sum_{n=1}^{\infty}(-1)^{n-1}\frac{A_n}{n-\frac{1}{2}} \sin\left[\left(n-\tfrac{1}{2}\right)\varphi\right] = 1, \qquad \gamma < \varphi \le \pi, \tag{4.2.12}$$

and γ is defined by $a = b\sin(\gamma/2)$, $0 < \gamma \le \pi$. If we now introduce the change of variables $\theta = \pi - \varphi$ and $c = \pi - \gamma$, we find that A_n is the solution of the following pair of dual series:

$$\sum_{n=1}^{\infty}\frac{A_n}{n-\frac{1}{2}} \cos\left[\left(n-\tfrac{1}{2}\right)\theta\right] = 1, \qquad 0 < \theta < c, \tag{4.2.13}$$

and

$$\sum_{n=1}^{\infty}A_n \cos\left[\left(n-\tfrac{1}{2}\right)\theta\right] = 0, \qquad c < \theta \le \pi. \tag{4.2.14}$$

Consequently, we have reduced three integral equations to two dual trigonometric series. Tranter[30] also analyzed dual trigonometric series of the form given by Equation 4.2.13 and Equation 4.2.14 and showed that in our particular case

$$A_n = \frac{P_{n-1}[\cos(c)]}{K(a/b)}, \tag{4.2.15}$$

where $K(\cdot)$ denotes the complete elliptic integral and $P_n(\cdot)$ is the Legendre polynomial of order n. Substituting Equation 4.2.15 into Equation 4.2.10 with $a = b\cos(c/2)$, we obtain

$$A(k) = \frac{2}{K(a/b)}\sum_{n=1}^{\infty}(-1)^{n-1}P_{n-1}[\cos(c)]J_{2n-1}(bk). \tag{4.2.16}$$

---

[29] Tranter, C. J., 1960: Some triple integral equations. *Proc. Glasgow Math. Assoc.*, **4**, 200–203. A more accessible analysis is given in Section 6.4 of Sneddon, I. N., 1966: *Mixed Boundary Value Problems in Potential Theory*. Wiley, 283 pp.

[30] Tranter, C. J., 1959: Dual trigonometric series. *Proc. Glasgow Math. Assoc.*, **4**, 49–57; Sneddon, op. cit., Section 5.4.5.

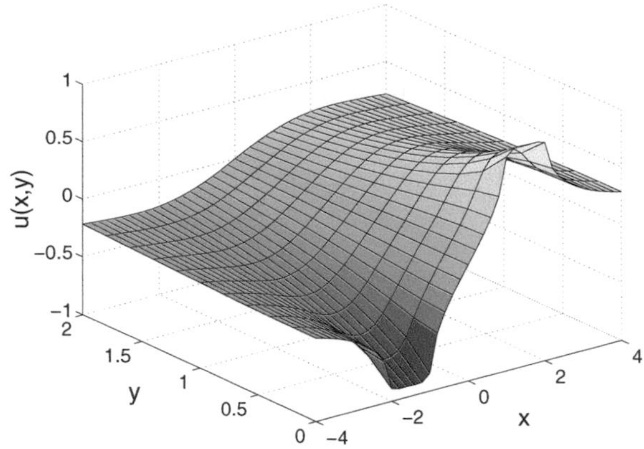

Figure 4.2.1: The solution for Equation 4.2.1 through Equation 4.2.5 when $a = 1$ and $b = 2$.

Figure 4.2.1 illustrates $u(x, y)$ when $a = 1$ and $b = 2$.

• Example 4.2.2

Let us solve Laplace's equation

$$\frac{\partial^2 u}{\partial x^2} + \frac{\partial^2 u}{\partial y^2} = 0, \qquad -\infty < x < \infty, \quad 0 < y < \pi, \tag{4.2.17}$$

with the boundary conditions

$$\lim_{|x|\to\infty} u(x, y) \to 0, \qquad 0 < y < \pi, \tag{4.2.18}$$

$$\begin{cases} u_y(x, 0) = 0, & |x| < a, \\ u(x, 0) = \operatorname{sgn}(x), & a < |x| < b, \\ u_y(x, 0) = 0, & b < |x|, \end{cases} \tag{4.2.19}$$

and

$$u(x, \pi) = 0, \qquad -\infty < x < \infty, \tag{4.2.20}$$

where $a < b$.

A quick check shows that

$$u(x, y) = \int_0^\infty A(k) \frac{\sinh[k(\pi - y)]}{\sinh(k\pi} \sin(kx) \, \frac{dk}{k} \tag{4.2.21}$$

satisfies Equation 4.2.17, Equation 4.2.18 and Equation 4.2.20. Upon substituting Equation 4.2.21 into Equation 4.2.19, we obtain three integral equations:

$$\int_0^\infty A(k) \coth(k\pi) \sin(kx) \, dk = 0, \qquad 0 < x < a, \tag{4.2.22}$$

$$\int_0^\infty A(k)\sin(kx)\,\frac{dk}{k} = 1, \qquad a < x < b, \tag{4.2.23}$$

and

$$\int_0^\infty A(k)\coth(k\pi)\sin(kx)\,dk = 0, \qquad b < x < \infty. \tag{4.2.24}$$

Singh[31] showed that the solution to

$$\int_0^\infty A(k)\coth(k\pi)\sin(k\eta)\,dk = F_1(\eta), \qquad 0 < \eta < a, \tag{4.2.25}$$

$$\int_0^\infty A(k)\sin(k\eta)\,\frac{dk}{k} = F_2(\eta), \qquad a < \eta < b, \tag{4.2.26}$$

and

$$\int_0^\infty A(k)\coth(k\pi)\sin(k\eta)\,dk = 0 \qquad b < \eta < \infty \tag{4.2.27}$$

is

$$\begin{aligned}\coth(k\pi)A(k) = {} & \frac{2}{\pi}\int_0^a F_1(\xi)\sin(k\xi)\,d\xi + \frac{2}{\pi}\int_a^b g[\cosh(\xi)]\sin(k\xi)\cosh(\xi/2)\,d\xi \\ & + \frac{2}{\pi}\int_b^\infty F_3(\xi)\sin(k\xi)\,d\xi,\end{aligned} \tag{4.2.28}$$

where

$$R(\eta) = \int_a^b g[\cosh(\xi)]\cosh(\xi/2)\ln\left|\frac{\sinh(\xi/2)+\sinh(\eta/2)}{\sinh(\xi/2)-\sinh(\eta/2)}\right|\,d\xi, \qquad a < \eta < b, \tag{4.2.29}$$

or

$$\begin{aligned}R(\eta) = \pi F_2(\eta) & - \int_0^a F_1(\xi)\ln\left|\frac{\sinh(\xi/2)+\sinh(\eta/2)}{\sinh(\xi/2)-\sinh(\eta/2)}\right|\,d\xi \\ & - \int_b^\infty F_3(\xi)\ln\left|\frac{\sinh(\xi/2)+\sinh(\eta/2)}{\sinh(\xi/2)-\sinh(\eta/2)}\right|\,d\xi,\end{aligned} \tag{4.2.30}$$

and

$$\begin{aligned}g[\cosh(\eta)] = {} & -\frac{2}{\pi^2}\sqrt{\frac{\cosh(\eta)-\cosh(a)}{\cosh(b)-\cosh(\eta)}} \\ & \times \int_a^b \sqrt{\frac{\cosh(b)-\cosh(\xi)}{\cosh(\xi)-\cosh(a)}}\,\frac{R'(\xi)\sinh(\xi/2)}{\cosh(\xi)-\cosh(\eta)}\,d\xi \\ & + \frac{C}{\sqrt{[\cosh(\eta)-\cosh(a)][\cosh(b)-\cosh(\eta)]}}.\end{aligned} \tag{4.2.31}$$

---

[31] Singh, B. M., 1973: On triple trigonometrical equations. *Glasgow Math. J.*, **14**, 174–178.

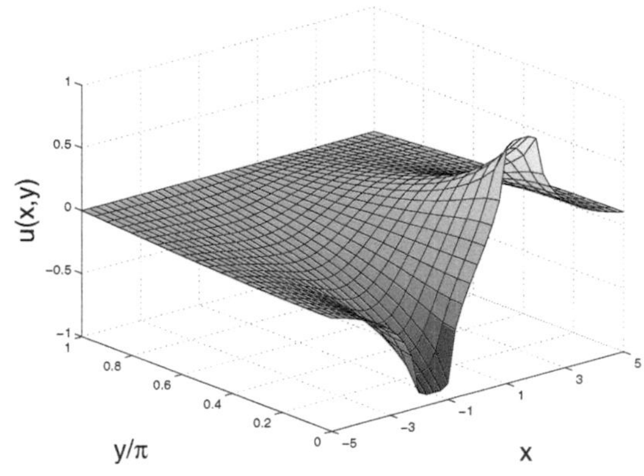

Figure 4.2.2: The solution for Equation 4.2.17 through Equation 4.2.20 when $a = 1$ and $b = 2$.

In the present case, $F_1(\eta) = F_3(\eta) = 0$ and $F_2(\eta) = 1$. Therefore,

$$g[\cosh(\eta)] = \frac{C}{\sqrt{[\cosh(\eta) - \cosh(a)][\cosh(b) - \cosh(\eta)]}}. \qquad \mathbf{(4.2.32)}$$

From Equation 4.2.29 and Equation 4.2.30, we find that

$$\int_a^b g[\cosh(\xi)] \cosh(\xi/2) \ln\left|\frac{\sinh(\xi/2) + \sinh(\eta/2)}{\sinh(\xi/2) - \sinh(\eta/2)}\right| d\xi = \pi, \qquad a < \eta < b. \qquad \mathbf{(4.2.33)}$$

Substituting Equation 4.2.32 into Equation 4.2.33 and evaluating the integral, we obtain

$$C = \frac{\sinh(b/2)}{K[\sinh(a/2)/\sinh(b/2)]}, \qquad \mathbf{(4.2.34)}$$

where $K(\cdot)$ denotes the complete elliptic integral. We then introduce this value of C into Equation 4.2.32 and find that

$$\coth(k\pi)A(k) = \frac{2}{\pi}\int_a^b g[\cosh(\xi)] \sin(k\xi) \cosh(\xi/2)\, d\xi. \qquad \mathbf{(4.2.35)}$$

Finally, the potential $u(x, y)$ follows from Equation 4.2.21. Figure 4.2.2 illustrates the present example.

• Example 4.2.3

A generalization[32] of the previous example is

$$\frac{\partial^2 u}{\partial x^2} + \frac{\partial^2 u}{\partial y^2} = 0, \qquad -\infty < x < \infty, \quad 0 < y < h, \qquad \mathbf{(4.2.36)}$$

---

[32] See Singh, B. M., and R. S. Dhaliwal, 1984: Closed form solutions to dynamic punch problems by integral transform method. *Z. Angew. Math. Mech.*, **64**, 31–34.

with the boundary conditions

$$\lim_{|x|\to\infty} u(x,y) \to 0, \qquad 0 < y < h, \tag{4.2.37}$$

$$\begin{cases} u_y(x,0) = 0, & |x| < a, \\ u(x,0) = f(x), & a < |x| < b, \\ u_y(x,0) = 0, & b < |x|, \end{cases} \tag{4.2.38}$$

and

$$u(x,h) = 0, \qquad -\infty < x < \infty, \tag{4.2.39}$$

where $a < b$.

A quick check shows that

$$u(x,y) = \frac{2}{\pi}\int_0^\infty A(k)\frac{e^{-ky} - e^{ky-2kh}}{1 + e^{-2kh}} \cos(kx)\, dk \tag{4.2.40}$$

satisfies Equation 4.2.36, Equation 4.2.37 and Equation 4.2.39. Upon substituting Equation 4.2.40 into Equation 4.2.38, we obtain three integral equations:

$$\frac{2}{\pi}\int_0^\infty kA(k)\cos(kx)\, dk = 0, \qquad 0 < x < a, \tag{4.2.41}$$

$$\frac{2}{\pi}\int_0^\infty \tanh(kh)A(k)\cos(kx)\, dk = f(x), \qquad a < x < b, \tag{4.2.42}$$

and

$$\frac{2}{\pi}\int_0^\infty kA(k)\cos(kx)\, dk = 0, \qquad b < x < \infty. \tag{4.2.43}$$

Let us introduce

$$\frac{2}{\pi}\int_0^\infty kA(k)\cos(kx)\, dk = g(x)\sinh(cx), \qquad a < x < b, \tag{4.2.44}$$

where $g(x)$ is an unknown function and $c = \pi/(2h)$. Using Fourier's inversion theorem,

$$kA(k) = \int_a^b g(\tau)\sinh(c\tau)\cos(k\tau)\, d\tau. \tag{4.2.45}$$

If we substitute Equation 4.2.45 into Equation 4.2.42, interchange the order of integration in the resulting equation, and use

$$\int_0^\infty \cos(kx)\cos(k\tau)\tanh(kh)\,\frac{dk}{k} = \frac{1}{2}\ln\left|\frac{\cosh(cx)+\cosh(c\tau)}{\cosh(cx)-\cosh(c\tau)}\right|, \tag{4.2.46}$$

we find that $g(\tau)$ is given by the integral equation

$$\int_a^b g(\tau)\sinh(c\tau)\ln\left|\frac{\cosh(cx)+\cosh(c\tau)}{\cosh(cx)-\cosh(c\tau)}\right| d\tau = \pi f(x), \qquad a < x < b. \tag{4.2.47}$$

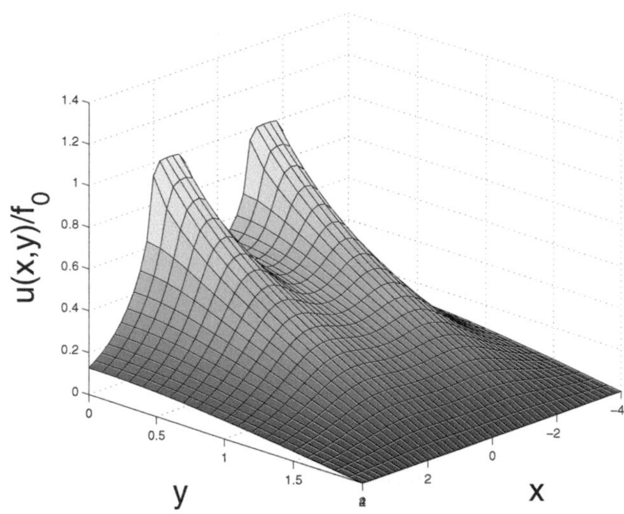

Figure 4.2.3: The solution to Equation 4.2.36 subject to the mixed boundary conditions given by Equation 4.2.37, Equation 4.2.38, and Equation 4.2.39 when $a = 1$, $b = 2$, $h = 2$ and $f(x) = f_0$.

Taking the derivative with respect to x of Equation 4.1.47, we find that

$$\int_a^b \frac{c\, g(\tau)\sinh(2c\tau)}{\cosh(2c\tau) - \cosh(2cx)}\, d\tau = \frac{\pi f'(x)}{2\sinh(cx)}, \qquad a < x < b. \tag{4.2.48}$$

The solution to Equation 4.1.48 is

$$\begin{aligned} g(\tau) = {} & -\frac{4}{\pi}\sqrt{\frac{\cosh(2c\tau) - \cosh(2ca)}{\cosh(2cb) - \cosh(2c\tau)}} \\ & \quad \times \int_a^b \frac{c\cosh(cx) f'(x)}{\cosh(2cx) - \cosh(2c\tau)} \sqrt{\frac{\cosh(2cb) - \cosh(2cx)}{\cosh(2cx) - \cosh(2ca)}}\, dx \\ & + \frac{B}{\sqrt{[\cosh(2c\tau) - \cosh(2ca)][\cosh(2cb) - \cosh(2c\tau)]}}. \end{aligned} \tag{4.2.49}$$

The constant B is found by substituting Equation 4.2.49 into Equation 4.2.47 and solving for B. Figure 4.2.3 illustrates this special case when $a = 1$, $b = 2$, $h = 2$ and $f(x) = f_0$.

In a similar manner, we can solve Equation 4.2.36, Equation 4.2.37 and Equation 4.2.39, plus the mixed boundary condition

$$\begin{cases} u_y(x,0) = 0, & |x| < a, \\ u(x,0) = \operatorname{sgn}(x) f(x), & a < |x| < b, \\ u_y(x,0) = 0, & b < |x|. \end{cases} \tag{4.2.50}$$

In the present case, the solution is given by

$$u(x,y) = \frac{2}{\pi}\int_0^\infty A(k)\frac{e^{-ky} - e^{ky-2kh}}{1+e^{-2kh}}\sin(kx)\,dk. \qquad (\mathbf{4.2.51})$$

Substituting Equation 4.2.51 into Equation 4.2.50,

$$\frac{2}{\pi}\int_0^\infty kA(k)\sin(kx)\,dk = 0, \qquad 0 < x < a, \quad b < x < \infty, \qquad (\mathbf{4.2.52})$$

and

$$\frac{2}{\pi}\int_0^\infty \tanh(kh)A(k)\sin(kx)\,dk = f(x), \qquad a < x < b. \qquad (\mathbf{4.2.53})$$

Let us introduce

$$\frac{2}{\pi}\int_0^\infty kA(k)\sin(kx)\,dk = \varphi(x)\cosh(cx), \qquad a < x < b, \qquad (\mathbf{4.2.54})$$

where $\varphi(x)$ is an unknown function and $c = \pi/(2h)$. Using Fourier's inversion theorem,

$$kA(k) = \int_a^b \varphi(\tau)\cosh(c\tau)\sin(k\tau)\,d\tau. \qquad (\mathbf{4.2.55})$$

If we substitute Equation 4.2.55 into Equation 4.2.53, interchange the order of integration in the resulting equation, and simplifying, $\varphi(\tau)$ is given by the integral equation

$$\int_a^b \varphi(\tau)\cosh(c\tau)\ln\left|\frac{\sinh(c\tau)+\sinh(cx)}{\sinh(c\tau)-\sinh(cx)}\right|\,d\tau = \pi f(x), \qquad a < x < b. \qquad (\mathbf{4.2.56})$$

Finally, taking the derivative with respect to x of Equation 4.2.56 and solving the resulting equation, we find that

$$\begin{aligned}\varphi(\tau) = {} & -\frac{4}{\pi}\sqrt{\frac{\cosh(2c\tau)-\cosh(2ca)}{\cosh(2cb)-\cosh(2c\tau)}} \\ & \times \int_a^b \frac{c\sinh(cx)f'(x)}{\cosh(2cx)-\cosh(2ca)}\sqrt{\frac{\cosh(2cb)-\cosh(2cx)}{\cosh(2cx)-\cosh(2ca)}}\,dx \\ & + \frac{B}{\sqrt{[\cosh(2c\tau)-\cosh(2ca)][\cosh(2cb)-\cosh(2c\tau)]}}. \end{aligned} \qquad (\mathbf{4.2.57})$$

The constant B is found by substituting Equation 4.2.57 into Equation 4.2.56 and solving for B. Figure 4.2.4 illustrates this special case when $a = 1$, $b = 2$, $h = 2$ and $f(x) = f_0$.

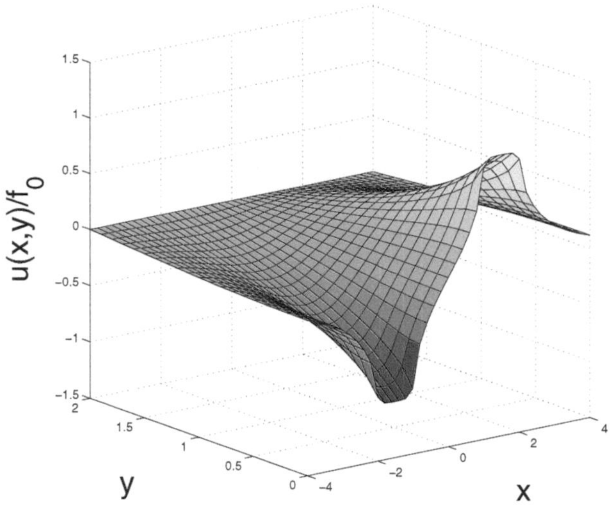

Figure 4.2.4: The solution to Equation 4.2.36 subject to the mixed boundary conditions given by Equation 4.2.37, Equation 4.2.50, and Equation 4.2.39 when $a = 1$, $b = 2$, $h = 2$ and $f(x) = f_0$.

4.3 DUAL FOURIER-BESSEL INTEGRALS

In the solution of mixed boundary value problems in domains where the radial direction extends to infinity, Hankel transforms are commonly used to solve these problems. Here we consider mixed boundary value problems that lead to dual integral equations.

• Example 4.3.1

One of the simplest mixed boundary value problems involving Hankel transforms is

$$\frac{\partial^2 u}{\partial r^2} + \frac{1}{r}\frac{\partial u}{\partial r} + \frac{\partial^2 u}{\partial z^2} = 0, \qquad 0 \le r < \infty, \quad 0 < z < \infty, \tag{4.3.1}$$

subject to the boundary conditions

$$\lim_{r\to 0} |u(r,z)| < \infty, \quad \lim_{r\to\infty} u(r,z) \to 0, \qquad 0 < z < \infty, \tag{4.3.2}$$

$$\lim_{z\to\infty} u(r,z) \to 0, \qquad 0 \le r < \infty, \tag{4.3.3}$$

and

$$\begin{cases} u(r,0) = f(r), & 0 \le r < 1, \\ u_z(r,0) = 0, & 1 \le r < \infty. \end{cases} \tag{4.3.4}$$

Using Hankel transforms, the solution to Equation 4.3.1 is

$$u(r,z) = \int_0^\infty A(k)e^{-kz} J_0(kr)k\,dk. \tag{4.3.5}$$

This solution satisfies not only Equation 4.3.1, but also Equation 4.3.2 and Equation 4.3.3. Substituting Equation 4.3.5 into Equation 4.3.4, we obtain the dual integral equations

$$\int_0^\infty A(k)k\, J_0(kr)\,dk = f(r), \qquad 0 \le r < 1, \tag{4.3.6}$$

and

$$\int_0^\infty A(k)k^2 J_0(kr)\,dk = 0, \qquad 1 < r < \infty. \tag{4.3.7}$$

Equation 4.3.6 and Equation 4.3.7 are special cases of a class of dual integral equations studied by Busbridge. See Equation 2.4.28 through Equation 2.4.30. Noting that $\nu = 0$ and $\alpha = -1$ and associating his $f(y)$ with $k^2 A(k)$, the solution to Equation 4.3.6 and Equation 4.3.7 is

$$\begin{aligned} kA(k) &= \frac{2}{\pi}\cos(k)\int_0^1 \frac{\eta f(\eta)}{\sqrt{1-\eta^2}}\,d\eta \\ &+ \frac{2}{\pi}\int_0^1 \frac{\eta}{\sqrt{1-\eta^2}}\left[\int_0^1 f(\eta\xi)k\xi\sin(k\xi)\,d\xi\right]d\eta. \end{aligned} \tag{4.3.8}$$

For the special case, $f(r) = C$, Equation 4.3.8 simplifies to

$$kA(k) = \frac{2C}{\pi}\frac{\sin(k)}{k}, \tag{4.3.9}$$

and

$$u(r,z) = \frac{2C}{\pi}\int_0^\infty \frac{\sin(k)}{k} e^{-kz} J_0(kr)\,dk. \tag{4.3.10}$$

An alternative form[33] of expressing the solution to Equation 4.3.1 through Equation 4.3.4 follows by rewriting Equation 4.3.5 as

$$u(r,z) = \int_0^\infty A(k)e^{-kz} J_0(kr)\,dk. \tag{4.3.11}$$

Substituting Equation 4.3.11 into Equation 4.3.4, we have that

$$\int_0^\infty A(k)J_0(kr)\,dk = f(r), \qquad 0 \le r < 1, \tag{4.3.12}$$

---

[33] See also Section 5.8 in Green, A. E., and W. Zerna, 1992: *Theoretical Elasticity*. Dover, 457 pp.; or Sneddon, op. cit., Section 7.5.

and

$$\int_0^\infty kA(k)J_0(kr)\,dk = 0, \qquad 1 < r < \infty. \tag{4.3.13}$$

Let us introduce

$$A(k) = \int_0^1 g(t)\cos(kt)\,dt = g(1)\frac{\sin(k)}{k} - \frac{1}{k}\int_0^1 g'(t)\sin(kt)\,dt. \tag{4.3.14}$$

Then,

$$\begin{aligned}\int_0^\infty kA(k)J_0(kr)\,dk &= g(1)\int_0^\infty \sin(kt)J_0(kr)\,dk \\ &\quad - \int_0^1 g'(t)\left[\int_0^\infty \sin(kt)J_0(kr)\,dk\right]dt.\end{aligned} \tag{4.3.15}$$

From Equation 1.4.13, these integrals vanish if $r > 1$ because $0 \le t \le 1$. Therefore, our choice for $A(k)$ given by Equation 4.3.14 satisfies Equation 4.3.13 identically. On the other hand, Equation 4.3.12 gives

$$\int_0^1 g(t)\left[\int_0^\infty \cos(kt)J_0(kr)\,dk\right]dt = f(r). \tag{4.3.16}$$

Using Equation 1.4.14, Equation 4.3.16 simplifies to

$$\int_0^r \frac{g(t)}{\sqrt{r^2 - t^2}}\,dt = f(r). \tag{4.3.17}$$

From Equation 1.2.13 and Equation 1.2.14, we obtain

$$g(t) = \frac{2}{\pi}\frac{d}{dt}\left[\int_0^t \frac{rf(r)}{\sqrt{t^2 - r^2}}\,dr\right]. \tag{4.3.18}$$

Next, substituting Equation 4.3.14 into Equation 4.3.11,

$$u(r,z) = \int_0^\infty \left[\int_0^1 g(t)\cos(kt)\,dt\right] e^{-kz}J_0(kr)\,dk \tag{4.3.19}$$

$$= \int_0^1 g(t)\left[\int_0^\infty e^{-kz}\cos(kz)J_0(kr)\,dk\right]dt \tag{4.3.20}$$

$$= \frac{1}{2}\int_0^1 \frac{g(t)}{\sqrt{r^2 + (z+it)^2}}\,dt + \frac{1}{2}\int_0^1 \frac{g(t)}{\sqrt{r^2 + (z-it)^2}}\,dt. \tag{4.3.21}$$

Recently, Fu et al.[34] showed that if $f(r)$ is a smooth function in $0 \le r < 1$, so that it can be experienced as the Maclaurin series

$$u(r,0) = h + \sum_{n=1}^\infty \frac{f^{(n)}(0)}{n!}r^n, \tag{4.3.22}$$

[34] Fu, G., T. Cao, and L. Cao, 2005: On the evaluation of the dopant concentration of a three-dimensional steady-state constant-source diffusion problem. *Mater. Lett.*, **59**, 3018–3020.

then

$$g(t) = \frac{2}{\sqrt{\pi}} \left[\frac{h}{\sqrt{\pi}} + \sum_{n=1}^{\infty} \frac{f^{(n)}(0)}{n!} \frac{\Gamma(1+n/2)}{\Gamma(1/2+n/2)} t^n \right]. \tag{4.3.23}$$

The solution of mixed boundary value problems in cylindrical coordinates often yields dual Fourier-Bessel integral equations of the form

$$\int_0^\infty G(k) A(k) J_\nu(kr)\, dk = r^\nu, \qquad 0 \le r < 1, \tag{4.3.24}$$

and

$$\int_0^\infty A(k) J_\nu(kr)\, dk = 0, \qquad 1 \le r < \infty, \tag{4.3.25}$$

where $G(k)$ is a known function of k. In 1956, Cooke[35] proved that the solution to Equation 4.3.24 and Equation 4.3.25 is

$$A(k) = \frac{2^\beta \Gamma(\nu+1)}{\Gamma(\nu-\beta+1)} k^{1+\beta} \int_0^1 f(t) t^{\alpha+1} J_{\nu-\beta}(kt)\, dt, \tag{4.3.26}$$

where $f(x)$ satisfies the integral equation

$$f(x) + x^{-\alpha} \int_0^1 t^{\alpha+1} f(t) \left\{ \int_0^\infty \left[a k^{2\beta} G(k) - 1 \right] k J_{\nu-\beta}(tk) J_{\nu-\beta}(xk)\, dk \right\} dt$$
$$= a x^{\nu-\alpha-\beta}. \tag{4.3.27}$$

Here, a, α and β are at our disposal as long as $0 < \Re(\beta) < 1$ and $-1 < \Re(\nu - \beta)$. Cooke suggested that we choose a and β so that $G(k)$ closely approximates $a^{-1}k^{-2\beta}$. The following examples illustrate the use of Cooke's method for solving Equation 4.3.24 and Equation 4.3.25 when they arise in mixed boundary value problems.

- **Example 4.3.2**

Let us solve[36]

$$\frac{\partial^2 u}{\partial r^2} + \frac{1}{r}\frac{\partial u}{\partial r} + \frac{\partial^2 u}{\partial z^2} = 0, \qquad 0 \le r < \infty, \quad 0 < z < a, \tag{4.3.28}$$

---

[35] Cooke, J. C., 1956: A solution of Tranter's dual integral equations problem. *Quart. J. Mech. Appl. Math.*, **9**, 103–110.

[36] See Leong, M. S., S. C. Choo, and K. H. Tay, 1976: The resistance of an infinite slab with a disc electrode as a mixed boundary value problem. *Solid-State Electron.*, **19**, 397–401. See also Belmont, B., and M. Shur, 1993: Spreading resistance of a round ohmic contact. *Solid-State Electron.*, **36**, 143–146.

subject to the boundary conditions

$$\lim_{r\to 0} |u(r,z)| < \infty, \quad \lim_{r\to\infty} u(r,z) \to 0, \qquad 0 < z < a, \tag{4.3.29}$$

$$\begin{cases} u(r,0) = 1, & 0 \le r < 1, \\ u_z(r,0) = 0, & 1 \le r < \infty, \end{cases} \tag{4.3.30}$$

and

$$u(r,a) = 0, \qquad 0 \le r < \infty. \tag{4.3.31}$$

Using Hankel transforms, the solution to Equation 4.3.28 is

$$u(r,z) = \int_0^\infty A(k) \frac{\sinh[k(a-z)]}{k \cosh(ak)} J_0(kr)\, dk. \tag{4.3.32}$$

This solution satisfies not only Equation 4.3.28, but also Equation 4.3.29 and Equation 4.3.31. Substituting Equation 4.3.32 into Equation 4.3.30, we obtain the dual integral equations

$$\int_0^\infty \frac{A(k)}{k} \tanh(ka) J_0(kr)\, dk = 1, \qquad 0 \le r < 1, \tag{4.3.33}$$

and

$$\int_0^\infty A(k) J_0(kr)\, dk = 0, \qquad 1 < r < \infty. \tag{4.3.34}$$

Because we can rewrite Equation 4.3.33 as

$$\int_0^\infty \frac{A(k)}{k} \left[1 + D(k)\right] J_0(kr)\, dk = 1, \qquad 0 \le r < 1, \tag{4.3.35}$$

where

$$D(k) = 2 \sum_{n=1}^\infty (-1)^n e^{-2nak}, \tag{4.3.36}$$

this suggests that we can apply Cooke's results if we set $a = 1$, $G(k) = [1 + D(k)]/k$, $\alpha = \beta = \frac{1}{2}$, and $\nu = 0$. From Equation 4.3.26, we have that

$$A(k) = \frac{2k}{\pi} \int_0^1 t \cos(kt) h(t)\, dt. \tag{4.3.37}$$

Turning to Equation 4.3.27, we find that

$$h(x) + \frac{4}{\pi x} \int_0^1 t\, h(t) \left[\int_0^\infty \sum_{n=1}^\infty (-1)^n e^{-2ank} \cos(kt) \cos(kx)\, dk \right] dt = \frac{1}{x}. \tag{4.3.38}$$

To derive Equation 4.3.38, we used the relation that

$$J_{-\frac{1}{2}}(z) = \sqrt{\frac{2}{\pi z}}\cos(z). \tag{4.3.39}$$

By interchanging the order of summation and integration in Equation 4.3.38, this equation simplifies to

$$h(x) + \int_0^1 h(\xi)K(x,\xi)\,d\xi = \frac{1}{x}, \tag{4.3.40}$$

where

$$K(x,\xi) = \frac{2\xi}{\pi x}\sum_{n=1}^{\infty}\left[\frac{2na(-1)^n}{4n^2a^2+(x-\xi)^2} + \frac{2na(-1)^n}{4n^2a^2+(x+\xi)^2}\right]. \tag{4.3.41}$$

Let $h(t) = f(t)/t$. Then, Equation 4.3.37 becomes

$$A(k) = \frac{2k}{\pi}\int_0^1 f(t)\cos(kt)\,dt. \tag{4.3.42}$$

Because $f(t)$ is an even function of t, we can rewrite Equation 4.3.40 in the more compact form of

$$f(t) + \frac{2}{\pi}\int_{-1}^1 f(\xi)\left[\sum_{n=1}^{\infty}\frac{2na(-1)^n}{4n^2a^2+(t-\xi)^2}\right]d\xi = 1. \tag{4.3.43}$$

Once we solve Equation 4.3.43, we substitute $f(t)$ into Equation 4.3.42 to obtain $A(k)$. This $A(k)$ can, in turn, be used in Equation 4.3.32 to find $u(r,z)$.

Equation 4.3.43 cannot be solved analytically and we must employ numerical techniques. Let us use MATLAB and show how this is done. We introduce nodal points at $t_j = (j - N/2)\Delta t$, where $j = 0, 1, \ldots, N$, and $\Delta t = 2/N$. Therefore, the first thing that we do in the MATLAB code is to compute the array for t_j:

```
% create arrays for t and ξ; N2 = N/2
for j = 0:N
t(j+1) = (j-N2)*dt; xi(j+1) = (j-N2)*dt; % dt = Δt
end
```

Next, using Simpson's rule, we replace Equation 4.3.43 with a set of $(N+1)\times(N+1)$ equations, which is expressed below as `AA*f=b'`. The corresponding MATLAB code is as follows:

```
for n = 0:N
tt = t(n+1);
b(n+1) = 1; % b' is the right side of the matrix equation
for m = 0:N
xxi = xi(m+1);
% ***************************************************************
% first term on the left side of Equation 4.3.43
% ***************************************************************
if (n==m) AA(n+1,m+1) = 1; else AA(n+1,m+1) = 0; end
% ***************************************************************
% summation inside of integral
% ***************************************************************
coeff = 0; sign = -1; sum = tt-xxi; sum2 = sum*sum;
for kk = 1:1000
anum = 2*kk*a;
coeff = coeff + sign*anum / (anum*anum+sum2);
sign = - sign;
end
% ***************************************************************
% approximate the integral by using Simpson's rule
% ***************************************************************
if ( (m>0) & (m<N) )
if ( mod(m+1,2)==0 )
AA(n+1,m+1) = AA(n+1,m+1) + 8*coeff*dt / (3*pi);
else
AA(n+1,m+1) = AA(n+1,m+1) + 4*coeff*dt / (3*pi);
end % end of inside logic loop
else
AA(n+1,m+1) = AA(n+1,m+1) + 2*coeff*dt / (3*pi);
end % end of outside logic loop
end % end of m loop
end % end of n loop
% ***************************************************************
% now find f(t_j), where t_j runs from -1 to 1
% ***************************************************************
f = AA\b';
```

Having computed $f(t_j)$, we are ready to compute $A(k)/k$ given by Equation 4.3.42. Note that we only need to retain those values of $f(t_j)$ where $t_j \geq 0$. This is done first. Because we plan to evaluate Equation 4.3.42 by using Simpson's rule, we also compute the coefficients and store them in the array `simpson`. The MATLAB code is as follows:

```
for n = 0:N2
t2(n+1) = n*dt; f2(n+1) = f(n+N2+1);
```

```
% ***********************************************************
% set up coefficients for Simpson's rule
% ***********************************************************
if ((n>0) & (n<N2))
if (mod(n+1,2)==0)
simpson(n+1) = 4*dt/3;
else
simpson(n+1) = 2*dt/3;
end
else
simpson(n+1) =   dt/3;
end; end
% ***********************************************************
% compute A(k)/k by using Simpson's rule
% ***********************************************************
for m = 0:M
A(m+1) = 0; k = m*dk;
for n = 0:N2
A(m+1) = A(m+1) + simpson(n+1)*f2(n+1)*cos(k*t2(n+1));
end
A(m+1) = 2*A(m+1)/pi;
end
```

We are now ready to compute the solution $u(r, z)$. For a given `r` and `z`, the solution `u` is computed as follows:

```
u = 0;
% use Simpson's rule to evaluate Equation 4.3.32
for m = 1:M
k = m*dk; factor = sinh(k*(a-z))*besselj(0,k*r)/cosh(k*a);
if (m<M)
if (mod(m+1,2) == 0)
u = u + 4*A(m+1)*factor;
else
u = u + 2*A(m+1)*factor;
end
else
u = u +   A(M+1)*factor;
end; end
u = dk*u/3;
```

Note that because the contribution from $k = 0$ is zero, we simply did not consider that case. Figure 4.3.1 illustrates $u(r, z)$ when $a = 1$. Schwarzbek

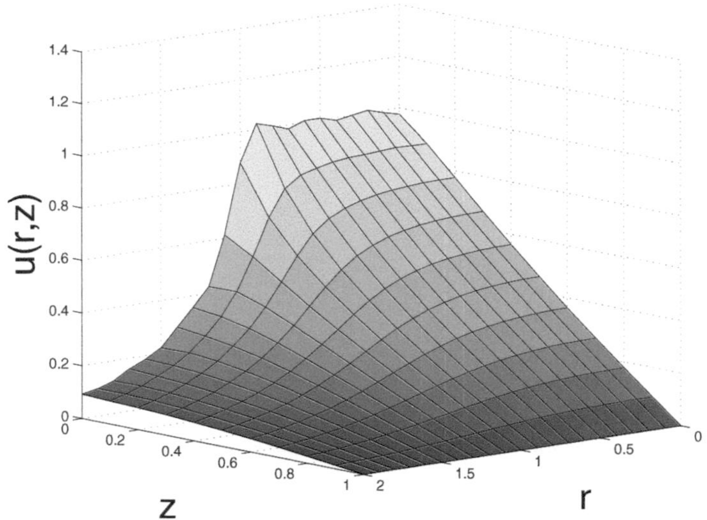

Figure 4.3.1: The solution $u(r, z)$ to the mixed boundary value problem governed by Equation 4.3.28 through Equation 4.3.31. The parameters used in this plot are $a = 1$, $M = 200$, $N = 20$, and $\Delta k = 0.1$.

and Ruggiero[37] employed this solution to calculate the effect of fringing fields on the measured resistance of a conducting film between two circular disks. Rossi and Nulman[38] used this analysis to model how a single circular flaw in a polymeric coated layer can reduce the protection to the underlying surface.

• Example 4.3.3

A similar problem[39] to the previous one is

$$\frac{\partial^2 u}{\partial r^2} + \frac{1}{r}\frac{\partial u}{\partial r} + \frac{\partial^2 u}{\partial z^2} = 0, \qquad 0 \le r < \infty, \quad 0 < z < a, \tag{4.3.44}$$

subject to the boundary conditions

$$\lim_{r\to 0} |u(r,z)| < \infty, \quad \lim_{r\to\infty} u(r,z) \to 0, \qquad 0 < z < a, \tag{4.3.45}$$

---

[37] Schwarzbek, S. M., and S. T. Ruggiero, 1986: The effect of fringing fields on the resistance of a conducting film. *IEEE Trans. Microwave Theory Tech.*, **MTT-34**, 977–981.

[38] Rossi, G., and M. Nulman, 1993: Effect of local flaws in polymeric permeation reducing barriers. *J. Appl. Phys.*, **74**, 5471–5475.

[39] See Chen, H., and J. C. M. Li, 2000: Anodic metal matrix removal rate in electrolytic in-process dressing. II: Protrusion effect and three-dimensional modeling. *J. Appl. Phys.*, **87**, 3159–3164. See also Yang, F.-Q., and J. C. M. Li, 1993: Impression creep of a thin film by vacancy diffusion. II. Cylindrical punch. *J. Appl. Phys.*, **74**, 4390–4397.

$$u(r,a) = 0, \qquad 0 \le r < \infty, \tag{4.3.46}$$

and

$$\begin{cases} u_z(r,0) = 1/a, & 0 \le r < 1, \\ u(r,0) = 0, & 1 \le r < \infty. \end{cases} \tag{4.3.47}$$

Using Hankel functions, the solution to Equation 4.3.44 is

$$u(r,z) = \int_0^\infty (ka)A(k)\frac{\sinh[k(z-a)]}{\sinh(ka)}J_0(kr)\,dk. \tag{4.3.48}$$

This solution satisfies not only Equation 4.3.44, but also Equation 4.3.45 and Equation 4.3.46. Substituting Equation 4.3.48 into Equation 4.3.47, we obtain the dual integral equations

$$\int_0^\infty (ka)^2 A(k)\coth(ka)J_0(kr)\,dk = 1, \qquad 0 \le r < 1, \tag{4.3.49}$$

and

$$\int_0^\infty (ka)A(k)J_0(kr)\,dk = 0, \qquad 1 < r < \infty. \tag{4.3.50}$$

Our solution of the dual integral equations, Equation 4.3.49 and Equation 4.3.50, begins by introducing the undetermined function $h(t)$ defined by

$$ka\,A(k) = \frac{2}{\pi a}\int_0^1 h(t)\sin(kt)\,dt, \qquad h(0) = 0. \tag{4.3.51}$$

The reason for introducing Equation 4.3.51 follows by substituting Equation 4.3.51 into Equation 4.3.50. We obtain

$$\begin{aligned} \frac{2}{\pi a}\int_0^\infty \left[\int_0^1 h(t)\sin(kt)\,dt\right] J_0(kr)\,dk & \\ &= \frac{2}{\pi a}\int_0^1 h(t)\left[\int_0^\infty \sin(kt)J_0(kr)\,dk\right]dt \qquad (4.3.52) \\ &= 0, \qquad (4.3.53) \end{aligned}$$

since the square bracketed term on the right side of Equation 4.3.52 vanishes. Therefore, Equation 4.3.50 is automatically satisfied.

To evaluate $h(t)$, we substitute Equation 4.3.51 into Equation 4.3.49 and find that

$$\frac{2}{\pi}\int_0^\infty k\coth(ka)\left[\int_0^1 h(t)\sin(kt)\,dt\right]J_0(kr)\,dk = 1, \tag{4.3.54}$$

or

$$\begin{aligned} \frac{2}{\pi}\int_0^\infty \left[\int_0^1 h(t)k\,\sin(kt)\,dt\right] J_0(kr)\,dk & \\ - \frac{2}{\pi}\int_0^\infty q(k)\left[\int_0^1 h(t)k\,\sin(kt)\,dt\right] J_0(kr)\,dk &= 1, \qquad (4.3.55) \end{aligned}$$

where $q(k) = 1 - \coth(ka)$. Because

$$\int_0^1 h(t)k\,\sin(kt)\,dt = -h(1)\cos(k) + \int_0^1 h'(t)\cos(kt)\,dt, \tag{4.3.56}$$

and

$$\int_0^\infty \cos(kt)J_0(kr)\,dk = \frac{H(r-t)}{\sqrt{r^2-t^2}}, \tag{4.3.57}$$

$$\int_0^\infty \left[\int_0^1 h(t)k\,\sin(kt)\,dt\right] J_0(kr)\,dk = \int_0^1 h'(t)\left[\int_0^\infty \cos(kt)J_0(kr)\,dk\right]dt \tag{4.3.58}$$

$$= \int_0^r \frac{h'(t)}{\sqrt{r^2-t^2}}\,dt. \tag{4.3.59}$$

Substituting Equation 4.3.59 into Equation 4.3.55,

$$\int_0^r \frac{h'(t)}{\sqrt{r^2-t^2}}\,dt - \int_0^1 h(t)\left[\int_0^\infty q(k)k\sin(kt)J_0(kr)\,dk\right]dt = \frac{\pi}{2}. \tag{4.3.60}$$

If we now define

$$f(r) = \int_0^r \frac{h'(t)}{\sqrt{r^2-t^2}}\,dt, \tag{4.3.61}$$

we have from Equation 1.2.13 and Equation 1.2.14 with $\alpha = \frac{1}{2}$ that

$$h'(t) = \frac{2}{\pi}\frac{d}{dt}\left[\int_0^t \frac{\eta f(\eta)}{\sqrt{t^2-\eta^2}}\,d\eta\right], \tag{4.3.62}$$

or

$$h(t) = \frac{2}{\pi}\int_0^t \frac{\eta f(\eta)}{\sqrt{t^2-\eta^2}}d\eta. \tag{4.3.63}$$

Therefore, Equation 4.3.60 simplifies to

$$f(r) - \int_0^1 h(t)\left[\int_0^\infty q(k)k\sin(kt)J_0(kr)\,dk\right]dt = \frac{\pi}{2}. \tag{4.3.64}$$

Setting r equal to η in Equation 4.3.64, multiplying it by $\eta/\sqrt{x^2-\eta^2}$ and integrating the resulting equation from 0 to x, we obtain

$$h(x) - \int_0^1 h(t)\left\{\frac{2}{\pi}\int_0^\infty q(k)k\sin(kt)\left[\int_0^x \frac{\eta J_0(k\eta)}{\sqrt{x^2-\eta^2}}\,d\eta\right]dk\right\}dt$$
$$= \int_0^x \frac{\eta}{\sqrt{x^2-\eta^2}}\,d\eta, \tag{4.3.65}$$

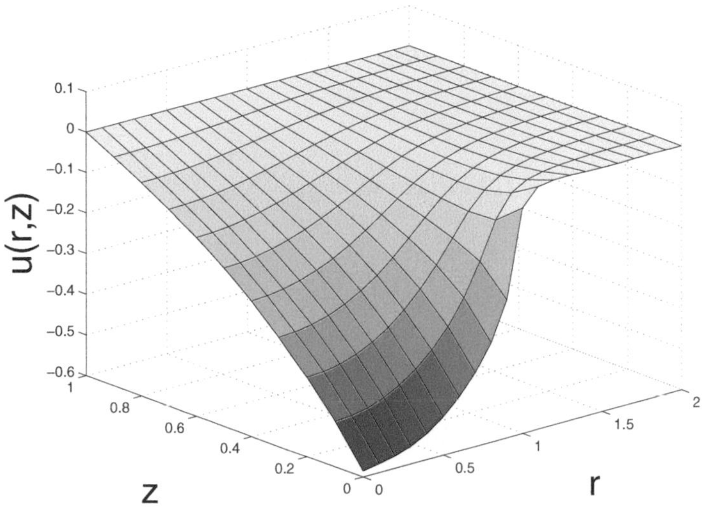

Figure 4.3.2: The solution $u(r,z)$ to the mixed boundary value problem governed by Equation 4.3.44 through Equation 4.3.47 with $a = 1$.

or

$$h(x) - \int_0^1 h(t) \left[\frac{2}{\pi} \int_0^\infty q(k) \sin(kx) \sin(kt)\, dk \right] dt = - \left. \sqrt{x^2 - \eta^2} \right|_0^x = x. \tag{4.3.66}$$

As in the previous example, we must solve for $h(x)$ numerically. The MATLAB code is very similar with the exception that the kernel in Equation 4.3.43 is replaced with a numerical integration of $\int_0^\infty q(k)\sin(kx)\sin(kt)\,dk$ by using Simpson's rule. This integral is easily evaluated due to the nature of $q(k)$. Figure 4.3.2 illustrates the solution $u(r,z)$ when $a = 1$.

Let us now generalize our results. We now wish to solve

$$\frac{\partial^2 u}{\partial r^2} + \frac{1}{r}\frac{\partial u}{\partial r} + \frac{\partial^2 u}{\partial z^2} = 0, \qquad 0 \le r < \infty, \quad 0 < z < h, \tag{4.3.67}$$

subject to the boundary conditions

$$\lim_{r\to 0} |u(r,z)| < \infty, \quad \lim_{r\to\infty} u(r,z) \to 0, \qquad 0 < z < h, \tag{4.3.68}$$

$$u(r,h) = 0, \qquad 0 \le r < \infty, \tag{4.3.69}$$

and

$$\begin{cases} u_z(r,0) = -g(r), & 0 \le r < a, \\ u(r,0) = 0, & a \le r < \infty. \end{cases} \tag{4.3.70}$$

The solution to Equation 4.3.67 through Equation 4.3.70 is

$$u(r,z) = \int_0^\infty A(k) \frac{\sinh[k(z-h)]}{\sinh(kh)} J_0(kr)\, dk, \tag{4.3.71}$$

with

$$\int_0^\infty kA(k) \coth(kh) J_0(kr)\, dk = g(r), \qquad 0 \le r < a, \tag{4.3.72}$$

and

$$\int_0^\infty A(k) J_0(kr)\, dk = 0, \qquad a < r < \infty. \tag{4.3.73}$$

We can rewrite Equation 4.3.72 as

$$\int_0^\infty kA(k)[1+G(k)] J_0(kr)\, dk = g(r), \qquad 0 \le r < a, \tag{4.3.74}$$

where $G(k) = 2\sum_{n=1}^\infty e^{-2nkh}$.

Our solution of the dual integral equations, Equation 4.3.73 and Equation 4.3.74, starts with the introduction of

$$A(k) = \int_0^a h(t) \sin(kt)\, dt, \qquad h(0) = 0, \tag{4.3.75}$$

or

$$kA(k) = -h(a)\cos(ka) + \int_0^a h'(t) \cos(kt)\, dt. \tag{4.3.76}$$

We can show that Equation 4.3.75 satisfies Equation 4.3.73 in the same manner as we did earlier. See Equation 4.3.52.

Now, Equation 4.3.74 can be rewritten

$$\int_0^\infty kA(k) J_0(kr)\, dk = g(r) - \int_0^\infty kA(k) G(k) J_0(kr)\, dk. \tag{4.3.77}$$

Substituting Equation 4.3.76 into Equation 4.3.77 and interchanging the order of integration, we have that

$$\begin{aligned} \int_0^a h'(t) \left[\int_0^\infty \cos(kt) J_0(kr)\, dk\right] dt - h(a) \int_0^\infty \cos(kt) J_0(kr)\, dk \\ = g(r) - \int_0^a h(t) \left[\int_0^\infty k\sin(kt) G(k) J_0(kr)\, dk\right] dt, \end{aligned} \tag{4.3.78}$$

for $0 < r < a$. If we employ Equation 1.4.14 to simplify Equation 4.3.78, then

$$\int_0^r \frac{h'(t)}{\sqrt{r^2-t^2}} dt = g(r) - \int_0^a h(\tau) \left[\int_0^\infty k\, G(k) \sin(k\tau) J_0(kr)\, dk\right] d\tau. \tag{4.3.79}$$

Applying Equation 1.2.13 and Equation 1.2.14,

$$h'(t) = \frac{2}{\pi}\frac{d}{dt}\left[\int_0^t \frac{r\,g(r)}{\sqrt{t^2-r^2}}\,dr\right] \tag{4.3.80}$$
$$-\frac{2}{\pi}\frac{d}{dt}\left(\int_0^t \left\{\int_0^a h(\tau)\left[\int_0^\infty kG(k)\sin(k\tau)J_0(kr)\,dk\right]d\tau\right\}\frac{r\,dr}{\sqrt{t^2-r^2}}\right).$$

Integrating Equation 4.3.80 with respect to t, we obtain the integral equation

$$h(t) = \frac{2}{\pi}\int_0^t \frac{rg(r)}{\sqrt{t^2-r^2}}\,dr - \frac{1}{\pi}\int_0^a K(t,\tau)h(\tau)\,d\tau, \tag{4.3.81}$$

where

$$K(t,\tau) = 2\int_0^t \left[\int_0^\infty kG(k)\sin(k\tau)J_0(kr)\,dk\right]\frac{r\,dr}{\sqrt{t^2-r^2}} \tag{4.3.82}$$
$$= 2\int_0^\infty kG(k)\left[\int_0^t \frac{rJ_0(kr)}{\sqrt{t^2-r^2}}\,dr\right]\sin(k\tau)\,dk \tag{4.3.83}$$
$$= 2\int_0^\infty G(k)\sin(kt)\sin(k\tau)\,dk \tag{4.3.84}$$
$$= \int_0^\infty G(k)\{\cos[k(t-\tau)] - \cos[k(t+\tau)]\}\,dk. \tag{4.3.85}$$

To compute $u(r,z)$, we must first solve the integral equation, Equation 4.3.81, then evaluate $A(k)$ using values of $h(t)$ via Equation 4.3.75, and finally employ Equation 4.3.71.

• Example 4.3.4

Let us solve[40]

$$\frac{\partial^2 u}{\partial r^2} + \frac{1}{r}\frac{\partial u}{\partial r} + \frac{\partial^2 u}{\partial z^2} = 0, \qquad 0 \le r < \infty, \quad 0 < z < \infty, \tag{4.3.86}$$

subject to the boundary conditions

$$\lim_{r\to 0}|u(r,z)| < \infty, \qquad \lim_{r\to\infty} u(r,z) \to 0, \qquad 0 < z < \infty, \tag{4.3.87}$$

$$\lim_{z\to\infty} u(r,z) \to 1, \qquad 0 \le r < \infty, \tag{4.3.88}$$

$$u(r,0) = 0, \qquad 0 \le r < \infty, \tag{4.3.89}$$

[40] See Lebedev, N. N., 1957: The electrostatic field of an immersion electron lens formed by two diaphragms. *Sov. Tech. Phys.*, **2**, 1943–1950.

and

$$\begin{cases} u_z(r,b^-) = u_z(r,b^+), & 0 \le r < a, \\ u(r,b) = 1, & a < r < \infty, \end{cases} \tag{4.3.90}$$

where b^- and b^+ denote points located slightly below and above the point $z = b > 0$.

Using transform methods or separation of variables, the general solution to Equation 4.3.86 through Equation 4.3.89 is

$$u(r,z) = \frac{z}{b} - \int_0^\infty A(k)\frac{\sinh(kz)}{\sinh(kb)} J_0(kr)\,dk, \quad 0 \le z \le b, \tag{4.3.91}$$

and

$$u(r,z) = 1 - \int_0^\infty A(k)e^{-k(z-b)} J_0(kr)\,dk, \qquad b \le z < \infty. \tag{4.3.92}$$

Substituting Equation 4.3.91 and Equation 4.3.92 into Equation 4.3.90, we have that

$$\int_0^\infty \frac{2kb}{1-e^{-2kb}} A(k)J_0(kr)\,dk = 1, \qquad 0 \le r < a, \tag{4.3.93}$$

and

$$\int_0^\infty A(k)J_0(kr)\,dk = 0, \qquad a < r < \infty. \tag{4.3.94}$$

To solve the dual integral equations, Equation 4.3.93 and Equation 4.3.94, we set

$$kA(k) = \frac{1}{2b}\int_0^a h(t)[\cos(kt) - \cos(ka)]\,dt. \tag{4.3.95}$$

We have chosen this definition for $A(k)$ because

$$\int_0^\infty A(k)J_0(kr)\,dk = \frac{1}{2b}\int_0^a h(t)\left\{\int_0^\infty [\cos(kt) - \cos(ka)]J_0(kr)\,\frac{dk}{k}\right\} dt = 0 \tag{4.3.96}$$

since $0 \le t \le a < r$. This follows from integrating Equation 1.4.13 with respect to t from 0 and a after setting $\nu = 0$ and noting that $a < r$.

Turning to Equation 4.3.93, we substitute Equation 4.3.95 into Equation 4.3.93. This yields

$$\int_0^a h(t)\left[\int_0^\infty \frac{\cos(kt) - \cos(ka)}{1-e^{-2kb}} J_0(kr)\,dk\right] dt = 1, \quad 0 \le r < a, \tag{4.3.97}$$

or

$$\begin{aligned} &\int_0^a h(t)\left\{\int_0^\infty [\cos(kt) - \cos(ka)]J_0(kr)\,dk\right\} dt \\ &\quad + \int_0^a h(t)\left\{\int_0^\infty \frac{e^{-2kb}}{1-e^{-2kb}}[\cos(kt) - \cos(ka)]J_0(kr)\,dk\right\} dt = 1. \end{aligned} \tag{4.3.98}$$

We can rewrite Equation 4.3.98 as

$$\int_0^r \frac{h(t)}{\sqrt{r^2-t^2}}\,dt + \int_0^{\pi/2}\int_0^a h(\tau)K[r\sin(\theta),\tau]\,d\tau\,d\theta = 1, \tag{4.3.99}$$

where

$$\begin{aligned} K[r\sin(\theta),\tau] &= \frac{2}{\pi}\int_0^{\pi/2}\int_0^\infty \frac{e^{-2kb}}{1-e^{-2kb}}[\cos(kt)-\cos(ka)]\cos[kr\sin(\theta)]\,dk\,d\theta \qquad (4.3.100)\\ &= \frac{2}{\pi}\int_0^{\pi/2}\int_0^\infty \frac{e^{-2kb}}{1-e^{-2kb}}\Big(\cos\{k[t-r\sin(\theta)]\}-\cos\{k[a-r\sin(\theta)]\}\\ &\qquad + \cos\{k[t+r\sin(\theta)]\}-\cos\{k[t+r\sin(\theta)]\}\Big)\,dk\,d\theta. \qquad (4.3.101)\end{aligned}$$

We have also used the integral representation[41] for $J_0(kr)$,

$$J_0(kr) = \frac{2}{\pi}\int_0^{\pi/2}\cos[kr\sin(\theta)]\,d\theta. \tag{4.3.102}$$

Introducing the logarithmic derivative of the gamma function,

$$\psi(z) = -\gamma + \int_0^\infty \frac{e^{-t}-e^{-tz}}{1-e^{-t}}\,dt, \qquad \Re(z)>0, \tag{4.3.103}$$

where γ denotes Euler's constant, we have that

$$\begin{aligned} K(t,\tau) = \frac{1}{2\pi b}\Re\Big\{&\psi\left[1+i\left(\frac{a-t}{2b}\right)\right]-\psi\left[1+i\left(\frac{\tau-t}{2b}\right)\right]\\ &+\psi\left[1+i\left(\frac{a+t}{2b}\right)\right]-\psi\left[1+i\left(\frac{\tau+t}{2b}\right)\right]\Big\}. \end{aligned} \tag{4.3.104}$$

Because

$$\int_0^r \frac{h(t)}{\sqrt{r^2-t^2}}\,dt = \int_0^{\pi/2} h[r\sin(\theta)]\,d\theta, \tag{4.3.105}$$

Equation 4.3.99 can be rewritten

$$\int_0^{\pi/2}\left\{h[r\sin(\theta)] + \int_0^a h(\tau)K[r\sin(\theta),\tau]\,d\tau\right\}d\theta = 1. \tag{4.3.106}$$

---

[41] Gradshteyn and Ryzhik, op. cit., Formula 8.411.1 with $n=0$.

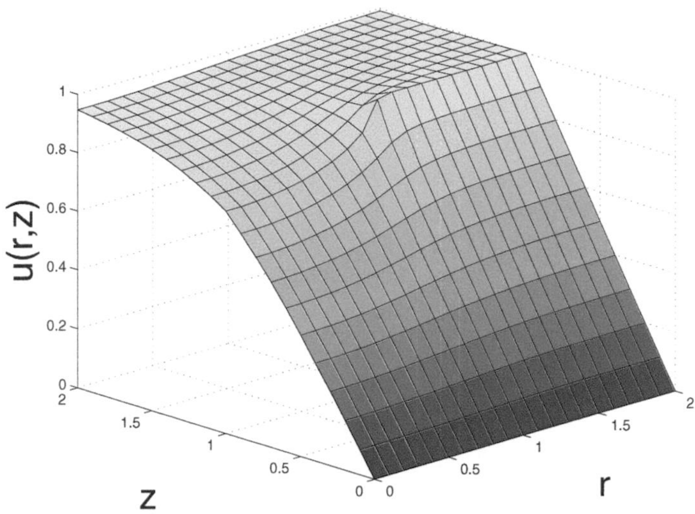

Figure 4.3.3: The solution $u(r,z)$ to the mixed boundary value problem governed by Equation 4.3.86 through Equation 4.3.90 with $a = b = 1$.

Equation 4.3.106 is satisfied if

$$h(t) + \int_0^a h(\tau)K(t,\tau)\,d\tau = \frac{2}{\pi}, \qquad 0 \le t \le a. \tag{4.3.107}$$

Figure 4.3.3 illustrates the solution $u(r,z)$ when $a = b = 1$. We first solve Equation 4.3.107 to find $h(t)$. Then Equation 4.3.95 gives $A(k)$ via $h(t)$. Finally, $u(r,z)$ follows from Equation 4.3.91 or Equation 4.3.92.

• Example 4.3.5

In the previous examples, the domain has been within a cylinder of given radius. Here we solve[42] Laplace's equation when the domain lies outside of a unit cylinder:

$$\frac{\partial^2 u}{\partial r^2} + \frac{1}{r}\frac{\partial u}{\partial r} + \frac{\partial^2 u}{\partial z^2} = 0, \qquad 1 \le r < \infty, \quad 0 < z < \infty, \tag{4.3.108}$$

subject to the boundary conditions

$$u_r(1,z) = 0, \qquad \lim_{r\to\infty} u(r,z) \to 0, \qquad 0 < z < \infty, \tag{4.3.109}$$

---

[42] See Srivastav, R. P., and P. Narain, 1966: Stress distribution due to pressurized exterior crack in an infinite isotropic elastic medium with coaxial cylindrical cavity. *Int. J. Engng. Sci.*, **4**, 689–697.

$$\lim_{z\to\infty} u(r,z) \to 0, \qquad 1 \le r < \infty, \tag{4.3.110}$$

and

$$\begin{cases} u_z(r,0) = 1, & 1 \le r < a, \\ u_{zz}(r,0) = f(r), & a < r < \infty. \end{cases} \tag{4.3.111}$$

Using transform methods or separation of variables, the general solution to Equation 4.3.108, Equation 4.3.109, and Equation 4.3.110 is

$$u(r,z) = \int_0^\infty A(k)e^{-kz} \frac{J_0(kr)Y_1(k) - Y_0(kr)J_1(k)}{Y_1^2(k) + J_1^2(k)} \frac{dk}{k}. \tag{4.3.112}$$

Substituting Equation 4.3.112 into Equation 4.3.111, we find that

$$\int_0^\infty A(k) \frac{J_0(kr)Y_1(k) - Y_0(kr)J_1(k)}{Y_1^2(k) + J_1^2(k)} \, dk = 0, \tag{4.3.113}$$

and

$$\int_0^\infty kA(k) \frac{J_0(kr)Y_1(k) - Y_0(kr)J_1(k)}{Y_1^2(k) + J_1^2(k)} \, dk = f(r). \tag{4.3.114}$$

To solve Equation 4.3.113 and Equation 4.3.114, let us introduce a $g(t)$ such that

$$\int_0^\infty A(k) \frac{J_0(kr)Y_1(k) - Y_0(kr)J_1(k)}{Y_1^2(k) + J_1^2(k)} \, dk = \int_a^r \frac{g(t)}{\sqrt{r^2 - t^2}} \, dt, \qquad a < r < \infty. \tag{4.3.115}$$

Now, from Weber's formula[43] the integral equation

$$F(y) = \int_1^\infty x f(x) [J_0(xy)Y_1(y) - Y_0(xy)J_1(y)] \, dx \tag{4.3.116}$$

has the solution

$$f(x) = \int_0^\infty yF(y) \frac{J_0(xy)Y_1(y) - Y_0(xy)J_1(y)}{Y_1^2(y) + J_1^2(y)} \, dy. \tag{4.3.117}$$

Therefore, from Equation 4.3.113 and Equation 4.3.115,

$$A(k) = \int_a^\infty g(\xi)[Y_1(k)\cos(k\xi) - J_1(k)\sin(k\xi)] \, d\xi. \tag{4.3.118}$$

If we multiply Equation 4.3.114 by $r/\sqrt{r^2 - t^2}$ and integrate from t to ∞, we obtain

$$\frac{\partial}{\partial t}\left[\int_0^\infty A(k) \frac{J_1(k)\cos(kt) + Y_1(k)\sin(kt)}{Y_1^2(k) + J_1^2(k)} \frac{dk}{k}\right] = \int_t^\infty \frac{rf(r)}{\sqrt{r^2 - t^2}} \, dr. \tag{4.3.119}$$

---

[43] Titchmarsh, E. C., 1946: *Eigenfunction Expansions Associated with Second Order Differential Equations. Part I.* Oxford, 203 pp. See Section 4.10.

Substituting for $A(k)$ from Equation 4.3.118 into Equation 4.3.119, we find after interchanging the order of integration that

$$\frac{\partial}{\partial t}\left[\int_a^\infty g(\xi) Q(\xi,t)\,d\xi\right] = \int_t^\infty \frac{r f(r)}{\sqrt{r^2-t^2}}\,dr, \tag{4.3.120}$$

where

$$\begin{aligned} Q(\xi,t) = \int_0^\infty & \frac{[J_1(k)\cos(kt) + Y_1(k)\sin(kt)]}{J_1^2(k)+Y_1^2(k)} \\ & \times [Y_1(k)\cos(k\xi) - J_1(k\xi)\sin(k\xi)]\,\frac{dk}{k}, \end{aligned} \tag{4.3.121}$$

or

$$\begin{aligned} Q(\xi,t) = -\frac{1}{2}\Bigg(&\int_0^\infty \sin[k(\xi-t)]\,\frac{dk}{k} - \int_0^\infty \sin[k(\xi+t)]\,\frac{dk}{k} \\ &+ \Im\left\{\int_0^\infty \left[\frac{H_1^{(2)}(k)}{H_1^{(1)}(k)} + 1\right] e^{i(\xi+t)k}\,\frac{dk}{k}\right\}\Bigg). \end{aligned} \tag{4.3.122}$$

Our final task is to evaluate the contour integral

$$\int_\Gamma \left[\frac{H_1^{(2)}(z)}{H_1^{(1)}(z)} + 1\right] e^{i(\xi+t)z}\,\frac{dz}{z},$$

where the contour Γ consists of the real axis from the origin to R, an arc in the first quadrant $|z| = R$, $0 \le \theta \le \pi/2$, and imaginary axis from iR to the origin. As $R \to \infty$, we find that

$$\Im\left\{\int_0^\infty \left[\frac{H_1^{(2)}(k)}{H_1^{(1)}(k)} + 1\right] e^{i(\xi+t)k}\,\frac{dk}{k}\right\} = -\pi\int_0^\infty \frac{I_1(k)}{K_1(k)} e^{-k(t+\xi)}\,\frac{dk}{k}. \tag{4.3.123}$$

Therefore, Equation 4.3.120 becomes

$$\begin{aligned} \frac{1}{2}\frac{\partial}{\partial t}\Bigg\{ & -\frac{\pi}{2}\int_a^t g(\xi)\,d\xi + \frac{\pi}{2}\int_t^\infty g(\xi)\,d\xi \\ & - \int_a^\infty g(\xi)\left[\pi\int_0^\infty \frac{I_1(k)}{K_1(k)} e^{-k(t+\xi)}\,\frac{dk}{k}\right] d\xi\Bigg\} = \int_t^\infty \frac{r f(r)}{\sqrt{r^2-t^2}}\,dr, \end{aligned} \tag{4.3.124}$$

or

$$g(t) + \int_a^\infty g(\xi)\left[\int_0^\infty \frac{I_1(k)}{K_1(k)} e^{-k(t+\xi)}\,dk\right] d\xi = \frac{2}{\pi}\int_t^\infty \frac{r f(r)}{\sqrt{r^2-t^2}}\,dr, \tag{4.3.125}$$

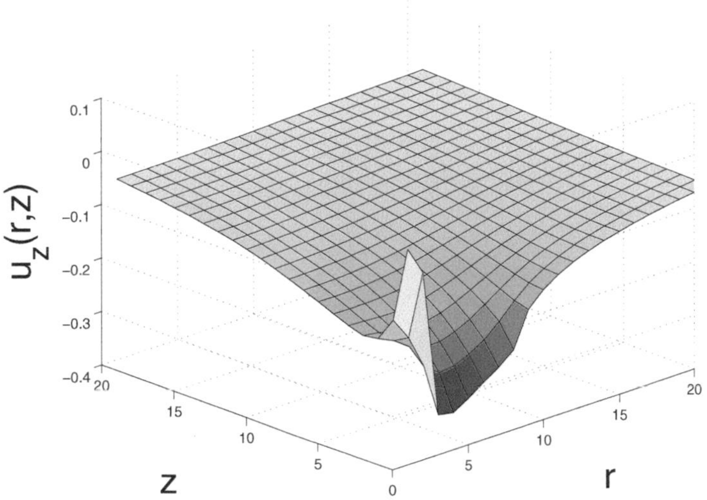

Figure 4.3.4: The solution $u(r,z)$ to the mixed boundary value problem governed by Equation 4.3.108 through Equation 4.3.111 with $a = 2$.

for $a < t < \infty$. Figure 4.3.4 illustrates the solution when $a = 2$.

• Example 4.3.6

Let us solve[44]

$$\frac{\partial^2 u}{\partial r^2} + \frac{1}{r}\frac{\partial u}{\partial r} + \frac{1}{z^p}\frac{\partial}{\partial z}\left(z^n \frac{\partial u}{\partial z}\right) = 0, \qquad 0 \le r < \infty, \quad 0 < z < \infty, \quad \textbf{(4.3.126)}$$

subject to the boundary conditions

$$\lim_{r\to 0} |u(r,z)| < \infty, \qquad \lim_{r\to\infty} u(r,z) \to 0, \qquad 0 < z < \infty, \qquad \textbf{(4.3.127)}$$

$$\lim_{z\to\infty} u(r,z) \to 0, \qquad 0 \le r < \infty, \qquad \textbf{(4.3.128)}$$

and

$$\begin{cases} u(r,0) = 1, & 0 \le r < 1, \\ z^n u_z(r,z)\big|_{z=0} = 0, & 1 < r < \infty, \end{cases} \qquad \textbf{(4.3.129)}$$

where $\kappa > 0$.

---

[44] Taken from Brutsaert, W., 1967: Evaporation from a very small water surface at ground level: Three-dimensional turbulent diffusion without convection. *J. Geophys. Res.*, **72**, 5631–5639. ©1967 American Geophysical Union. Reproduced/modified by permission of American Geophysical Union.

Using transform methods or separation of variables, the general solution to Equation 4.3.126, Equation 4.3.127, and Equation 4.3.128 is

$$u(r,z) = z^{(1-n)/2} \int_0^\infty A(k) K_\nu \left[\frac{2\nu k}{1-n} z^{(1-n)/(2\nu)} \right] J_0(kr)\, dk, \qquad \textbf{(4.3.130)}$$

where $\nu = (1-n)/(p-n+2)$. Substituting Equation 4.3.130 into Equation 4.3.129, we have that

$$\int_0^\infty A(k) J_0(kr) \frac{dk}{k^\nu} = C, \qquad 0 \le r < 1, \qquad \textbf{(4.3.131)}$$

and

$$\int_0^\infty k^\nu A(k) J_0(kr)\, dk = 0, \qquad 1 < r < \infty, \qquad \textbf{(4.3.132)}$$

where

$$C = \frac{2}{\Gamma(\nu)} \left(\frac{\nu}{1-n} \right)^\nu. \qquad \textbf{(4.3.133)}$$

If we now restrict ν so that it lies between 0 and $\frac{1}{4}$, then

$$A(k) = \frac{(2k)^\nu C}{\Gamma(1-\nu)} \left\{ k^{1-\nu} J_{-\nu}(k) \int_0^1 \frac{\eta}{(1-\eta^2)^\nu}\, d\eta \right.$$
$$\left. + \left[\int_0^1 \frac{\zeta}{(1-\zeta^2)^\nu}\, d\zeta \right] \left[\int_0^1 (k\eta)^{2-\nu} J_{1-\nu}(k\eta)\, d\eta \right] \right\} \qquad \textbf{(4.3.134)}$$

$$= \frac{2^{\nu-1} k\, C}{\Gamma(2-\nu)} \left[J_{-\nu}(k) + J_{2-\nu}(k) \right] \qquad \textbf{(4.3.135)}$$

$$= 2^{\nu+1} \left(\frac{\nu}{1-n} \right)^\nu \frac{\sin(\nu\pi)}{\pi} J_{1-\nu}(k). \qquad \textbf{(4.3.136)}$$

Consequently, the final solution is

$$u(r,z) = \left[2^{\nu+1} \left(\frac{\nu}{1-n} \right)^\nu \frac{\sin(\nu\pi)}{\pi} \right] z^{(1-n)/2}$$
$$\times \int_0^\infty K_\nu \left[\frac{2\nu k}{1-n} z^{(1-n)/(2\nu)} \right] J_0(kr) J_{1-\nu}(k)\, dk. \qquad \textbf{(4.3.137)}$$

Figure 4.3.5 illustrates this solution when $n = \frac{1}{2}$ and $p = 1$.

• Example 4.3.7

Let us solve[45]

$$\frac{\partial^2 u}{\partial r^2} + \frac{1}{r}\frac{\partial u}{\partial r} - \frac{u}{r^2} + \frac{\partial^2 u}{\partial z^2} = \kappa^2 u, \qquad 0 \le r < \infty, \quad 0 < z < \infty, \qquad \textbf{(4.3.138)}$$

---

[45] A simplified version of a problem solved by Borodachev, N. M., and Yu. A. Mamteyew, 1969: Unsteady torsional oscillations of an elastic half-space. *Mech. Solids*, **4(1)**, 79–83.

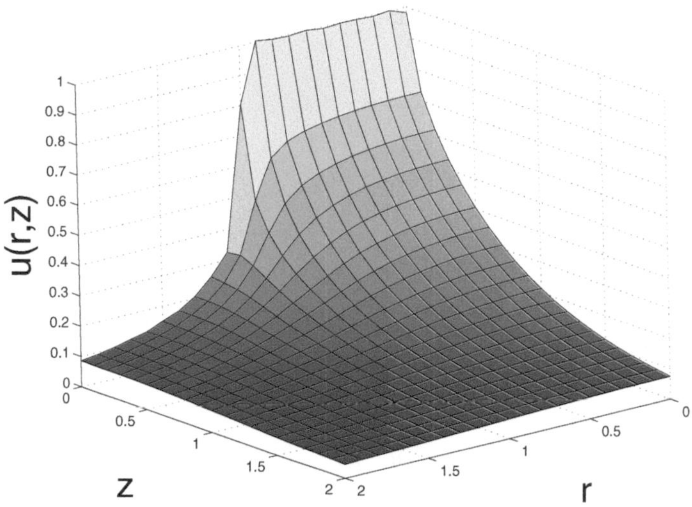

Figure 4.3.5: The solution $u(r,z)$ to the mixed boundary value problem governed by Equation 4.3.126 through Equation 4.3.129 with $n = \frac{1}{2}$ and $p = 1$.

subject to the boundary conditions

$$\lim_{r\to 0} |u(r,z)| < \infty, \qquad \lim_{r\to\infty} u(r,z) \to 0, \qquad 0 < z < \infty, \tag{4.3.139}$$

$$\lim_{z\to\infty} u(r,z) \to 0, \qquad 0 \le r < \infty, \tag{4.3.140}$$

and

$$\begin{cases} u(r,0) = r, & 0 \le r < a, \\ u_z(r,0) = 0, & a < r < \infty, \end{cases} \tag{4.3.141}$$

where $\kappa > 0$.

Using transform methods or separation of variables, the general solution to Equation 4.3.138, Equation 4.3.139, and Equation 4.3.140 is

$$u(r,z) = \int_0^\infty A(k) J_1(kr) e^{-z\sqrt{k^2+\kappa^2}}\, dk. \tag{4.3.142}$$

Substituting Equation 4.3.142 into Equation 4.3.141, we have that

$$\int_0^\infty A(k) J_1(kr)\, dk = r, \qquad 0 \le r < a, \tag{4.3.143}$$

and

$$\int_0^\infty \sqrt{k^2+\kappa^2}\, A(k) J_1(kr)\, dk = 0, \qquad a < r < \infty. \tag{4.3.144}$$

Setting $x = r/a$, $\xi = ka$, and $g(\xi) = \sqrt{\xi^2 + (\kappa a)^2}\, A(\xi)$ in Equation 4.3.143 and Equation 4.3.144, we find that

$$\int_0^\infty \frac{g(\xi)}{\sqrt{\xi^2 + (\kappa a)^2}} J_1(\xi x)\, d\xi = x, \qquad 0 \le x < 1, \tag{4.3.145}$$

and

$$\int_0^\infty g(\xi) J_1(\xi x)\, d\xi = 0, \qquad 1 < x < \infty. \tag{4.3.146}$$

By comparing our problem with the canonical form given by Equation 4.3.26 through Equation 4.3.27, then $\nu = 1$ and $G(\xi) = \left[\xi^2 + (\kappa a)^2\right]^{-1/2}$. Selecting $a = 1$, $\alpha = -\frac{1}{2}$, and $\beta = \frac{1}{2}$, then

$$g(\xi) = \frac{4\xi}{\pi} \int_0^1 h(t) \sin(\xi t)\, dt, \tag{4.3.147}$$

and

$$h(t) + \int_0^1 K(t, \eta) h(\eta)\, d\eta = t, \qquad 0 \le t \le 1, \tag{4.3.148}$$

where

$$K(t,\eta) = \frac{2}{\pi} \int_0^1 \left[1 - \frac{\xi}{\sqrt{\xi^2 + (\kappa a)^2}}\right] \sin(t\xi)\sin(\eta\xi)\, d\xi \tag{4.3.149}$$

$$= \frac{2}{\pi} \int_0^1 \left[\frac{\sqrt{\xi^2 - (\kappa a)^2} - \xi}{\sqrt{\xi^2 + (\kappa a)^2}}\right] \sin(t\xi)\sin(\eta\xi)\, d\xi \tag{4.3.150}$$

$$= \frac{2}{\pi}(\kappa a)^2 \int_0^1 \frac{\sin(t\xi)\sin(\eta\xi)}{\sqrt{\xi^2 + (\kappa a)^2}\left[\xi + \sqrt{\xi^2 + (\kappa a)^2}\right]}\, d\xi \tag{4.3.151}$$

$$= \frac{\kappa a}{2} \{ L_1[\kappa a(\eta + t)] - I_1[\kappa a(\eta + t)] - L_1[\kappa a|\eta - t|\,] + I_1[\kappa a|\eta - t|\,]\}, \tag{4.3.152}$$

where $L_1(\cdot)$ denotes a modified Struve function of the first kind. Vasudevaiah and Majhi[46] showed how to evaluate the integral in Equation 4.3.151.

As in the previous examples, we must solve for $h(x)$ numerically. Then $g(\xi)$ is computed from Equation 4.3.147. Finally, Equation 4.3.142 gives $u(r, z)$. Figure 4.3.6 illustrates this solution when $\kappa a = 1$.

---

[46] Vasudevaiah, M., and S. N. Majhi, 1981: Viscous impulsive rotation of two finite coaxial disks. *Indian J. Pure Appl. Math.*, **12**, 1027–1042.

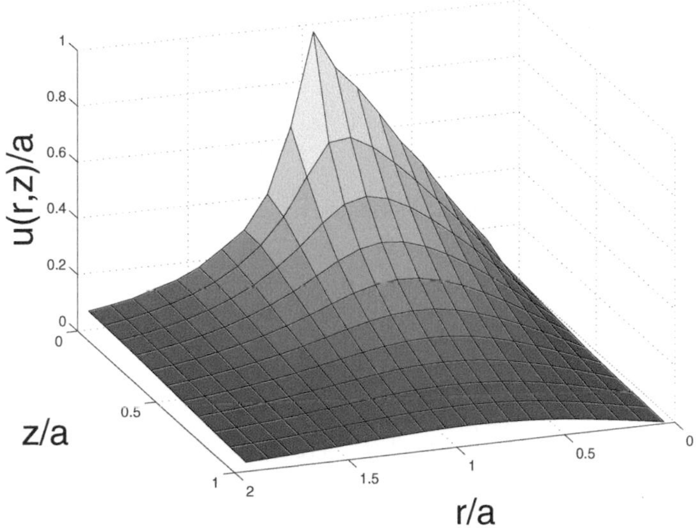

Figure 4.3.6: The solution $u(r,z)$ to the mixed boundary value problem governed by Equation 4.3.138 through Equation 4.3.141 with $\kappa a = 1$.

• Example 4.3.8

Let us solve[47]

$$\frac{\partial^2 u}{\partial r^2} + \frac{1}{r}\frac{\partial u}{\partial r} + \frac{\partial^2 u}{\partial z^2} + \alpha^2 u = 0, \qquad 0 \le r < \infty, \quad -\infty < z < \infty, \quad \textbf{(4.3.153)}$$

subject to the boundary conditions

$$\lim_{r\to 0} |u(r,z)| < \infty, \qquad \lim_{r\to\infty} u(r,z) \to 0, \qquad -\infty < z < \infty, \qquad \textbf{(4.3.154)}$$

$$\lim_{|z|\to\infty} u(r,z) \to 0, \qquad 0 \le r < \infty, \qquad \textbf{(4.3.155)}$$

and

$$\begin{cases} u_z(r,0^-) = u_z(r,0^+) = 1, & 0 \le r < a, \\ u(r,0^-) = u(r,0^+), & a < r < \infty. \end{cases} \qquad \textbf{(4.3.156)}$$

Using transform methods or separation of variables, the general solution to Equation 4.3.153, Equation 4.3.154, and Equation 4.3.155 is

$$u(r,z) = \mp \int_0^\infty A(k) J_0(kr) e^{-|z|\sqrt{k^2-\alpha^2}}\, dk. \qquad \textbf{(4.3.157)}$$

---

[47] See Lebedev, N. N., and I. P. Skal'skaya, 1959: A new method for solving the problem of the diffraction of electromagnetic waves by a thin conducting disk. *Sov. Tech. Phys.*, **4**, 627–637.

Substituting Equation 4.3.157 into Equation 4.3.156, we have that

$$\int_0^\infty \sqrt{k^2-\alpha^2}\,A(k)J_0(kr)\,dk = 1, \qquad 0 \le r < a, \tag{4.3.158}$$

and

$$\int_0^\infty A(k)J_0(kr)\,dk = 0, \qquad a < r < \infty. \tag{4.3.159}$$

To solve the dual integral equations, Equation 4.3.158 and Equation 4.3.159, we set

$$kA(k) = \frac{2}{\pi}\int_0^a h(t)[\cos(kt)-\cos(ka)]\,dt. \tag{4.3.160}$$

We chose this definition for $A(k)$ because

$$\int_0^\infty A(k)J_0(kr)\,dk = \frac{2}{\pi}\int_0^a h(t)\left\{\int_0^\infty [\cos(kt)-\cos(ka)]J_0(kr)\,\frac{dk}{k}\right\}dt = 0, \tag{4.3.161}$$

where we have integrated Equation 1.4.13 with respect to t from 0 and a after setting $\nu = 0$ and noted that $0 \le t \le a < r$.

Turning to Equation 4.3.158, we substitute Equation 4.3.160 into Equation 4.3.158. This yields

$$\int_0^a h(t)\left\{\frac{2}{\pi}\int_0^\infty \sqrt{k^2-\alpha^2}\,[\cos(kt)-\cos(ka)]J_0(kr)\,\frac{dk}{k}\right\}dt = 1, \ 0 \le r < a, \tag{4.3.162}$$

or

$$\begin{aligned}\int_0^a h(t)&\left\{\frac{2}{\pi}\int_0^\infty [\cos(kt)-\cos(ka)]J_0(kr)\,dk\right\}dt \\ &-\int_0^a h(t)\left\{\frac{2}{\pi}\int_0^\infty\left(1-\frac{\sqrt{k^2-\alpha^2}}{k}\right)[\cos(kt)-\cos(ka)]J_0(kr)\,dk\right\}dt = 1.\end{aligned} \tag{4.3.163}$$

Let us evaluate

$$\begin{aligned}&\frac{2}{\pi}\int_0^\infty\left(1-\frac{\sqrt{k^2-\alpha^2}}{k}\right)[\cos(kt)-\cos(ka)]J_0(kr)\,dk \\ &\quad= \frac{4}{\pi^2}\int_0^{\pi/2}\int_0^\infty\left(1-\frac{\sqrt{k^2-\alpha^2}}{k}\right)[\cos(k\tau)-\cos(ka)]\cos[kr\sin(\theta)]\,dk\,d\theta\end{aligned} \tag{4.3.164}$$

$$\begin{aligned}&\quad= \frac{2}{\pi^2}\int_0^{\pi/2}\int_0^\infty\left(1-\frac{\sqrt{k^2-\alpha^2}}{k}\right)\Big(\cos\{k[\tau-r\sin(\theta)]\}-\cos\{k[a-r\sin(\theta)]\} \\ &\qquad+\cos\{k[t+r\sin(\theta)]\}-\cos\{k[a+r\sin(\theta)]\}\Big)\,dk\,d\theta.\end{aligned} \tag{4.3.165}$$

We used the integral definition of $J_0(kr)$ to obtain Equation 4.3.164.

Consider now the integral

$$L = \frac{2}{\pi}\int_0^\infty \left(1 - \frac{\sqrt{k^2-\kappa^2}}{k}\right)[\cos(k\alpha) - \cos(k\beta)]\,dk, \quad \alpha,\beta > 0. \quad \textbf{(4.3.166)}$$

Then

$$\frac{\partial L}{\partial \alpha} = -\frac{2}{\pi}\int_0^\infty \left(k - \sqrt{k^2-\kappa^2}\right)\sin(k\alpha)\,dk \quad \textbf{(4.3.167)}$$

$$= \frac{2}{\pi\alpha}\int_0^\infty \left(k - \sqrt{k^2-\kappa^2}\right)d[\cos(k\alpha)] \quad \textbf{(4.3.168)}$$

$$= \frac{2}{\pi\alpha}\left[i\kappa - \int_0^\infty \left(1 - \frac{k}{\sqrt{k^2-\kappa^2}}\right)\cos(k\alpha)\,dk\right] \quad \textbf{(4.3.169)}$$

$$= \frac{2}{\pi\alpha}\left\{i\kappa - \frac{d}{d\alpha}\left[\int_0^\infty \frac{\sin(k\alpha)}{k}\,dk\right] + \frac{d}{d\alpha}\left[\int_0^\infty \frac{\sin(k\alpha)}{\sqrt{k^2-\kappa^2}}\,dk\right]\right\} \quad \textbf{(4.3.170)}$$

$$= \frac{2}{\pi\alpha}\left\{i\kappa + \frac{d}{d\alpha}\left[\int_0^\infty \frac{\sin(k\alpha)}{\sqrt{k^2-\kappa^2}}\,dk\right]\right\}. \quad \textbf{(4.3.171)}$$

Using the integral representation for the Bessel and Struve functions[48]

$$J_0(x) = \frac{2}{\pi}\int_1^\infty \frac{\sin(xt)}{\sqrt{t^2-1}}\,dt, \qquad H_0(x) = \frac{2}{\pi}\int_0^1 \frac{\sin(xt)}{\sqrt{1-t^2}}\,dt, \quad \textbf{(4.3.172)}$$

with $J_0'(x) = -J_1(x)$ and $H_0'(x) = 2/\pi - H_1(x)$, we obtain the final result that

$$\frac{\partial L}{\partial \alpha} = -\frac{\kappa}{\alpha}\left[J_1(\kappa\alpha) - iH_1(\kappa\alpha)\right]. \quad \textbf{(4.3.173)}$$

Upon integrating Equation 4.3.173 with respect to α and noting that $L = 0$ when $\alpha = \beta$, we find that

$$\frac{2}{\pi}\int_0^\infty \left(1 - \frac{\sqrt{k^2-\kappa^2}}{k}\right)[\cos(k\alpha) - \cos(k\beta)]\,dk = \kappa\Big\{Ji_1(\kappa\beta) - Ji_1(\kappa\alpha) - i\left[Hi_1(\kappa\beta) - Hi_1(\kappa\alpha)\right]\Big\}, \quad \textbf{(4.3.174)}$$

if $\alpha, \beta > 0$, where

$$Ji_1(x) = \int_0^x J_1(y)\,\frac{dy}{y}, \quad \text{and} \quad Hi_1(x) = \int_0^x H_1(y)\,\frac{dy}{y}. \quad \textbf{(4.3.175)}$$

---

[48] Gradshteyn and Ryzhik, op. cit., Formula 8.411.9 and Formula 8.551.1.

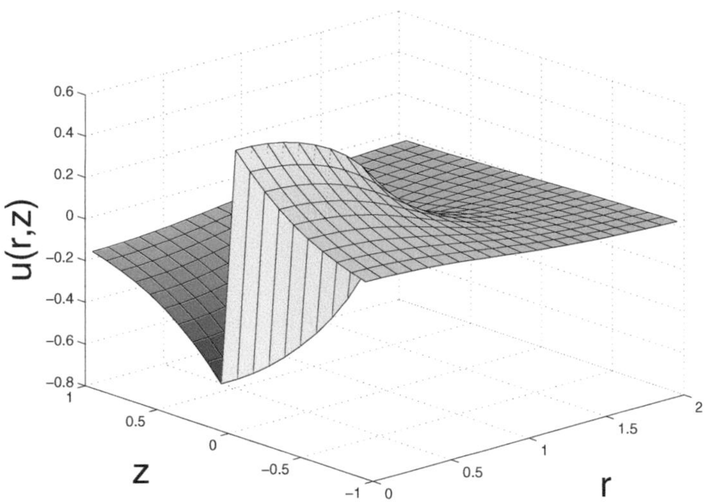

Figure 4.3.7: The solution $u(r,z)$ to the mixed boundary value problem governed by Equation 4.3.153 through Equation 4.3.156 with $a = 1$ and $\alpha = 0.1$.

Applying these results to Equation 4.3.163, we have

$$\frac{2}{\pi}\int_0^r \frac{h(t)}{\sqrt{r^2-t^2}}\,dt - \frac{2\alpha}{\pi}\int_0^{\pi/2}\int_0^a K[r\sin(\theta),\tau]h(\tau)\,d\tau\,d\theta = 1 \qquad (\mathbf{4.3.176})$$

with

$$K(r,\tau) = \tfrac{1}{2}\Big\{ Ji_1[\alpha(a-r)] - Ji_1[\alpha|t-r|] + Ji_1[\alpha(a+r)] - Ji_1[\alpha(t+r)] \\ - iHi_1[\alpha(a-r)] + iHi_1[\alpha|t-r|] - iHi_1[\alpha(a+r)] + iHi_1[\alpha(t+r)]\Big\}. \qquad (\mathbf{4.3.177})$$

Because

$$\int_0^r \frac{h(t)}{\sqrt{r^2-t^2}}\,dt = \int_0^{\pi/2} h[r\sin(\theta)]\,d\theta, \qquad (\mathbf{4.3.178})$$

Equation 4.3.176 can be rewritten

$$\frac{2}{\pi}\int_0^{\pi/2}\left\{h[r\sin(\theta)] - \alpha\int_0^a K[r\sin(\theta),\tau]h(\tau)\,d\tau\right\}d\theta = 1, \qquad 0 \le r < a. \qquad (\mathbf{4.3.179})$$

Equation 4.3.179 is satisfied if

$$h(x) - \alpha\int_0^a K(x,\tau)h(\tau)\,d\tau = 1, \qquad 0 \le x \le a. \qquad (\mathbf{4.3.180})$$

Figure 4.3.7 illustrates $u(r,z)$ when $a = 1$ and $\alpha = 0.1$.

In a similar manner,[49] we can solve

$$\frac{\partial^2 u}{\partial r^2} + \frac{1}{r}\frac{\partial u}{\partial r} + \frac{\partial^2 u}{\partial z^2} + \left(\alpha^2 - \frac{1}{r^2}\right) u = 0, \qquad 0 \le r < \infty, \quad -\infty < z < \infty, \tag{4.3.181}$$

subject to the boundary conditions

$$\lim_{r\to 0} |u(r,z)| < \infty, \qquad \lim_{r\to\infty} u(r,z) \to 0, \qquad -\infty < z < \infty, \tag{4.3.182}$$

$$\lim_{|z|\to\infty} u(r,z) \to 0, \qquad 0 \le r < \infty, \tag{4.3.183}$$

and

$$\begin{cases} u(r,0^-) = u(r,0^+) = r, & 0 \le r < a, \\ u_z(r,0^-) = u_z(r,0^+), & a < r < \infty. \end{cases} \tag{4.3.184}$$

Using transform methods or separation of variables, the general solution to Equation 4.3.181, Equation 4.3.182, and Equation 4.3.183 is

$$u(r,z) = \int_0^\infty A(k) J_1(kr) e^{-|z|\sqrt{k^2-\alpha^2}}\, dk. \tag{4.3.185}$$

Substituting Equation 4.3.185 into Equation 4.3.184, we have that

$$\int_0^\infty A(k) J_1(kr)\, dk = r, \qquad 0 \le r < a, \tag{4.3.186}$$

and

$$\int_0^\infty \sqrt{k^2-\alpha^2}\, A(k) J_1(kr)\, dk = 0, \qquad a < r < \infty. \tag{4.3.187}$$

We can satisfy Equation 4.3.187 identically if we set

$$A(k) = \frac{2k}{\pi\sqrt{k^2-\alpha^2}} \int_0^a h(t) \sin(kt)\, dt, \tag{4.3.188}$$

because

$$\int_0^\infty \sqrt{k^2-\alpha^2}\, A(k) J_1(kr)\, dk = \int_0^a h(t) \left[\int_0^\infty k \sin(kt) J_1(kr)\, dk\right] dt \tag{4.3.189}$$

$$= -\int_0^1 h(t) \frac{d}{dr}\left[\int_0^\infty \sin(kt) J_0(kr)\, dk\right] dt = 0 \tag{4.3.190}$$

---

[49] See also Ufliand, Ia. S., 1961: On torsional vibrations of half-space. *J. Appl. Math. Mech.*, **25**, 228–233.

since the integral within the square brackets vanishes in Equation 4.3.190 when $0 \le t \le a < r$.

Turning Equation 4.3.186, we substitute Equation 4.3.188 into it. This yields

$$\int_0^a h(t) \left[\frac{2}{\pi} \int_0^\infty \frac{k}{\sqrt{k^2 - \alpha^2}} \sin(kt) J_1(kr)\, dk \right] dt = r, \quad 0 \le r < a, \tag{4.3.191}$$

or

$$\int_0^a h(t) \left[\frac{2}{\pi} \int_0^\infty \sin(kt) J_1(kr)\, dk \right] dt \tag{4.3.192}$$
$$- \int_0^a h(t) \left[\frac{2}{\pi} \int_0^\infty \left(1 - \frac{k}{\sqrt{k^2 - \alpha^2}}\right) \sin(kt) J_1(kr)\, dk \right] dt = r.$$

Let us evaluate

$$\frac{2}{\pi} \int_0^\infty \left(1 - \frac{k}{\sqrt{k^2 - \alpha^2}}\right) \sin(kt) J_1(kr)\, dk$$
$$= \frac{4}{\pi^2} \int_0^{\pi/2} \sin(\theta) \left\{ \int_0^\infty \left(1 - \frac{k}{\sqrt{k^2 - \alpha^2}}\right) \sin(kt) \sin[kr\sin(\theta)]\, dk \right\} d\theta \tag{4.3.193}$$
$$= \frac{2}{\pi^2} \int_0^{\pi/2} \sin(\theta) \left[\int_0^\infty \left(1 - \frac{k}{\sqrt{k^2 - \alpha^2}}\right) \left(\cos\{k[\tau - r\sin(\theta)]\} - \cos\{k[t + r\sin(\theta)]\} \right) dk \right] d\theta. \tag{4.3.194}$$

We used the integral definition of $J_1(kr)$ to obtain Equation 4.3.193.

Consider now the integral

$$L = \frac{2}{\pi} \int_0^\infty \left(1 - \frac{k}{\sqrt{k^2 - \kappa^2}}\right) \cos(k\alpha)\, dk, \quad \alpha > 0. \tag{4.3.195}$$

Then,

$$L = \frac{d}{d\alpha} \left[\frac{2}{\pi} \int_0^\infty \left(1 - \frac{k}{\sqrt{k^2 - \kappa^2}}\right) \sin(k\alpha)\, \frac{dk}{k} \right] \tag{4.3.196}$$
$$= -\frac{d}{d\alpha} \left[\frac{2}{\pi} \int_0^\infty \frac{\sin(k\alpha)}{\sqrt{k^2 - \kappa^2}}\, dk \right] \tag{4.3.197}$$
$$= \kappa \left[J_1(\kappa\alpha) - iH_1(\kappa\alpha) + \frac{2i}{\pi} \right]. \tag{4.3.198}$$

Applying these results to Equation 4.3.194, we find

$$\frac{2}{\pi r} \int_0^r \frac{t\, h(t)}{\sqrt{r^2 - t^2}}\, dt - \frac{2\alpha}{\pi} \int_0^{\pi/2} \left\{ \int_0^a K[r\sin(\theta), \tau] h(\tau)\, d\tau \right\} \sin(\theta)\, d\theta = r \tag{4.3.199}$$

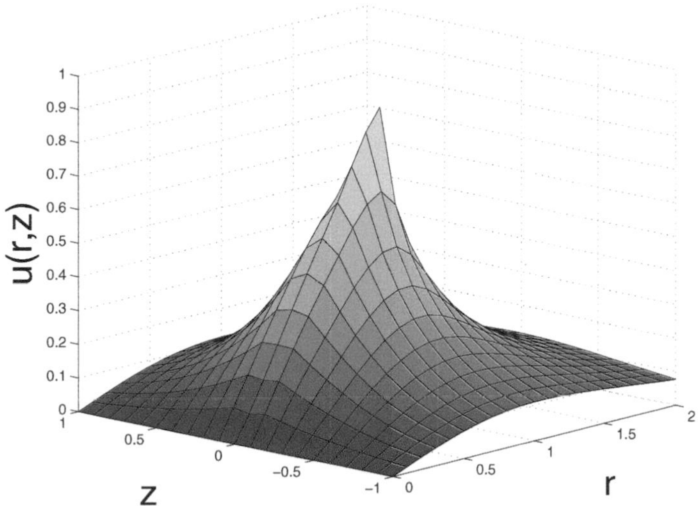

Figure 4.3.8: The solution $u(r,z)$ to the mixed boundary value problem governed by Equation 4.3.181 through Equation 4.3.184 with $a = 1$ and $\alpha = 0.1$.

with

$$K(r,\tau) = \tfrac{1}{2}\left\{J_1[\alpha|\tau - r|] - J_1[\alpha(t+r)] - iH_1[\alpha|t-r|] + iH_1[\alpha(t+r)]\right\}. \quad \textbf{(4.3.200)}$$

Because

$$\frac{2}{\pi r}\int_0^r \frac{t\,h(t)}{\sqrt{r^2-t^2}}\,dt = \frac{2}{\pi}\int_0^{\pi/2} h[r\sin(\theta)]\sin(\theta)\,d\theta, \quad \textbf{(4.3.201)}$$

Equation 4.3.199 can be written

$$\frac{2}{\pi}\int_0^{\pi/2}\left\{h[r\sin(\theta)] - \alpha\int_0^a K[r\sin(\theta),\tau]h(\tau)\,d\tau\right\}\sin(\theta)\,d\theta = r \quad \textbf{(4.3.202)}$$

with $0 \le r < a$. If we set

$$h(x) - \alpha\int_0^a K(x,\tau)h(\tau)\,d\tau = f(x), \qquad 0 \le x \le a, \quad \textbf{(4.3.203)}$$

then Equation 4.3.202 becomes

$$\frac{2}{\pi}\int_0^{\pi/2} r\sin(\theta)f[r\sin(\theta)]\,d\theta = r^2 \quad \textbf{(4.3.204)}$$

which has the solution $f(x) = 2x$. Therefore, $h(t)$ is given by

$$h(x) - \alpha\int_0^a K(x,\tau)h(\tau)\,d\tau = 2x, \qquad 0 \le x \le a. \quad \textbf{(4.3.205)}$$

Figure 4.3.8 illustrates $u(r,z)$ when $a = 1$ and $\alpha = 0.1$.

• **Example 4.3.9**

Let us solve[50]

$$\frac{\partial^2 u}{\partial r^2} + \frac{1}{r}\frac{\partial u}{\partial r} + \frac{\partial^2 u}{\partial z^2} - \alpha^2 u = 0, \qquad 0 \le r < \infty, \quad -h < z < \infty, \quad (\mathbf{4.3.206})$$

subject to the boundary conditions

$$\lim_{r\to 0} |u(r,z)| < \infty, \qquad \lim_{r\to\infty} u(r,z) \to 0, \qquad -h < z < \infty, \quad (\mathbf{4.3.207})$$

$$\lim_{z\to\infty} u(r,z) \to 0, \qquad 0 \le r < \infty, \quad (\mathbf{4.3.208})$$

$$u_r(r,-h) = 0, \qquad 0 \le r < \infty, \quad (\mathbf{4.3.209})$$

$$u_r(r,0^-) = u_r(r,0^+), \qquad 0 \le r < \infty, \quad (\mathbf{4.3.210})$$

and

$$\begin{cases} u_r(r,0^-) = u_r(r,0^+) = -r, & 0 \le r < 1, \\ u_{rz}(r,0^-) = u_{rz}(r,0^+), & 1 < r < \infty. \end{cases} \quad (\mathbf{4.3.211})$$

Using transform methods or separation of variables, the general solution to Equation 4.3.206 through Equation 4.3.210 is

$$u(r,z) = \int_0^\infty A(k) e^{-\lambda z} J_0(kr)\,\frac{dk}{k}, \qquad 0 < z < \infty, \quad (\mathbf{4.3.212})$$

and

$$u(r,z) = \int_0^\infty A(k) \frac{\sinh[\lambda(z+h)]}{\sinh(\lambda h)} J_0(kr)\,\frac{dk}{k}, \qquad -h < z < 0, \quad (\mathbf{4.3.213})$$

where $\lambda = \sqrt{k^2+\alpha^2}$ with $\Re(\lambda) > 0$. Substituting Equation 4.3.212 and Equation 4.3.213 into Equation 4.3.211, we have that

$$\int_0^\infty B(k)\left(1 - e^{-2\lambda h}\right) J_1(kr)\,\frac{dk}{\lambda} = -2r, \qquad 0 \le r < 1, \quad (\mathbf{4.3.214})$$

and

$$\int_0^\infty B(k) J_1(kr)\,dk = 0, \qquad 1 \le r < \infty, \quad (\mathbf{4.3.215})$$

where

$$B(k) = -\frac{\lambda e^{\lambda h} A(k)}{\sinh(\lambda h)}. \quad (\mathbf{4.3.216})$$

---

[50] See Chu, J. H., and M.-U. Kim, 2004: Oscillatory Stokes flow due to motions of a circular disk parallel to an infinite plane wall. *Fluid Dyn. Res.*, **34**, 77–97.

To solve the dual integral equations, Equation 4.3.214 and Equation 4.3.215, we introduce

$$B(k) = k \int_0^1 h(t) \sin(kt)\, dt. \qquad \textbf{(4.3.217)}$$

Following Equation 4.3.189 and Equation 4.3.190, we can show that this choice satisfies Equation 4.3.215 identically.

Turning to Equation 4.3.214, we substitute Equation 4.3.217 into Equation 4.3.214. This yields

$$\begin{aligned} &\int_0^1 h(t) \left[\int_0^\infty \sin(kt) J_1(kr)\, dk \right] dt \qquad \textbf{(4.3.218)} \\ &\quad - \int_0^1 h(t) \left[\sin(kt) J_1(kr) - \frac{k}{\lambda} \left(1 - e^{-2\lambda h}\right) \sin(kt) J_1(kr)\, dk \right] dt = -2r. \end{aligned}$$

From integral tables,[51]

$$\int_0^\infty J_1(\alpha x) \sin(\beta x)\, dx = \begin{cases} \frac{\beta}{\alpha\sqrt{\alpha^2-\beta^2}}, & \alpha > \beta, \\ 0, & \alpha < \beta, \end{cases} \qquad \textbf{(4.3.219)}$$

we can evaluate the first term in Equation 4.3.218 and this equation now reads

$$\begin{aligned} &\int_0^r \frac{t\, h(t)}{r\sqrt{r^2 - t^2}}\, dt \qquad \textbf{(4.3.220)} \\ &\quad = \int_0^1 h(\tau) \left[\int_0^\infty \left(1 - \frac{k}{\lambda} + \frac{k}{\lambda} e^{-2\lambda h}\right) \sin(k\tau) J_1(kr)\, dk \right] d\tau - 2r. \end{aligned}$$

Upon applying the results from Equation 1.2.13 and Equation 1.2.14,

$$\begin{aligned} r\, h(r) = \frac{2}{\pi} \int_0^1 h(\tau) \Bigg\{ \int_0^\infty & \left(1 - \frac{k}{\lambda} + \frac{k}{\lambda} e^{-2\lambda h}\right) \\ & \times \sin(k\tau) \frac{d}{dr} \left[\int_0^r \frac{\xi^2 J_1(k\xi)}{\sqrt{r^2 - \xi^2}}\, d\xi \right] dk \Bigg\}\, d\tau \\ - \frac{4}{\pi} \frac{d}{dr} \left(\int_0^r \frac{\xi^3}{\sqrt{r^2 - \xi^2}}\, d\xi \right). & \qquad \textbf{(4.3.221)} \end{aligned}$$

Now

$$-\frac{4}{\pi} \frac{d}{dr} \left(\int_0^r \frac{\xi^3}{\sqrt{r^2 - \xi^2}}\, d\xi \right) = -\frac{8r^2}{\pi}, \qquad \textbf{(4.3.222)}$$

---

[51] Gradshteyn and Ryzhik, op. cit., Formula 6.671.

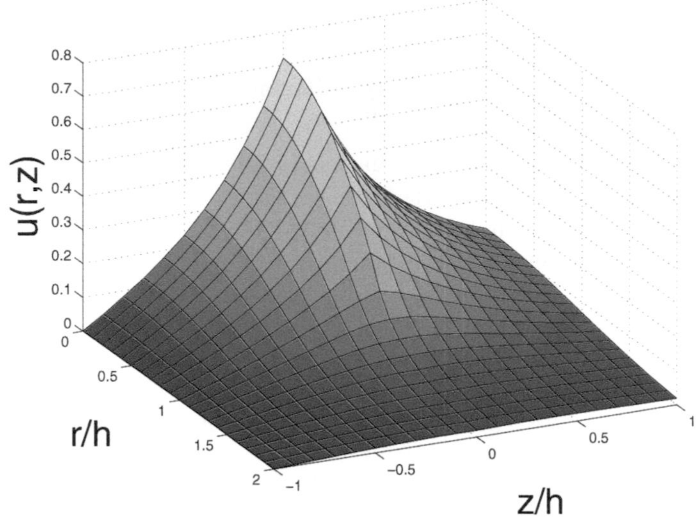

Figure 4.3.9: The solution $u(r,z)$ to the mixed boundary value problem governed by Equation 4.3.206 through Equation 4.3.211 with $h = 1$ and $\alpha = 1$.

and

$$\frac{d}{dr}\left[\int_0^r \frac{\xi^2 J_1(k\xi)}{\sqrt{r^2-\xi^2}}\,d\xi\right] = r\sin(kr) \tag{4.3.223}$$

after using integral tables.[52] Substituting these results into Equation 4.3.221 and dividing by r,

$$h(r) - \frac{2}{\pi}\int_0^1 h(\tau)\left[\int_0^\infty \left(1 - \frac{k}{\lambda} + \frac{k}{\lambda}e^{-2\lambda h}\right)\sin(k\tau)\sin(kr)\,dk\right]\,d\tau = -\frac{8r}{\pi}. \tag{4.3.224}$$

Figure 4.3.9 illustrates $u(r,z)$ when $h = 1$ and $\alpha = 1$.

• Example 4.3.10

In the previous examples, the boundary condition was $u(r,0) = 0$ or $u_z(r,0) = 0$ for $0 < a < r < \infty$. In this example we consider the other situation where $u(r,0) = 0$ applies when $0 < r < 1$. In particular, we find the solution[53] to

$$\frac{\partial^2 u}{\partial r^2} + \frac{1}{r}\frac{\partial u}{\partial r} + \frac{\partial^2 u}{\partial z^2} = 0, \qquad 0 \le r < \infty, \quad 0 < z < h, \tag{4.3.225}$$

---

52 Ibid., Formula 6.567.1 with $\nu = 1$ and $\mu = -\frac{1}{2}$.

53 Taken from Dhaliwal, R. S., 1967: An axisymmetric mixed boundary value problem for a thick slab. *SIAM J. Appl. Math.*, **15**, 98–106. ©1967 Society for Industrial and Applied Mathematics. Reprinted with permission.

subject to the boundary conditions

$$\lim_{r\to 0} |u(r,z)| < \infty, \qquad \lim_{r\to\infty} u(r,z) \to 0, \qquad 0 < z < h, \tag{4.3.226}$$

$$u_z(r,h) = 0, \qquad 0 \le r < \infty, \tag{4.3.227}$$

and

$$\begin{cases} u(r,0) = 0, & 0 \le r < 1, \\ u_z(r,0) = -f(r), & 1 < r < \infty. \end{cases} \tag{4.3.228}$$

Using transform methods or separation of variables, the general solution to Equation 4.3.225, Equation 4.3.226, and Equation 4.3.227 is

$$u(r,z) = \int_0^\infty A(k) \frac{\cosh[k(z-h)]}{\cosh(kh)} J_0(kr)\, dk. \tag{4.3.229}$$

Substituting Equation 4.3.229 into Equation 4.3.228, we have that

$$\int_0^\infty A(k) J_0(kr)\, dk = 0, \qquad 0 \le r < 1, \tag{4.3.230}$$

and

$$\int_0^\infty kA(k) \tanh(kh) J_0(kr)\, dk = f(r), \qquad 1 < r < \infty. \tag{4.3.231}$$

To solve Equation 4.3.230 and Equation 4.3.231, we set

$$A(k) = \int_1^\infty g(t) \cos(kt)\, dt, \tag{4.3.232}$$

where $\lim_{t\to\infty} g(t) \to 0$. We did this because

$$\int_0^\infty A(k) J_0(kr)\, dk = \int_1^\infty g(t) \left[\int_0^\infty \cos(kt) J_0(kr)\, dk \right] dt = 0 \tag{4.3.233}$$

from Equation 1.4.14 with $0 < r < 1 \le t < \infty$. Turning to Equation 4.3.231, the substitution of Equation 4.3.232 yields

$$\begin{aligned} &\int_0^\infty J_0(kr) \left[\int_1^\infty k\, g(t) \cos(kt)\, dt \right] dk \\ &\qquad - \int_0^\infty kM(kh) J_0(kr) \left[\int_1^\infty g(t) \cos(kt)\, dt \right] dk = f(r), \end{aligned} \tag{4.3.234}$$

where $M(kh) = 2/\left(1 + e^{2kh}\right)$.

We now simplify Equation 4.3.234 in two ways. In the first term we integrate by parts the integral within the square brackets and apply Equation 1.4.13. We then replace $J_0(kr)$ by its integral representation.[54] This gives

$$-\int_r^\infty \frac{g'(t)}{\sqrt{t^2-r^2}}\,dt - \frac{2}{\pi}\int_0^\infty kM(kh)\left[\int_r^\infty \frac{\sin(kt)}{\sqrt{t^2-r^2}}\,dt\right] \times \left[\int_1^\infty g(x)\cos(kx)\,dx\right] dk = f(r). \tag{4.3.235}$$

Interchanging the order of integration and using the trigonometric product formula, Equation 4.3.235 becomes

$$\int_r^\infty \left\{g'(t) - \frac{1}{\pi h}\int_1^\infty g(x)\left[G'(t+x)+G'(t-x)\right]dx\right\}\frac{dt}{\sqrt{t^2-r^2}} = -f(r), \tag{4.3.236}$$

where $G(\xi) = \int_0^\infty M(\eta)\cos(\xi\eta/h)\,d\eta$. Viewing Equation 4.3.236 as an integral equation of the Abel type, Equation 1.2.15 and Equation 1.2.16 yield

$$g'(t) - \frac{1}{\pi h}\int_1^\infty g(x)\left[G'(t+x)+G'(t-x)\right]dx = \frac{2}{\pi}\frac{d}{dt}\left[\int_t^\infty \frac{rf(r)}{\sqrt{r^2-t^2}}\,dr\right]. \tag{4.3.237}$$

Integrating Equation 4.3.237 with respect to t,

$$g(t) - \frac{1}{\pi h}\int_1^\infty g(x)\left[G(t+x)+G(t-x)\right]dx = \frac{2}{\pi}\int_t^\infty \frac{rf(r)}{\sqrt{r^2-t^2}}\,dr. \tag{4.3.238}$$

For the special case

$$f(r) = \begin{cases} 1, & 1 < r < a, \\ 0, & a < r < \infty, \end{cases} \tag{4.3.239}$$

Equation 4.3.238 becomes

$$g(t) - \frac{1}{\pi h}\int_1^\infty g(x)K(x,t)\,dx = \frac{2}{\pi}\sqrt{a^2-t^2}, \qquad 1 \le t \le a, \tag{4.3.240}$$

and $g(t) = 0$ for $a < t < \infty$, where

$$K(x,t) = 2\int_0^\infty M(\xi)\cos\left(\frac{\xi x}{h}\right)\cos\left(\frac{\xi t}{h}\right)d\xi. \tag{4.3.241}$$

---

[54] Gradshteyn and Ryzhik, op. cit., Formula 8.41.9

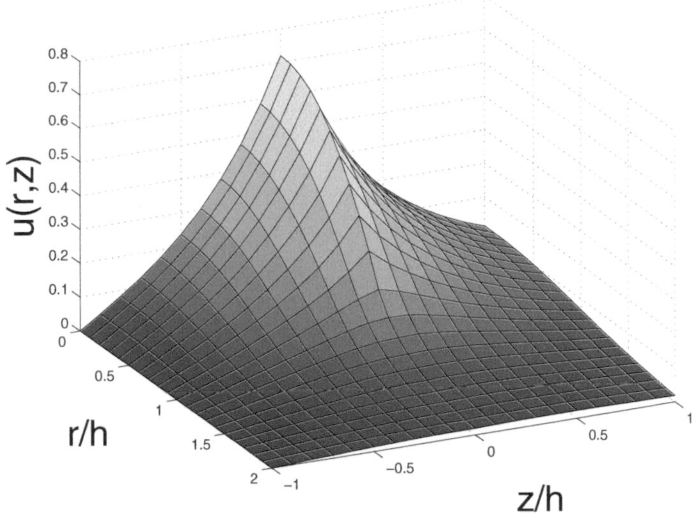

Figure 4.3.10: The solution $u(r,z)$ to the mixed boundary value problem governed by Equation 4.3.225 through Equation 4.3.228 with $a = h = 2$.

Figure 4.3.10 illustrates this solution when $a = h = 2$.

• **Example 4.3.11**

Let us solve[55]

$$\frac{\partial^2 u}{\partial r^2} + \frac{1}{r}\frac{\partial u}{\partial r} + \frac{\partial^2 u}{\partial z^2} = 0, \qquad 0 \le r < \infty, \quad 0 < z < \infty, \tag{4.3.242}$$

subject to the boundary conditions

$$\lim_{r\to 0} |u(r,z)| < \infty, \qquad \lim_{r\to\infty} u(r,z) \to 0, \qquad 0 < z < \infty, \tag{4.3.243}$$

$$\lim_{z\to\infty} u(r,z) \to 0, \qquad 0 \le r < \infty, \tag{4.3.244}$$

and

$$\begin{cases} \alpha u_z(r,0) - \beta u(r,0) = -f(r), & 0 \le r < 1, \\ \gamma u_z(r,0) - \delta u(r,0) = 0, & 1 < r < \infty. \end{cases} \tag{4.3.245}$$

All of the coefficients in Equation 4.3.245 are nonzero.

Using transform methods or separation of variables, the general solution to Equation 4.3.242, Equation 4.3.243, and Equation 4.3.244 is

$$u(r,z) = \int_0^\infty A(k) J_0(kr) e^{-kz}\, dk. \tag{4.3.246}$$

[55] See Kuz'min, Yu. N., 1966: Some axially symmetric problems in heat flow with mixed boundary conditions. *Sov. Tech. Phys.*, **11**, 169–173.

Substituting Equation 4.3.246 into Equation 4.3.245, we have that

$$\int_0^\infty (\alpha k + \beta) A(k) J_0(kr)\, dk = f(r), \qquad 0 \le r < 1, \tag{4.3.247}$$

and

$$\int_0^\infty (\gamma k + \delta) A(k) J_0(kr)\, dk = 0, \qquad 1 < r < \infty; \tag{4.3.248}$$

or

$$\int_0^\infty M(k)[1 + g(k)] J_0(kr)\, dk = f(r), \qquad 0 \le r < 1, \tag{4.3.249}$$

and

$$\int_0^\infty M(k) J_0(kr)\, dk = 0, \qquad 1 < r < \infty, \tag{4.3.250}$$

where

$$M(k) = \alpha(\gamma k + \delta) A(k)/\gamma, \quad \text{and} \quad g(k) = \frac{\beta\gamma - \alpha\delta}{\alpha(\gamma k + \delta)}. \tag{4.3.251}$$

Let us now introduce the function $M(k)$, where

$$M(k) = \int_0^1 h'(t) \sin(kt)\, dt. \tag{4.3.252}$$

Then, by integration by parts,

$$M(k) = h(1)\sin(k) + k \int_0^1 h(t)\cos(kt)\, dt. \tag{4.3.253}$$

Therefore,

$$\int_0^\infty M(k) J_0(kr)\, dk = \int_0^1 h'(t) \left[\int_0^\infty \sin(kt) J_0(kr)\, dk\right] dt = 0 \tag{4.3.254}$$

if $r > 1$ by Equation 1.4.13; our choice of $M(k)$ satisfies Equation 4.3.250 identically.

Turning to Equation 4.3.249,

$$\int_0^\infty \left[\int_0^1 h'(t)\sin(kt)\, dt\right] J_0(kr)\, dk = f(r) - \int_0^\infty k \left[\int_0^1 h(t)\cos(kt)\, dt\right] g(k) J_0(kr)\, dk \tag{4.3.255}$$

if $h(1) = 0$; or,

$$\int_0^1 h'(t) \left[\int_0^\infty \sin(kt) J_0(kr)\, dk\right] dt = f(r) - \int_0^1 h(t) \left[\int_0^\infty k\, g(k)\cos(kt) J_0(kr)\, dk\right] dt. \tag{4.3.256}$$

Upon applying Equation 1.4.13 to the integral within the square brackets on the left side of Equation 4.3.256,

$$\int_r^1 \frac{h'(t)}{\sqrt{t^2-r^2}}\,dt = f(r) - \int_0^1 h(\tau)\left[\int_0^\infty k\,g(k)\cos(k\tau)J_0(kr)\,dk\right]dt. \tag{4.3.257}$$

From Equation 1.2.15 and Equation 1.2.16,

$$h'(t) = -\frac{2}{\pi}\frac{d}{dt}\left[\int_t^1 \frac{rf(r)}{\sqrt{r^2-t^2}}\,dr\right] \tag{4.3.258}$$

$$+\frac{2}{\pi}\frac{d}{dt}\left(\int_0^1 h(\tau)\left\{\int_t^1 \frac{r}{\sqrt{r^2-t^2}}\left[\int_0^\infty k\,g(k)\cos(k\tau)J_0(kr)\,dk\right]dr\right\}d\tau\right);$$

or

$$h(t) = -\frac{2}{\pi}\int_t^1 \frac{r\,f(r)}{\sqrt{r^2-t^2}}\,dr + \frac{2}{\pi}\int_0^1 K(t,\tau)h(\tau)\,d\tau, \tag{4.3.259}$$

where

$$K(t,\tau) = \int_t^1 \frac{r}{\sqrt{r^2-t^2}}\left[\int_0^\infty k\,g(k)\cos(k\tau)J_0(kr)\,dk\right]dr \tag{4.3.260}$$

$$= \frac{\beta\gamma-\alpha\delta}{\alpha\gamma}\int_t^1 \frac{r}{\sqrt{r^2-t^2}}\left[\int_0^\infty \frac{\gamma k}{\gamma k+\delta}\cos(k\tau)J_0(kr)\,dk\right]dr \tag{4.3.261}$$

$$= \frac{\beta\gamma-\alpha\delta}{\alpha\gamma}\int_t^1 \frac{r}{\sqrt{r^2-t^2}}\left[\int_0^\infty \cos(k\tau)J_0(kr)\,dk\right]dr$$

$$-\frac{\delta(\beta\gamma-\alpha\delta)}{\alpha\gamma^2}\int_t^1 \frac{r}{\sqrt{r^2-t^2}}\left[\int_0^\infty \frac{\cos(k\tau)J_0(kr)}{k+\lambda}\,dk\right]dr \tag{4.3.262}$$

$$= \frac{\beta\gamma-\alpha\delta}{\alpha\gamma}\left\{\ln\left[\frac{\sqrt{1-t^2}+\sqrt{1-\tau^2}}{\sqrt{|t^2-\tau^2|}}\right]\right.$$

$$\left.+\frac{\delta}{\pi\gamma}\int_0^1 \ln\left[\frac{\sqrt{1-t^2}+\sqrt{1-\eta^2}}{\sqrt{|t^2-\eta^2|}}\right]R(\tau,\eta,\delta/\gamma)\,d\eta\right\}, \tag{4.3.263}$$

$$R(\tau,t,k) = \sin[k(t+\tau)]\,\mathrm{si}[k(t+\tau)] + \cos[k(t+\tau)]\,\mathrm{ci}[k(t+\tau)]$$
$$+\sin[k|t-\tau|]\,\mathrm{si}[k|t-\tau|] + \cos[k|t-\tau|]\,\mathrm{ci}[k|t-\tau|], \tag{4.3.264}$$

$\lambda = \delta/\gamma$ and $\mathrm{si}(\cdot)$ and $\mathrm{ci}(\cdot)$ are the sine and cosine integrals.

• Example 4.3.12

Consider[56] the axisymmetric Laplace equation

$$\frac{1}{r}\frac{\partial}{\partial r}\left(r\frac{\partial u}{\partial r}\right) + \frac{\partial^2 u}{\partial z^2} = 0, \qquad 0 \le r < \infty, \quad 0 < z < 1, \tag{4.3.265}$$

---

[56] Reprinted from *J. Theor. Biol.*, **81**, A. Nir and R. Pfeffer, Transport of macromolecules across arterial wall in the presence of local endothial injury, 685–711, ©1979, with permission from Elsevier.

subject to the boundary conditions

$$\lim_{r\to 0} |u(r,z)| < \infty, \qquad \lim_{r\to\infty} |u(r,z)| < \infty, \qquad u(r,0) = 0, \tag{4.3.266}$$

and

$$\begin{cases} u(r,1) = 1, & 0 < r \le a, \\ u(r,1) + \dfrac{u_z(r,1)}{\sigma} = 1, & a < r < \infty. \end{cases} \tag{4.3.267}$$

The interesting aspect of this example is the mixture of boundary conditions along the boundary $z = 1$. For $0 < r < a$, we have a Dirichlet boundary condition that becomes a Robin boundary condition when $a < r < \infty$.

Applying Hankel transforms, the solution to Equation 4.3.265 and the boundary conditions given by Equation 4.3.266 is

$$u(r,z) = \frac{\sigma z}{1+\sigma} + \frac{a}{1+\sigma}\int_0^\infty A(k)\sinh(kz)J_0(kr)\,dk. \tag{4.3.268}$$

Substitution of Equation 4.3.268 into Equation 4.3.267 leads to the dual integral equations:

$$a\int_0^\infty A(k)\sinh(k)J_0(kr)\,dk = 1, \qquad 0 < r \le a, \tag{4.3.269}$$

and

$$\int_0^\infty A(k)\left[\sinh(k) + \frac{k\cosh(k)}{\sigma}\right] J_0(kr)\,dk = 0, \qquad a < r < \infty. \tag{4.3.270}$$

A procedure for solving Equation 4.3.269 and Equation 4.3.270 was developed by Tranter[57] who proved that dual integral equations of the form

$$\int_0^\infty G(\lambda)f(\lambda)J_0(\lambda a)\,d\lambda = g(a), \tag{4.3.271}$$

and

$$\int_0^\infty f(\lambda)J_0(\lambda a)\,d\lambda = 0 \tag{4.3.272}$$

have the solution

$$f(\lambda) = \lambda^{1-\kappa}\sum_{n=0}^\infty A_n J_{2m+\kappa}(\lambda), \tag{4.3.273}$$

if $G(\lambda)$ and $g(a)$ are known. The value of κ is chosen so that the difference $G(\lambda) - \lambda^{2\kappa-2}$ is fairly small. In the present case, $f(\lambda) = \sinh(\lambda)A(\lambda,a)$, $g(a) = 1$ and $G(\lambda) = 1 + \lambda\coth(\lambda)/\sigma$.

---

[57] Tranter, C. J., 1950: On some dual integral equations occurring in potential problems with axial symmetry. *Quart. J. Mech. Appl. Math.*, **3**, 411–419.

Figure 4.3.11: Educated at Queen's College, Oxford, Clement John Tranter, CBE, (1909–1991) excelled both as a researcher and educator, primarily at the Military College of Science at Woolrich and then Shrivenham. His mathematical papers fall into two camps: (a) the solution of boundary value problems by classical and transform methods and (b) the solution of dual integral equations and series. He is equally well known for a series of popular textbooks on integral transforms and Bessel functions. (Portrait provided by kind permission of the Defense College of Management and Technology Library's Heritage Centre.)

What is the value of κ here? Clearly, we would like our solution to be valid for a wide range of σ. Because $G(\lambda) \to 1$ as $\sigma \to \infty$, a reasonable choice is $\kappa = 1$. Therefore, we take

$$\sinh(k)A(k) = \sum_{n=1}^{\infty} \frac{A_n}{1 + k\coth(k)/\sigma} J_{2n-1}(ka). \tag{4.3.274}$$

Our final task remains to find A_n.

We begin by writing

$$\frac{A_n}{1 + k\coth(k)/\sigma} J_{2n-1}(ka) = \sum_{m=1}^{\infty} B_{mn}\, J_{2m-1}(ka), \tag{4.3.275}$$

where B_{mn} depends only on a and σ. Multiplying Equation 4.3.275 by $dk/k \times J_{2p-1}(ka)$ and integrating

$$\int_0^{\infty} \frac{A_n}{1 + k\coth(k)/\sigma}\, J_{2n-1}(ka)\, J_{2p-1}(ka)\, \frac{dk}{k}$$
$$= \int_0^{\infty} \sum_{m=1}^{\infty} B_{mn}\, J_{2m-1}(ka)\, J_{2p-1}(ka)\, \frac{dk}{k}. \tag{4.3.276}$$

Because[58]

$$\int_0^\infty J_{2n-1}(ka)\, J_{2p-1}(ka)\, \frac{dk}{k} = \frac{\delta_{mp}}{2(2m-1)}, \tag{4.3.277}$$

where δ_{mp} is the Kronecker delta:

$$\delta_{mp} = \begin{cases} 1, & m = p, \\ 0, & m \neq p, \end{cases} \tag{4.3.278}$$

Equation 4.3.276 reduces to

$$A_n \int_0^\infty \frac{J_{2n-1}(ka) J_{2m-1}(ka)}{1 + k\coth(k)/\sigma}\, \frac{dk}{k} = \frac{B_{mn}}{2(2m-1)}. \tag{4.3.279}$$

If we define

$$S_{mn} = \int_0^\infty \frac{J_{2n-1}(ka)\, J_{2m-1}(ka)}{1 + k\coth(k)/\sigma}\, \frac{dk}{k}, \tag{4.3.280}$$

then we can rewrite Equation 4.3.279 as

$$A_n S_{mn} = \frac{B_{mn}}{2(2m-1)}. \tag{4.3.281}$$

Because[59]

$$a \int_0^\infty J_0(kr)\, J_{2m-1}(ka)\, dk = P_{m-1}\left(1 - \frac{2r^2}{a^2}\right), \qquad r < a, \tag{4.3.282}$$

where $P_m(\cdot)$ is the Legendre polynomial of order m, Equation 4.3.282 can be rewritten

$$\sum_{n=1}^\infty \sum_{m=1}^\infty B_{mn} P_{m-1}\left(1 - \frac{2r^2}{a^2}\right) = 1. \tag{4.3.283}$$

Equation 4.3.283 follows from the substitution of Equation 4.3.274 into Equation 4.3.269 and then using Equation 4.3.282. Multiplying Equation 4.3.283 by $P_{m-1}(\xi)\, d\xi$, integrating between -1 and 1, and using the orthogonality properties of the Legendre polynomial, we have

$$\sum_{n=1}^\infty B_{mn} \int_{-1}^1 \left[P_{m-1}(\xi)\right]^2 d\xi = \int_{-1}^1 P_{m-1}(\xi)\, d\xi = \int_{-1}^1 P_0(\xi) P_{m-1}(\xi)\, d\xi, \tag{4.3.284}$$

---

[58] Gradshteyn and Ryzhik, op. cit., Formula 6.538.2.

[59] Ibid., Formula 6.512.4.

Table 4.3.1: The Convergence of the Coefficients A_n Given by Equation 4.3.287 Where S_{mn} Has Nonzero Values for $1 \leq m, n \leq N$

N	A_1	A_2	A_3	A_4	A_5	A_6	A_7	A_8
1	2.9980							
2	3.1573	−1.7181						
3	3.2084	−2.0329	1.5978					
4	3.2300	−2.1562	1.9813	−1.4517				
5	3.2411	−2.2174	2.1548	−1.8631	1.3347			
6	3.2475	−2.2521	2.2495	−2.0670	1.7549	−1.2399		
7	3.2515	−2.2738	2.3073	−2.1862	1.9770	−1.6597	1.1620	
8	3.2542	−2.2882	2.3452	−2.2626	2.1133	−1.8925	1.5772	−1.0972

which shows that only $m = 1$ yields a nontrivial sum. Thus,

$$\sum_{n=1}^{\infty} B_{mn} = 2(2m-1) \sum_{n=1}^{\infty} A_n S_{mn} = 0, \quad 2 \leq m, \tag{4.3.285}$$

and

$$\sum_{n=1}^{\infty} B_{1n} = 2 \sum_{n=1}^{\infty} A_n S_{1n} = 1, \tag{4.3.286}$$

or

$$\sum_{n=1}^{\infty} S_{mn} A_n = \tfrac{1}{2} \delta_{m1}. \tag{4.3.287}$$

Thus, we reduced the problem to the solution of an infinite number of linear equations that yield A_n. Selecting some maximum value for n and m, say N, each term in the matrix S_{mn}, $1 \leq m, n \leq N$, is evaluated numerically for a given value of a and σ. By inverting Equation 4.3.287, we obtain the coefficients A_n for $n = 1, \ldots, N$. Because we solved a truncated version of Equation 4.3.287, they will only be approximate. To find more accurate values, we can increase N by 1 and again invert Equation 4.3.287. In addition to the new A_{N+1}, the previous coefficients will become more accurate. We can repeat this process of increasing N until the coefficients converge to their correct values. This is illustrated in Table 4.3.1 when $\sigma = a = 1$.

Once we have computed the coefficients A_n necessary for the desired accuracy, we use Equation 4.3.274 to find $A(k)$ and then obtain $u(r, z)$ from Equation 4.3.268 via numerical integration. Figure 4.3.12 illustrates the solution when $\sigma = 1$ and $a = 2$.

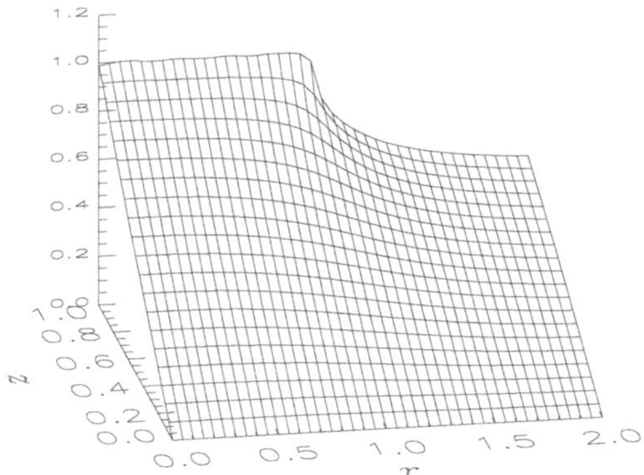

Figure 4.3.12: The solution of the axisymmetric Laplace equation, Equation 4.3.265, with $u(r, 0) = 0$ and the mixed boundary condition given by Equation 4.3.267. Here we have chosen $\sigma = 1$ and $a = 2$.

• Example 4.3.13

Let us now examine the case of systems of partial differential equations where one of the boundary conditions is mixed. Consider[60]

$$\frac{\partial^2 u_1}{\partial r^2} + \frac{1}{r}\frac{\partial u_1}{\partial r} + \frac{\partial^2 u_1}{\partial z^2} = 0, \qquad 0 \le r < \infty, \quad 0 < z < b, \tag{4.3.288}$$

$$\frac{\partial^2 u_2}{\partial r^2} + \frac{1}{r}\frac{\partial u_2}{\partial r} + \frac{\partial^2 u_2}{\partial z^2} = 0, \qquad 0 \le r < \infty, \quad z < 0, \tag{4.3.289}$$

subject to the boundary conditions

$$\lim_{r\to 0} |u_1(r, z)| < \infty, \quad \lim_{r\to\infty} u_1(r, z) \to 0, \qquad 0 < z < b, \tag{4.3.290}$$

$$\lim_{r\to 0} |u_2(r, z)| < \infty, \quad \lim_{r\to\infty} u_2(r, z) \to 0, \qquad z < 0, \tag{4.3.291}$$

$$u_1(r, b) = 0, \qquad 0 \le r < \infty, \tag{4.3.292}$$

$$\lim_{z\to -\infty} u_2(r, z) \to 0, \qquad 0 \le r < \infty, \tag{4.3.293}$$

and

$$\begin{cases} u_1(r, 0) = 1, & 0 \le r < a, \\ \epsilon\, \partial u_1(r, 0)/\partial z = \epsilon_0\, \partial u_2(r, 0)/\partial z, & a \le r < \infty. \end{cases} \tag{4.3.294}$$

Using Hankel transforms, the solutions to Equation 4.3.288 and Equation 4.3.289 are

$$u_1(r, z) = \int_0^\infty A(k) \frac{\sinh[k(b - z)/a]}{\cosh(bk/a)} J_0(kr/a)\, \frac{dk}{k}, \tag{4.3.295}$$

[60] See Gelmont, B., M. S. Shur, and R. J. Mattauch, 1995: Disk and stripe capacitances. *Solid-State Electron.*, **38**, 731–734.

and

$$u_2(r,z) = \int_0^\infty A(k)\tanh(kb/a)e^{kz/a}J_0(kr/a)\,\frac{dk}{k}. \qquad \textbf{(4.3.296)}$$

Equation 4.3.295 satisfies not only Equation 4.3.288, but also Equation 4.3.290 and Equation 4.3.292. Similarly, Equation 4.3.296 satisfies not only Equation 4.3.289, but also Equation 4.3.291 and Equation 4.3.293. Substituting Equation 4.3.295 and Equation 4.3.296 into Equation 4.3.294, we obtain the dual integral equations

$$\int_0^\infty A(k)\tanh(kb/a)J_0(kr/a)\,\frac{dk}{k} = 1, \qquad 0 \le r < a, \qquad \textbf{(4.3.297)}$$

and

$$\int_0^\infty A(k)\left[1+\epsilon_0\tanh(kb/a)/\epsilon\right]J_0(kr/a)\,dk = 0, \qquad a < r < \infty. \quad \textbf{(4.3.298)}$$

If we define $A(k)$ by

$$\left[1+\epsilon_0\tanh(kb/a)/\epsilon\right]A(k) = k\int_0^1 f(t)\cos(kt)\,dt, \qquad \textbf{(4.3.299)}$$

then direct substitution of Equation 4.3.299 into Equation 4.3.298 shows that it is satisfied identically. We next substitute Equation 4.3.299 into Equation 4.3.297 and interchange the order of integration. This yields

$$\int_0^\infty \frac{\tanh(kb/a)J_0(kr/a)}{1+\epsilon_0\tanh(kb/a)/\epsilon}\left[\int_0^1 f(t)\cos(kt)\,dt\right]dk = 1, \qquad 0 < t < 1. \qquad \textbf{(4.3.300)}$$

From Equation 1.4.9, we find that

$$\frac{d}{dt}\left[\int_0^{at}\frac{r\,J_0(kr)}{\sqrt{t^2-r^2/a^2}}\,dr\right] = a^2\cos(kt). \qquad \textbf{(4.3.301)}$$

Why did we derive Equation 4.3.301? If we multiply both sides of Equation 4.3.300 by $r\,dr/\sqrt{t^2-r^2/a^2}$, integrate from 0 to at, differentiate with respect to t, and use Equation 4.3.301, we obtain the following integral equation that gives $f(t)$:

$$\int_0^1 f(\tau)\left[\int_0^\infty \frac{\tanh(kb/a)}{1+\epsilon_0\tanh(kb/a)/\epsilon}\cos(kt)\cos(k\tau)\,dk\right]d\tau = 1; \qquad \textbf{(4.3.302)}$$

or,

$$\frac{\pi}{2}f(t) - \int_0^1 f(\tau)\left[\int_0^\infty \frac{1-\tanh(kb/a)}{1+\epsilon_0\tanh(kb/a)/\epsilon}\cos(kt)\cos(k\tau)\,dk\right]d\tau = \left(1+\frac{\epsilon_0}{\epsilon}\right), \qquad \textbf{(4.3.303)}$$

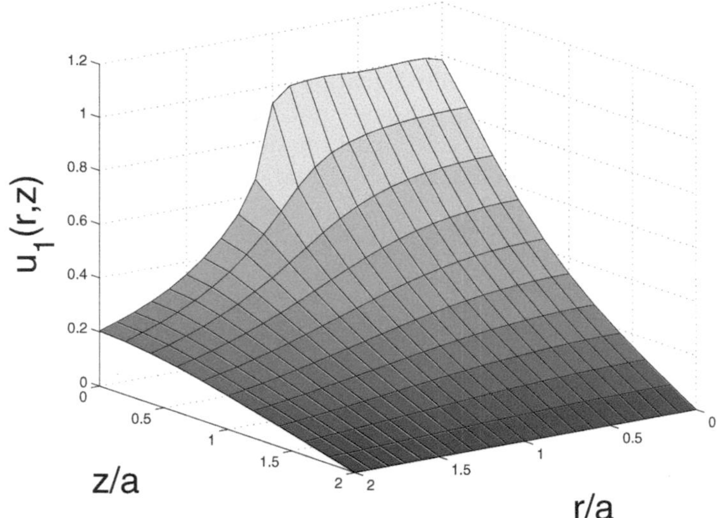

Figure 4.3.13: The solution $u_1(r,z)$ to the mixed boundary value problem governed by Equation 4.3.288 through Equation 4.3.294 when $\epsilon = 3\epsilon_0$.

if $0 < t < 1$.

At this point we must solve Equation 4.3.303 numerically to compute $f(t)$. Before we do that, there are two limiting cases of interest. When $\epsilon = \epsilon_0$, we have the same problem that we solved in Section 2.2 on the disc capacitor. The second limit is $\epsilon_0 \ll \epsilon$. In this case $u_2(r,z) \to 0$ and $u_1(r,z)$ is given by the solution to Example 4.3.2. Figure 4.3.13 shows the solution somewhere between these two limits with $\epsilon = 3\epsilon_0$.

• Example 4.3.14

During their study of a circular disk in a Brinkman medium, Feng et al.[61] solved a system of mixed boundary value problems. We join their problem midway in progress where they derived the following governing partial differential equations and boundary conditions:

$$\frac{\partial^2 u_1}{\partial r^2} + \frac{1}{r}\frac{\partial u_1}{\partial r} - \frac{\partial^2 u_1}{\partial z^2} = 0, \qquad 0 \le r < \infty, \quad -\infty < z < \infty, \qquad \textbf{(4.3.304)}$$

$$\frac{\partial^2 u_2}{\partial r^2} + \frac{1}{r}\frac{\partial u_2}{\partial r} - \frac{\partial^2 u_2}{\partial z^2} - \gamma^2 u_2 = 0, \qquad 0 \le r < \infty, \quad -\infty < z < \infty, \qquad \textbf{(4.3.305)}$$

subject to the boundary conditions

$$\lim_{r\to 0} |u_1(r,z)| < \infty, \quad \lim_{r\to\infty} u_1(r,z) \to 0, \qquad -\infty < z < \infty, \qquad \textbf{(4.3.306)}$$

---

[61] Feng, J., P. Ganatos, and S. Weinbaum, 1998: The general motion of a circular disk in a Brinkman medium. *Phys. Fluids*, **10**, 2137–2146.

$$\lim_{r\to 0} |u_2(r,z)| < \infty, \quad \lim_{r\to\infty} u_2(r,z) \to 0, \qquad -\infty < z < \infty, \tag{4.3.307}$$

$$\lim_{|z|\to\infty} u_1(r,z) \to 0, \qquad 0 \le r < \infty, \tag{4.3.308}$$

$$\lim_{|z|\to\infty} u_2(r,z) \to 0, \qquad 0 \le r < \infty, \tag{4.3.309}$$

$$\left.\frac{\partial u_1}{\partial z} + \frac{\partial u_2}{\partial z}\right|_{z=0^-} = \left.\frac{\partial u_1}{\partial z} + \frac{\partial u_2}{\partial z}\right|_{z=0^+}, \tag{4.3.310}$$

and

$$\begin{cases} \left.\dfrac{\partial u_1}{\partial r} + \dfrac{\partial u_2}{\partial r}\right|_{z=0^-} = \left.\dfrac{\partial u_1}{\partial r} + \dfrac{\partial u_2}{\partial r}\right|_{z=0^+} = r, & 0 \le r < 1, \\ p(r,0^-) = p(r,0^+), & 1 < r < \infty, \end{cases} \tag{4.3.311}$$

where

$$\frac{\partial p}{\partial z} = -\frac{1}{r}\frac{\partial u_1}{\partial r} \qquad \text{and} \qquad \frac{\partial p}{\partial r} = \frac{1}{r}\frac{\partial u_1}{\partial z}. \tag{4.3.312}$$

Using Hankel transforms, the solutions to Equation 4.3.304 and Equation 4.3.305 are

$$u_1(r,z) = \int_0^\infty A(k)e^{-k|z|}\, rJ_1(kr)\, dk, \tag{4.3.313}$$

and

$$u_2(r,z) = \int_0^\infty B(k)e^{-|z|\sqrt{k^2+\gamma^2}} rJ_1(kr)\, dk. \tag{4.3.314}$$

Equation 4.3.313 satisfies not only Equation 4.3.304, but also Equation 4.3.306 and Equation 4.3.308. Similarly, Equation 4.3.314 satisfies not only Equation 4.3.305, but also Equation 4.3.307 and Equation 4.3.309. Substituting Equation 4.3.313 and Equation 4.3.314 into Equation 4.3.310, we find that

$$\int_0^\infty \left[-kA(k) - \sqrt{k^2+\gamma^2}\, B(k)\right] rJ_1(kr)\, dk = 0, \qquad 0 \le r < \infty. \tag{4.3.315}$$

Hence,

$$B(k) = -\frac{k\, A(k)}{\sqrt{k^2+\gamma^2}}. \tag{4.3.316}$$

Let us now turn to the equation involving $p(r,z)$ in the mixed boundary condition Equation 4.3.311. Now,

$$\frac{\partial p}{\partial z} = -\frac{1}{r}\int_0^\infty A(k)e^{-k|z|}\frac{d}{dr}\left[rJ_1(kr)\right]\, dk = k\int_0^\infty A(k)e^{-k|z|}J_0(kr)\, dk. \tag{4.3.317}$$

Therefore,

$$p(r,z) = f(r) + \int_0^\infty A(k)\,\mathrm{sgn}(z)e^{-k|z|}J_0(kr)\, dk, \tag{4.3.318}$$

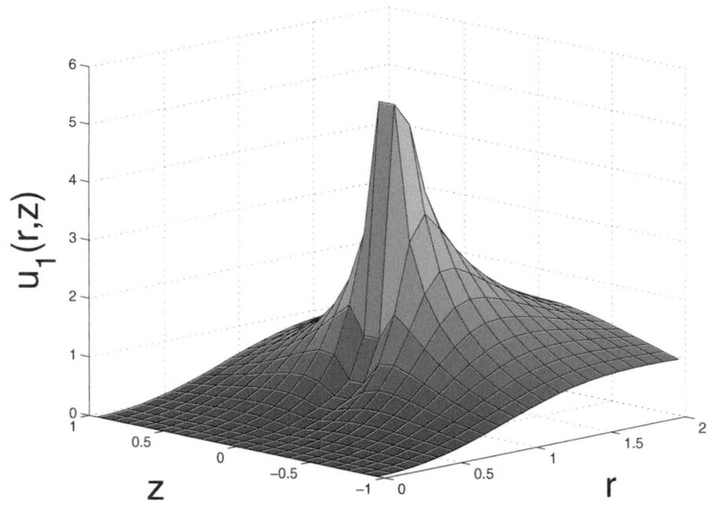

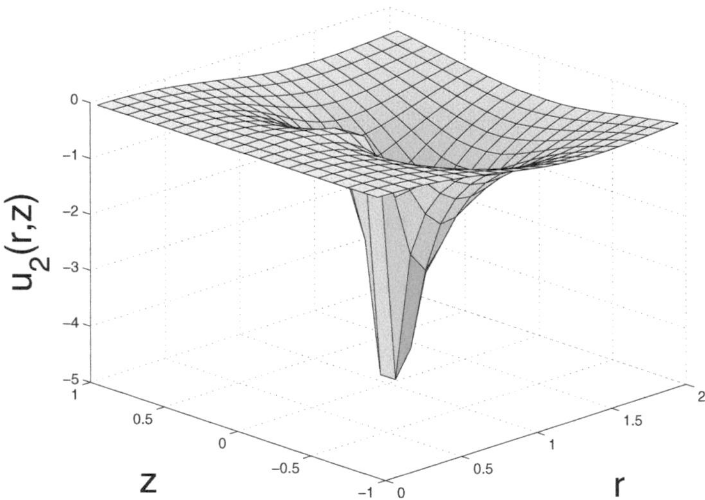

Figure 4.3.14: The solution to the mixed boundary value problem governed by Equation 4.3.304 through Equation 4.3.311 when $\gamma = 1$.

where $f(r)$ is an arbitrary function of r. The mixed boundary condition, Equation 4.3.311, then gives

$$p(r, 0^+) - p(r, 0^-) = 2 \int_0^\infty A(k) J_0(kr)\, dk = 0. \qquad \textbf{(4.3.319)}$$

Consider now the first condition in Equation 4.3.311. Substitution of

Equation 4.3.313 and Equation 4.3.314 yields

$$\int_0^\infty k\left(1-\frac{k}{\sqrt{k^2+\gamma^2}}\right)A(k)J_0(kr)\,dk=1,\qquad 0\le r<1,\tag{4.3.320}$$

or

$$\int_0^\infty \frac{kJ_0(kr)}{\sqrt{k^2+\gamma^2}\left(k+\sqrt{k^2+\gamma^2}\right)}A(k)\,dk=\frac{1}{\gamma^2},\qquad 0\le r<1;\tag{4.3.321}$$

or

$$\int_0^\infty [1-F(k)]\,C(k)J_0(kr)\,\frac{dk}{k}=1,\qquad 0\le r<1,\tag{4.3.322}$$

where

$$A(k)=\frac{2C(k)}{\gamma^2},\quad\text{and}\quad F(k)=1-\frac{2k^2}{\sqrt{k^2+\gamma^2}\left(k+\sqrt{k^2+\gamma^2}\right)}.\tag{4.3.323}$$

Using Cooke's results given in Equation 4.3.24 through Equation 4.3.27 with $a=1$, $\alpha=\beta=\frac{1}{2}$, $\nu=0$, and $G(k)=[1-F(k)]/k$, we have that

$$A(k)=k\int_0^1 h(t)\cos(kt)\,dt,\tag{4.3.324}$$

where

$$h(t)=\frac{4}{\pi\gamma^2}+\int_0^1 h(\tau)K(t,\tau)\,d\tau,\tag{4.3.325}$$

and

$$K(x,\tau)=\frac{2}{\pi}\int_0^\infty F(k)\cos(xk)\cos(\tau k)\,dk.\tag{4.3.326}$$

At this point we must solve Equation 4.3.325 numerically to compute $h(t)$. Once, we have $h(t)$, then $A(k)$ and $B(k)$ follow from Equation 4.3.324 and Equation 4.3.316, respectively. Finally, Equation 4.3.313 and Equation 4.3.314 yield $u(r,z)$. Figure 4.3.14 illustrates the solution when $\gamma=1$

• Example 4.3.15

Consider the mixed boundary value problem

$$\frac{\partial^2 u}{\partial r^2}+\frac{1}{r}\frac{\partial u}{\partial r}+\frac{\partial^2 u}{\partial z^2}=0,\qquad 0\le r<\infty,\quad 0<z<\infty,\tag{4.3.327}$$

subject to the boundary conditions

$$\lim_{r\to 0}|u(r,z)|<\infty,\qquad \lim_{r\to\infty}u(r,z)\to 0,\qquad 0<z<\infty,\tag{4.3.328}$$

$$\lim_{z\to\infty} u(r,z) \to 0, \qquad 0 \le r < \infty, \tag{4.3.329}$$

and

$$\begin{cases} \alpha u_z(r,0) - \beta u(r,0) = -f(r), & 0 \le r < 1, \\ u(r,0) = 0, & 1 < r < \infty. \end{cases} \tag{4.3.330}$$

Both α and β are nonzero.

In line with previous examples the solution that satisfies Equation 4.3.327 through Equation 4.3.329 is

$$u(r,z) = \int_0^\infty A(k)e^{-kz} J_0(kr)\, dk. \tag{4.3.331}$$

Substituting $u(r,z)$ into the mixed boundary condition,

$$\int_0^\infty (\alpha k + \beta)A(k)J_0(kr)\, dk = f(r), \qquad 0 \le r < 1, \tag{4.3.332}$$

and

$$\int_0^\infty A(k)J_0(kr)\, dk = 0, \qquad 1 \le r < \infty. \tag{4.3.333}$$

At this point we introduce an integral definition for $A(k)$,

$$A(k) = \int_0^1 h(t)\sin(kt)\, dt, \qquad h(0) = 0. \tag{4.3.334}$$

The demonstration that this definition of $A(k)$ satisfies Equation 4.3.333 is left as part of Problem 6. Turning to Equation 4.3.332, we substitute $A(k)$ into Equation 4.3.332 and find that

$$\begin{aligned} \alpha \int_0^1 h(t) \left[\int_0^\infty k\sin(kt)J_0(kr)\, dk\right] dt & \\ + \beta \int_0^1 h(t)\left[\int_0^\infty \sin(kt)J_0(kr)\, dk\right] dt = f(r), & \qquad 0 \le r < 1. \end{aligned} \tag{4.3.335}$$

At this point, we would normally manipulate Equation 4.3.335 into a Fredholm integral equation. This is left as an exercise in Problem 6. Here we introduce an alternative method developed by Gladwell et al.[62] The derivation begins by showing that

$$\int_0^x \left[t\int_0^\infty kA(k)J_0(kt)\, dk\right] \frac{dt}{\sqrt{x^2-t^2}} = \int_0^\infty A(k)\sin(kx)\, dk. \tag{4.3.336}$$

---

[62] Taken from Gladwell, G. M. L., J. R. Barber, and Z. Olesiak, 1983: Thermal problems with radiation boundary conditions. *Quart. J. Mech. Appl. Math.*, **36**, 387–401 by permission of Oxford University Press; see also Lemczyk, T. F., and M. M. Yovanovich, 1988: Thermal constriction resistance with convective boundary conditions–1. Half-space contacts. *Int. J. Heat Mass Transfer*, **31**, 1861–1872.

This follows from interchanging the order of integration and applying Equation 1.4.9. Next, we view the quantity within the square brackets on the left side of Equation 4.3.336 as the unknown in an integral equation of the Abel type. From Equation 1.2.13 and Equation 1.2.14, we have that

$$t\int_0^\infty kA(k)J_0(kt)\,dk = \frac{2}{\pi}\frac{d}{dt}\left\{\int_0^t\left[\int_0^\infty A(k)\sin(k\tau)\,dk\right]\frac{\tau\,d\tau}{\sqrt{t^2-\tau^2}}\right\} \tag{4.3.337}$$

$$= \frac{2}{\pi}\frac{d}{dt}\left\{\int_0^\infty A(k)\left[\int_0^t \frac{\tau\sin(k\tau)}{\sqrt{t^2-\tau^2}}\,d\tau\right]dk\right\} \tag{4.3.338}$$

$$= \frac{d}{dt}\left[t\int_0^\infty A(k)J_1(kt)\,dk\right] \tag{4.3.339}$$

$$= \frac{d}{dt}\left\{t\int_0^1 h(\tau)\left[\int_0^\infty \sin(k\tau)J_1(kt)\,dk\right]d\tau\right\} \tag{4.3.340}$$

$$= \frac{d}{dt}\left[\int_0^t \frac{\tau\,h(\tau)}{\sqrt{t^2-\tau^2}}\,d\tau\right]. \tag{4.3.341}$$

If we divide the left side of Equation 4.3.341 by t, we have the first term on the left side of Equation 4.3.335. The second term can be evaluated from integral tables.[63] Consequently Equation 4.3.335 becomes

$$\frac{\alpha}{r}\frac{d}{dr}\left[\int_0^r \frac{t\,h(t)}{\sqrt{r^2-t^2}}\,dt\right] + \beta\int_r^1 \frac{h(t)}{\sqrt{t^2-r^2}}\,dt = f(r), \quad 0 \le r < 1. \tag{4.3.342}$$

To solve Equation 4.3.342, let

$$h(t) = \psi(\theta) = \sum_{n=0}^\infty A_n \cos[(2n+1)\theta], \qquad 0 < \theta < \pi/2, \tag{4.3.343}$$

and

$$\sin(\theta)\psi(\theta) = \sum_{n=0}^\infty B_n \cos[(2n+1)\theta], \qquad 0 < \theta < \pi/2, \tag{4.3.344}$$

where $t = \cos(\theta)$. If $r = \cos(\varphi)$, then

$$\int_0^r \frac{t\,h(t)}{\sqrt{r^2-t^2}}\,dt = {\sum_{n=0}^\infty}' A_n \int_\varphi^\pi \frac{\sin(2\theta)\cos[(2n+1)\theta]}{\sqrt{2[\cos(2\varphi)-\cos(2\theta)]}}\,d\theta \tag{4.3.345}$$

$$= \frac{\pi}{8}{\sum_{n=0}^\infty}' A_n\left\{P_{n+1}[\cos(2\varphi)] - P_{n-1}[\cos(2\varphi)]\right\} \tag{4.3.346}$$

63 Gradshteyn and Ryzhik, op. cit., Formula 4.671.1.

from Equation 1.3.4 and Equation 1.3.5, where the prime denotes that whenever $P_{-1}(\cdot)$ occurs, then it is replaced by $P_0(\cdot)$. Similarly,

$$\int_r^1 \frac{h(t)}{\sqrt{r^2-t^2}}\,dt = 2\sum_{n=0}^{\infty} B_n \int_0^{\varphi} \frac{\cos[(2n+1)\theta]}{\sqrt{2[\cos(2\varphi)-\cos(2\theta)]}}\,d\theta \tag{4.3.347}$$

$$= \frac{\pi}{2}\sum_{n=0}^{\infty} B_n P_n[\cos(2\varphi)]. \tag{4.3.348}$$

Because

$$\frac{P_n(x)-P_{n+2}(x)}{1-x^2} = \frac{(2n+3)}{2(n+1)(n+2)}\sum_{k=0}^{n}\left[1+(-1)^{n+k}\right](2k+1)P_k(x) \tag{4.3.349}$$

and using the derivative rule for Legendre polynomials, we find that

$$\frac{1}{r}\frac{d}{dr}\left[P_{n+1}(2r^2-1)-P_{n-1}(2r^2-1)\right] = 4(2n+1)P_n(2r^2-1). \tag{4.3.350}$$

Therefore, Equation 4.3.342 becomes

$$\frac{\alpha\pi}{2}\sum_{n=0}^{\infty}(2n+1)A_nP_n(2r^2-1)+\frac{\beta\pi}{2}\sum_{n=0}^{\infty}B_nP_n(2r^2-1)=f(r). \tag{4.3.351}$$

If we reexpress $f(r)$ as the Fourier-Legendre expansion

$$f(r) = \frac{\pi}{2}\sum_{n=0}^{\infty}(2n+1)C_nP_n(2r^2-1), \tag{4.3.352}$$

then

$$\alpha A_n + \frac{\beta}{2n+1}B_n = C_n, \qquad n=0,1,2,\ldots. \tag{4.3.353}$$

In addition to Equation 4.3.353, we also have from Equation 4.3.344 that

$$B_n = \sum_{m=0}^{\infty} c_{m,n}A_n, \tag{4.3.354}$$

where

$$c_{m,n} = \frac{2}{\pi}\int_0^{\pi}\sin(\theta)\cos[(2m+1)\theta]\cos[(2n+1)\theta]\,d\theta \tag{4.3.355}$$

$$= \frac{1}{\pi}\left(\frac{1}{2n+2m+3}-\frac{1}{2n+2m+1}+\frac{1}{2n-2m+1}-\frac{1}{2n-2m-1}\right). \tag{4.3.356}$$

We can now solve for A_n via Equation 4.3.353 and Equation 4.3.354 after we truncate the set of infinite equations to a finite number. Then $h(t)$ follows from Equation 4.3.343. Finally, $A(k)$ is computed from Equation 4.3.334 while $u(r, z)$ is obtained from Equation 4.3.331.

Problems

1. Solve the potential problem[64]

$$\frac{\partial^2 u}{\partial r^2} + \frac{1}{r}\frac{\partial u}{\partial r} + \frac{\partial^2 u}{\partial z^2} = 0, \quad 0 \le r < \infty, \quad 0 < z < \infty,$$

subject to the boundary conditions

$$\lim_{r\to 0} |u(r,z)| < \infty, \qquad \lim_{r\to\infty} |u(r,z)| < \infty, \qquad 0 < z < \infty,$$

$$\lim_{z\to\infty} u(r,z) \to u_\infty, \qquad 0 \le r < \infty,$$

and

$$\begin{cases} u(r,0) = u_\infty - \Delta u, & 0 \le r < a, \\ u_z(r,0) = 0, & a \le r < \infty, \end{cases}$$

where u_∞ and Δu are constants.

Step 1: Show that

$$u(r,z) = u_\infty - \int_0^\infty A(k) e^{-kz} J_0(kr)\, dk$$

satisfies the partial differential equation and the boundary conditions as $r \to 0$, $r \to \infty$, and $z \to \infty$.

Step 2: Show that

$$\int_0^\infty k A(k) J_0(kr)\, dk = 0, \qquad a < r < \infty.$$

Step 3: Using the relationship[65]

$$\int_0^\infty \sin(ka) J_0(kr)\, dk = \begin{cases} (a^2 - r^2)^{-\frac{1}{2}}, & r < a, \\ 0, & r > a, \end{cases}$$

---

[64] See Fleischmann, M., and S. Pons, 1987: The behavior of microdisk and microring electrodes. *J. Electroanal. Chem.*, **222**, 107–115; Gupta, S. C., 1957: Slow broad side motion of a flat plate in a viscous liquid. *Z. Angew. Math. Phys.*, **8**, 257–261.

[65] Gradshteyn and Ryzhik, op. cit., Formula 6.671.7.

show that $kA(k) = C\sin(ka)$.

Step 4: Using the relationship[66]

$$\int_0^\infty \sin(ka)J_0(kr)\,\frac{dk}{k} = \begin{cases} \pi/2, & r \le a, \\ \arcsin(a/r), & r \ge a, \end{cases}$$

show that

$$\begin{aligned} u(r,z) &= u_\infty - \frac{2\Delta u}{\pi}\int_0^\infty e^{-kz}\sin(ka)J_0(kr)\,\frac{dk}{k} \\ &= u_\infty - \frac{2\Delta u}{\pi}\arcsin\left[\frac{2a}{\sqrt{(r-a)^2+z^2}+\sqrt{(r+a)^2+z^2}}\right]. \end{aligned}$$

Wiley and Webster[67] used this solution in an improved design for a circular electrosurgical dispersive electrode.

2. Solve Laplace's equation[68]

$$\frac{1}{r}\frac{\partial}{\partial r}\left(r\frac{\partial u}{\partial r}\right) + \frac{\partial^2 u}{\partial z^2} = 0, \qquad 0 \le r < \infty, \quad 0 < z < \infty,$$

with the boundary conditions

$$\lim_{r\to 0}|u(r,z)| < \infty, \qquad \lim_{r\to\infty} u(r,z) \to 0, \qquad 0 < z < \infty,$$

$$\lim_{z\to\infty} u(r,z) \to 0, \qquad 0 \le r < \infty,$$

and

$$\begin{cases} u_z(r,0) = 1, & 0 \le r < 1, \\ u(r,0) = 0, & 1 < r < \infty. \end{cases} \tag{1}$$

See Problem 5 for a generalization of this problem.

Step 1: Using separation of variables or transform methods, show that the general solution to the problem is

$$u(r,z) = \int_0^\infty kA(k)e^{-kz}J_0(kr)\,dk.$$

---

[66] Ibid., Formula 6.693.1 with $\nu = 0$.

[67] Wiley, J. D., and J. G. Webster, 1982: Analysis and control of the current distribution under circular dispersive electrode. *IEEE Trans. Biomed. Engng.*, **BME-29**, 381–385.

[68] See Yang, F.-Q., and J. C. M. Li, 1995: Impression and diffusion creep of anisotropic media. *J. Appl. Phys.*, **77**, 110–117. This problem also appears while finding the temperature field in a paper by Florence, A. L., and J. N. Goodier, 1963: The linear thermoelastic problem of uniform heat flow disturbed by a penny-shaped insulated crack. *Int. J. Engng. Sci.*, **1**, 533–540.

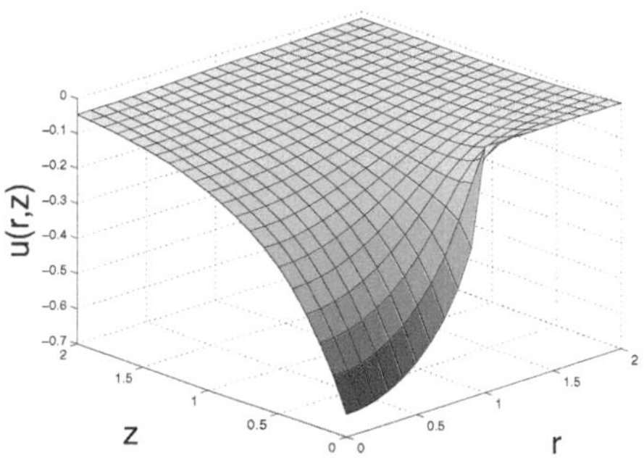

Problem 2

Step 2: Using boundary condition (1), show that $A(k)$ satisfies the dual integral equations

$$\int_0^\infty k^2 A(k) J_0(kr)\, dk = -1, \qquad 0 \le r < 1,$$

and

$$\int_0^\infty k A(k) J_0(kr)\, dk = 0, \qquad 1 < r < \infty.$$

Step 3: Using Equation 1.4.13 and the result[69] from integral tables that

$$\int_0^\infty J_\nu(ax) \sin(bx) \frac{dx}{x} = \begin{cases} \sin[\nu \arcsin(b/a)]/\nu, & b \le a, \\ \dfrac{a^\nu \sin(\nu\pi/2)}{\nu \left(b + \sqrt{b^2 - a^2}\right)^\nu} & b \ge a, \end{cases}$$

if $\Re(\nu) > -1$, show that

$$A(k) = -\frac{2}{k\pi} \int_0^1 t \sin(kt)\, dt = -\frac{2}{\pi} \left[\frac{\sin(k)}{k^3} - \frac{\cos(k)}{k^2} \right]$$

satisfies both integral equations given in Step 2.

Step 4: Show that the solution to this problem is

$$u(r, z) = -\frac{2}{\pi} \int_0^\infty \left[\frac{\sin(k)}{k^2} - \frac{\cos(k)}{k} \right] e^{-kz} J_0(kr)\, dk.$$

In particular, show that $u(r, 0) = -2\sqrt{1 - r^2}/\pi$ if $0 \le r < 1$. The figure labeled Problem 2 illustrates the solution $u(r, z)$.

---

[69] Gradshteyn and Ryzhik, op. cit., Formula 6.693.1.

3. Solve the potential problem[70]

$$\frac{1}{r}\frac{\partial}{\partial r}\left(r\frac{\partial u}{\partial r}\right)+\frac{\partial^2 u}{\partial z^2}=0, \qquad 0 \le r < \infty, \quad 0 < z < \infty,$$

subject to the boundary conditions

$$\lim_{r\to 0}|u(r,z)|<\infty, \qquad \lim_{r\to\infty} u(r,z)\to 0, \qquad 0<z<\infty,$$

$$\lim_{z\to\infty} u(r,z)\to 0, \qquad 0 \le r < \infty,$$

and

$$\begin{cases} u(r,0)=U_0, & r<a, \\ u_z(r,0)=0, & r>a. \end{cases}$$

Step 1: By using either separation of variables or transform methods, show that the general solution to partial differential equation is

$$u(r,z)=\frac{2}{\pi}\int_0^\infty A(k)e^{-kz}J_0(kr)\,dk.$$

Note that this solution also satisfies the first three boundary conditions.

Step 2: Setting

$$A(k)=\int_0^a f(\xi)\cos(k\xi)\,d\xi,$$

show that $u_z(r,0)=0$ if $r>a$.

Step 3: Show that

$$\frac{2}{\pi}\int_0^\infty J_0(kr)\left[\int_0^a f(\xi)\cos(k\xi)\,d\xi\right]dk=U_0, \qquad 0<r<a.$$

Step 4: By replacing r by η in the integral equation given in Step 3, multiplying the resulting equation by $\eta\,d\eta/\sqrt{r^2-\eta^2}$, integrating from 0 to r, and taking the derivative with respect to r, show that

$$\frac{2}{\pi}\frac{d}{dr}\left[\int_0^r \frac{\eta}{\sqrt{r^2-\eta^2}}\left\{\int_0^\infty J_0(k\eta)\left[\int_0^a f(\xi)\cos(k\xi)\,d\xi\right]dk\right\}d\eta\right]=U_0.$$

Step 5: By interchanging the order of the η and k integrations and using the relationship that

$$\int_0^r \frac{\eta\,J_0(k\eta)}{\sqrt{r^2-\eta^2}}\,d\eta=\frac{\sin(kr)}{k},$$

[70] See Laporte, O., and R. G. Fowler, 1967: Weber's mixed boundary value problem in electrodynamics. *J. Math. Phys.*, **8**, 518–522.

show that the integral equation in Step 4 simplifies to

$$\frac{2}{\pi}\frac{d}{dr}\left\{\int_0^a f(\xi)\left[\int_0^\infty \sin(kr)\cos(k\xi)\,\frac{dk}{k}\right]\,d\xi\right\} = U_0.$$

Step 6: Use the results from Step 5 to show that if $f(\xi) = U_0$, then $A(k) = U_0\sin(ka)/k$, and

$$\begin{aligned} u(r,z) &= \frac{2U_0}{\pi}\int_0^\infty \frac{\sin(ka)}{k}e^{-kz}J_0(kr)\,dk \\ &= \frac{2U_0}{\pi}\arcsin\left[\frac{2a}{\sqrt{(r-a)^2+z^2}+\sqrt{(r+a)^2+z^2}}\right]. \end{aligned}$$

4. Solve Laplace's equation[71]

$$\frac{\partial^2 u}{\partial r^2} + \frac{1}{r}\frac{\partial u}{\partial r} + \frac{\partial^2 u}{\partial z^2} = 0, \qquad 0 \le r < \infty, \quad 0 < z < \infty,$$

subject to the boundary conditions

$$\lim_{r\to 0}|u(r,z)| < \infty, \qquad \lim_{r\to\infty} u(r,z) \to 0, \qquad 0 < z < \infty,$$

$$\lim_{z\to\infty} u(r,z) \to 0, \qquad 0 \le r < \infty,$$

and

$$\begin{cases} u_z(r,0) = 1, & 0 \le r < a, \\ u(r,0) = 0, & a < r < \infty. \end{cases}$$

Step 1: Show that

$$u(r,z) = \int_0^\infty kA(k)e^{-kz}J_0(kr)\,dk$$

satisfies the partial differential equation and the boundary conditions provided that $A(k)$ satisfies the dual integral equations

$$\int_0^\infty k^2A(k)J_0(kr)\,dk = -1, \qquad 0 \le r < a, \tag{1}$$

and

$$\int_0^\infty kA(k)J_0(kr)\,dk = 0, \qquad a \le r < \infty. \tag{2}$$

---

[71] See Shindo, Y., 1986: The linear magnetoelastic problem of a uniform current flow disturbed by a penny-shaped crack in a constant axial magnetic field. *Eng. Fract. Mech.*, **23**, 977–982.

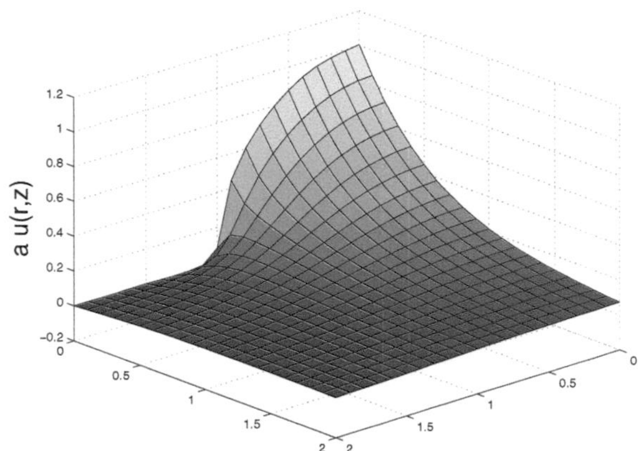

Problem 4

Step 2: By introducing

$$kA(k) = \int_0^a h(t)\sin(kt)\,dt, \qquad h(0) = 0,$$

show that Equation (2) is automatically satisfied.

Step 3: By using Equation (1), show that $h(t) = -2t/\pi$ and

$$kA(k) = -\frac{2}{\pi k^2}\left[\sin(ka) - ka\cos(ka)\right].$$

Step 4: Show that the solution to the problem is

$$u(r,z) = -\frac{2}{\pi}\int_0^\infty \left[\sin(ka) - ka\cos(ka)\right] e^{-kz} J_0(kr)\,\frac{dk}{k^2}.$$

The figure labeled Problem 4 illustrates this solution $u(r,z)$.

5. Solve Laplace's equation[72]

$$\frac{\partial^2 u}{\partial r^2} + \frac{1}{r}\frac{\partial u}{\partial r} + \frac{\partial^2 u}{\partial z^2} = 0, \qquad 0 \le r < \infty, \quad 0 < z < \infty,$$

subject to the boundary conditions

$$\lim_{r\to 0} |u(r,z)| < \infty, \qquad \lim_{r\to\infty} u(r,z) \to 0, \qquad 0 < z < \infty,$$

[72] See Riffert, H., 1980: Pulsating X-ray sources: The oblique dipole configuration. *Astrophys. Space Sci.*, **71**, 195–201.

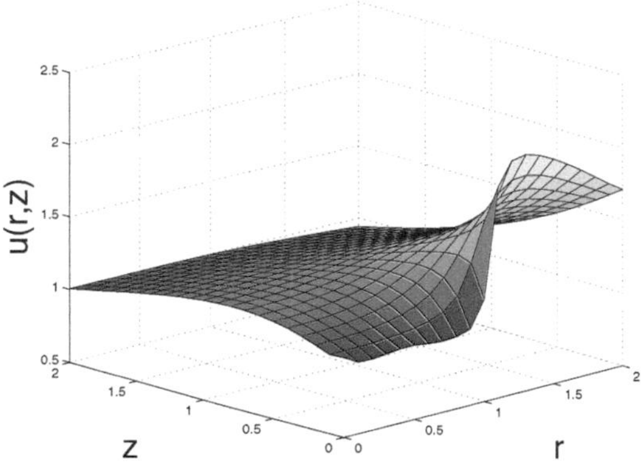

Problem 5

$$\lim_{z \to \infty} u(r, z) \to 0, \qquad 0 \le r < \infty,$$

and

$$\begin{cases} u(r, 0) = C_1, & 0 \le r < 1, \\ u_z(r, 0) = -C_2/r^3, & 1 < r < \infty. \end{cases}$$

Step 1: Show that

$$u(r, z) = \int_0^\infty A(k) e^{-kz} J_0(kr)\, dk$$

satisfies the partial differential equation and the boundary conditions provided that $A(k)$ satisfies the dual integral equations

$$\int_0^\infty A(k) J_0(kr)\, dk = C_1, \qquad 0 \le r < 1,$$

and

$$\int_0^\infty kA(k) J_0(kr)\, dk = C_2/r^3, \qquad 1 \le r < \infty.$$

Step 2: Show that

$$A(k) = \frac{2}{\pi}(C_1 - C_2)\frac{\sin(k)}{k} + \frac{4C_2}{\pi k} \int_1^\infty \sin(k\xi)\, \frac{d\xi}{\xi^3}.$$

The figure labeled Problem 5 illustrates this solution $u(r, z)$ when $C_1 = 1$ and $C_5 = 5$.

6. Solve the partial differential equation[73]

$$\frac{\partial^2 u}{\partial r^2} + \frac{1}{r}\frac{\partial u}{\partial r} - \frac{u}{r^2} + \frac{\partial^2 u}{\partial z^2} = 0, \qquad 0 \le r < \infty, \quad 0 < z < \infty,$$

subject to the boundary conditions

$$\lim_{r\to 0} |u(r,z)| < \infty, \qquad \lim_{r\to\infty} u(r,z) \to 0, \qquad 0 < z < \infty,$$

$$\lim_{z\to\infty} u(r,z) \to 0, \qquad 0 \le r < \infty,$$

and

$$\begin{cases} u_z(r,0) = 0, & 0 \le r < a, \\ u(r,0) = a/r, & a < r < \infty. \end{cases}$$

Step 1: Show that

$$u(r,z) = \int_0^\infty A(k) e^{-kz} J_1(kr)\, dk$$

satisfies the partial differential equation and the boundary conditions provided that $A(k)$ satisfies the dual integral equations

$$\int_0^\infty k\, A(k) J_1(kr)\, dk = 0, \qquad 0 \le r < a,$$

and

$$\int_0^\infty A(k) J_1(kr)\, dk = a/r, \qquad a \le r < \infty.$$

Step 2: Show that

$$A(k) = \frac{\sin(ka)}{k}$$

satisfies the integral equations. The figure labeled Problem 6 illustrates this solution.

7. A generalization of a problem originally suggested by Popova[74] was given by Kuz'min[75] who solved

$$\frac{\partial^2 u}{\partial r^2} + \frac{1}{r}\frac{\partial u}{\partial r} + \frac{\partial^2 u}{\partial z^2} = 0, \qquad 0 \le r < \infty, \quad 0 < z < \infty,$$

---

[73] Suggested by a problem solved in Appendix A of Raynolds, J. E., B. A. Munk, J. B. Pryor, and R. J. Marhefka, 2003: Ohmic loss in frequency-selective surfaces. *J. Appl. Phys.*, **93**, 5346–5358.

[74] Popova, A. P., 1973: Nonstationary mixed problem of thermal conductivity for the half-space. *J. Engng. Phys.*, **25**, 934–935.

[75] Kuz'min, Yu. N., 1966: Some axially symmetric problems in heat flow with mixed boundary conditions. *Sov. Tech. Phys.*, **11**, 169–173.

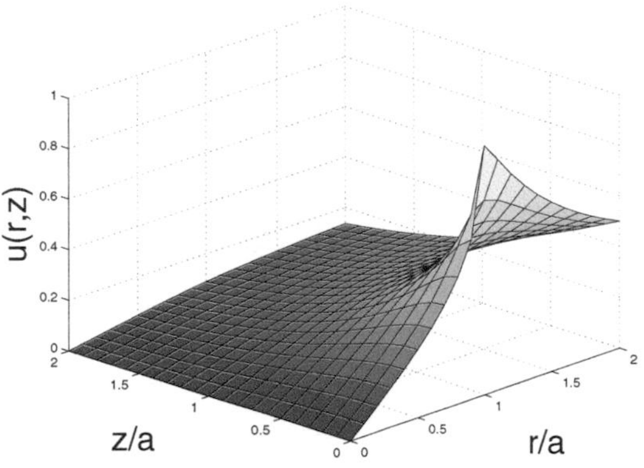

Problem 6

subject to the boundary conditions

$$\lim_{r\to 0} |u(r,z)| < \infty, \qquad \lim_{r\to\infty} u(r,z) \to 0, \qquad 0 < z < \infty,$$

$$\lim_{z\to\infty} u(r,z) \to 0, \qquad 0 \le r < \infty,$$

and

$$\begin{cases} \alpha u_z(r,0) - \beta u(r,0) = -f(r), & 0 < r < 1, \\ u(r,0) = 0, & 1 < r < \infty. \end{cases}$$

Step 1: Show that

$$u(r,z) = \int_0^\infty A(k) e^{-kz} J_0(kr)\, dk$$

satisfies the partial differential equation and the boundary conditions provided that $A(k)$ satisfies the dual integral equations

$$\int_0^\infty (\alpha k + \beta) A(k) J_0(kr)\, dk = f(r), \qquad 0 \le r < 1, \tag{1}$$

and

$$\int_0^\infty A(k) J_0(kr)\, dk = 0, \qquad 1 \le r < \infty. \tag{2}$$

Step 2: By introducing

$$A(k) = \int_0^1 h(t) \sin(kt)\, dt, \qquad h(0) = 0,$$

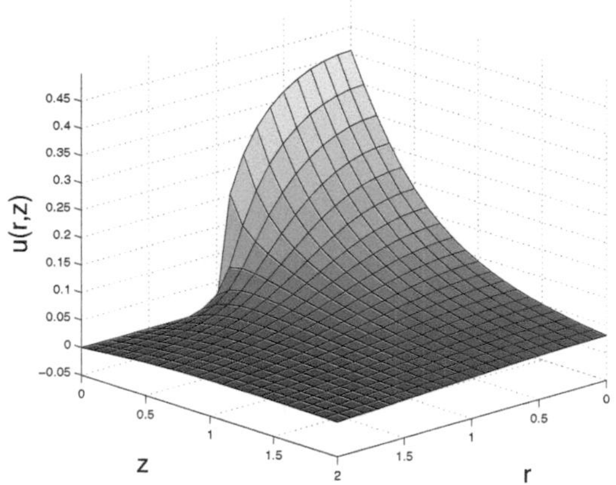

Problem 7

show that Equation (2) is automatically satisfied.

Step 3: By using Equation (1) and Equation 1.4.14 and noting that

$$kA(k) = -h(1)\cos(k) + \int_0^1 h'(t)\cos(kt)\,dt,$$

show that

$$\alpha \int_0^r \frac{h'(t)}{\sqrt{r^2 - t^2}}\,dt = f(r) - \beta \int_0^1 h(\tau)\left[\int_0^\infty \sin(k\tau) J_0(kr)\,dk\right]d\tau.$$

Step 4: Using Equation 1.2.13, Equation 1.2.14 and Equation 1.4.9, show that

$$\alpha h(t) = \frac{2}{\pi}\int_0^t \frac{rf(r)}{\sqrt{t^2 - r^2}}\,dr - \frac{2\beta}{\pi}\int_0^1 K(t,\tau)h(\tau)\,d\tau,$$

where

$$K(t,\tau) = \int_0^\infty \sin(kt)\sin(k\tau)\frac{dk}{k} = \frac{1}{2}\ln\left|\frac{t+\tau}{t-\tau}\right|.$$

The figure labeled Problem 7 illustrates the solution $u(r,z)$ when $\alpha = \beta = 1$ and $f(r) = 1$.

8. Solve[76] Laplace's equation

$$\frac{\partial^2 u}{\partial r^2} + \frac{1}{r}\frac{\partial u}{\partial r} + \frac{\partial^2 u}{\partial z^2} = 0, \qquad 0 \le r < \infty, \quad 0 < z < \infty,$$

[76] A generalization of a problem solved by Mahalanabis, R. K., 1967: A mixed boundary-value problem of thermoelasticity for a half-space. *Quart. J. Mech. Appl. Math.*, **20**, 127–134.

subject to the boundary conditions

$$\lim_{r\to 0} |u(r,z)| < \infty, \qquad \lim_{r\to\infty} u(r,z) \to 0, \qquad 0 < z < \infty,$$

$$\lim_{z\to\infty} u(r,z) \to 0, \qquad 0 \le r < \infty,$$

and

$$\begin{cases} u(r,0) = f(r), & 0 \le r < 1, \\ \gamma u_z(r,0) = \delta u(r,0), & 1 < r < \infty. \end{cases}$$

Step 1: Show that

$$u(r,z) = \int_0^\infty A(k) e^{-kz} J_0(kr)\, dk$$

satisfies the partial differential equation and the boundary conditions provided that $A(k)$ satisfies the dual integral equations

$$\int_0^\infty A(k) J_0(kr)\, dk = f(r), \qquad 0 \le r < 1, \tag{1}$$

and

$$\int_0^\infty (\gamma k + \delta) A(k) J_0(kr)\, dk = 0, \qquad 1 \le r < \infty. \tag{2}$$

Step 2: By introducing

$$(\gamma k + \delta) A(k) = \gamma k \int_0^1 h(t) \cos(kt)\, dt = \gamma h(1) \sin(k) - \gamma \int_0^1 h'(t) \sin(kt)\, dt,$$

show that Equation (2) is automatically satisfied.

Step 3: By using Equation (1) and Equation 1.4.14 and noting that

$$A(k) = \int_0^1 h(t) \cos(kt)\, dt - \frac{\delta}{\gamma k + \delta} \int_0^1 h(t) \cos(kt)\, dt,$$

show that

$$\int_0^r \frac{h(t)}{\sqrt{r^2 - t^2}}\, dt = f(r) + \delta \int_0^1 h(\tau) \left[\int_0^\infty \cos(k\tau) J_0(kr) \frac{dk}{\gamma k + \delta} \right] d\tau.$$

Step 4: Using Equation 1.2.13, Equation 1.2.14 and Equation 1.4.9, show that

$$h(t) = \frac{2}{\pi} \frac{d}{dt} \left[\int_0^t \frac{r f(r)}{\sqrt{t^2 - r^2}}\, dr \right] - \frac{2}{\pi} \int_0^1 K(t,\tau) h(\tau)\, d\tau,$$

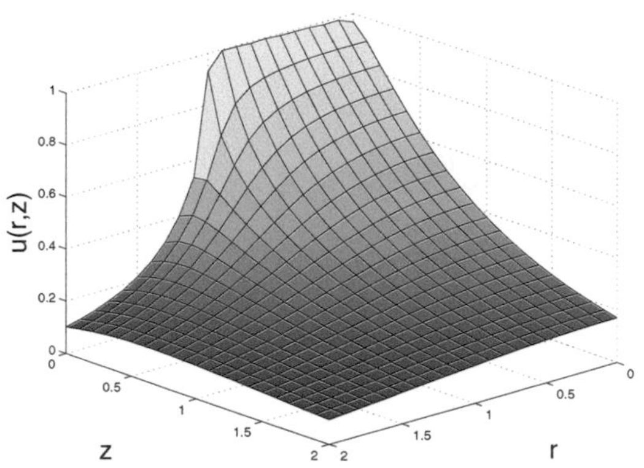

Problem 8

where

$$K(t,\tau) = -\delta \int_0^\infty \cos(kt)\cos(k\tau)\frac{dk}{\gamma k + \delta}$$
$$= \frac{\delta}{2\gamma}\{\sin[\lambda(t+\tau)]\,\mathrm{si}[\lambda(t+\tau)] + \cos[\lambda(t+\tau)]\,\mathrm{ci}[\lambda(t+\tau)]$$
$$+ \sin[\lambda|t-\tau|]\,\mathrm{si}[\lambda|t-\tau|] + \cos[\lambda|t-\tau|]\,\mathrm{ci}[\lambda|t-\tau|]\},$$

$\lambda = \delta/\gamma$ and $\mathrm{si}(\cdot)$ and $\mathrm{ci}(\cdot)$ are the sine and cosine integrals. The figure labeled Problem 8 illustrates the solution $u(r,z)$ when $\gamma = \delta = 1$ and $f(r) = 1$.

9. Solve Laplace's equation

$$\frac{\partial^2 u}{\partial r^2} + \frac{1}{r}\frac{\partial u}{\partial r} + \frac{\partial^2 u}{\partial z^2} = 0, \qquad 0 \le r < \infty, \quad 0 < z < \infty,$$

subject to the boundary conditions

$$\lim_{r\to 0} |u(r,z)| < \infty, \qquad \lim_{r\to\infty} u(r,z) \to 0, \qquad 0 < z < \infty,$$

$$\lim_{z\to\infty} u(r,z) \to 0, \qquad 0 \le r < \infty,$$

and

$$\begin{cases} \alpha u_z(r,0) - \beta u(r,0) = -f(r), & 0 \le r < 1, \\ u_z(r,0) = 0, & 1 < r < \infty. \end{cases}$$

Step 1: Show that

$$u(r,z) = \int_0^\infty A(k)e^{-kz}J_0(kr)\,dk$$

satisfies the partial differential equation and the boundary conditions provided that $A(k)$ satisfies the dual integral equations

$$\int_0^\infty (\alpha k + \beta) A(k) J_0(kr)\, dk = f(r), \qquad 0 \le r < 1, \tag{1}$$

and

$$\int_0^\infty k A(k) J_0(kr)\, dk = 0, \qquad 1 \le r < \infty. \tag{2}$$

Step 2: By introducing

$$A(k) = \int_0^1 h(t) \cos(kt)\, dt, \qquad h(1) = 0,$$

or

$$kA(k) = -\int_0^1 h'(t) \sin(kt)\, dt,$$

show that Equation (2) is automatically satisfied.

Step 3: By using Equation (1) and Equation 1.4.13, show that

$$\alpha \int_r^1 \frac{h'(t)}{\sqrt{t^2 - r^2}}\, dt = f(r) - \beta \int_0^1 h(\tau) \left[\int_0^\infty \cos(k\tau) J_0(kr)\, dk \right] d\tau.$$

Step 4: Using Equation 1.2.15, Equation 1.2.16 and Equation 1.4.14, show that

$$\alpha h(t) = -\frac{2}{\pi} \int_t^1 \frac{r\, f(r)}{\sqrt{r^2 - t^2}}\, dr + \frac{2\beta}{\pi} \int_0^1 K(t, \tau) h(\tau)\, d\tau,$$

where

$$\begin{aligned} K(t, \tau) &= \int_t^1 \frac{r}{\sqrt{r^2 - t^2}} \left[\int_0^\infty \cos(k\tau) J_0(kr)\, dk \right] dr \\ &= \ln \left[\frac{\sqrt{1 - t^2} + \sqrt{1 - \tau^2}}{\sqrt{|t^2 - \tau^2|}} \right]. \end{aligned}$$

The figure labeled Problem 9 illustrates the solution $u(r, z)$ when $\alpha = \beta = 1$ and $f(r) = 1$.

10. Solve Laplace's equation

$$\frac{1}{r} \frac{\partial}{\partial r} \left(r \frac{\partial u}{\partial r} \right) + \frac{\partial^2 u}{\partial z^2} = 0, \qquad 0 \le r < \infty, \quad 0 < z < \infty,$$

with the boundary conditions

$$\lim_{r \to 0} |u(r, z)| < \infty, \qquad \lim_{r \to \infty} u(r, z) \to 0, \qquad 0 < z < \infty,$$

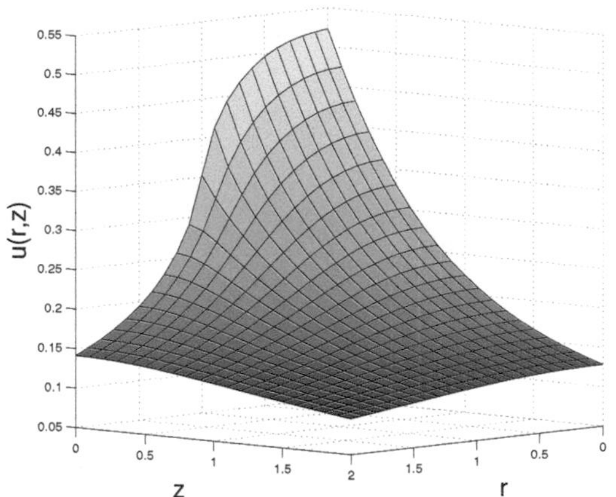

Problem 9

$$\lim_{z\to\infty} u(r,z) \to 0, \qquad 0 \le r < \infty,$$

and

$$\begin{cases} u_z(r,0) = -f(r), & 0 \le r < 1, \\ u(r,0) = 0, & 1 < r < \infty. \end{cases} \tag{1}$$

Step 1: Using separation of variables or transform methods, show that the general solution to the problem is

$$u(r,z) = \int_0^\infty A(k) e^{-kz} J_0(kr)\, dk.$$

Step 2: Using boundary condition (1), show that $A(k)$ satisfies the dual integral equations

$$\int_0^\infty k A(k) J_0(kr)\, dk = f(r), \qquad 0 \le r < 1,$$

and

$$\int_0^\infty A(k) J_0(kr)\, dk = 0, \qquad 1 < r < \infty.$$

Step 3: If

$$A(k) = \int_0^1 g(t) \sin(kt)\, dt,$$

show that $A(k)$ satisfies the second integral equation in Step 2 identically.

Step 4: Using the first integral equation in Step 2, show that

$$g(t) = \frac{2}{\pi} \int_0^t \frac{r f(r)}{\sqrt{t^2 - r^2}}\, dr.$$

Step 5: Because

$$k\, A(k) = -g(1)\cos(k) + \int_0^1 g'(t)\cos(kt)\, dt, \qquad g(0) = 0,$$

show that

$$\int_0^r \frac{g'(t)}{\sqrt{r^2 - t^2}}\, dt = f(r),$$

or

$$g(t) = \frac{2}{\pi} \int_0^t \frac{r f(r)}{\sqrt{t^2 - r^2}}\, dr.$$

11. Let us solve Laplace's equation[77]

$$\frac{1}{r}\frac{\partial}{\partial r}\left(r\frac{\partial u}{\partial r}\right) + \frac{\partial^2 u}{\partial z^2} = 0, \qquad 0 \le r < \infty, \quad 0 < z < \infty,$$

with the boundary conditions

$$\lim_{r\to 0} |u(r,z)| < \infty, \qquad \lim_{r\to\infty} u(r,z) \to 0, \qquad 0 < z < \infty,$$

$$\lim_{z\to\infty} u(r,z) \to 0, \qquad 0 \le r < \infty,$$

and

$$\begin{cases} u_z(r,0) = 0, & 0 \le r < 1, \\ u(r,0) = f(r), & 1 < r < \infty. \end{cases} \tag{1}$$

Step 1: Using separation of variables or transform methods, show that the general solution to the problem is

$$u(r,z) = \int_0^\infty A(k) e^{-kz} J_0(kr)\, \frac{dk}{k}.$$

Step 2: Using boundary condition (1), show that $A(k)$ satisfies the dual integral equations

$$\int_0^\infty A(k) J_0(kr)\, dk = 0, \qquad 0 \le r < 1,$$

[77] Reprinted from *Int. J. Engng. Sci.*, **9**, B. R. Das, Some axially symmetric thermal stress distributions in elastic solids containing cracks – I. An external crack in an infinite solid, 469–478, ©1971, with permission Elsevier.

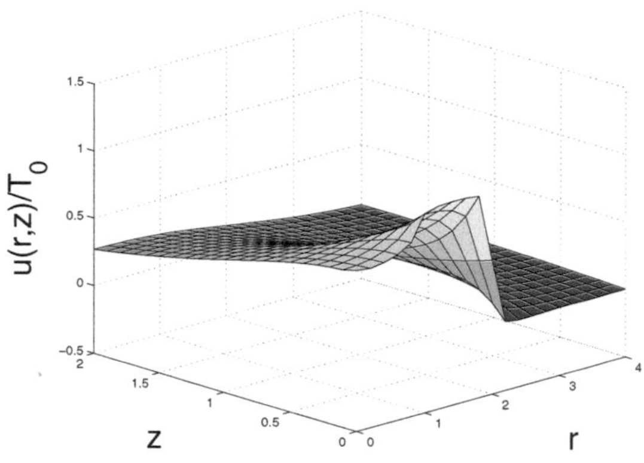

Problem 11

and

$$\int_0^\infty A(k) J_0(kr)\,\frac{dk}{k} = f(r), \qquad 1 < r < \infty.$$

Step 3: If

$$A(k) = k\int_1^\infty h(t)\sin(kt)\,dt = h(1)\cos(k) + \int_1^\infty h'(t)\cos(kt)\,dt,$$

with $\lim_{t\to\infty} h(t) \to 0$, show that $A(k)$ satisfies the second integral equation in Step 2 identically.

Step 4: Using the first integral equation in Step 2, show that

$$\int_0^\infty A(k) J_0(kr)\frac{dk}{k} = \int_r^\infty \frac{h(t)}{\sqrt{t^2 - r^2}}\,dt = f(r), \qquad 1 < r < \infty.$$

Step 5: Solving the integral equation in Step 4, show that

$$h(t) = -\frac{2}{\pi}\frac{d}{dt}\left[\int_t^\infty \frac{r f(r)}{\sqrt{t^2 - r^2}}\,dr\right].$$

The figure entitled Problem 11 illustrates this solution when $f(r) = -T_0 H(a - r)$ with $a = 2$.

12. Let us solve the Poisson equation[78]

$$\frac{\partial^2 u}{\partial r^2} + \frac{1}{r}\frac{\partial u}{\partial r} + \frac{\partial^2 u}{\partial z^2} = \kappa^2 u, \qquad 0 \le r < \infty, \quad -\infty < z < \infty,$$

[78] See Agra, R., E. Trizac, and L. Bocquet, 2004: The interplay between screening properties and colloid anisotropy: Towards a reliable pair potential for disc-like charged particles. *Eur. Phys. J., Ser. E*, **15**, 345–357.

subject to the boundary conditions

$$\lim_{r\to 0} |u(r,z)| < \infty, \qquad \lim_{r\to\infty} u(r,z) \to 0, \qquad -\infty < z < \infty,$$

$$\begin{cases} u(r,0) = u_0, & 0 \le r < a, \\ u_z(r,0) = 0, & a < r < \infty, \end{cases}$$

and

$$\lim_{|z|\to\infty} u(r,z) \to 0, \qquad 0 \le r < \infty,$$

where $\kappa > 0$. The case $\kappa = 0$ was considered already in Problem 1.

Step 1: Show that

$$u(r,z) = \int_0^\infty A(k) J_0(kr) e^{-\sqrt{k^2+\kappa^2}\,|z|}\, dk$$

satisfies the partial differential equation and the boundary conditions provided that $A(k)$ satisfies the dual integral equations

$$\int_0^\infty A(k) J_0(kr)\, dk = u_0, \qquad 0 \le r < a,$$

and

$$\int_0^\infty \sqrt{k^2+\kappa^2} A(k) J_0(kr)\, dk = 0, \qquad a < r < \infty.$$

Step 2: Setting $x = r/a$, $\xi = ka$ and $g(\xi) = \sqrt{\xi^2 + (\kappa a)^2} A(\xi)/u_0$, show that dual integral equations in Step 1 become

$$\int_0^\infty \frac{g(\xi)}{\sqrt{\xi^2+(\kappa a)^2}} J_0(x\xi)\, d\xi = 1, \qquad 0 \le x < 1,$$

and

$$\int_0^\infty g(\xi) J_0(x\xi)\, d\xi = 0, \qquad 1 < x < \infty.$$

Step 3: Using Equation 4.3.24 through Equation 4.3.27, show that

$$g(\xi) = \frac{2\xi}{\pi} \int_0^1 h(t) \cos(\xi t)\, dt,$$

where $h(t)$ is given by the integral equation

$$h(x) + \int_0^1 h(t) \left\{ \frac{2}{\pi} \int_0^\infty \left[\frac{k}{\sqrt{k^2+(\kappa a)^2}} - 1 \right] \cos(tk)\cos(xk)\, dk \right\} dt = 1,$$

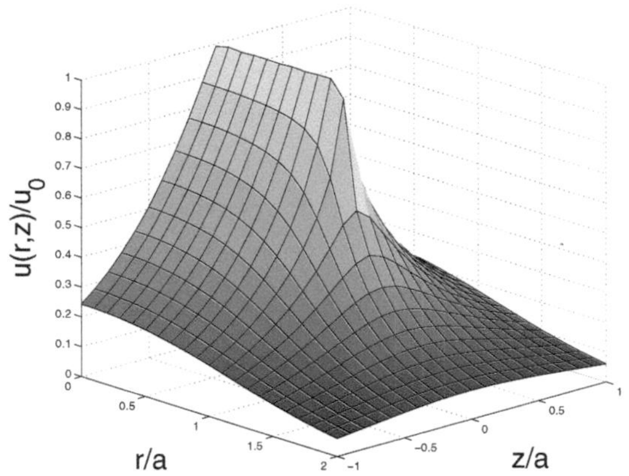

Problem 12

for $0 \le x \le 1$.

Step 4: Show that

$$\frac{2}{\pi}\int_0^\infty \left[\frac{k}{\sqrt{k^2+(\kappa a)^2}} - 1\right] \cos(tk)\cos(xk)\,dk$$
$$= -\frac{(\kappa a)^2}{\pi}\int_0^\infty \frac{\cos[(t+x)k]+\cos[(t-x)k]}{\left[k+\sqrt{k^2+(\kappa a)^2}\right]\sqrt{k^2+(\kappa a)^2}}\,dk.$$

Given κa, this integral is evaluated numerically and the result is substituted into the integral equation given in Step 3. The computed values of $h(x)$ are used to find $g(\xi)$ using a numerical scheme developed by Ehrenmark[79] for integrands that oscillate rapidly. Finally, the potential is computed. The figure labeled Problem 12 illustrates an example when $\kappa a = 1$.

13. A problem similar to Example 4.3.2 involves solving Laplace's equation[80]

$$\frac{\partial^2 u}{\partial r^2} + \frac{1}{r}\frac{\partial u}{\partial r} + \frac{\partial^2 u}{\partial z^2} = 0, \qquad 0 \le r < \infty, \quad 0 < z < L,$$

subject to the boundary conditions

$$\lim_{r\to 0} |u(r,z)| < \infty, \qquad \lim_{r\to\infty} u(r,z) \to 0, \qquad 0 < z < L,$$

---

[79] Ehrenmark, U. T., 1988: A three-point formula for numerical quadrature of oscillatory integrals with variable frequency. *J. Comput. Appl. Math.*, **21**, 87–99.

[80] See Yang, F.-Q., and J. C. M. Li, 1993: Impression creep of a thin film by vacancy diffusion. II. Cylindrical punch. *J. Appl. Phys.*, **74**, 4390–4397.

$$\begin{cases} u_z(r,0) = 1, & 0 \le r < 1, \\ u(r,0) = 0, & 1 < r < \infty, \end{cases}$$

and

$$u_z(r,L) = 0, \qquad 0 \le r < \infty.$$

Step 1: Show that

$$u(r,z) = \int_0^\infty A(k) \frac{\cosh[k(z-L)]}{\cosh(kL)} J_0(kr)\, dk$$

satisfies the partial differential equation and the boundary conditions provided that $A(k)$ satisfies the dual integral equations

$$\int_0^\infty kA(k)\tanh(kL) J_0(kr)\, dk = -1, \qquad 0 \le r < 1, \tag{1}$$

and

$$\int_0^\infty A(k) J_0(kr)\, dk = 0, \qquad 1 < r < \infty. \tag{2}$$

Step 2: Introducing

$$A(k) = \frac{2}{\pi} \int_0^1 f(t) \sin(kt)\, dt,$$

where $f(t)$ is a real and continuous function, show that Equation (2) is automatically satisfied. Hint: Use Equation 1.4.13.

Step 3: Using Equation (1), show that $f(t)$ is given by the integral equation

$$f(x) - \frac{2}{\pi} \int_0^1 f(t) \left\{ \int_0^\infty [1 - \tanh(kL)] \sin(kx) \sin(kt)\, dk \right\} dt = -x.$$

This integral equation must be solved numerically. The values of $f(t)$ are then used to compute $A(k)$. Finally, the values of $A(k)$ are substituted into the solution of $u(r,z)$. The figure labeled Problem 13 illustrates this solution $u(r,z)$ when $L = 1$.

14. Solve Laplace's equation[81] in cylindrical coordinates:

$$\frac{\partial^2 u}{\partial r^2} + \frac{1}{r}\frac{\partial u}{\partial r} + \frac{\partial^2 u}{\partial z^2} = 0, \qquad 0 \le r < \infty, \quad 0 < z < L,$$

---

[81] Reprinted from *Solid-State Electron.*, **42**, L. S. Tan, M. S. Leong, and S. C. Choo, Theory for the determination of backside contact resistance of semiconductor wafers from surface potential measurements, 589–594, ©1998 with the permission from Elsevier.

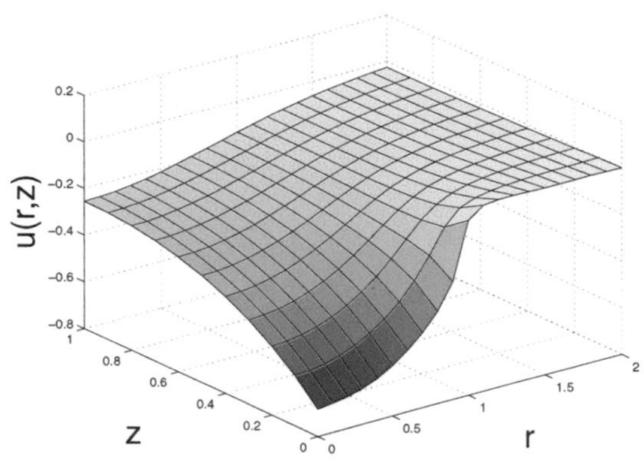

Problem 13

subject to the boundary conditions

$$\lim_{r\to 0}|u(r,z)| < \infty, \qquad \lim_{r\to\infty} u(r,z) \to 0, \qquad 0 < z < L,$$

$$\begin{cases} u(r,0) = 1, & 0 \le r < 1, \\ u_z(r,0) = 0, & 1 < r < \infty, \end{cases}$$

and

$$u(r,L) = -\gamma u_z(r,L), \qquad 0 \le r < \infty.$$

Step 1: Show that

$$u(r,z) = \int_0^\infty \frac{A(k)}{k}\left\{[1+D(k)]\cosh(kz) - \sinh(kz)\right\} J_0(kr)\, dk,$$

with $D(k) = [k\gamma + \tanh(kL)]/[1 + k\gamma\tanh(kL)] - 1$, satisfies the partial differential equation and the boundary conditions provided that $A(k)$ satisfies the dual integral equations

$$\int_0^\infty \frac{A(k)}{k}[1+D(k)]\, J_0(kr)\, dk = 1, \qquad 0 \le r < 1, \tag{1}$$

and

$$\int_0^\infty A(k) J_0(kr)\, dk = 0, \qquad 1 < r < \infty. \tag{2}$$

Step 2: By introducing

$$A(k) = \frac{2k}{\pi}\int_0^1 f(t)\cos(kt)\, dt,$$

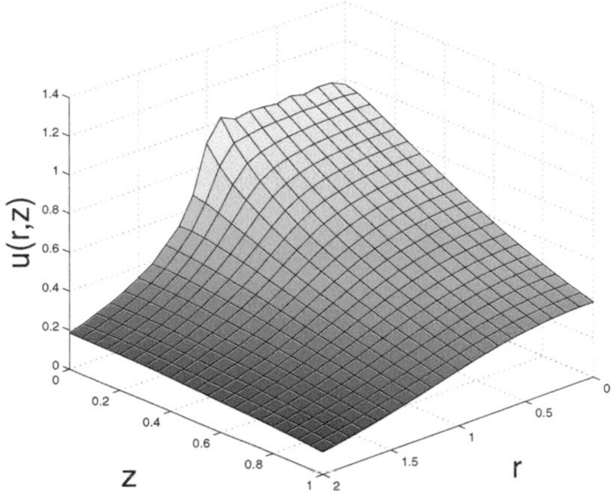

Problem 14

where $f(t)$ is a real, even and continuous function, show that Equation (2) is automatically satisfied. Hint: See Section 2.2.

Step 3: Using Equation (1), show that $f(t)$ is given by the integral equation

$$f(t) + \frac{1}{\pi}\int_0^1 f(\eta)\left\{\int_0^\infty D(k)\cos[(t-\eta)k]\,dk + \int_0^\infty D(k)\cos[(t+\eta)k]\,dk\right\}d\eta = 1.$$

This integral equation must be solved numerically. Because the integrals involving $\cos[(t-\eta)k]$ and $\cos[(t+\eta)k]$ can oscillate rapidly, we use a numerical scheme by Ehrenmark[82] for their evaluation. $A(k)$ and $B(k)$ then follow from $f(t)$. Finally, the solution $u(r,z)$ involves a numerical integration where the integrand includes both $A(k)$ and $B(k)$. The figure labeled Problem 14 illustrates this solution $u(r,z)$ when $\gamma = L = 1$.

14. Solve Laplace's equation in cylindrical coordinates:

$$\frac{\partial^2 u}{\partial r^2} + \frac{1}{r}\frac{\partial u}{\partial r} + \frac{\partial^2 u}{\partial z^2} = 0, \qquad 0 \le r < \infty, \quad -h < z < \infty,$$

subject to the boundary conditions

$$\lim_{r\to 0}|u(r,z)| < \infty, \qquad \lim_{r\to\infty} u(r,z) \to 0, \qquad -h < z < \infty, \tag{1}$$

$$u(r,-h) = 0, \qquad 0 \le r < \infty, \tag{2}$$

[82] Ehrenmark, op. cit.

$$\begin{cases} u(r, 0^-) = u(r, 0^+) = 1, & 0 \le r < 1, \\ u(r, 0^-) = u(r, 0^+), \quad u_z(r, 0^-) = u_z(r, 0^+), & 1 < r < \infty, \end{cases} \tag{3}$$

and

$$\lim_{z \to \infty} u(r, z) \to 0. \tag{4}$$

Step 1: Show that

$$u(r, z) = \int_0^\infty A(k) \frac{\sinh[k(z+h)]}{\sinh(kh)} J_0(kr)\, dk, \qquad -h < z < 0,$$

and

$$u(r, z) = \int_0^\infty A(k) e^{-kz} J_0(kr)\, dk, \qquad 0 < z < \infty,$$

satisfy the partial differential equation and the boundary conditions given by Equation (1), Equation (2), Equation (4), and $u(r, 0^-) = u(r, 0^+)$.

Step 2: Show that the boundary condition given by Equation (3) yields the dual integral equations

$$\int_0^\infty C(k) \left[1 - e^{-2kh}\right] J_0(kr) \frac{dk}{k} = 2, \qquad 0 \le r < 1, \tag{a}$$

and

$$\int_0^\infty C(k) J_0(kr)\, dk = 0, \qquad 1 \le r < \infty, \tag{b}$$

where $C(k) = kA(k)[1 + \coth(kh)]$.

Step 3: Verify that

$$C(k) = \frac{2k}{\pi} \int_0^1 \cos(kt) g(t)\, dt$$

satisfies Equation (b).

Step 4: Show that $g(t)$ is governed by

$$g(t) - \frac{2}{\pi} \int_0^1 g(\tau) F(t, \tau)\, d\tau = \frac{4}{\pi},$$

where

$$\begin{aligned} F(t, \tau) &= \int_0^\infty e^{-2hk} \cos(kt) \cos(k\tau)\, dk \\ &= \frac{4h}{\pi} \left[\frac{1}{4h^2 + (t+\tau)^2} + \frac{1}{4h^2 + (t-\tau)^2} \right]. \end{aligned}$$

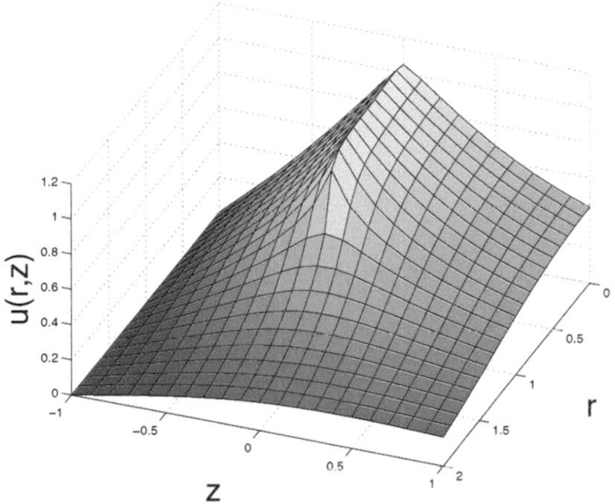

Problem 15

The figure labeled Problem 15 illustrates this solution when $h = 1$.

16. Solve

$$\frac{\partial^2 u}{\partial r^2} + \frac{1}{r}\frac{\partial u}{\partial r} - \frac{u}{r^2} + \frac{\partial^2 u}{\partial z^2} = 0, \qquad 0 \le r < \infty, \quad 0 < z < \infty,$$

subject to the boundary conditions

$$\lim_{r\to 0} |u(r,z)| < \infty, \qquad \lim_{r\to\infty} u(r,z) \to 0, \qquad 0 < z < \infty,$$

$$\lim_{z\to\infty} u(r,z) \to 0, \qquad 0 \le r < \infty,$$

and

$$\begin{cases} u(r,0) = r, & 0 \le r < a, \\ u_z(r,0) = 0, & a < r < \infty. \end{cases}$$

Step 1: Show that

$$u(r,z) = \int_0^\infty A(k) e^{-kz} J_1(kr)\, dk$$

satisfies the partial differential equation and the first three boundary conditions.

Step 2: Show that the last boundary condition yields the dual integral equations

$$\int_0^\infty A(k) J_1(kr)\, dk = r, \qquad 0 \le r < a,$$

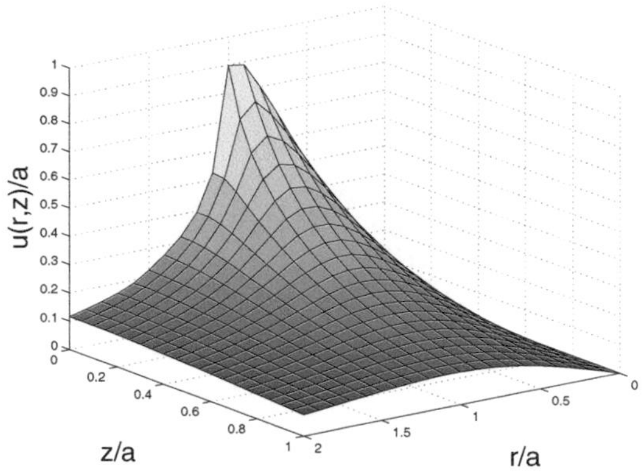

Problem 16

and

$$\int_0^\infty kA(k)J_1(kr)\,dk = 0, \qquad a < r < \infty.$$

Step 3: Using Busbridge's results (see Equation 2.4.48 and Equation 2.4.49), show that

$$A(k) = \frac{4a^2}{\pi}\left[\frac{\sin(ka)}{k^2a^2} - \frac{\cos(ka)}{ka}\right].$$

Step 4: Show that the solution to the problem is

$$u(r,z) = \frac{4a}{\pi}\int_0^\infty \left[\frac{\sin(ka)}{ka} - \cos(ka)\right] e^{-kz} J_1(kr)\,\frac{dk}{k}.$$

The figure labeled Problem 16 illustrates this solution.[83]

4.4 TRIPLE AND HIGHER FOURIER-BESSEL INTEGRALS

In Section 4.3 we examined in detail mixed boundary value problems which yielded dual integral equations. Here we extend our studies where we obtain triple integral equations.

---

[83] For an alternative derivation, see Ray, M., 1936: Application of Bessel functions to the solution of problem of motion of a circular disk in viscous liquid. *Philos. Mag., Ser. 7*, **21**, 546–564.

• Example 4.4.1

Let us solve Laplace's equation[84]

$$\frac{\partial^2 u}{\partial r^2} + \frac{1}{r}\frac{\partial u}{\partial r} + \frac{\partial^2 u}{\partial z^2} = 0, \qquad 0 \le r < \infty, \quad 0 < z < \infty, \tag{4.4.1}$$

subject to the boundary conditions

$$\lim_{r\to 0} |u(r,z)| < \infty, \quad \lim_{r\to\infty} u(r,z) \to 0, \qquad 0 < z < \infty, \tag{4.4.2}$$

$$\lim_{z\to\infty} u(r,z) \to 0, \qquad 0 \le r < \infty, \tag{4.4.3}$$

and

$$\begin{cases} u(r,0) = 1, & 0 \le r < a, \\ u_z(r,0) = 0, & a < r < 1, \\ u(r,0) = 0, & 1 < r < \infty, \end{cases} \tag{4.4.4}$$

where $a < 1$.

Using transform methods or separation of variables, the general solution to Equation 4.4.1, Equation 4.4.2, and Equation 4.4.3 is

$$u(r,z) = \int_0^\infty A(k) J_0(kr) e^{-kz}\, dk. \tag{4.4.5}$$

Substituting Equation 4.4.5 into Equation 4.4.4, we find that

$$\int_0^\infty A(k) J_0(kr)\, dk = 1, \qquad 0 \le r < a, \tag{4.4.6}$$

$$\int_0^\infty kA(k) J_0(kr)\, dk = 0, \qquad a < r < 1, \tag{4.4.7}$$

and

$$\int_0^\infty A(k) J_0(kr)\, dk = 0, \qquad 1 < r < \infty. \tag{4.4.8}$$

To solve this set of integral equations, let us introduce the unknown functions $f(r)$ and $g(r)$ such that

$$\int_0^\infty kA(k) J_0(kr)\, dk = f(r), \qquad 0 \le r < a, \tag{4.4.9}$$

---

[84] Taken from Tartakovsky, D. M., J. D. Moulton, and V. A. Zlotnik, 2000: Kinematic structure of minipermeameter flow. *Water Resourc. Res.*, **36**, 2433–2442. ©2000 American Geophysical Union. Reproduced/modified by permission of American Geophysical Union.

$$\int_0^\infty kA(k)J_0(kr)\,dk = 0, \qquad a < r < 1, \tag{4.4.10}$$

and

$$\int_0^\infty kA(k)J_0(kr)\,dk = g(r), \qquad 1 < r < \infty. \tag{4.4.11}$$

Invoking the inversion theorem as it applies to Hankel transforms,

$$A(k) = \int_0^a r\,f(r)J_0(kr)\,dr + \int_1^\infty r\,g(r)J_0(kr)\,dr. \tag{4.4.12}$$

Substituting Equation 4.4.12 into Equation 4.4.6 and interchanging the order of integration, we find that

$$1 - \int_1^\infty kg(k)\vartheta(r,k)\,dk = \frac{2}{\pi}\int_0^r \int_\xi^a \frac{kf(k)}{\sqrt{k^2-\xi^2}\sqrt{r^2-\xi^2}}\,dk\,d\xi, \tag{4.4.13}$$

where

$$\vartheta(r,k) = \int_0^\infty J_0(r\xi)J_0(k\xi)\,d\xi = \frac{2}{\pi}\int_0^{\min(k,r)} \frac{d\xi}{\sqrt{k^2-\xi^2}\sqrt{r^2-\xi^2}}. \tag{4.4.14}$$

Equation 4.4.13 can be viewed as an integral equation of the Abel type. Applying Equation 1.2.13 and Equation 1.2.14, we have that

$$\sqrt{a^2-k^2}\,f(k) = \frac{2}{\pi} - \frac{2}{\pi}\int_1^\infty \frac{t\sqrt{t^2-a^2}}{t^2-k^2}g(t)\,dt, \quad 0 \le k \le a. \tag{4.4.15}$$

In a similar manner, substituting Equation 4.4.12 into Equation 4.4.8, interchanging the order of integration and introducing

$$\theta(r,k) = \frac{2}{\pi}\int_{\max(k,r)}^\infty \frac{d\xi}{\sqrt{\xi^2-k^2}\sqrt{\xi^2-r^2}}, \tag{4.4.16}$$

we have

$$-\int_0^a kf(k)\theta(r,k)\,dk = \frac{2}{\pi}\int_r^\infty \int_1^\xi \frac{k\,g(k)}{\sqrt{\xi^2-k^2}\sqrt{\xi^2-r^2}}\,dk\,d\xi. \tag{4.4.17}$$

Again, Equation 4.4.17 can be viewed as an integral equation of the Abel type so that

$$\sqrt{t^2-1}\,g(t) = -\frac{2}{\pi}\int_0^a \frac{\xi\sqrt{1-\xi^2}}{t^2-\xi^2}f(\xi)\,d\xi, \qquad 1 \le t < \infty. \tag{4.4.18}$$

Next, we rewrite Equation 4.4.15 and Equation 4.4.18 as

$$a\sqrt{1-\lambda^2}\,f(a\lambda) = \frac{2}{\pi} - \frac{2}{\pi}\int_0^1 \frac{\sqrt{1-a^2\tau^2}}{1-a^2\lambda^2\tau^2}\,g\!\left(\frac{1}{\tau}\right)\frac{d\tau}{\tau^2}, \tag{4.4.19}$$

and

$$\frac{1}{\mu^2}\sqrt{\frac{1}{\mu^2}-1}\, g\left(\frac{1}{\mu}\right) = -\frac{2a^2}{\pi}\int_0^1 \frac{\sigma\sqrt{1-a^2\sigma^2}}{1-a^2\mu^2\sigma^2}\, f(a\sigma)\, d\sigma. \tag{4.4.20}$$

Introducing the functions

$$\phi(\lambda) = a\sqrt{1-\lambda^2}\, f(a\lambda), \tag{4.4.21}$$

and

$$\psi(\mu) = \frac{1}{\mu^2}\sqrt{\frac{1}{\mu^2}-1}\, g\left(\frac{1}{\mu}\right), \tag{4.4.22}$$

we obtain

$$A(k) = a\int_0^1 \frac{r\,\phi(r)}{\sqrt{1-r^2}} J_0(akr)\, dr + \int_0^1 \frac{\psi(r)}{\sqrt{1-r^2}} J_0\left(\frac{k}{r}\right)\, dr, \tag{4.4.23}$$

$$\phi(r) = \frac{2}{\pi} - \frac{2}{\pi}\int_0^1 K(r,\sigma)\psi(\sigma)\, d\sigma, \qquad 0 \le r \le 1, \tag{4.4.24}$$

and

$$\psi(r) = -\frac{2a}{\pi}\int_0^1 K(r,\tau)\phi(\tau)\, d\tau, \qquad 0 \le r \le 1, \tag{4.4.25}$$

where

$$K(x,y) = \frac{y\sqrt{1-a^2y^2}}{(1-a^2x^2y^2)\sqrt{1-y^2}}. \tag{4.4.26}$$

Belyaev[85] gave an alternative approach to this problem. Again, we wish to solve Laplace's equation

$$\frac{\partial^2 u}{\partial r^2} + \frac{1}{r}\frac{\partial u}{\partial r} + \frac{\partial^2 u}{\partial z^2} = 0, \qquad 0 \le r < \infty, \quad 0 < z < \infty, \tag{4.4.27}$$

subject to the boundary conditions

$$\lim_{r\to 0} |u(r,z)| < \infty, \quad \lim_{r\to\infty} u(r,z) \to 0, \qquad 0 < z < \infty, \tag{4.4.28}$$

$$\lim_{z\to\infty} u(r,z) \to 0, \qquad 0 \le r < \infty, \tag{4.4.29}$$

and

$$\begin{cases} u(r,0) = V, & 0 \le r < a, \\ u_z(r,0) = 0, & a < r < 1, \\ u(r,0) = 0, & 1 < r < \infty, \end{cases} \tag{4.4.30}$$

---

[85] See Belyaev, S. Yu., 1980: Electrostatic problem of a disk in a coaxial circular aperature in a conducting plane. *Sov. Tech. Phys.*, **25**, 12–16.

where $a < 1$.

Using transform methods or separation of variables, the general solution to Equation 4.4.27, Equation 4.4.28, and Equation 4.4.29 is

$$u(r,z) = \int_0^\infty A(k) J_0(kr) e^{-kz}\, dk. \tag{4.4.31}$$

Substituting Equation 4.4.31 into Equation 4.4.30, we find that

$$\int_0^\infty A(k) J_0(kr)\, dk = V, \qquad 0 \le r < a, \tag{4.4.32}$$

$$\int_0^\infty kA(k) J_0(kr)\, dk = 0, \qquad a < r < 1, \tag{4.4.33}$$

and

$$\int_0^\infty A(k) J_0(kr)\, dk = 0, \qquad 1 < r < \infty. \tag{4.4.34}$$

To solve this set of integral equations, we let $A(k) = B(k) + D(k)$. Then, Equation 4.4.32 through Equation 4.4.34 can be rewritten

$$\int_0^\infty B(k) J_0(kr)\, dk = f(r), \qquad 0 \le r < a, \tag{4.4.35}$$

$$\int_0^\infty kB(k) J_0(kr)\, dk = 0, \qquad a < r < \infty, \tag{4.4.36}$$

$$\int_0^\infty kD(k) J_0(kr)\, dk = 0, \qquad 0 \le r < 1, \tag{4.4.37}$$

and

$$\int_0^\infty D(k) J_0(kr)\, dk = g(r), \qquad 1 < r < \infty, \tag{4.4.38}$$

where

$$f(r) = V - \int_0^\infty D(k) J_0(kr)\, dk, \tag{4.4.39}$$

and

$$g(r) = -\int_0^\infty B(k) J_0(kr)\, dk. \tag{4.4.40}$$

Equation 4.4.36 and Equation 4.4.37 are automatically satisfied if we define $B(k)$ and $D(k)$ as follows:

$$B(k) = \int_0^a \phi(t) \cos(kt)\, dt, \qquad D(k) = \int_1^\infty \psi(\tau) \sin(k\tau)\, d\tau. \tag{4.4.41}$$

If we substitute Equation 4.4.41 into Equation 4.4.35, we have that

$$\int_0^a \phi(t) \left[\int_0^\infty \cos(kt) J_0(kr)\, dk\right] dt = f(r) \tag{4.4.42}$$

after we interchange the order of integration. Using Equation 1.4.14, Equation 4.4.42 simplifies to

$$\int_0^r \frac{\phi(t)}{\sqrt{r^2 - t^2}}\, dt = f(r). \tag{4.4.43}$$

From Equation 1.2.13 and Equation 1.2.14, we obtain

$$\phi(t) = \frac{2}{\pi} \frac{d}{dt}\left[\int_0^t \frac{r f(r)}{\sqrt{t^2 - r^2}}\, dr\right]. \tag{4.4.44}$$

In a similar manner, we find that

$$\psi(\tau) = -\frac{2}{\pi} \frac{d}{d\tau}\left[\int_\tau^\infty \frac{r\, g(r)}{\sqrt{r^2 - \tau^2}}\, dr\right]. \tag{4.4.45}$$

Next, we substitute for $D(k)$ in Equation 4.4.39 and find that

$$f(r) = V - \int_0^\infty \left[\int_1^\infty \psi(\tau) \sin(k\tau)\, d\tau\right] J_0(kr)\, dk \tag{4.4.46}$$

$$= V - \int_1^\infty \psi(\tau) \left[\int_0^\infty \sin(k\tau) J_0(kr)\, dk\right] d\tau \tag{4.4.47}$$

$$= V - \int_1^\infty \frac{\psi(\tau)}{\sqrt{\tau^2 - r^2}}\, d\tau \tag{4.4.48}$$

for $0 \le r < a$. Here we have used Equation 1.4.13. In a similar manner, it is readily shown that

$$g(r) = -\int_0^a \frac{\phi(t)}{\sqrt{r^2 - t^2}}\, dt, \qquad 1 < r < \infty. \tag{4.4.49}$$

Finally, we substitute Equation 4.4.48 into Equation 4.4.44 and find that

$$\phi(t) = \frac{2}{\pi} \frac{d}{dt}\left[\int_0^t \frac{rV}{\sqrt{t^2 - r^2}}\, dr\right] \tag{4.4.50}$$

$$- \frac{2}{\pi} \frac{d}{dt}\left\{\int_0^t \left[\int_1^\infty \frac{\psi(\tau)}{\sqrt{\tau^2 - r^2}}\, d\tau\right] \frac{r\, dr}{\sqrt{t^2 - r^2}}\right\}$$

$$= \frac{2V}{\pi} - \frac{2}{\pi} \int_1^\infty \psi(\tau) \frac{d}{dt}\left[\int_0^t \frac{r\, dr}{\sqrt{\tau^2 - r^2}\sqrt{t^2 - r^2}}\right] d\tau \tag{4.4.51}$$

$$= \frac{2V}{\pi} - \frac{2}{\pi} \int_1^\infty \frac{\tau\, \psi(\tau)}{t^2 - \tau^2}\, d\tau. \tag{4.4.52}$$

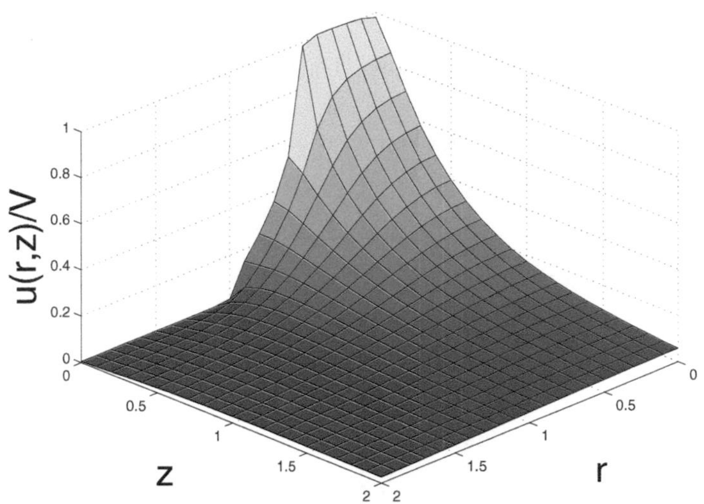

Figure 4.4.1: The solution to Equation 4.4.27 subject to the mixed boundary conditions given by Equation 4.4.28 through Equation 4.4.30 when $a = 0.5$.

In a similar manner, we also find that

$$\psi(\tau) = -\frac{2}{\pi}\int_0^a \frac{\tau\,\phi(t)}{\tau^2 - t^2}\,dt. \tag{4.4.53}$$

If we introduce $t = ax$, $\tau = ay$, $\Phi(x) = \pi\phi(t)/(2V)$, and $\Psi(y) = \pi\psi(t)/(2V)$, Equation 4.4.52 and Equation 4.4.53 can be combined to yield

$$\Phi(x) = 1 + \frac{2}{\pi^2}\int_0^1 \Phi(\xi)K(x,\xi)\,d\xi, \qquad 0 < x < 1, \tag{4.4.54}$$

where

$$K(x,\xi) = 2a\int_1^\infty \frac{\eta^2}{(\eta^2 - a^2x^2)(\eta^2 - a^2\xi^2)}\,d\eta \tag{4.4.55}$$

$$= \frac{1}{\xi^2 - x^2}\left[\xi\ln\left(\frac{1+a\xi}{1-a\xi}\right) - x\ln\left(\frac{1+ax}{1-ax}\right)\right]. \tag{4.4.56}$$

For the special case of $\xi = x$, we use L'Hospital rule and find that

$$K(x,x) = \frac{1}{2x}\ln\left(\frac{1+ax}{1-ax}\right) + \frac{a}{1-a^2x^2}. \tag{4.4.57}$$

Our computations begin by finding $\phi(t)$ from Equation 4.4.54. The function $\psi(\tau)$ follows from Equation 4.4.53. With $\phi(t)$ and $\psi(\tau)$, we compute $B(k)$, $D(k)$ and $A(k)$. Finally, Equation 4.4.31 gives the potential $u(r,z)$. Figure 4.4.1 illustrates this potential when $a = 0.5$.

Finally Selvadurai[86] solved this problem as a system of integral equations. In the present case we have

$$\frac{\partial^2 u}{\partial r^2} + \frac{1}{r}\frac{\partial u}{\partial r} + \frac{\partial^2 u}{\partial z^2} = 0, \qquad 0 \le r < \infty, \quad 0 < z < \infty, \tag{4.4.58}$$

subject to the boundary conditions

$$\lim_{r\to 0} |u(r,z)| < \infty, \quad \lim_{r\to\infty} u(r,z) \to 0, \qquad 0 < z < \infty, \tag{4.4.59}$$

$$\lim_{z\to\infty} u(r,z) \to 0, \qquad 0 \le r < \infty, \tag{4.4.60}$$

and

$$\begin{cases} u(r,0) = 1, & 0 \le r < a, \\ u_z(r,0) = 0, & a < r < b, \\ u(r,0) = 0, & b < r < \infty, \end{cases} \tag{4.4.61}$$

where $b > a$.

Using transform methods or separation of variables, the general solution to Equation 4.4.58, Equation 4.4.59, and Equation 4.4.60 is

$$u(r,z) = \int_0^\infty A(k) J_0(kr) e^{-kz}\, dk. \tag{4.4.62}$$

Substituting Equation 4.4.62 into Equation 4.4.61, we find that

$$\int_0^\infty A(k) J_0(kr)\, dk = 1, \qquad 0 \le r < a, \tag{4.4.63}$$

$$\int_0^\infty kA(k) J_0(kr)\, dk = 0, \qquad a < r < b, \tag{4.4.64}$$

and

$$\int_0^\infty A(k) J_0(kr)\, dk = 0, \qquad b < r < \infty. \tag{4.4.65}$$

To solve this set of integral equations, we introduce two new functions $f(r)$ and $g(r)$ such that

$$\int_0^\infty kA(k) J_0(kr)\, dk = f(r), \qquad 0 \le r < a, \tag{4.4.66}$$

and

$$\int_0^\infty kA(k) J_0(kr)\, dk = g(r), \qquad b < r < \infty. \tag{4.4.67}$$

---

[86] Reprinted from *Mech. Res. Commun.*, **23**, A. P. S. Selvadurai, On the problem of an electrified disc located at the central opening of a coplanar earthed sheet, 621–624, ©1996, with permission from Elsevier.

Then, because

$$A(k) = \int_0^a r\, f(r) J_0(kr)\, dr + \int_b^\infty r\, g(r) J_0(kr)\, dr, \tag{4.4.68}$$

Equation 4.4.63 and Equation 4.4.65 become

$$\int_0^a \tau\, f(\tau) L(\tau, r)\, d\tau + \int_b^\infty \tau\, g(\tau) L(\tau, r)\, d\tau = 1, \qquad 0 \le r < a, \tag{4.4.69}$$

and

$$\int_0^a \tau\, f(\tau) L(\tau, r)\, d\tau + \int_b^\infty \tau\, g(\tau) L(\tau, r)\, d\tau = 0, \qquad b \le r < \infty \tag{4.4.70}$$

after interchanging the order of integration, where

$$L(\tau, r) = \int_0^\infty J_0(k\tau) J_0(kr)\, dk. \tag{4.4.71}$$

Because[87]

$$L(\tau, r) = \int_0^{\min(\tau,r)} \frac{ds}{\sqrt{(\tau^2 - s^2)(r^2 - s^2)}} \tag{4.4.72}$$

$$= \int_{\max(\tau,r)}^\infty \frac{ds}{\sqrt{(s^2 - \tau^2)(s^2 - r^2)}}, \tag{4.4.73}$$

Equation 4.4.69 can be rewritten

$$\begin{aligned} &\frac{2}{\pi} \int_0^r \tau f(\tau) \left[\int_0^\tau \frac{ds}{\sqrt{(\tau^2 - s^2)(r^2 - s^2)}} \right] d\tau \\ &\quad + \frac{2}{\pi} \int_r^a \tau f(\tau) \left[\int_0^r \frac{ds}{\sqrt{(\tau^2 - s^2)(r^2 - s^2)}} \right] d\tau \\ &\quad + \frac{2}{\pi} \int_b^\infty \tau g(\tau) \left[\int_\tau^\infty \frac{ds}{\sqrt{(s^2 - \tau^2)(s^2 - r^2)}} \right] d\tau = 1, \end{aligned} \tag{4.4.74}$$

or, after interchanging the order of integration,

$$\frac{2}{\pi} \int_0^r \left[\int_s^a \frac{\tau\, f(\tau)}{\sqrt{\tau^2 - s^2}}\, d\tau \right] \frac{ds}{\sqrt{r^2 - s^2}} + \frac{2}{\pi} \int_b^\infty \left[\int_b^s \frac{\tau\, g(\tau)}{\sqrt{s^2 - \tau^2}}\, d\tau \right] \frac{ds}{\sqrt{s^2 - r^2}} = 1. \tag{4.4.75}$$

---

[87] Cooke, op. cit.

Setting

$$F(s) = \int_s^a \frac{\tau f(\tau)}{\sqrt{\tau^2 - s^2}}\, d\tau, \qquad 0 \le s \le a, \tag{4.4.76}$$

and

$$G(s) = \int_b^s \frac{\tau g(\tau)}{\sqrt{s^2 - \tau^2}}\, d\tau, \qquad b \le s \le \infty, \tag{4.4.77}$$

Equation 4.4.75 simplifies to

$$\int_0^r \frac{F(s)}{\sqrt{r^2 - s^2}}\, ds = \frac{\pi}{2} - \int_b^\infty \frac{G(s)}{\sqrt{s^2 - r^2}}\, ds. \tag{4.4.78}$$

Applying Equation 1.2.13 and Equation 1.2.14 to Equation 4.4.78, we have that

$$F(s) = -\frac{d}{ds}\left[\sqrt{s^2 - r^2}\Big|_0^s\right] - \frac{2}{\pi}\int_b^\infty G(t)\left\{\frac{d}{ds}\left[\int_0^s \frac{dy}{\sqrt{t^2 - s^2 + y^2}}\right]\right\} dt, \tag{4.4.79}$$

where we interchanged the order of integration and set $s^2 - r^2 = y^2$. Carrying out the integration in y and simplifying, we finally obtain

$$F(r) + \frac{2}{\pi}\int_b^\infty \frac{t\, G(t)}{t^2 - r^2}\, dt = 1, \qquad 0 \le r < a. \tag{4.4.80}$$

In a similar manner, for Equation 4.4.70, we have that

$$\begin{aligned}\int_0^a \tau f(\tau) &\left[\int_0^\tau \frac{ds}{\sqrt{(\tau^2 - s^2)(r^2 - s^2)}}\right] d\tau \\ &+ \int_b^r \tau g(\tau)\left[\int_r^\infty \frac{ds}{\sqrt{(s^2 - \tau^2)(s^2 - r^2)}}\right] d\tau \\ &+ \int_r^\infty \tau g(\tau)\left[\int_\tau^\infty \frac{ds}{\sqrt{(s^2 - \tau^2)(s^2 - r^2)}}\right] d\tau = 0,\end{aligned} \tag{4.4.81}$$

or

$$\int_0^a \left[\int_s^a \frac{\tau\, f(\tau)}{\sqrt{\tau^2 - s^2}}\, d\tau\right] \frac{ds}{\sqrt{r^2 - s^2}} + \int_r^\infty \left[\int_b^s \frac{\tau\, g(\tau)}{\sqrt{s^2 - \tau^2}}\, d\tau\right] \frac{ds}{\sqrt{s^2 - r^2}} = 0. \tag{4.4.82}$$

Equation 4.4.82 simplifies to

$$\int_r^\infty \frac{G(s)}{\sqrt{s^2 - r^2}}\, ds = -\int_0^a \frac{F(s)}{\sqrt{r^2 - s^2}}\, ds. \tag{4.4.83}$$

Applying Equation 1.2.15 and Equation 1.2.16 to Equation 4.4.83, we have that

$$G(s) = \frac{2}{\pi}\int_0^a F(t)\left\{\frac{d}{ds}\left[\int_s^\infty \frac{r}{\sqrt{(r^2 - t^2)(r^2 - s^2)}}\, dr\right]\right\} dt. \tag{4.4.84}$$

Carrying out the integration and differentiation within the wavy brackets, we obtain

$$G(r) + \frac{2s}{\pi}\int_0^a \frac{F(t)}{s^2 - t^2}\,dt = 0, \qquad b < r < \infty. \tag{4.4.85}$$

Upon solving the dual Fredholm integral equations, Equation 4.4.80 and Equation 4.4.85, we have $F(r)$ and $G(r)$. Next, we invert Equation 4.4.76 and Equation 4.4.77 to find $f(r)$ and $g(r)$. The Fourier coefficient $A(k)$ follows from Equation 4.4.68 while Equation 4.4.62 yields $u(r,z)$.

• **Example 4.4.2**

Let us now solve Laplace's equation[88]

$$\frac{\partial^2 u}{\partial r^2} + \frac{1}{r}\frac{\partial u}{\partial r} + \frac{\partial^2 u}{\partial z^2} = 0, \qquad 0 \le r < \infty, \quad 0 < z < \infty, \tag{4.4.86}$$

when the boundary conditions are

$$\lim_{r\to 0} |u(r,z)| < \infty, \quad \lim_{r\to\infty} u(r,z) \to 0, \qquad 0 < z < \infty, \tag{4.4.87}$$

$$\lim_{z\to\infty} u(r,z) \to 0, \qquad 0 \le r < \infty, \tag{4.4.88}$$

and

$$\begin{cases} u_z(r,0) = 0, & 0 \le r < a, \\ u(r,0) = 1, & a < r < b, \\ u_z(r,0) = 0, & b < r < \infty, \end{cases} \tag{4.4.89}$$

where $b > a > 0$.

Using transform methods or separation of variables, the general solution to Equation 4.4.86, Equation 4.4.87, and Equation 4.4.88 is

$$u(r,z) = \int_0^\infty A(k) J_0(kr) e^{-kz}\,\frac{dk}{k}. \tag{4.4.90}$$

Substituting Equation 4.4.90 into Equation 4.4.89, we find that

$$\int_0^\infty A(k) J_0(kr)\,dk = 0, \qquad 0 \le r < a, \tag{4.4.91}$$

$$\int_0^\infty A(k) J_0(kr)\,\frac{dk}{k} = 1, \qquad a < r < b, \tag{4.4.92}$$

and

$$\int_0^\infty A(k) J_0(kr)\,dk = 0, \qquad b < r < \infty. \tag{4.4.93}$$

The mixed boundary condition, Equation 4.4.89, has led to three integral equations involving Fourier-Bessel integrals. Our remaining task is to find the Fourier-Bessel coefficient $A(k)$. Can we find some general result that might assist us in solving these triple Fourier-Bessel equations?

In 1963 Cooke[89] studied how to find $A(k)$ governed by the following

---

[88] Similar to a problem by Borodachev, N. M., and F. N. Borodacheva, 1966: Pentration of an annular stamp into an elastic half-space. *Mech. Solids*, **1(4)**, 101–103.

[89] Cooke, op. cit.

integral equations:

$$\int_0^\infty A(k) J_\nu(kr)\, dk = 0, \qquad 0 \le r < a, \tag{4.4.94}$$

$$\int_0^\infty k^p A(k) J_\nu(kr)\, dk = f(r), \qquad a < r < b, \tag{4.4.95}$$

and

$$\int_0^\infty A(k) J_\nu(kr)\, dk = 0, \qquad b < r < \infty, \tag{4.4.96}$$

where $p = \pm 1$, $\nu > -\frac{1}{2}$, and $b > a > 0$. For $p = -1$, he proved that

$$A(k) = k \int_a^b r g(r) J_\nu(kr)\, dr, \tag{4.4.97}$$

where

$$g(r) = -\frac{2}{\pi} r^{\nu-1} \frac{d}{dr}\left[\int_r^b \frac{\eta\, h(\eta)}{\sqrt{\eta^2 - r^2}}\, d\eta\right], \qquad a < r < b, \tag{4.4.98}$$

$$\eta^{2\nu} h(\eta) = \frac{d}{d\eta}\left[\int_a^\eta \frac{x^{\nu+1} f(x)}{\sqrt{\eta^2 - x^2}}\, dx\right] - \frac{4}{\pi^2} \frac{\eta}{\sqrt{\eta^2 - a^2}} \int_a^b \frac{t\, h(t)}{\sqrt{t^2 - a^2}} K(\eta, t)\, dt, \tag{4.4.99}$$

$$K(\eta, t) = \int_0^a \frac{y^{2\nu}(a^2 - y^2)}{(\eta^2 - y^2)(t^2 - y^2)}\, dy, \tag{4.4.100}$$

and $a < \eta < b$.

We can use Cooke's results if we set $\nu = 0$. Then, from Equation 4.4.97 through Equation 4.4.100, we have that

$$A(k) = k \int_a^b r\, g(r) J_0(kr)\, dr, \tag{4.4.101}$$

where

$$g(r) = -\frac{2}{\pi r} \frac{d}{dr}\left[\int_r^b \frac{\eta\, h(\eta)}{\sqrt{\eta^2 - r^2}}\, d\eta\right], \tag{4.4.102}$$

$$h(\eta) = \frac{d}{d\eta}\left[\int_a^\eta \frac{x}{\sqrt{\eta^2 - x^2}}\, dx\right] - \frac{4}{\pi^2} \frac{\eta}{\sqrt{\eta^2 - a^2}} \int_a^b \frac{t\, h(t)}{\sqrt{t^2 - a^2}} K(\eta, t)\, dt \tag{4.4.103}$$

$$= \frac{\eta}{\sqrt{\eta^2 - a^2}} - \frac{4}{\pi^2} \frac{\eta}{\sqrt{\eta^2 - a^2}} \int_a^b \frac{t\, h(t)}{\sqrt{t^2 - a^2}} K(\eta, t)\, dt, \tag{4.4.104}$$

and

$$K(\eta,t) = \int_0^a \frac{a^2 - y^2}{(\eta^2 - y^2)(t^2 - y^2)}\,dy \tag{4.4.105}$$

$$= \frac{1}{2(\eta^2 - t^2)}\left[\frac{\eta^2 - a^2}{\eta}\ln\left(\frac{\eta + a}{\eta - a}\right) - \frac{t^2 - a^2}{t}\ln\left(\frac{t + a}{t - a}\right)\right]. \tag{4.4.106}$$

In the special case when $t = \eta$, we employ L'Hospital rule and find that

$$K(\eta,\eta) = -\frac{a}{2\eta^2} + \frac{\eta^2 + a^2}{4\eta^3}\ln\left(\frac{\eta + a}{\eta - a}\right). \tag{4.4.107}$$

If we introduce

$$\eta h(\eta) = \sqrt{\eta^2 - a^2}\,\chi(\eta), \tag{4.4.108}$$

then Equation 4.102 and Equation 4.4.104 become

$$g(r) = -\frac{2}{\pi r}\frac{d}{dr}\left[\int_r^b \sqrt{\frac{\eta^2 - a^2}{\eta^2 - r^2}}\,\chi(\eta)\,d\eta\right], \tag{4.4.109}$$

and

$$\frac{\eta^2 - a^2}{\eta^2}\chi(\eta) = 1 - \frac{4}{\pi^2}\int_a^b K(\eta,t)\chi(t)\,dt. \tag{4.4.110}$$

To compute $u(r,z)$, we solve Equation 4.4.110 by replacing the integral with its representation from the midpoint rule. Setting `d_eta = (b-a) / N`, the MATLAB code for computing $\chi(\eta)$ is:

```
for j = 1:N
xi(j) = (j-0.5)*d_eta + a;
eta(j) = (j-0.5)*d_eta + a;
end

for m = 1:N

for K = 0:K_max
k = K*dK; factor(K+1,m) = bessel(0,k*eta(m));
end; end

for n = 1:N % rows loop (top to bottom in the matrix)
x = xi(n); b(n) = 1; % right side of the integral equation
for m = 1:N % columns loop (left to right in the matrix)
t = eta(m);
% start setting up Equation 4.4.110
if (n==m) AA(n,m) = (x-a)*(x+a)/(x*x); % first term on left side
else AA(n,m) = 0; end
```

```
% introduce the integral in Equation 4.4.110
temp1 = (x-a)*(x+a)/x; temp2 = (t-a)*(t+a)/t;
if (t == x)
integrand = -a/(2*x*x)+(x*x+a*a)*log((x+a)/(x-a))/(4*x*x*x);
else
integrand = temp1*log((x+a)/(x-a))-temp2*log((t+a)/(t-a));
integrand = integrand/(2*(x*x-t*t));
end
AA(n,m) = AA(n,m) + 4*integrand*d_eta/(pi*pi);
end
end
% compute χ(η) and call it f
f = AA\b'
```

Equation 4.4.109 gives $g(r)$. First use the midpoint rule to compute the integral and put it in `F(n)`. Then compute the derivative to find `g(n)`. The MATLAB code for this is:

```
for n = 1:N
r = a + (n-1)*d_eta; F(n) = 0;
for m = n:N
sq = xi(m)*xi(m); temp1 = sqrt((sq-a*a)/(sq-r*r));
F(n) = F(n) + temp1*f(m)*d_eta;
end; end

F(N+1) = 0;
for n = 1:N
g(n) = -2*(F(n+1)-F(n))/(pi*xi(n)*d_eta);
end
```

Finally, combining Equation 4.4.90 and Equation 4.4.101,

$$u(r,z) = \int_a^b \eta g(\eta) \left[\int_0^\infty J_0(k\eta) J_0(kr) e^{-kz}\, dk \right] d\eta, \tag{4.4.111}$$

after the order of integration is interchanged. The integral within the square brackets is evaluated using Simpson's rule. The MATLAB code is:

```
for j = 1:21
z = 0.1*(j-1);

for K = 0:K_max
k = K*dK; Z(K+1) = exp(-k*z);
end
```

```
for i = 1:31
r = 0.1*(i-1); u(i,j) = 0;

for K = 0:K_max
k = K*dK; R(K+1) = besselj(0,k*r);
end

for m = 1:N
integral = 0;
for K = 0:K_max
k = K*dK; integrand = factor(K+1,m)*R(K+1)*Z(K+1);
if ( (K>0) & (K<K_max))
if (mod(K+1,2) == 0)
integral = integral + 4*integrand;
else
integral = integral + 2*integrand;
end
else
integral = integral +   integrand;
end
end
integral = integral*dK/3;
u(i,j) = u(i,j) + g(m)*integral*eta(m)*d_eta;
end; end; end
```

Figure 4.4.2 illustrates this potential when $a = 1$ and $b = 2$.

• Example 4.4.3

Let us solve Laplace's equation:[90]

$$\frac{\partial^2 u}{\partial r^2} + \frac{1}{r}\frac{\partial u}{\partial r} + \frac{\partial^2 u}{\partial z^2} = 0, \qquad 0 \le r < \infty, \quad 0 < z < \infty, \qquad \textbf{(4.4.112)}$$

subject to the boundary conditions

$$\lim_{r\to 0} |u(r,z)| < \infty, \quad \lim_{r\to\infty} u(r,z) \to 0, \qquad 0 < z < \infty, \qquad \textbf{(4.4.113)}$$

$$\lim_{z\to\infty} u(r,z) \to 0, \qquad 0 \le r < \infty, \qquad \textbf{(4.4.114)}$$

---

[90] Taken with permission from Sibgatullin, N. R., I. N. Sibgatullin, A. A. Garcia, and V. S. Manko, 2004: Magnetic fields of pulsars surrounded by accretion disks of finite extension. *Astron. Astrophys.*, **422**, 587–590.

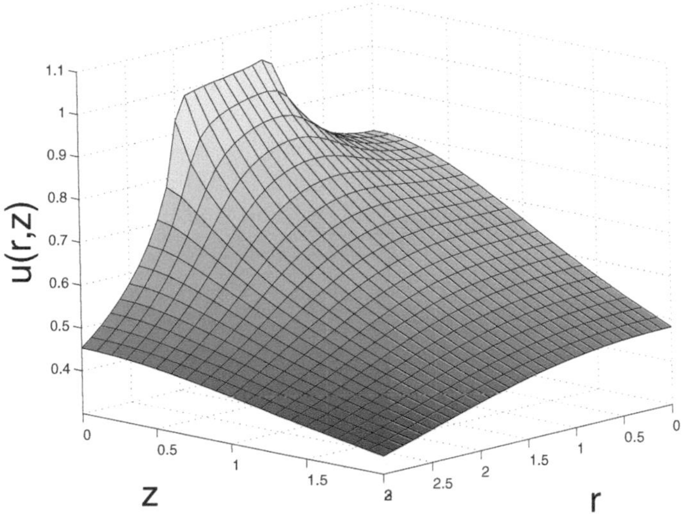

Figure 4.4.2: The solution to Equation 4.4.86 subject to the mixed boundary conditions given by Equation 4.4.87 through Equation 4.4.89 when $a = 1$ and $b = 2$.

and

$$\begin{cases} u(r,0) = K, & 0 \le r < a, \\ u_z(r,0) = A/r^3, & a < r < b, \\ u(r,0) = 0, & b < r < \infty, \end{cases} \tag{4.4.115}$$

where $b > a > 0$.

Using separation of variables or transform methods, the general solution to Equation 4.4.112 is

$$u(r,z) = \int_0^\infty F(k) e^{-kz} J_0(kr)\, dk. \tag{4.4.116}$$

This solution satisfies not only Equation 4.4.112, but also Equation 4.4.113 and Equation 4.4.114. Substituting Equation 4.2.116 into Equation 4.2.115, we obtain the triple integral equations

$$\int_0^\infty F(k) J_0(kr)\, dk = K, \qquad 0 \le r < a, \tag{4.4.117}$$

$$\int_0^\infty k F(k) J_0(kr)\, dk = -A/r^3, \qquad a < r < b, \tag{4.4.118}$$

and

$$\int_0^\infty F(k) J_0(kr)\, dk = 0, \qquad b < r < \infty. \tag{4.4.119}$$

We begin our solution of Equation 4.4.117 through Equation 4.4.119 by introducing

$$F(k) = \int_0^a \mu_1(\xi)\sin(k\xi)\,d\xi + \int_a^b \mu_2(\xi)\sin(k\xi)\,d\xi. \tag{4.4.120}$$

Note that $\mu_1(\xi)$ is defined over the interval $[0,a]$ while $\mu_2(\xi)$ is defined over the interval $[a,b]$. Substituting Equation 4.4.120 into Equation 4.4.119 and interchanging the order of integration, we find that

$$\begin{aligned}\int_0^\infty F(k)J_0(kr)\,dk = {} & \int_0^a \mu_1(\xi)\left[\int_0^\infty \sin(k\xi)J_0(kr)\,dk\right]d\xi \\ & + \int_a^b \mu_2(\xi)\left[\int_0^\infty \sin(k\xi)J_0(kr)\,dk\right]d\xi\end{aligned} \tag{4.4.121}$$

and Equation 4.4.119 is satisfied because the integrals within the square brackets vanish according to Equation 1.4.13 since $r > b > a$. In a similar manner, substituting Equation 4.4.120 into Equation 4.4.117 yields

$$\int_r^a \frac{\mu_1(\xi)}{\sqrt{\xi^2 - r^2}}\,d\xi + \int_a^b \frac{\mu_2(\xi)}{\sqrt{\xi^2 - r^2}}\,d\xi = K, \qquad 0 \le r < a, \tag{4.4.122}$$

again by using Equation 1.4.13 and noting that $r < a < b$.

To solve Equation 4.4.118, we first note that

$$\begin{aligned}kF(k) = {} & \int_0^a k\,\mu_1(\xi)\sin(k\xi)\,d\xi \\ & + \mu_2(a)\cos(ka) - \mu_2(b)\cos(kb) + \int_a^b \mu_2'(\xi)\cos(k\xi)\,d\xi\end{aligned} \tag{4.4.123}$$

by integrating the second integral in Equation 4.4.120 by parts. Substituting Equation 4.4.123 into Equation 4.4.118 and interchanging the order of integration,

$$\begin{aligned}&\int_0^a \mu_1(\xi)\left[\int_0^\infty k\sin(k\xi)J_0(kr)\,dk\right]d\xi + \mu_2(a)\int_0^\infty \cos(ka)J_0(kr)\,dk \\ &- \mu_2(b)\int_0^\infty \cos(kb)J_0(kr)\,dk + \int_a^b \mu_2'(\xi)\left[\int_0^\infty \cos(k\xi)J_0(kr)\,dk\right]d\xi = -\frac{A}{r^3}.\end{aligned} \tag{4.4.124}$$

Now, by integration by parts,

$$\int_0^\infty k\sin(k\xi)J_0(kr)\,dk = \frac{k}{r}\sin(k\xi)J_1(kr)\bigg|_0^\infty - \frac{\xi}{r}\int_0^\infty \cos(\xi k)J_1(rk)k\,dk \tag{4.4.125}$$

$$= -\frac{\xi\,H(r-\xi)}{(r^2-\xi^2)^{3/2}}, \tag{4.4.126}$$

where we used tables[91] to simplify the integral on the right side of Equation 4.4.125 and $d[z^n J_n(z)] = z^n J_{n-1}(z)\,dz, n = 1, 2, \ldots$. Using Equation 1.4.13 and Equation 4.4.126, Equation 4.4.124 becomes

$$-\int_0^a \frac{\xi r \mu_1(\xi)}{(r^2-\xi^2)^{3/2}}\,d\xi + \frac{r\mu_2(a)}{\sqrt{r^2-a^2}} + \int_a^r \frac{r\mu_2'(\xi)}{\sqrt{r^2-\xi^2}}\,d\xi = -\frac{A}{r^2}, \qquad \textbf{(4.4.127)}$$

or

$$\frac{\partial}{\partial r}\left[\int_0^a \frac{\xi\mu_1(\xi)}{\sqrt{r^2-\xi^2}}\,d\xi + \int_a^r \frac{\xi\mu_2(\xi)}{\sqrt{r^2-\xi^2}}\,d\xi\right] = -\frac{A}{r^2}, \qquad a < r < b. \quad \textbf{(4.4.128)}$$

Upon integrating Equation 4.4.128, we finally obtain

$$\int_0^a \frac{\xi\mu_1(\xi)}{\sqrt{r^2-\xi^2}}\,d\xi + \int_a^r \frac{\xi\mu_2(\xi)}{\sqrt{r^2-\xi^2}}\,d\xi = \frac{A}{r} + \frac{aAC}{a^2}, \qquad a < r < b, \quad \textbf{(4.4.129)}$$

where C is an arbitrary constant.

Because we can write Equation 4.4.129 as an integral equation of the Abel type:

$$\int_a^r \frac{\xi\mu_2(\xi)}{\sqrt{r^2-\xi^2}}\,d\xi = \frac{A}{r} + \frac{aAC}{a^2} - \int_0^a \frac{\xi\mu_1(\xi)}{\sqrt{r^2-\xi^2}}\,d\xi, \qquad a < r < b, \quad \textbf{(4.4.130)}$$

we can solve for $r\mu_2(r)$ using Equation 1.2.14. This yields

$$r\mu_2(r) = \frac{2}{\pi}\frac{d}{dr}\left\{\int_a^r \frac{\tau}{\sqrt{r^2-\tau^2}}\left[\frac{A}{\tau} + \frac{aAC}{a^2} - \int_0^a \frac{\xi\mu_1(\xi)}{\sqrt{\tau^2-\xi^2}}\,d\xi\right]d\tau\right\}. \qquad \textbf{(4.4.131)}$$

Carrying out the τ integration and the r differentiation, we have that

$$r\mu_2(r) = \frac{2}{\pi}\left[\frac{aA}{r\sqrt{r^2-a^2}} + \frac{aACr}{a^2\sqrt{r^2-a^2}} - \frac{r}{\sqrt{r^2-a^2}}\int_0^a \frac{\xi\sqrt{a^2-\xi^2}}{r^2-\xi^2}\mu_1(\xi)\,d\xi\right]. \qquad \textbf{(4.4.132)}$$

In a similar manner, we can write Equation 4.4.122 as an integral equation of the Abel type:

$$\int_r^a \frac{\mu_1(\xi)}{\sqrt{\xi^2-r^2}}\,d\xi = K - \int_a^b \frac{\mu_2(\xi)}{\sqrt{\xi^2-r^2}}\,d\xi, \qquad 0 \le r < a. \qquad \textbf{(4.4.133)}$$

Its solution yields

$$\mu_1(r) = -\frac{2}{\pi}\frac{d}{dr}\left\{\int_r^a \frac{\tau}{\sqrt{\tau^2-r^2}}\left[K - \int_a^b \frac{\mu_2(\xi)}{\sqrt{\xi^2-\tau^2}}\,d\xi\right]d\tau\right\}. \qquad \textbf{(4.4.134)}$$

---

[91] Gradshteyn and Ryzhik, op. cit., Formula 6.699.6.

Carrying out the τ integration and then taking the r derivative,

$$\mu_1(r) = \frac{2}{\pi}\left[\frac{Kr}{\sqrt{a^2-r^2}} - \frac{r}{\sqrt{a^2-r^2}}\int_a^b \frac{\sqrt{\xi^2-a^2}}{\xi^2-r^2}\mu_2(\xi)\,d\xi\right]. \qquad \textbf{(4.4.135)}$$

Upon substituting Equation 4.4.135 into Equation 4.4.131 and interchanging the order of integration,

$$\sqrt{r^2-a^2}\,\mu_2(r) = \frac{2aA}{\pi r^2} + \frac{2aAC}{\pi a^2} - \frac{4K}{\pi^2}\int_0^a \frac{\xi^2}{r^2-\xi^2}\,d\xi \qquad \textbf{(4.4.136)}$$

$$+ \frac{4}{\pi^2}\int_a^b \sqrt{\eta^2-a^2}\,\mu_2(\eta)\left[\int_0^a \frac{\xi^2}{(r^2-\xi^2)(\eta^2-\xi^2)}\,d\xi\right]d\eta$$

$$= \frac{2aA}{\pi r^2} + \frac{2aAC}{\pi a^2} + \frac{2aK}{\pi^2}\left[2 + \frac{r}{a}\ln\left(\frac{r-a}{r+a}\right)\right] \qquad \textbf{(4.4.137)}$$

$$+ \frac{2}{\pi^2}\int_a^b \frac{\sqrt{\eta^2-a^2}}{\eta^2-r^2}\mu_2(\eta)\left[\eta\ln\left(\frac{\eta-a}{\eta+a}\right) - r\ln\left(\frac{r-a}{r+a}\right)\right]d\eta.$$

Introducing the variables $x = r^2/a^2$, $t = \eta^2/a^2$, $\beta = b^2/a^2$, $\mu_2(r) = Ay(t)/a^2$, and $G = a^2K/A$, Equation 4.4.137 simplifies to

$$\sqrt{x-1}\;y(x) = \frac{2}{\pi x} + C + \frac{2G}{\pi^2}\left[2 + \sqrt{x}\ln\left(\frac{\sqrt{x}-1}{\sqrt{x}+1}\right)\right] \qquad \textbf{(4.4.138)}$$

$$+ \frac{1}{\pi^2}\int_1^\beta \frac{\sqrt{t-1}\,y(t)}{\sqrt{t}(t-x)}\left[\sqrt{t}\ln\left(\frac{\sqrt{t}-1}{\sqrt{t}+1}\right) - \sqrt{x}\ln\left(\frac{\sqrt{x}-1}{\sqrt{x}+1}\right)\right]dt$$

$$= \frac{2}{\pi x} + C' + \frac{1}{\pi^2}\int_1^\beta \frac{y(t)}{\sqrt{t}\sqrt{t-1}(t-x)}\,dt \qquad \textbf{(4.4.139)}$$

$$\times\left[(t-1)\sqrt{t}\ln\left(\frac{\sqrt{t}-1}{\sqrt{t}+1}\right) - (x-1)\sqrt{x}\ln\left(\frac{\sqrt{x}-1}{\sqrt{x}+1}\right)\right],$$

where $C' = C + 4G/\pi^2$. Equation 4.4.139 has the advantage that its kernel is not singular. We obtained it from Equation 4.4.138 by using the relationship that

$$\int_1^\beta \frac{y(t)}{\sqrt{t}\sqrt{t-1}}\,dt = 2G, \qquad \textbf{(4.4.140)}$$

which follows from Equation 4.4.133 in the limit of $r \to a$.

The potential is computed as follows: For a specific C', we find $y(x)$ from Equation 4.4.139. The corresponding value of G follows from Equation 4.4.140. By varying C', we can compute the $y(x)$ for a desired G. We compute the function $\mu_1(\xi)$ from Equation 4.4.135. Finally, combining Equation 4.4.116 and Equation 4.4.120, we have that

$$u(r,z) = \int_0^a \mu_1(\xi)\left[\int_0^\infty e^{-kz}\sin(k\xi)J_0(kr)\,dk\right]d\xi$$

$$+ \int_a^b \mu_2(\xi)\left[\int_0^\infty e^{-kz}\sin(k\xi)J_0(kr)\,dk\right]d\xi. \qquad \textbf{(4.4.141)}$$

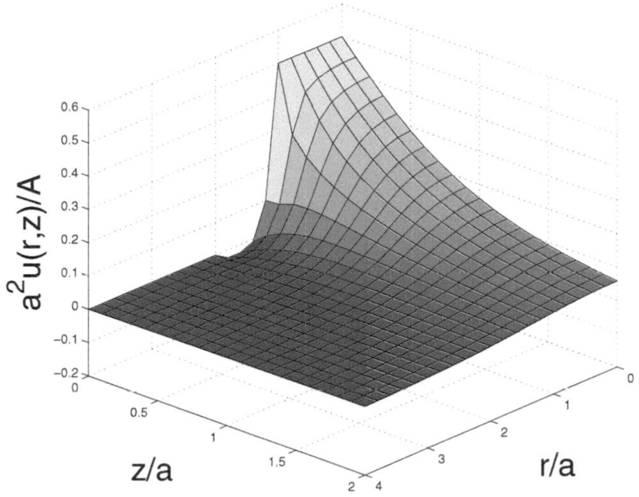

Figure 4.4.3: The solution to Equation 4.4.112 subject to the mixed boundary conditions given by Equation 4.4.113 through Equation 4.4.115 when $b/a = 2$ and $a^2K/A = 0.5$.

We evaluate numerically the integrals inside of the square brackets (except for the case when $z = 0$ where there is an exact expression) and then we compute the ξ integration. Figure 4.4.3 illustrates the solution for $b/a = 2$ and $G = 0.5$.

• Example 4.4.4

Cooke's results are also useful in solving[92]

$$\frac{\partial^2 u}{\partial r^2} + \frac{1}{r}\frac{\partial u}{\partial r} - \frac{u}{r^2} + \frac{\partial^2 u}{\partial z^2} = 0, \qquad 0 \le r < \infty, \quad 0 < z < \infty, \tag{4.4.142}$$

subject to the boundary conditions

$$\lim_{r\to 0} |u(r,z)| < \infty, \quad \lim_{r\to\infty} u(r,z) \to 0, \qquad 0 < z < \infty, \tag{4.4.143}$$

$$\lim_{z\to\infty} u(r,z) \to 0, \qquad 0 \le r < \infty, \tag{4.4.144}$$

and

$$\begin{cases} u_z(r,0) = 0, & 0 \le r < a, \\ u(r,0) = r, & a < r < b, \\ u_z(r,0) = 0, & b < r < \infty, \end{cases} \tag{4.4.145}$$

[92] See Borodachev, N. M., and F. N. Borodacheva, 1966: Twisting of an elastic half-space by the rotation of a ring-shaped punch. *Mech. Solids*, **1(1)**, 63–66.

where $b > a > 0$.

Using transform methods or separation of variables, the general solution to Equation 4.4.142, Equation 4.4.143, and Equation 4.4.144 is

$$u(r,z) = \int_0^\infty A(k) J_1(kr) e^{-kz} \frac{dk}{k}. \tag{4.4.146}$$

Substituting Equation 4.4.146 into Equation 4.4.145, we find that

$$\int_0^\infty A(k) J_1(kr)\, dk = 0, \qquad 0 \le r < a, \tag{4.4.147}$$

$$\int_0^\infty A(k) J_1(kr)\, \frac{dk}{k} = r, \qquad a < r < b, \tag{4.4.148}$$

and

$$\int_0^\infty A(k) J_1(kr)\, dk = 0, \qquad b < r < \infty. \tag{4.4.149}$$

We can use Cooke's results if we set $\nu = 1$. Then, from Equation 4.4.97 through Equation 4.4.100, we have that

$$A(k) = k \int_a^b r\, g(r) J_1(kr)\, dr, \tag{4.4.150}$$

where

$$g(r) = -\frac{2}{\pi}\frac{d}{dr}\left[\int_r^b \frac{\eta\, h(\eta)}{\sqrt{\eta^2 - r^2}}\, d\eta\right], \tag{4.4.151}$$

$$\eta^2 h(\eta) = \frac{d}{d\eta}\left[\int_a^\eta \frac{x^3}{\sqrt{\eta^2 - x^2}}\, dx\right] - \frac{4}{\pi^2}\frac{\eta}{\sqrt{\eta^2 - a^2}} \int_a^b \frac{t\, h(t)}{\sqrt{t^2 - a^2}} K(\eta, t)\, dt \tag{4.4.152}$$

$$= \frac{(2\eta^2 - a^2)\eta}{\sqrt{\eta^2 - a^2}} - \frac{4}{\pi^2}\frac{\eta}{\sqrt{\eta^2 - a^2}} \int_a^b \frac{t\, h(t)}{\sqrt{t^2 - a^2}} K(\eta, t)\, dt, \tag{4.4.153}$$

and

$$K(\eta, t) = \int_0^a \frac{y^2(a^2 - y^2)}{(\eta^2 - y^2)(t^2 - y^2)}\, dy \tag{4.4.154}$$

$$= \frac{1}{2(\eta^2 - t^2)}\left[\eta(\eta^2 - a^2)\ln\left(\frac{\eta + a}{\eta - a}\right) - t(t^2 - a^2)\ln\left(\frac{t + a}{t - a}\right)\right] - a. \tag{4.4.155}$$

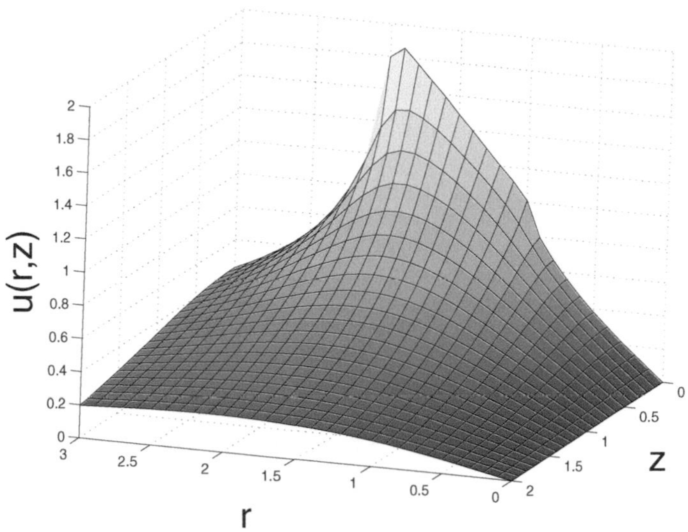

Figure 4.4.4: The solution to Equation 4.4.142 subject to the mixed boundary conditions given by Equation 4.4.143 through Equation 4.4.145 when $a = 1$ and $b = 2$.

In the special case when $t = \eta$, we employ L'Hospital rule and find that

$$K(\eta, \eta) = \frac{3\eta^2 - a^2}{4\eta} \ln\left(\frac{\eta + a}{\eta - a}\right) - \frac{3a}{2}. \qquad \textbf{(4.4.156)}$$

If we introduce

$$\eta\, h(\eta) = \sqrt{\eta^2 - a^2}\, \chi(\eta), \qquad \textbf{(4.4.157)}$$

then Equation 4.4.151 and Equation 4.4.153 become

$$g(r) = -\frac{2}{\pi}\frac{d}{dr}\left[\int_r^b \sqrt{\frac{\eta^2 - a^2}{\eta^2 - r^2}}\, \chi(\eta)\, d\eta\right], \qquad \textbf{(4.4.158)}$$

and

$$(\eta^2 - a^2)\chi(\eta) = 2\eta^2 - a^2 - \frac{4}{\pi^2}\int_a^b K(\eta, t)\chi(t)\, dt. \qquad \textbf{(4.4.159)}$$

The evaluation of $u(r, z)$ begins by solving Equation 4.4.159 by replacing the integral with its representation from the midpoint rule. With those values of $\chi(\eta)$, Equation 4.4.158 gives $g(r)$. Finally, combining Equation 4.4.146 and Equation 4.4.158, we find that

$$u(r, z) = \int_a^b \eta g(\eta) \left[\int_0^\infty J_1(k\eta)J_1(kr)e^{-kz}\, dk\right] d\eta, \qquad \textbf{(4.4.160)}$$

after we interchange the order of integration. We use Simpson's rule to evaluate the integral within the square brackets. Figure 4.4.4 illustrates this potential when $a = 1$ and $b = 2$.

• Example 4.4.5

Kim and Kim solved Laplace's equation[93]

$$\frac{\partial^2 u}{\partial r^2} + \frac{1}{r}\frac{\partial u}{\partial r} + \frac{\partial^2 u}{\partial z^2} = 0, \qquad 0 \le r < \infty, \quad 0 < z < \infty, \tag{4.4.161}$$

subject to the boundary conditions

$$\lim_{r\to 0} |u(r,z)| < \infty, \quad \lim_{r\to\infty} u(r,z) \to 0, \qquad 0 < z < \infty, \tag{4.4.162}$$

$$\lim_{z\to\infty} u(r,z) \to 0, \qquad 0 \le r < \infty, \tag{4.4.163}$$

and

$$\begin{cases} u_{rz}(r,0) = 0, & 0 \le r < a, \\ u_r(r,0) = -r, & a < r < b, \\ u_{rz}(r,0) = 0, & b < r < \infty. \end{cases} \tag{4.4.164}$$

Using transform methods or separation of variables, the general solution to Equation 4.4.161, Equation 4.4.162, and Equation 4.4.163 is

$$u(r,z) = \int_0^\infty A(k) J_0(kr) e^{-kz} \frac{dk}{k^2}. \tag{4.4.165}$$

Substituting Equation 4.4.165 into Equation 4.4.164, we find that

$$\int_0^\infty A(k) J_1(kr)\, dk = 0, \qquad 0 \le r < a, \tag{4.4.166}$$

$$\int_0^\infty A(k) J_1(kr) \frac{dk}{k} = r, \qquad a < r < b, \tag{4.4.167}$$

and

$$\int_0^\infty A(k) J_1(kr)\, dk = 0, \qquad b < r < \infty. \tag{4.4.168}$$

To solve this set of integral equations, we introduce

$$F(r) = \int_0^\infty A(k) J_1(kr)\, dk, \qquad a < r < b. \tag{4.4.169}$$

Therefore, the Hankel transform of Equation 4.4.169 is

$$\frac{A(k)}{k^2} = \int_a^b \xi F(\xi) \frac{J_1(k\xi)}{k}\, d\xi. \tag{4.4.170}$$

---

[93] Taken with permission from Kim, M.-U., and J.-U. Kim, 1985: Slow rotation of an annular disk in a viscous fluid. *J. Phys. Soc. Japan*, **54**, 3337–3341.

Then, from Equation 4.4.167,

$$\int_a^b \xi F(\xi) \left[\int_0^\infty J_1(k\xi) J_1(kr)\, dk \right] d\xi = r. \tag{4.4.171}$$

Now,[94]

$$\int_0^\infty J_1(k\xi) J_1(kr)\, dk = \frac{2}{\pi \xi r} \int_0^{\min(r,\xi)} \frac{\tau^2}{\sqrt{r^2 - \tau^2}\sqrt{\xi^2 - \tau^2}}\, d\tau. \tag{4.4.172}$$

Because

$$\int_a^b \int_0^{\min(r,\xi)} (\cdots)\, d\tau\, d\xi = \int_a^r \int_\tau^b (\cdots)\, d\xi\, d\tau + \int_0^a \int_a^b (\cdots)\, d\xi\, d\tau, \tag{4.4.173}$$

Equation 4.4.171 becomes

$$\int_a^r \frac{\tau^2 G(\tau)}{\sqrt{r^2 - \tau^2}}\, d\tau = \frac{\pi r^2}{2} - \int_0^a \frac{\tau^2}{\sqrt{r^2 - \tau^2}} \left[\int_a^b \frac{F(\xi)}{\sqrt{\xi^2 - \tau^2}}\, d\xi \right] d\tau, \tag{4.4.174}$$

where

$$G(\tau) = \int_\tau^b \frac{F(\xi)}{\sqrt{\xi^2 - \tau^2}}\, d\xi, \qquad a < \tau < b. \tag{4.4.175}$$

Solving for $F(\xi)$ from Equation 4.4.175 using Equation 1.2.15 and Equation 1.2.16,

$$F(\xi) = -\frac{2}{\pi} \frac{d}{d\xi} \left[\int_\xi^b \frac{\tau\, G(\tau)}{\sqrt{\tau^2 - \xi^2}}\, d\tau \right]. \tag{4.4.176}$$

Turning to Equation 4.4.174, we treat the right side as a known. Then, from Equation 1.2.13 and Equation 1.2.14,

$$G(\tau) = \left(\frac{\sqrt{\tau^2 - a^2}}{\tau} + \frac{\tau}{\sqrt{\tau^2 - a^2}} \right) - \frac{4}{\pi^2 \tau \sqrt{\tau^2 - a^2}} \int_a^b \frac{\xi\, G(\xi)}{\sqrt{\xi^2 - a^2}} K(\tau, \xi)\, d\xi, \tag{4.4.177}$$

where

$$K(\tau, \xi) = \frac{1}{2(\tau^2 - \xi^2)} \left[\tau(\tau^2 - a^2) \ln\left(\frac{\tau + a}{\tau - a} \right) - \xi(\xi^2 - a^2) \ln\left(\frac{\xi + a}{\xi - a} \right) \right] - a. \tag{4.4.178}$$

Our numerical calculation begins by setting $\tau = a \sec(\sigma)$ and $\xi = a \sec(\zeta)$. Equation 4.4.177 is then finite differenced and solved to yield $G(\tau)$. Then,

---

94 Cooke, op. cit.

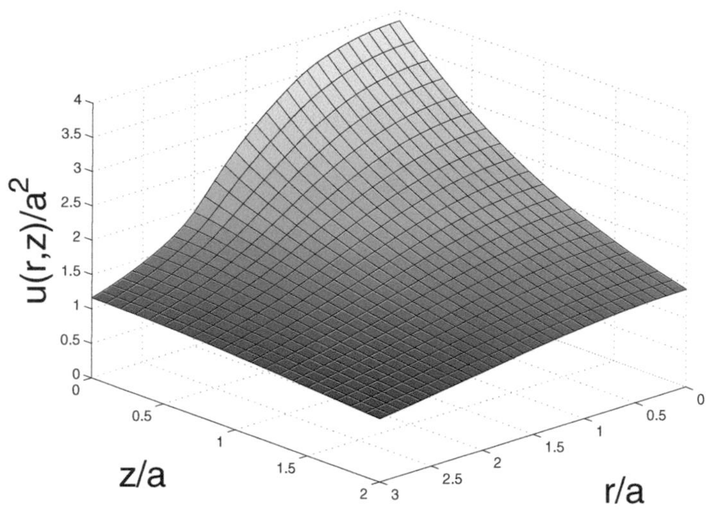

Figure 4.4.5: The solution to Equation 4.4.161 subject to the mixed boundary conditions given by Equation 4.4.162 through Equation 4.4.164 when $b/a = 2$

$F(\xi)$ and $A(k)$ follow from Equation 4.4.176 and Equation 4.4.170, respectively. Finally Equation 4.4.165 provides $u(r, z)$. Figure 4.4.5 illustrates this solution when $b/a = 2$.

• Example 4.4.6

Let us solve[95] Laplace's equation over a slab of thickness h. The governing equation is

$$\frac{\partial^2 u}{\partial r^2} + \frac{1}{r}\frac{\partial u}{\partial r} + \frac{\partial^2 u}{\partial z^2} = 0, \qquad 0 \le r < \infty, \quad 0 < z < h, \tag{4.4.179}$$

subject to the boundary conditions

$$\lim_{r\to 0} |u(r, z)| < \infty, \quad \lim_{r\to\infty} u(r, z) \to 0, \qquad 0 < z < h, \tag{4.4.180}$$

and

$$\begin{cases} u(r, 0) = f(r), \quad u_z(r, h) = g(r), & 0 \le r < 1, \\ \quad u_z(r, 0) = u(r, h) = 0, & 1 < r < \infty. \end{cases} \tag{4.4.181}$$

---

[95] Taken from Dhaliwal, R. S., 1966: Mixed boundary value problem of heat conduction for infinite slab. *Appl. Sci. Res.*, **16**, 228–240 with kind permission from Springer Science and Business Media.

Using Hankel transforms, the solution to Equation 4.4.179 and Equation 4.4.180 is

$$u(r,z) = \int_0^\infty [A(k)\cosh(kz) + B(k)\sinh(kz)]\, J_0(kr)\, dk. \qquad \textbf{(4.4.182)}$$

Substituting Equation 4.4.182 into the mixed boundary conditions, Equation 4.4.181, we find that

$$\int_0^\infty A(k) J_0(kr)\, dk = f(r), \qquad 0 \le r < 1, \qquad \textbf{(4.4.183)}$$

$$\int_0^\infty [A(k)\sinh(kh) + B(k)\cosh(kh)]\, k\, J_0(kr)\, dk = g(r), \qquad 0 \le r < 1, \qquad \textbf{(4.4.184)}$$

$$\int_0^\infty k\, B(k) J_0(kr)\, dk = 0, \qquad 1 < r < \infty, \qquad \textbf{(4.4.185)}$$

and

$$\int_0^\infty [A(k)\cosh(kh) + B(k)\sinh(kh)]\, J_0(kr)\, dk = 0, \qquad 1 < r < \infty. \qquad \textbf{(4.4.186)}$$

To solve Equation 4.4.183 through Equation 4.4.186, we introduce

$$B(k) = \int_0^1 \psi_1(t)\cos(kt)\, dt, \qquad \textbf{(4.4.187)}$$

and

$$A(k)\cosh(kh) + B(k)\sinh(kh) = \int_0^1 \psi_2(t)\sin(kt)\, dt, \quad \psi_2(0) = 0. \qquad \textbf{(4.4.188)}$$

We introduced these relations because Equation 4.4.185 and Equation 4.4.186 are automatically satisfied. Simple algebra yields

$$A(k) = \frac{1}{\cosh(kh)} \int_0^1 \psi_2(t)\sin(kt)\, dt - \left[1 - \frac{e^{-kh}}{\cosh(kh)}\right] \int_0^1 \psi_1(t)\cos(kt)\, dt, \qquad \textbf{(4.4.189)}$$

and

$$A(k)\sinh(kh) + B(k)\cosh(kh) = \frac{1}{\cosh(kh)} \int_0^1 \psi_1(t)\cos(kt)\, dt + \left[1 - \frac{e^{-kh}}{\cosh(kh)}\right] \int_0^1 \psi_2(t)\sin(kt)\, dt. \qquad \textbf{(4.4.190)}$$

Introducing Equation 4.4.187 and Equation 4.4.189 into Equation 4.4.183 and Equation 4.4.184, multiplying the resulting equations by $d\eta/\sqrt{r^2-\eta^2}$ and integrating from 0 to r, we obtain

$$\begin{aligned}\int_0^r \frac{d\eta}{\sqrt{r^2-\eta^2}}\bigg\{&-\psi_1(\eta) - \frac{1}{\pi h}\int_0^1 \psi_1(t)\left[G_1(\eta+t)+G_1(\eta-t)\right] dt \\ &+ \frac{1}{\pi h}\int_0^1 \psi_2(t)\left[G_2(\eta+t)-G_2(\eta-t)\right] dt\bigg\} = f(r), \quad 0 \le r < 1,\end{aligned} \tag{4.4.191}$$

and

$$\begin{aligned}\int_0^r \frac{d\eta}{\sqrt{r^2-\eta^2}}\bigg\{&\psi_2'(\eta) + \frac{1}{\pi h}\int_0^1 \psi_1(t)\left[G_2'(\eta+t)+G_2'(\eta-t)\right] dt \\ &- \frac{1}{\pi h}\int_0^1 \psi_2(t)\left[G_1'(\eta+t)-G_1'(\eta-t)\right] dt\bigg\} = g(r), \quad 0 \le r < 1,\end{aligned} \tag{4.4.192}$$

where

$$G_1(\xi) = -\int_0^\infty \frac{e^{-\eta}}{\cosh(\eta)}\cos(\xi\eta/h)\, d\eta \quad \text{and} \quad G_2(\xi) = \int_0^\infty \frac{\sin(\xi\eta/h)}{\cosh(\eta)}\, d\eta. \tag{4.4.193}$$

In deriving Equation 4.4.191 and Equation 4.4.192, we used Equation 1.4.14 and

$$J_0(kr) = \frac{2}{\pi}\int_0^r \frac{\cos(kx)}{\sqrt{r^2-x^2}}\, dx. \tag{4.4.194}$$

Equation 4.4.191 and Equation 4.4.192 are integral equations of the Abel type. If the quantities inside the large brackets are treated as unknowns, then Equation 1.2.14 gives

$$\psi_1(\eta)+\frac{1}{\pi h}\int_0^1 \left[K_1(\eta,t)\psi_1(t) - K_2(\eta,t)\psi_2(t)\right] dt = -\frac{2}{\pi}\frac{d}{d\eta}\left[\int_0^\eta \frac{r\, f(r)}{\sqrt{\eta^2-r^2}}\, dr\right], \tag{4.4.195}$$

and

$$\psi_2(\eta) + \frac{1}{\pi h}\int_0^1 \left[K_3(\eta,t)\psi_1(t) + K_4(\eta,t)\psi_2(t)\right] dt = \frac{2}{\pi}\int_0^\eta \frac{r\, g(r)}{\sqrt{\eta^2-r^2}}\, dr, \tag{4.4.196}$$

if $0 \le \eta < 1$, where

$$K_1(\eta,t) = G_1(\eta+t)+G_1(\eta-t) = -2\int_0^\infty e^{-\xi}\frac{\cos(\eta\xi/h)\cos(t\xi/h)}{\cosh(\xi)}\, d\xi, \tag{4.4.197}$$

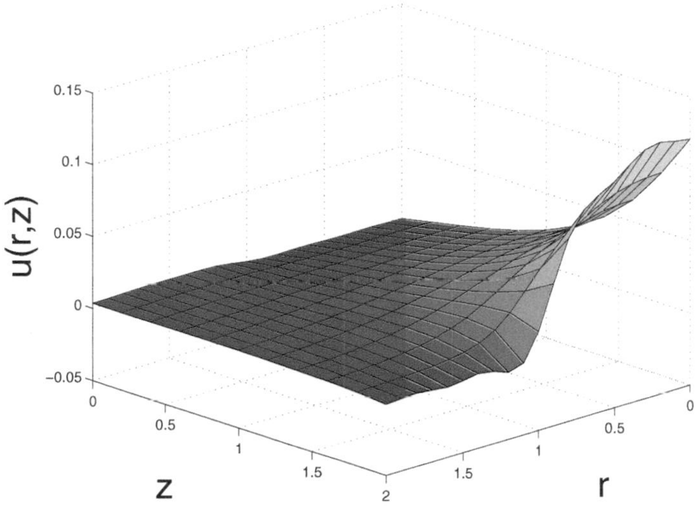

Figure 4.4.6: The solution $u(r,z)$ to the mixed boundary value problem governed by Equation 4.4.179 through Equation 4.4.181.

$$K_2(\eta,t) = G_2(\eta+t) - G_2(\eta-t) = 2\int_0^\infty \frac{\cos(\eta\xi/h)\sin(t\xi/h)}{\cosh(\xi)}\,d\xi, \quad \textbf{(4.4.198)}$$

$$K_3(\eta,t) = G_2(\eta+t) + G_2(\eta-t) = 2\int_0^\infty \frac{\sin(\eta\xi/h)\cos(t\xi/h)}{\cosh(\xi)}\,d\xi, \quad \textbf{(4.4.199)}$$

and

$$K_4(\eta,t) = G_1(\eta-t) - G_1(\eta+t) = -2\int_0^\infty e^{-\xi}\frac{\sin(\eta\xi/h)\sin(t\xi/h)}{\cosh(\xi)}\,d\xi. \quad \textbf{(4.4.200)}$$

We compute $u(r,z)$ as follows: First, we find $\psi_1(\eta)$ and $\psi_2(\eta)$ via Equation 4.4.195 and Equation 4.4.196. Next, $A(k)$ and $B(k)$ follow from Equation 4.4.187 and Equation 4.4.189, respectively. Finally, Equation 4.4.182 yields $u(r,z)$. Figure 4.4.6 illustrates the solution when $f(r) = 0$, $g(r) = 1$, and $h = 2$.

• Example 4.4.7

Let us solve Laplace's equation:[96]

$$\frac{\partial^2 u}{\partial r^2} + \frac{1}{r}\frac{\partial u}{\partial r} + \frac{\partial^2 u}{\partial z^2} = 0, \qquad 0 \le r < \infty, \quad -\infty < z < \infty, \qquad \textbf{(4.4.201)}$$

[96] See Kuz'min, Yu. N., 1972: Electrostatic field of a circular disk near a plane containing an aperture. *Sov. Tech. Phys.*, **17**, 473–476.

subject to the boundary conditions

$$\lim_{r\to 0}|u(r,z)| < \infty, \quad \lim_{r\to\infty} u(r,z) \to 0, \qquad -\infty < z < \infty, \tag{4.4.202}$$

$$\lim_{|z|\to\infty} u(r,z) \to 0, \qquad 0 \le r < \infty, \tag{4.4.203}$$

$$u(r,0^-) = u(r,0^+), \qquad 0 \le r < \infty, \tag{4.4.204}$$

$$\begin{cases} u_z(r,0^-) = u_z(r,0^+), & 0 \le r < b, \\ u(r,0) = 0, & b < r < \infty, \end{cases} \tag{4.4.205}$$

$$u(r,h^-) = u(r,h^+), \qquad 0 \le r < \infty, \tag{4.4.206}$$

and

$$\begin{cases} u(r,h) = V, & 0 < r < a, \\ u(r,h^-) = u(r,h^+), & a < r < \infty, \end{cases} \tag{4.4.207}$$

where $b > a$.

Using Hankel transforms, the solution to Equation 4.4.201 through Equation 4.4.204 and Equation 4.4.206 is

$$u(r,z) = \int_0^\infty B(k)e^{kz} J_0(kr)\,dk, \qquad -\infty < z \le 0, \tag{4.4.208}$$

$$u(r,z) = \int_0^\infty \frac{A(k)\sinh(kz) + B(k)\sinh[k(h-z)]}{\sinh(kh)} J_0(kr)\,dk, \quad 0 \le z \le h, \tag{4.4.209}$$

and

$$u(r,z) = \int_0^\infty A(k)e^{k(h-z)} J_0(kr)\,dk, \qquad h \le z < \infty. \tag{4.4.210}$$

Substituting Equation 4.4.208 through Equation 4.4.210 into Equation 4.4.205 and Equation 4.4.207, we find that

$$\int_0^\infty A(k) J_0(kr)\,dk = V, \qquad 0 \le r < a, \tag{4.4.211}$$

$$\int_0^\infty k\frac{e^{kh}A(k) - B(k)}{\sinh(kh)} J_0(kr)\,dk = 0, \qquad a < r < \infty, \tag{4.4.212}$$

$$\int_0^\infty k\frac{e^{kh}B(k) - A(k)}{\sinh(kh)} J_0(kr)\,dk = 0, \qquad 0 \le r < b, \tag{4.4.213}$$

and

$$\int_0^\infty B(k) J_0(kr)\,dk = 0, \qquad b < r < \infty. \tag{4.4.214}$$

If we introduce

$$C(k) = \frac{k\left[A(k)e^{kh} - B(k)\right]}{\sinh(kh)}, \quad \text{or} \quad A(k) = e^{-kh}\left[B(k) + \frac{\sinh(kh)}{k}C(k)\right], \tag{4.4.215}$$

then Equation 4.4.211 through Equation 4.4.214 become

$$\int_0^\infty C(k)J_0(kr)\,\frac{dk}{k} = f(r), \qquad 0 \le r < a, \tag{4.4.216}$$

$$\int_0^\infty C(k)J_0(kr)\,dk = 0, \qquad a < r < \infty, \tag{4.4.217}$$

$$\int_0^\infty k\,B(k)J_0(kr)\,dk = g(r), \qquad 0 \le r < b, \tag{4.4.218}$$

and

$$\int_0^\infty B(k)J_0(kr)\,dk = 0, \qquad b < r < \infty, \tag{4.4.219}$$

where

$$f(r) = 2V + \int_0^\infty C(k)e^{-2kh}J_0(kr)\,\frac{dk}{k} - 2\int_0^\infty B(k)e^{-kh}J_0(kr)\,dk, \tag{4.4.220}$$

and

$$g(r) = \frac{1}{2}\int_0^\infty C(k)e^{-kh}J_0(kr)\,dk. \tag{4.4.221}$$

Equation 4.4.217 and Equation 4.4.219 are satisfied identically if we introduce a $\phi(t)$ and $\psi(t)$ such that

$$C(k)/k = \int_0^a \phi(\tau)\cos(k\tau)\,d\tau, \tag{4.4.222}$$

and

$$B(k) = \int_0^b \psi(\tau)\sin(k\tau)\,d\tau. \tag{4.4.223}$$

Substituting for $B(k)$ and $C(k)$ in Equation 4.4.216 and interchanging the order of integration, we have that

$$\begin{aligned} \int_0^a \phi(\tau)\left[\int_0^\infty \cos(k\tau)J_0(kr)\,dk\right]d\tau & \\ = 2V + \int_0^a \phi(\tau)&\left[\int_0^\infty e^{-2kh}\cos(k\tau)J_0(kr)\,dk\right]d\tau \\ - 2\int_0^b \psi(\tau)&\left[\int_0^\infty e^{-kh}\sin(k\tau)J_0(kr)\,dk\right]d\tau, \end{aligned} \tag{4.4.224}$$

or

$$\int_0^r \frac{\phi(\tau)}{\sqrt{r^2-\tau^2}}\,d\tau = 2V + \int_0^a \phi(\tau)\left[\int_0^\infty e^{-2kh}\cos(k\tau)J_0(kr)\,dk\right]d\tau$$
$$-2\int_0^b \psi(\tau)\left[\int_0^\infty e^{-kh}\sin(k\tau)J_0(kr)\,dk\right]d\tau. \quad \textbf{(4.4.225)}$$

Solving this integral equation of the Abel type,

$$\phi(t) = \frac{4V}{\pi}\frac{d}{dt}\left[\int_0^t \frac{\tau}{\sqrt{t^2-\tau^2}}\,d\tau\right] \quad \textbf{(4.4.226)}$$
$$+\frac{2}{\pi}\int_0^a \phi(\tau)\left\{\int_0^\infty e^{-2kh}\cos(k\tau)\frac{d}{dt}\left[\int_0^t \frac{rJ_0(kr)}{\sqrt{t^2-r^2}}\,dr\right]dk\right\}d\tau$$
$$-\frac{4}{\pi}\int_0^b \psi(\tau)\left\{\int_0^\infty e^{-kh}\sin(k\tau)\frac{d}{dt}\left[\int_0^t \frac{rJ_0(kr)}{\sqrt{t^2-r^2}}\,dr\right]dk\right\}d\tau.$$

Evaluating the first integral on the right side of Equation 4.4.226 and employing Equation 1.4.9, we finally obtain

$$\phi(t) = \frac{4V}{\pi} + \frac{2}{\pi}\int_0^a K(t,\tau)\phi(\tau)\,d\tau - \frac{4}{\pi}\int_0^b M(t,\tau)\psi(\tau)\,d\tau, \quad \textbf{(4.4.227)}$$

where

$$K(t,\tau) = \int_0^\infty e^{-2kh}\cos(k\tau)\cos(kt)\,dk = \frac{h}{4h^2+(t+\tau)^2} + \frac{h}{4h^2+(t-\tau)^2}, \quad \textbf{(4.4.228)}$$

and

$$M(t,\tau) = \int_0^\infty e^{-kh}\sin(k\tau)\cos(kt)\,dk = \frac{1}{2}\left[\frac{t+\tau}{4h^2+(t+\tau)^2} - \frac{t-\tau}{4h^2+(t-\tau)^2}\right]. \quad \textbf{(4.4.229)}$$

Turning to Equation 4.4.218 and substituting for $B(k)$, we find that

$$\int_0^b \psi(\tau)\left[\int_0^\infty k\sin(k\tau)J_0(kr)\,dk\right]d\tau = g(r). \quad \textbf{(4.4.230)}$$

Multiplying Equation 4.4.230 by $r\,dr/\sqrt{t^2-r^2}$, integrating from 0 to t, and using Equation 1.4.9, Equation 4.4.230 transforms into

$$\int_0^\infty \psi(\tau)\left[\int_0^\infty \sin(k\tau)\sin(kt)\,dk\right]d\tau = \int_0^t \frac{r\,g(r)}{\sqrt{t^2-r^2}}\,dr, \quad \textbf{(4.4.231)}$$

or

$$\psi(t) = \frac{2}{\pi}\int_0^t \frac{r\,g(r)}{\sqrt{t^2-r^2}}\,dr. \quad \textbf{(4.4.232)}$$

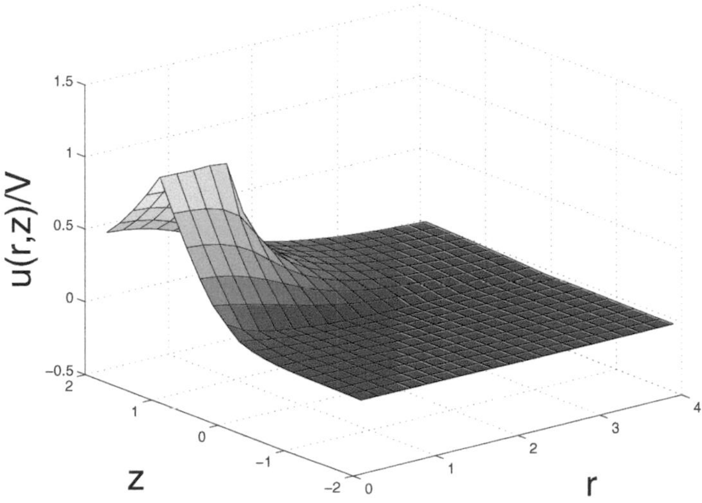

Figure 4.4.7: The solution $u(r, z)$ to the mixed boundary value problem governed by Equation 4.4.201 through Equation 4.4.207.

Upon substituting Equation 4.4.221, Equation 4.4.232 becomes

$$\psi(t) = \frac{1}{\pi}\int_0^\infty C(k)e^{-kh}\left[\int_0^t \frac{rJ_0(kr)}{\sqrt{t^2-r^2}}\,dr\right]dk \tag{4.4.233}$$

$$= \frac{1}{\pi}\int_0^\infty C(k)e^{-kh}\sin(kt)\frac{dk}{k} \tag{4.4.234}$$

$$= \frac{1}{\pi}\int_0^a \phi(\tau)\left[\int_0^\infty e^{-kh}\sin(kt)\cos(k\tau)\,dk\right]d\tau \tag{4.4.235}$$

$$= \frac{1}{\pi}\int_0^a M(\tau,t)\phi(\tau)\,d\tau. \tag{4.4.236}$$

In summary, once we find $\phi(t)$ and $\psi(t)$ by solving the simultaneous integral equations, Equation 4.4.227 and Equation 4.4.236, respectively, we can compute $A(k)$ and $B(k)$ via Equation 4.4.222, Equation 4.4.223, and Equation 4.4.215. Finally $u(r, z)$ follows from Equation 4.4.208 through Equation 4.4.210. Figure 4.4.7 illustrates the solution when $a = 1$, $b = 2$, and $h = 1$.

Problems

1. If $0 < a < 1$, solve[97]

$$\frac{\partial^2 u}{\partial r^2} + \frac{1}{r}\frac{\partial u}{\partial r} + \frac{\partial^2 u}{\partial z^2} = 0, \qquad 0 \le r < \infty, \quad -\infty < z < \infty,$$

[97] See Davis, A. M. J., 1991: Slow viscous flow due to motion of an annular disk; pressure-driven extrusion through an annular hole in a wall. *J. Fluid Mech.*, **231**, 51–71.

subject to the boundary conditions

$$\lim_{r\to 0} |u(r,z)| < \infty, \quad \lim_{r\to\infty} u(r,z) \to 0, \qquad -\infty < z < \infty,$$

$$\lim_{|z|\to\infty} u(r,z) \to 0, \qquad 0 \le r < \infty,$$

and

$$\begin{cases} u_z(r,0) = 0, & 0 \le r < a, \\ u(r,0) = 1, & a < r < 1, \\ u_z(r,0) = 0, & 1 < r < \infty. \end{cases}$$

Step 1: Using separation of variables or transform methods, show that the general solution to the partial differential equation and the first two boundary conditions is

$$u(r,z) = \frac{2}{\pi}\int_0^\infty A(k)e^{-k|z|}J_0(kr)\,dk.$$

Step 2: Using the mixed boundary condition, show that $A(k)$ satisfies the triple equation

$$\int_0^\infty kA(k)J_0(kr)\,dk = 0, \qquad 0 \le r < a, 1 < r < \infty, \tag{1}$$

and

$$\int_0^\infty A(k)J_0(kr)\,dk = \frac{\pi}{2}, \qquad a \le r \le 1. \tag{2}$$

Step 3: Given[98]

$$\int_0^\infty J_0(kr)\sin(k)\frac{dk}{k} = \begin{cases} \pi/2, & 0 \le r < 1, \\ \arcsin(1/r), & 1 < r < \infty, \end{cases}$$

and setting

$$A(k) = \frac{\sin(k)}{k} - \int_0^a \frac{F(t)}{t}\sin(kt)\,dt - \int_1^\infty G(t)\cos(kt)\,dt,$$

show that Equation (2) is satisfied. Hint: Use Equation 1.4.13 and Equation 1.4.14.

Step 4: Using[99]

$$\int_a^\infty \frac{r\,J_0(kr)}{\sqrt{r^2-a^2}}\,dr = \frac{\cos(ka)}{k}, \qquad \text{and} \qquad \int_0^a \frac{r\,J_0(kr)}{\sqrt{a^2-r^2}}\,dr = \frac{\sin(ka)}{k},$$

---

[98] Gradshteyn and Ryzhik, op. cit., Formula 6.693.

[99] Ibid., Formula 6.554.3 and Formula 6.554.2.

show that

$$\int_0^\infty A(k)\cos(kr)\,dk = 0, \qquad r \ge 1,$$

and

$$\int_0^\infty A(k)\sin(kr)\,dk = 0, \qquad r \le a.$$

Step 5: Substituting $A(k)$ into the results from Step 4, show that

$$\int_0^a \frac{F(t)}{r^2 - t^2}\,dt = \frac{\pi}{2} G(r), \qquad 1 \le r,$$

and

$$\frac{1}{2}\ln\left(\frac{1+r}{1-r}\right) = \frac{\pi\, F(r)}{2r} - r\int_1^\infty \frac{G(t)}{t^2 - r^2}\,dt, \qquad 0 \le r \le a.$$

Step 6: Eliminating $G(r)$, show that

$$F(r) - \frac{2r^2}{\pi^2}\int_0^a \left[\frac{1}{r}\ln\left(\frac{1+r}{1-r}\right) - \frac{1}{t}\ln\left(\frac{1+t}{1-t}\right)\right]\frac{F(t)}{r^2 - t^2}\,dt = \frac{r}{\pi}\ln\left(\frac{1+r}{1-r}\right),$$

for $0 \le r \le a$.

Step 7: Using the relationship[100] that

$$\ln\left(\frac{1+x}{1-x}\right) = 2\sum_{n=1}^\infty \frac{x^{2n-1}}{2n-1}, \qquad |x| < 1,$$

show that $F(0) = 0$.

Step 8: Show that the potential is given by

$$\begin{aligned} u(r,z) = {} & \frac{2}{\pi}\arcsin\left[\frac{2}{\sqrt{z^2 + (1+r)^2} + \sqrt{z^2 + (1-r)^2}}\right] \\ & - \frac{2}{\pi}\int_0^a F(t)\left[\int_0^\infty e^{-k|z|}\frac{\sin(kt)}{t} J_0(kr)\,dk\right] dt \\ & - \frac{2}{\pi}\int_1^\infty G(t)\left[\int_0^\infty e^{-k|z|}\cos(kt) J_0(kr)\,dk\right] dt. \end{aligned}$$

---

[100] Ibid., Formula 1.513.1.

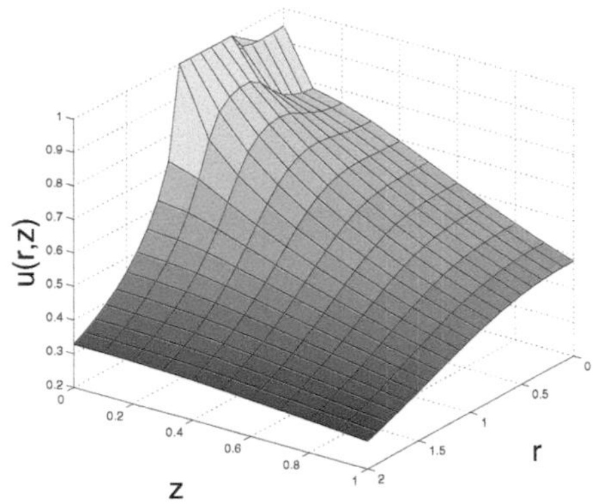

Problem 1

The figure labeled Problem 1 illustrates this solution when $a = 0.5$.

4.5 JOINT TRANSFORM METHODS

In the previous sections we sought to separate problems according to whether the kernels of dual or triple integral equations contained trigonometrical or Bessel functions. Such clear-cut lines of demarcation are not always possible and we conclude with examples where the analysis includes both Fourier and Hankel transforms as well as Fourier and Fourier-Bessel series.

• Example 4.5.1

In Section 4.1 we solved Laplace's equation on an infinite strip. See Equation 4.1.84 through Equation 4.1.100. Here, we again solve[101] Laplace's equation but on a semi-infinite domain which contains two regions with different properties:

$$\frac{\partial^2 u}{\partial x^2} + \frac{\partial^2 u}{\partial y^2} = 0, \qquad 0 < x < \infty, \quad 0 < y < L, \tag{4.5.1}$$

subject to the boundary conditions

$$u_x(0, y) = 0, \qquad 0 < y < L, \tag{4.5.2}$$

$$\lim_{x\to\infty} u(x, y) \to 0, \qquad 0 < y < L, \tag{4.5.3}$$

$$u(x, L) = 0, \qquad 0 < x < \infty, \tag{4.5.4}$$

[101] See Shindo, Y., and A. Atsumi, 1975: Thermal stresses in a laminate composite with infinite row of parallel cracks normal to the interfaces. *Int. J. Engng. Sci.*, **13**, 25–42.

$$u(h^-,y) = u(h^+,y), \quad \epsilon_1 u_x(h^-,y) = \epsilon_2 u_x(h^+,y), \qquad 0 < y < L, \tag{4.5.5}$$

and

$$\begin{cases} u_y(x,0) = -1, & 0 < x < a, \\ u(x,0) = 0, & a < x < \infty, \end{cases} \tag{4.5.6}$$

where $h > a$. The effect of these two different regions introduces an interfacial condition, Equation 4.5.5.

The solution to Equation 4.5.1 to Equation 4.5.4 is

$$u(x,y) = \int_0^\infty A(k) \frac{\sinh[k(L-y)]}{\sinh(kL)} \cos(kx)\,dk + \sum_{n=1}^\infty A_n \cosh\left(\frac{n\pi x}{L}\right) \sin\left(\frac{n\pi y}{L}\right) \tag{4.5.7}$$

for $0 < x < h$; and

$$u(x,y) = \sum_{n=1}^\infty B_n \exp\left(-\frac{n\pi x}{L}\right) \sin\left(\frac{n\pi y}{L}\right) \tag{4.5.8}$$

for $h < x < \infty$. An interesting aspect of this problem is that Equation 4.5.7 contains both a Fourier cosine transform and a Fourier sine series. Substituting Equation 4.5.7 and Equation 4.5.8 into Equation 4.5.6 yields the dual integral equations

$$\int_0^\infty kA(k)[1+M(kL)] \cos(kx)\,dk - \sum_{n=1}^\infty \left(\frac{n\pi}{L}\right) A_n \cosh\left(\frac{n\pi x}{L}\right) = 1, \quad 0 < x < a, \tag{4.5.9}$$

and

$$\int_0^\infty A(k) \cos(kx)\,dk = 0, \qquad a < x < \infty, \tag{4.5.10}$$

where $M(\mu) = e^{-\mu}/\sinh(\mu)$. To solve this set of dual integral equations, we introduce

$$A(k) = \int_0^a g(t) J_0(kt)\,dt. \tag{4.5.11}$$

We chose this definition for $A(k)$ because

$$\int_0^\infty A(k) \cos(kx)\,dk = \int_0^a g(t) \left[\int_0^\infty \cos(kx) J_0(kt)\,dk\right] dt = 0, \tag{4.5.12}$$

where we used Equation 1.4.14. Note that $0 \le t \le x < \infty$.

Turning to Equation 4.5.9, we first integrate it with respect to x and obtain

$$\int_0^\infty A(k)[1+M(kL)] \sin(kx)\,dk - \sum_{n=1}^\infty A_n \sinh\left(\frac{n\pi x}{L}\right) = x, \quad 0 < x < a; \tag{4.5.13}$$

or

$$\int_0^a g(t)\left[\int_0^\infty \sin(kx)J_0(kt)\,dk\right]dt + \int_0^a g(t)\left[\int_0^\infty M(kL)\sin(kx)J_0(kt)\,dk\right]dt - \sum_{n=1}^\infty A_n \sinh\left(\frac{n\pi x}{L}\right) = x, \quad 0 < x < a. \tag{4.5.14}$$

Using Equation 1.4.13, the first term in Equation 4.5.14 can be simplified and we find that

$$\int_0^x \frac{g(t)}{\sqrt{x^2-t^2}}\,dt + \int_0^a g(\tau)\left[\int_0^\infty M(kL)\sin(kx)J_0(k\tau)\,dk\right]d\tau - \sum_{n=1}^\infty A_n \sinh\left(\frac{n\pi x}{L}\right) = x, \quad 0 < x < a. \tag{4.5.15}$$

Solving for $g(t)$ in the first term by using Equation 1.2.13 and Equation 1.2.14,

$$g(x) + \frac{2}{\pi}\int_0^a g(\tau)\left\{\int_0^\infty M(kL)\frac{d}{dx}\left[\int_0^x \frac{t\,\sin(kt)}{\sqrt{x^2-t^2}}\,dt\right]J_0(k\tau)\,dk\right\}d\tau - \frac{2}{\pi}\sum_{n=1}^\infty A_n \frac{d}{dx}\left[\int_0^x \frac{t\sinh(n\pi x/L)}{\sqrt{x^2-t^2}}\,dt\right] = \frac{2\tau_0}{\pi}\frac{d}{dx}\left[\frac{t^2}{\sqrt{x^2-t^2}}\,dt\right] \tag{4.5.16}$$

when $0 < x < a$. Using Equation 4.1.98, Equation 4.1.99, and the fact[102] that

$$\int_0^x \frac{t\sinh(n\pi t)}{\sqrt{x^2-t^2}}\,dt = \frac{\pi x}{2}I_1\left(\frac{n\pi x}{L}\right), \tag{4.5.17}$$

we finally obtain

$$g(x) + x\int_0^a g(\tau)\left[\int_0^\infty k\,M(kL)J_0(kx)J_0(k\tau)\,dk\right]d\tau - \sum_{n=1}^\infty A_n\left(\frac{n\pi x}{L}\right)I_0\left(\frac{n\pi x}{L}\right) = x \tag{4.5.18}$$

for $0 < x < a$.

Before we can solve Equation 4.5.18, we must eliminate A_n from this equation. To that end, we apply the interfacial condition Equation 4.5.5 and find that

$$-\cosh\left(\frac{n\pi h}{L}\right)A_n + \exp\left(-\frac{n\pi h}{L}\right)B_n = \frac{2}{L}\int_0^\infty A(k)\frac{n\pi/L}{k^2+n^2\pi^2/L^2}\cos(kh)\,dk \tag{4.5.19}$$

---

102 Gradshteyn and Ryzhik, op. cit., Formula 3.365.1 and Formula 3.389.3.

and

$$\sinh\left(\frac{n\pi h}{L}\right) A_n + \frac{\epsilon_2}{\epsilon_1} \exp\left(-\frac{n\pi h}{L}\right) B_n = \frac{2}{L} \int_0^\infty A(k) \frac{k \sin(kh)}{k^2 + n^2\pi^2/L^2}\, dk. \tag{4.5.20}$$

Solving for A_n and B_n,

$$A_n = \frac{2(1 - \epsilon_2/\epsilon_1)}{L[\sinh(n\pi h/L) + (\epsilon_2/\epsilon_1)\cosh(n\pi h/L)]} F(n\pi/L), \tag{4.5.21}$$

and

$$B_n = \frac{2[\sinh(n\pi h/L) + \cosh(n\pi h/L)]}{L[\sinh(n\pi h/L) + (\epsilon_2/\epsilon_1)\cosh(n\pi h/L)]} F(n\pi/L), \tag{4.5.22}$$

where

$$F(n\pi/L) = \int_0^\infty A(k) \frac{k \sin(kh)}{k^2 + n^2\pi^2/L^2}\, dk \tag{4.5.23}$$

$$= \int_0^\infty A(k) \frac{n\pi L \cos(kh)}{k^2 L^2 + n^2\pi^2}\, dk \tag{4.5.24}$$

$$= \frac{\pi}{2} e^{-n\pi h/L} \int_0^a g(t) I_0\left(\frac{n\pi t}{L}\right) dt. \tag{4.5.25}$$

Finally, we can bring everything into a nondimensional form by introducing $\xi = x/a$, $\eta = \tau/a$, $\rho = L/a$, $\kappa = a/h$, and $G(\xi) = h(a\xi)/(a\sqrt{\xi}\,)$. This gives

$$G(\xi) + \int_0^1 G(\eta) K(\xi, \eta)\, d\eta = \sqrt{\xi}, \tag{4.5.26}$$

where

$$K(\xi, \eta) = \frac{\sqrt{\xi\eta}}{\rho^2} \left[\int_0^\infty \mu M(\mu) J_0(\xi\mu/\rho) J_0(\eta\mu/\rho)\, d\mu \right. \tag{4.5.27}$$

$$\left. - \pi^2 \sum_{n=1}^\infty \frac{n(1 - \epsilon_2/\epsilon_1) e^{-n\pi/(\kappa\rho)}}{\sinh[n\pi/(\kappa\rho)] + \epsilon_2 \cosh[n\pi/(\kappa\rho)]/\epsilon_1} I_0\left(\frac{n\pi\xi}{\rho}\right) I_0\left(\frac{n\pi\eta}{\rho}\right) \right].$$

Once $G(\xi)$ is computed via Equation 4.5.26, we find A_n, B_n and $A(k)$ from Equation 4.5.21, Equation 4.5.22, and Equation 4.5.11, respectively. The solution $u(x, y)$ follows from Equation 4.5.7 and Equation 4.5.8. Figure 4.5.1 illustrates this solution when $h/a = \frac{1}{2}$ and $a/L = 1$.

• Example 4.5.2

Let us solve[103]

$$\frac{\partial^2 u}{\partial r^2} + \frac{1}{r}\frac{\partial u}{\partial r} - \frac{u}{r^2} + \frac{\partial^2 u}{\partial z^2} = 0, \qquad 0 \le r < a, \quad -\infty < z < \infty, \tag{4.5.28}$$

[103] Taken with permission from Kim, M.-U., 1981: Slow rotation of a disk in a fluid-filled circular cylinder. *J. Phys. Soc. Japan*, **50**, 4063–4067.

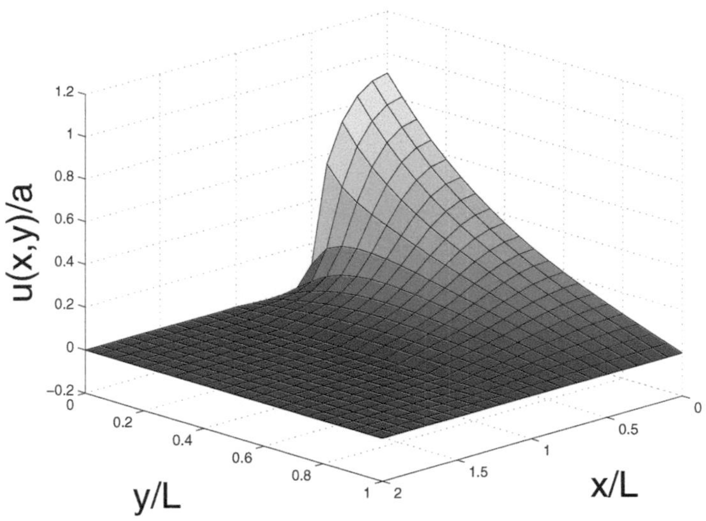

Figure 4.5.1: The solution $u(r,z)$ to the mixed boundary value problem governed by Equation 4.5.1 through Equation 4.5.6 when $h/a = \frac{1}{2}$ and $a/L = 1$.

subject to the boundary conditions

$$\lim_{r\to 0} |u(r,z)| < \infty, \quad u(a,z) = 0, \qquad -\infty < z < \infty, \tag{4.5.29}$$

$$\lim_{|z|\to\infty} u(r,z) \to 0, \qquad 0 \le r < a, \tag{4.5.30}$$

and

$$\begin{cases} u(r,0) = r, & 0 \le r < 1, \\ u_z(r,0^-) = u_z(r,0^+), & 1 < r < a, \end{cases} \tag{4.5.31}$$

where $a > 1$.

Using separation of variables, the general solution to Equation 4.5.28 is

$$u(r,z) = \int_0^\infty A(k) e^{-k|z|} J_1(kr)\, dk + \int_0^\infty B(k) I_1(kr) \cos(kz)\, dk. \tag{4.5.32}$$

This solution satisfies not only Equation 4.5.28, but also Equation 4.5.29 and Equation 4.5.30. Substituting Equation 4.5.32 into Equation 4.5.31, we obtain the dual integral equations

$$\int_0^\infty A(k) J_1(kr)\, dk + \int_0^\infty B(k) I_1(kr)\, dk = r, \qquad 0 \le r < 1, \tag{4.5.33}$$

and

$$\int_0^\infty k A(k) J_1(kr)\, dk = 0, \qquad 1 < r < a. \tag{4.5.34}$$

The interesting aspect of this problem is that the dual integral equations contain both Fourier and Fourier-Bessel transforms.

We begin our solution of Equation 4.5.33 and Equation 4.5.34 by introducing

$$A(k) = \int_0^1 h(t)\sin(kt)\,dt. \tag{4.5.35}$$

If we substitute Equation 4.5.35 into Equation 4.5.34, we can show that this definition of $A(k)$ satisfies Equation 4.5.34 identically. On the other hand, Equation 4.5.33 yields

$$\int_0^1 h(t)\left[\int_0^\infty \sin(kt)J_1(kt)\,dk\right]dt = r - \int_0^\infty B(k)I_1(kr)\,dk. \tag{4.5.36}$$

Because

$$\int_0^\infty \sin(kt)J_1(kr)\,dk = \frac{t\,H(r-t)}{r\sqrt{r^2-t^2}}, \tag{4.5.37}$$

Equation 4.5.36 becomes

$$\int_0^r \frac{t\,h(t)}{\sqrt{r^2-t^2}}\,dt = r^2 - r\int_0^\infty B(k)I_1(kr)\,dk. \tag{4.5.38}$$

We now use results from Chapter 1, namely Equation 1.2.7 with $\alpha = \frac{1}{2}$, that the solution to Equation 4.5.38 is

$$h(t) = \frac{2}{\pi t}\frac{d}{dt}\left\{\int_0^t \left[\xi^3 - \int_0^\infty B(k)\xi^2 I_1(k\xi)\,dk\right]\frac{d\xi}{\sqrt{t^2-\xi^2}}\right\}. \tag{4.5.39}$$

Using the relationship that

$$\int_0^t \frac{\xi^2 I_1(k\xi)}{\sqrt{t^2-\xi^2}}\,d\xi = \frac{kt\cosh(kt)-\sinh(kt)}{k^2}, \tag{4.5.40}$$

we can simplify Equation 4.5.39 to

$$h(t) = \frac{2}{\pi}\left[2t - \int_0^\infty B(k)\sinh(kt)\,dk\right]. \tag{4.5.41}$$

To evaluate $B(k)$, we first substitute Equation 4.5.32 into the boundary condition given by Equation 4.5.31. This yields

$$\int_0^\infty B(k)I_1(ka)\cos(kz)\,dk = -\int_0^\infty A(\eta)J_1(\eta a)e^{-\eta|z|}\,d\eta. \tag{4.5.42}$$

Recognizing that the left side of Equation 4.5.42 is the Fourier cosine transform representation of the right side, $B(k)$ is given by

$$I_1(ka)B(k) = -\frac{2}{\pi}\int_0^\infty \left\{\int_0^\infty J_1(\eta a)e^{-\eta z}\left[\int_0^1 h(\tau)\sin(\eta\tau)\,d\tau\right]d\eta\right\}\cos(kz)\,dz \tag{4.5.43}$$

$$= -\frac{2}{\pi}\int_0^1 h(\tau)\left\{\int_0^\infty \sin(\eta\tau)J_1(\eta a)\left[\int_0^\infty e^{-\eta z}\cos(kz)\,dz\right]d\eta\right\}d\tau \tag{4.5.44}$$

$$= -\frac{2}{\pi}\int_0^1 h(\tau)\left[\int_0^\infty \sin(\eta\tau)J_1(\eta a)\frac{\eta}{\eta^2+k^2}\,d\eta\right]d\tau \tag{4.5.45}$$

$$= -\frac{2}{\pi}\int_0^1 h(\tau)\sinh(k\tau)K_1(ka)\,d\tau, \tag{4.5.46}$$

where we used integral tables[104] for the η integration. Substituting Equation 4.5.46 into Equation 4.5.41, we obtain

$$h(t) = \frac{2}{\pi}\left\{2t + \frac{2}{\pi}\int_0^1 h(\tau)\left[\int_0^\infty \frac{K_1(ka)}{I_1(ka)}\sinh(kt)\sinh(k\tau)\,dk\right]d\tau\right\}; \tag{4.5.47}$$

or

$$h(t) - \frac{4}{\pi^2}\int_0^1 h(\tau)\left[\int_0^\infty \frac{K_1(ka)}{I_1(ka)}\sinh(kt)\sinh(k\tau)\,dk\right]d\tau = \frac{4t}{\pi}, \tag{4.5.48}$$

for $0 < t < 1$. To compute $u(r,z)$, we first solve Equation 4.5.48 to find $h(t)$. Next, we use Simpson's rule to evaluate $A(k)$ and $B(k)$ from Equation 4.5.35 and Equation 4.5.46, respectively. Finally, Equation 4.5.32 gives $u(r,z)$. Figure 4.5.2 illustrates the solution when $a = 2$.

• Example 4.5.3

Given $a < 1$ and denoting a nonnegative integer by n, let us solve[105]

$$\frac{\partial^2 u}{\partial r^2} + \frac{1}{r}\frac{\partial u}{\partial r} - \frac{n^2}{r^2}u + \frac{\partial^2 u}{\partial z^2} = 0, \qquad 0 \le r < 1, \quad 0 \le z < \infty, \tag{4.5.49}$$

subject to the boundary conditions

$$\lim_{r\to 0}|u(r,z)| < \infty, \quad u(1,z) = g(z), \qquad 0 \le z < \infty, \tag{4.5.50}$$

---

[104] Gradshteyn and Ryzhik, op. cit., Formula 6.718.

[105] See Rusia, K. C., 1968: On certain asymmetric mixed boundary value problems of an electrified circular disc situated inside a coaxial infinite hollow cylinder. *Indian J. Pure Appl. Phys.*, **6**, 44–46.

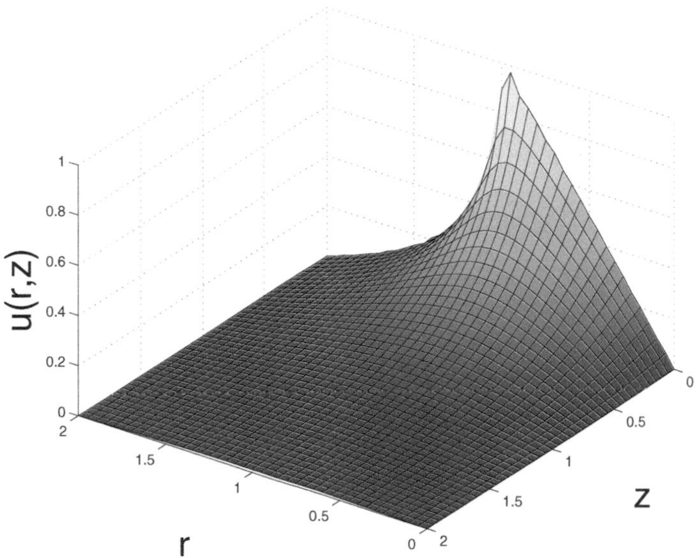

Figure 4.5.2: The solution $u(r, z)$ to the mixed boundary value problem governed by Equation 4.5.28 through Equation 4.5.31 with $a = 2$.

$$\lim_{z\to\infty} u(r, z) \to 0, \qquad 0 \le r < 1, \tag{4.5.51}$$

and

$$\begin{cases} u_z(r, 0) = f(r), & 0 \le r < a, \\ u(r, 0) = 0, & a < r < 1. \end{cases} \tag{4.5.52}$$

Using separation of variables, the general solution to Equation 4.5.49 is

$$u(r, z) = \int_0^\infty A(k)e^{-kz} J_n(kr)\, dk + \int_0^\infty B(k)I_n(kr)\sin(kz)\, dk. \tag{4.5.53}$$

This solution satisfies not only Equation 4.5.49, but also Equation 4.5.51 and the first part of Equation 4.5.50. Substituting Equation 4.2.53 into Equation 4.2.52, we obtain the dual integral equations

$$\int_0^\infty kA(k)J_n(kr)\, dk = \int_0^\infty kB(k)I_n(kr)\, dk - f(r), \qquad 0 \le r < a, \tag{4.5.54}$$

and

$$\int_0^\infty A(k)J_n(kr)\, dk = 0, \qquad a < r < 1. \tag{4.5.55}$$

The interesting aspect of this problem is that the dual integral equations contain both Fourier and Fourier-Bessel transforms.

We begin our solution of Equation 4.5.54 and Equation 4.5.55 by introducing

$$A(k) = \sqrt{k} \int_0^a h(t) t^{-(n-\frac{1}{2})} J_{n+\frac{1}{2}}(kt)\, dt. \tag{4.5.56}$$

If we substitute Equation 4.5.56 into Equation 4.5.55 and interchange the order of integration,

$$\int_0^\infty A(k) J_n(kr)\, dk = \int_0^a h(t) t^{-(n-\frac{1}{2})} \left[\int_0^\infty \sqrt{k}\, J_n(kr) J_{n+\frac{1}{2}}(kt)\, dk \right] dt. \tag{4.5.57}$$

From tables[106] and noting that $r > t$, the value of the integral within the square brackets equals zero and this choice of $A(k)$ satisfies Equation 4.5.55.

If we integrate Equation 4.5.56 by parts and assuming that $h(t)t^{-(n-\frac{1}{2})}$ tends to zero as $t \to 0$,

$$A(k) = -\frac{h(a) J_{n-\frac{1}{2}}(ka)}{\sqrt{k}\, a^{n-\frac{1}{2}}} + \frac{1}{\sqrt{k}} \int_0^a h'(t) t^{-(n-\frac{1}{2})} J_{n-\frac{1}{2}}(kt)\, dt. \tag{4.5.58}$$

Substituting Equation 4.5.58 into Equation 4.5.54, interchanging the order of integration, and carrying out the k-integration,

$$-\frac{h(a)}{a^{n-\frac{1}{2}}} \int_0^\infty \sqrt{k}\, J_{n-\frac{1}{2}}(ka) J_n(kr)\, dk + \int_0^a \frac{h'(t)}{t^{n-\frac{1}{2}}} \left[\int_0^\infty \sqrt{k}\, J_{n-\frac{1}{2}}(kt) J_n(kr)\, dk \right] dt$$
$$= \int_0^\infty k B(k) I_n(kr)\, dk - f(r). \tag{4.5.59}$$

In Equation 4.5.59, the first integral on the left side vanishes while the integral inside the square brackets can be evaluated using tables. The end result is

$$\int_0^r \frac{h'(t)}{\sqrt{r^2 - t^2}}\, dt = \sqrt{\frac{\pi}{2}}\, r^n \left[\int_0^\infty k B(k) I_n(kr)\, dk - f(r) \right], \qquad 0 \le r < a. \tag{4.5.60}$$

Applying the results from Equation 1.2.13 and Equation 1.2.14, we find for $h(t)$ that

$$h(t) = \frac{2}{\pi} \int_0^t \frac{r\, dr}{\sqrt{t^2 - r^2}} \left\{ \sqrt{\frac{\pi}{2}}\, r^n \left[\int_0^\infty k B(k) I_n(kr)\, dk - f(r) \right] \right\} \tag{4.5.61}$$

$$= \sqrt{\frac{2}{\pi}} \int_0^t \frac{r^{n+1}}{\sqrt{t^2 - r^2}} \left[\int_0^\infty k B(k) I_n(kr)\, dk - f(r) \right] dr \tag{4.5.62}$$

---

[106] Gradshteyn and Ryzhik, op. cit., Formula 6.575.1. Note that this formula has a typo; the condition should read $\Re(\nu + 1) > \Re(\mu) > -1$. See p. 100 in Magnus, W., F. Oberhettinger, and R. P. Soni, 1966: *Formulas and Theorems for the Special Functions of Mathematical Physics.* Springer-Verlag, 508 pp.

$$= -\sqrt{\frac{2}{\pi}} \int_0^t \frac{r^{n+1} f(r)}{\sqrt{t^2 - r^2}}\, dr + \sqrt{\frac{2}{\pi}} \int_0^\infty k B(k) \left[\int_0^t \frac{r^{n+1} I_n(kr)}{\sqrt{t^2 - r^2}}\, dr \right] dk \tag{4.5.63}$$

$$= -\sqrt{\frac{2}{\pi}} \int_0^t \frac{r^{n+1} f(r)}{\sqrt{t^2 - r^2}}\, dr + t^{n+\frac{1}{2}} \int_0^\infty \sqrt{k}\, B(k) I_{n+\frac{1}{2}}(kt)\, dk, \tag{4.5.64}$$

where we used[107]

$$\int_0^t \frac{r^{n+1} I_n(kr)}{\sqrt{t^2 - r^2}}\, dr = \sqrt{\frac{\pi}{2}} \frac{t^{n+\frac{1}{2}}}{\sqrt{k}} I_{n+\frac{1}{2}}(kt), \qquad \Re(n) > -1. \tag{4.5.65}$$

Turning to the last boundary condition, the second part of Equation 4.5.50,

$$\int_0^\infty A(k) J_n(k) e^{-kz}\, dk + \int_0^\infty B(k) I_n(k) \sin(kz)\, dk = g(z), \qquad 0 < z < \infty. \tag{4.5.66}$$

Noting that Equation 4.5.66 is a Fourier sine transform with $B(k)$ as the Fourier coefficient,

$$I_n(k)B(k) = \frac{2}{\pi} \int_0^\infty g(z) \sin(kz)\, dz - \frac{2}{\pi} \int_0^\infty \left[\int_0^\infty A(t) J_n(t) e^{-tz}\, dt \right] \sin(kz)\, dz \tag{4.5.67}$$

$$= \frac{2}{\pi} \int_0^\infty g(z) \sin(kz)\, dz - \frac{2k}{\pi} \int_0^\infty \frac{A(t) J_n(t)}{t^2 + k^2}\, dt. \tag{4.5.68}$$

Substituting for $A(k)$ the expression given by Equation 4.5.56,

$$I_n(k)B(k) = \frac{2}{\pi} \int_0^\infty g(z) \sin(kz)\, dz - \frac{2k}{\pi} \int_0^\infty \frac{J_n(t)}{t^2 + k^2} \left[\sqrt{t} \int_0^a \frac{h(\xi)}{\xi^{n-\frac{1}{2}}} J_{n+\frac{1}{2}}(t\xi)\, d\xi \right] dt \tag{4.5.69}$$

$$= \frac{2}{\pi} \int_0^\infty g(z) \sin(kz)\, dz - \frac{2k}{\pi} \int_0^a \frac{h(\xi)}{\xi^{n-\frac{1}{2}}} \left[\int_0^\infty \frac{\sqrt{t}}{t^2 + k^2} J_n(t) J_{n+\frac{1}{2}}(\xi t)\, dt \right] d\xi \tag{4.5.70}$$

$$= \frac{2}{\pi} \int_0^\infty g(z) \sin(kz)\, dz - \frac{2}{\pi} \sqrt{k}\, K_n(k) \int_0^a \frac{h(\xi)}{\xi^{n-\frac{1}{2}}} I_{n+\frac{1}{2}}(k\xi)\, d\xi. \tag{4.5.71}$$

---

[107] Sneddon [Sneddon, op. cit., p. 30.] showed how Sonine's first integral can be rewritten as

$$\int_0^t \xi^{\mu+1} (t^2 - \xi^2)^\nu J_\mu(y\xi)\, d\xi = 2^\nu t^{\mu+\nu+1} y^{-\nu-1} \Gamma(\nu+1) J_{\nu+\nu+1}(ty).$$

Equation 4.5.65 follows by setting $\xi = r$, $\mu = n$, $\nu = -\frac{1}{2}$, and $y = ik$.

Substituting Equation 4.5.71 into Equation 4.5.64 and interchanging the order of integration,

$$h(t) = -\sqrt{\frac{2}{\pi}} \int_0^t \frac{r^{n+1} f(r)}{\sqrt{t^2 - r^2}}\, dr + t^{n+\frac{1}{2}} \int_0^\infty \frac{\sqrt{k}\, I_{n+\frac{1}{2}}(kt)}{I_n(k)}\, dk \tag{4.5.72}$$

$$\times \left[\frac{2}{\pi} \int_0^\infty g(z) \sin(kz)\, dz - \frac{2}{\pi}\sqrt{k}\, K_n(k) \int_0^a \frac{h(\xi)}{\xi^{n-\frac{1}{2}}} I_{n+\frac{1}{2}}(k\xi)\, d\xi \right]$$

$$= -\sqrt{\frac{2}{\pi}} \int_0^t \frac{r^{n+1} f(r)}{\sqrt{t^2 - r^2}}\, dr$$

$$+ \frac{2}{\pi} t^{n+\frac{1}{2}} \int_0^\infty \frac{\sqrt{k}\, I_{n+\frac{1}{2}}(kt)}{I_n(k)} \left[\int_0^\infty g(z) \sin(kz)\, dz \right] dk$$

$$- \frac{2}{\pi} t^{n+\frac{1}{2}} \int_0^\infty \frac{h(\xi)}{\xi^{n-\frac{1}{2}}} \left[\int_0^\infty \frac{k K_n(k) I_{n+\frac{1}{2}}(kt) I_{n+\frac{1}{2}}(k\xi)}{I_n(k)}\, dk \right] d\xi, \tag{4.5.73}$$

where $0 < t < a$.

From Equation 4.5.73 we can compute $h(t)$, and consequently $A(k)$ and $B(k)$. The potential $u(r, z)$ then follows from

$$u(r, z) = \int_0^\infty A(k) e^{-kz} J_n(kr)\, dk + \int_0^\infty B(k) I_n(kr) \sin(kz)\, dk \tag{4.5.74}$$

$$= \int_0^a \frac{h(t)}{t^{n-\frac{1}{2}}} \left[\int_0^\infty \sqrt{k}\, e^{-kz} J_{n+\frac{1}{2}}(kt) J_n(kr)\, dk \right] dt$$

$$+ \frac{2}{\pi} \int_0^\infty \frac{I_n(kr)}{I_n(k)} \left[\int_0^\infty g(t) \sin(kt)\, dt \right] \sin(kz)\, dk \tag{4.5.75}$$

$$- \frac{2}{\pi} \int_0^a \frac{h(t)}{t^{n-\frac{1}{2}}} \left[\int_0^\infty \sqrt{k}\, \frac{K_n(k)}{I_n(k)} I_n(kr) I_{n+\frac{1}{2}}(kt) \sin(kz)\, dk \right] dt.$$

Figure 4.5.3 illustrates Equation 4.5.75 when $a = 0.5$, $n = 1$, $g(z) = 0$ and $f(r) = 1$.

We can also use this technique to solve Equation 4.5.49 through Equation 4.5.52 when Equation 4.5.50 reads

$$\lim_{r \to 0} |u(r, z)| < \infty, \quad u_r(1, z) = g(z), \qquad 0 \le z < \infty. \tag{4.5.76}$$

The analysis is identical to our earlier problem except for computing $B(k)$. Equation 4.5.71 now becomes

$$I_n'(k) B(k) = \frac{2}{\pi k} \int_0^\infty g(z) \sin(kz)\, dz - \frac{2}{\pi}\sqrt{k}\, K_n'(k) \int_0^a \frac{h(\xi)}{\xi^{n-\frac{1}{2}}} I_{n+\frac{1}{2}}(k\xi)\, d\xi \tag{4.5.77}$$

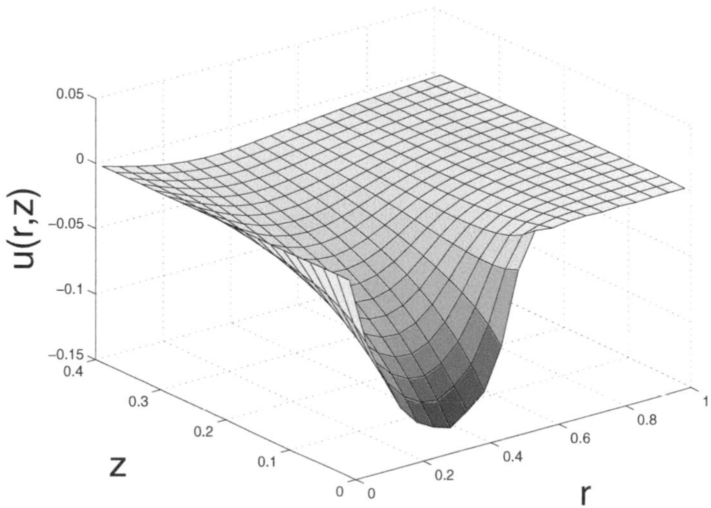

Figure 4.5.3: The solution $u(r,z)$ to the mixed boundary value problem governed by Equation 4.5.49 through Equation 4.5.52 with $a = 0.5$, $n = 1$, $g(z) = 0$ and $f(r) = 1$.

and the integral equation governing $h(t)$ now reads

$$
\begin{aligned}
h(t) &= -\sqrt{\frac{2}{\pi}} \int_0^t \frac{r^{n+1} f(r)}{\sqrt{t^2 - r^2}}\, dr \\
&\quad + \frac{2}{\pi} t^{n+\frac{1}{2}} \int_0^\infty \frac{I_{n+\frac{1}{2}}(kt)}{\sqrt{k}\, I_n'(k)} \left[\int_0^\infty g(z) \sin(kz)\, dz \right] dk \\
&\quad - \frac{2}{\pi} t^{n+\frac{1}{2}} \int_0^a \frac{h(\xi)}{\xi^{n-\frac{1}{2}}} \left[\int_0^\infty \frac{k K_n'(k)}{I_n'(k)} I_{n+\frac{1}{2}}(kt) I_{n+\frac{1}{2}}(k\xi)\, dk \right] d\xi, \qquad \textbf{(4.5.78)}
\end{aligned}
$$

where $0 < t < a$. Figure 4.5.4 illustrates the potential when the modified boundary condition Equation 4.5.76 occurs and $a = 0.8$, $n = 1$, $g(z) = 0$ and $f(r) = 1$. Note that Equation 4.5.75 must be modified to include the new form of $B(k)$.

• Example 4.5.4

Let us solve[108]

$$
\frac{\partial^2 u}{\partial r^2} + \frac{1}{r} \frac{\partial u}{\partial r} - \frac{u}{r^2} + \frac{\partial^2 u}{\partial z^2} = 0, \qquad 0 \le r < \infty, \quad 0 < z < \infty, \qquad \textbf{(4.5.79)}
$$

[108] Reprinted from *Int. J. Engng. Sci.*, **9**, N. J. Freeman and L. M. Keer, On the breaking of an embedded fibre in torsion, 1007–1017, ©1971, with permission from Elsevier.

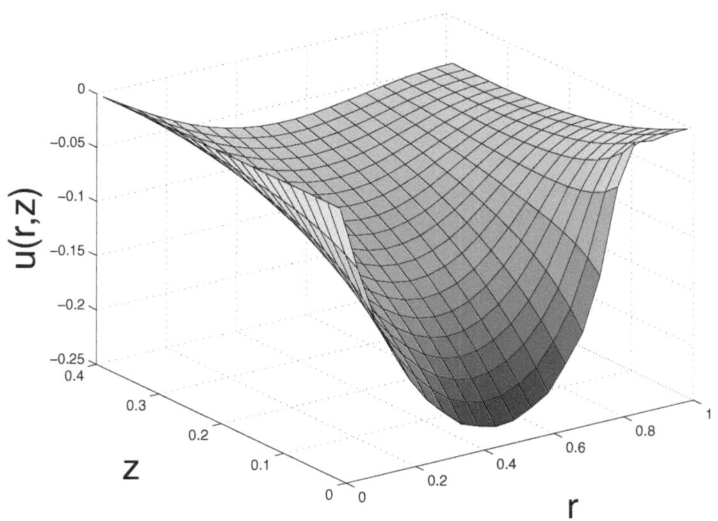

Figure 4.5.4: The solution $u(r,z)$ to the mixed boundary value problem governed by Equation 4.5.49, Equation 4.5.76, Equation 4.5.51 and Equation 4.5.52 with $a = 0.8$, $n = 1$, $g(z) = 0$ and $f(r) = 1$.

subject to the boundary conditions

$$\lim_{r\to 0} |u(r,z)| < \infty, \quad \lim_{r\to\infty} u(r,z) \to 0, \qquad 0 < z < \infty, \tag{4.5.80}$$

$$\lim_{z\to\infty} u(r,z) \to 0, \qquad 0 \le r < \infty, \tag{4.5.81}$$

$$u(1^-,z) = u(1^+,z), \qquad 0 < z < \infty, \tag{4.5.82}$$

$$\epsilon_1 \frac{\partial}{\partial r}\left[\frac{u(r,z)}{r}\right]\bigg|_{r=1^-} = \epsilon_2 \frac{\partial}{\partial r}\left[\frac{u(r,z)}{r}\right]\bigg|_{r=1^+}, \qquad 0 < z < \infty, \tag{4.5.83}$$

and

$$\begin{cases} u_z(r,0) = f(r), & 0 \le r < c, \\ u(r,0) = 0, & c < r < \infty, \end{cases} \tag{4.5.84}$$

where $c < 1$.

Using separation of variables, the general solution to Equation 4.5.79 is

$$u(r,z) = \int_0^\infty A(k)e^{-kz} J_1(kr)\, dk + \int_0^\infty B(k) I_1(kr) \sin(kz)\, dk, \qquad 0 \le r < 1, \tag{4.5.85}$$

and

$$u(r,z) = \int_0^\infty C(k) K_1(kr) \sin(kz)\, dk, \qquad 1 < r < \infty, \tag{4.5.86}$$

where

$$A(k) = \sqrt{k} \int_0^c t^{3/2} J_{3/2}(kt) g(t)\, dt. \tag{4.5.87}$$

This choice for $A(k)$ ensures that Equation 4.5.85 satisfies Equation 4.5.84 for $c < r < 1$.

To satisfy the remaining boundary conditions, we introduce the relationship[109]

$$\int_0^\infty k^{\pm 1/2} e^{-kz} J_n(kr) J_{n\pm\frac{1}{2}}(kt)\, dk = \frac{2}{\pi} \int_0^\infty k^{\pm 1/2} K_n(kr) I_{n\pm\frac{1}{2}}(kt) \cos(kz)\, dk, \tag{4.5.88}$$

if $z > 0$. Then, combining Equation 4.5.82 and Equation 4.5.83 with Equation 4.5.85 and Equation 4.5.86, we find that

$$\begin{aligned} \frac{2}{\pi} \int_0^c t^{3/2} g(t) \left[\int_0^\infty k^{1/2} K_1(k) I_{3/2}(kt) \sin(kz)\, dk \right] dt + \int_0^\infty B(k) I_1(k) \sin(kz)\, dk \\ = \int_0^\infty C(k) K_1(k) \sin(kz)\, dk, \end{aligned} \tag{4.5.89}$$

and

$$\begin{aligned} \frac{2}{\pi} \int_0^c t^{3/2} g(t) \left[\int_0^\infty k^{3/2} K_2(k) I_{3/2}(kt) \sin(kz)\, dk \right] dt - \int_0^\infty k B(k) I_2(k) \sin(kz)\, dk \\ = \beta \int_0^\infty k\, C(k) K_2(k) \sin(kz)\, dk, \end{aligned} \tag{4.5.90}$$

where $\beta = \epsilon_2/\epsilon_1$. Noting that $B(k)$ and $C(k)$ are Fourier coefficients of Fourier sine transforms

$$B(k) = -\frac{2(\beta - 1)}{\pi \Delta(k)} k^{1/2} K_1(k) K_2(k) \int_0^c t^{3/2} g(t) I_{3/2}(kt)\, dt, \tag{4.5.91}$$

and

$$C(k) = \frac{2}{\pi\, k^{1/2} \Delta(k)} \int_0^c t^{3/2} g(t) I_{3/2}(kt)\, dt, \tag{4.5.92}$$

where $\Delta(k) = I_2(k) K_1(k) + \beta I_1(k) K_2(k)$.

Turning now to Equation 4.5.84,

$$\begin{aligned} \sqrt{\frac{2}{\pi}} \frac{1}{r^2} \frac{d}{dr} \left[\int_0^r \frac{t^3 g(t)}{\sqrt{r^2 - t^2}}\, dt \right] + \frac{2}{\pi}(\beta - 1) \int_0^c t^{3/2} g(t) \\ \times \left[\int_0^\infty \frac{k^{3/2}}{\Delta(k)} K_1(k) K_2(k) I_{3/2}(kt) I_1(kr)\, dk \right] dt = -f(r), \quad 0 \le r \le c. \end{aligned} \tag{4.5.93}$$

---

[109] Eason, G., B. Noble, and I. N. Sneddon, 1955: On certain integrals of Lipschitz-Hankel type involving products of Bessel functions. *Philos. Trans. R. Soc. London, Ser. A*, **247**, 529–551.

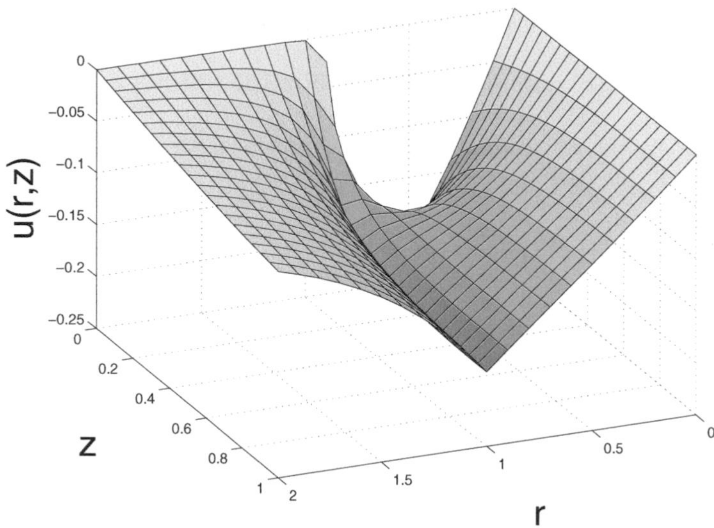

Figure 4.5.5: The solution $u(r,z)$ to the mixed boundary value problem governed by Equation 4.5.79 through Equation 4.5.84 with $\beta = 0$ and $c = 0.9$.

Solving this integral equation of the Abel type,

$$t\,g(t) + \frac{2}{\pi}(\beta - 1)\int_0^c L(\tau,t)\tau g(\tau)\,d\tau = -\sqrt{\frac{2}{\pi t^2}}\int_0^t \frac{\tau^2 f(\tau)}{\sqrt{t^2-\tau^2}}\,d\tau, \quad 0 \le t,\tau \le c, \tag{4.5.94}$$

where

$$L(\tau,t) = \frac{2}{\pi}\int_0^\infty \frac{K_1(k)K_2(k)}{\Delta(k)}\left[\cosh(k\tau) - \frac{\sinh(k\tau)}{k\tau}\right] \times \left[\cosh(kt) - \frac{\sinh(kt)}{kt}\right] dk. \tag{4.5.95}$$

For the special case $f(r) = r$ and defining

$$G(t) = \frac{3\sqrt{2\pi}}{4}\,t\,g(t), \tag{4.5.96}$$

Equation 4.5.95 can be rewritten

$$G(t) - \frac{2(1-\beta)}{\pi}\int_0^c L(\tau,t)G(\tau)\,d\tau = -t^2. \tag{4.5.97}$$

Figure 4.5.5 illustrates this special case when $\beta = 0$ and $c = 0.9$.

• **Example 4.5.5**

Let us solve the biharmonic equation[110]

$$\left(\frac{\partial^2}{\partial r^2} + \frac{1}{r}\frac{\partial}{\partial r} + \frac{\partial^2}{\partial z^2}\right)^2 u = 0, \qquad 0 \le r < 1, \quad 0 < z < \infty, \tag{4.5.98}$$

subject to the boundary conditions

$$\lim_{r\to 0} |u(r,z)| < \infty, \quad u(1,z) = 1, \quad u_r(1,z) = 0, \qquad 0 < z < \infty, \tag{4.5.99}$$

$$\lim_{z\to\infty} u(r,z) \to 2\left(r^2 - r^4/2\right), \qquad 0 \le r < 1, \tag{4.5.100}$$

and

$$\begin{cases} u_z(r,0) = 0, & 0 \le r \le a, \\ u(r,0) = 1, & a \le r \le 1, \end{cases} \tag{4.5.101}$$

where $a < 1$.

Using separation of variables, the general solution to Equation 4.5.98 is

$$u(r,z) = 2\left(r^2 - \frac{r^4}{2}\right) + \int_0^\infty A(k)\cos(kz)\left[rI_0(k)I_1(kr) - r^2 I_1(k)I_0(kr)\right] dk$$
$$+ \sum_{n=1}^\infty B_n r J_1(k_n r)(1 + \alpha_n z)e^{-\alpha_n z}, \tag{4.5.102}$$

where α_n is the nth positive root of $J_1(\alpha) = 0$. Let us now apply the condition $u_r(1,z) = 0$. This yields

$$\int_0^\infty A(k)\{2I_1(k)I_0(k) + k[I_1^2(k) - I_0^2(k)]\}\cos(kz)\,dk$$
$$= \sum_{n=1}^\infty k_n B_n J_0(k_n)(1 + k_n z)e^{-k_n z}. \tag{4.5.103}$$

Solving for $A(k)$,

$$\{2I_1(k)I_0(k) - k[I_0^2(k) - I_1^2(k)]\}A(k)$$
$$= \frac{2}{\pi}\sum_{n=1}^\infty k_n B_n J_0(k_n)\int_0^\infty (1 + k_n z)e^{-k_n z}\cos(kz)\,dz, \tag{4.5.104}$$

or

$$\pi\left\{2I_0(k)I_1(k) - k[I_0^2(k) - I_1^2(k)]\right\} A(k) = 4\sum_{n=1}^\infty \frac{k_n^4 J_0(k_n)}{(k^2 + k_n^2)^2} B_n. \tag{4.5.105}$$

---

[110] See Jeong, J.-T., and S.-R. Choi, 2005: Axisymmetric Stokes flow through a circular orifice in a tube. *Phys. Fluids*, **17**, Art. No. 053602.

Substituting $A(k)$ into Equation 4.5.102 and carrying out the integration over k, we have

$$u(r,z) = 2\left(r^2 - \frac{r^4}{2}\right) + 4\Re\left\{\sum_{n=1}^{\infty} k_n^4 J_0(k_n) B_n \right. \tag{4.5.106}$$
$$\left. \times \sum_{m=1}^{\infty} \frac{rJ_0(\beta_m)J_1(\beta_m r) - r^2 J_1(\beta_m)J_0(\beta_m r)}{J_1^2(\beta_m)(\alpha_n^2 - \beta_m^2)^2} e^{-\beta_m z}\right\},$$

where β_m is mth (complex) root of $2J_0(\beta)J_1(\beta) - \beta\left[J_0^2(\beta) + J_1^2(\beta)\right]$ with positive real and imaginary parts. Asymptotic analysis shows that these roots vary as

$$\beta_m \approx \left(m + \tfrac{1}{2}\right)\pi + \frac{i}{2}\ln\left[(2m+1)\pi + \sqrt{(2m+1)^2\pi^2 - 1}\right], \qquad m \gg 1. \tag{4.5.107}$$

Our remaining task is to compute B_n. Substituting Equation 4.5.106 into Equation 4.5.101, we have the dual series:

$$\Re\left\{\sum_{n=1}^{\infty} k_n^4 J_0(k_n) B_n \sum_{m=1}^{\infty} \frac{\beta_m[1 - J_0(\beta_m r)]}{J_1(\beta_m)(\alpha_n^2 - \beta_m^2)^2}\right\} = 0 \tag{4.5.108}$$

for $0 \le r \le a$;

$$\Re\left\{\sum_{n=1}^{\infty} k_n^4 J_0(k_n) B_n \sum_{m=1}^{\infty} \frac{rJ_0(\beta_m)J_1(\beta_m r) - r^2 J_1(\beta_m)J_0(\beta_m r)}{J_1^2(\beta_m)(\alpha_n^2 - \beta_m^2)^2}\right\} = \tfrac{1}{4}(1 - r^2)^2 \tag{4.5.109}$$

for $a \le r \le 1$.

The procedure for computing B_n is as follows: We truncate Equation 4.5.108 and Equation 4.5.109 to N linear equations with $r_n = \left(n - \frac{1}{2}\right)\Delta r$ with $\Delta r = 1/N$; we truncate the m summation to M terms. Inverting the $N \times N$ system of linear equations, we employ B_n in Equation 4.5.106 to find $u(r,z)$. Figure 4.5.6 illustrates the results when $N = 50$ and $M = 100$ when $a = \frac{1}{2}$.

• Example 4.5.6: Change of variables

Consider the following two-dimensional heat conduction problem that arises during the manufacture of p-n junctions:

$$\frac{\partial u}{\partial t} = a^2\left(\frac{\partial^2 u}{\partial x^2} + \frac{\partial^2 u}{\partial y^2}\right), \quad -\infty < x < \infty,\ 0 < y < \infty,\ 0 < t, \tag{4.5.110}$$

subject to the mixed boundary value conditions

$$\lim_{|x|\to\infty} |u(x,y,t)| < \infty, \qquad 0 < y < \infty, \quad 0 < t, \tag{4.5.111}$$

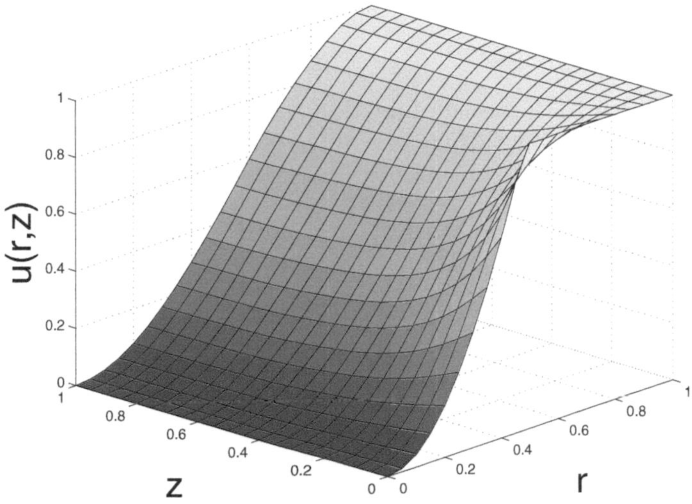

Figure 4.5.6: The solution $u(r, z)$ to the mixed boundary value problem governed by Equation 4.5.79 through Equation 4.5.84 with $a = \frac{1}{2}$.

$$\lim_{y\to\infty} |u(x, y, t)| < \infty, \qquad -\infty < x < \infty, \quad 0 < t, \tag{4.5.112}$$

$$\begin{cases} u_y(x, 0, t) = 0, & 0 < x < \infty, \\ u(x, 0, t) = U_0, & -\infty < x < 0, \end{cases} \quad 0 < t, \tag{4.5.113}$$

and the initial condition

$$u(x, y, 0) = 0, \qquad -\infty < x < \infty, \quad 0 < y < \infty. \tag{4.5.114}$$

Kennedy and O'Brien[111] solved this mixed boundary value problem by reformulating it in polar coordinates; or,

$$\frac{\partial u}{\partial t} = a^2 \left(\frac{\partial^2 u}{\partial r^2} + \frac{1}{r}\frac{\partial u}{\partial r} + \frac{1}{r^2}\frac{\partial^2 u}{\partial \theta^2} \right), \qquad 0 \le r < \infty, \quad 0 < \theta < \pi, \quad 0 < t, \tag{4.5.115}$$

subject to the boundary conditions

$$\lim_{r\to 0} |u(r, \theta, t)| < \infty, \quad \lim_{r\to\infty} |u(r, \theta, t)| < \infty, \qquad 0 < \theta < \pi, \quad 0 < t, \tag{4.5.116}$$

---

[111] Kennedy, D. P., and R. R. O'Brien, 1965: Analysis of the impurity atom distribution near the diffusion mask for a planar *p-n* junction. *IBM J. Res. Develop.*, **9**, 179–186. For an alternative derivation, see Cherednichenko, D. I., H. Gruenberg, and T. K. Sarkar, 1974: Solution to a diffusion problem with mixed boundary conditions. *Solid-State Electron.*, **17**, 315–318.

$$u_\theta(r,0,t) = 0, \quad u(r,\pi,t) = U_0, \qquad 0 \le r < \infty, \quad 0 < t, \tag{4.5.117}$$

and

$$u(r,\theta,0) = 0, \qquad 0 \le r < \infty, \quad 0 < \theta < \pi. \tag{4.5.118}$$

Equation 4.5.115 through Equation 4.5.117 can be solved via transform methods. We begin by taking the Laplace transform of these equations and find that

$$\frac{\partial^2 U}{\partial r^2} + \frac{1}{r}\frac{\partial U}{\partial r} + \frac{1}{r^2}\frac{\partial^2 U}{\partial \theta^2} - \frac{s}{a^2}U = 0, \qquad 0 \le r < \infty, \quad 0 < \theta < \pi, \tag{4.5.119}$$

subject to the boundary conditions

$$\lim_{r\to 0} |U(r,\theta,s)| < \infty, \quad \lim_{r\to\infty} |U(r,\theta,s)| < \infty, \qquad 0 < \theta < \pi, \tag{4.5.120}$$

and

$$U_\theta(r,0,s) = 0, \quad U(r,\pi,s) = \frac{U_0}{s}, \qquad 0 \le r < \infty. \tag{4.5.121}$$

Let

$$U(r,\theta,s) = \frac{U_0}{s} + V(r,\theta,s). \tag{4.5.122}$$

Then, Equation 4.5.119 through Equation 4.5.121 become

$$\frac{\partial^2 V}{\partial r^2} + \frac{1}{r}\frac{\partial V}{\partial r} + \frac{1}{r^2}\frac{\partial^2 V}{\partial \theta^2} - \frac{s}{a^2}V = \frac{U_0}{a^2}, \qquad 0 \le r < \infty, \quad 0 < \theta < \pi, \tag{4.5.123}$$

with

$$\lim_{r\to 0} |V(r,\theta,s)| < \infty, \quad \lim_{r\to\infty} V(r,\theta,s) \to 0, \qquad 0 < \theta < \pi, \tag{4.5.124}$$

and

$$V_\theta(r,0,s) = V(r,\pi,s) = 0, \qquad 0 \le r < \infty. \tag{4.5.125}$$

We next express the solution to Equation 4.5.123 as the Fourier series

$$V(r,\theta,s) = \sum_{n=0}^{\infty} V_n(r,s)\cos\left[\left(n+\tfrac{1}{2}\right)\theta\right]. \tag{4.5.126}$$

Note that Equation 4.5.126 satisfies the boundary condition given by Equation 4.5.125. Each Fourier coefficient $V_n(r,s)$ is governed by

$$\frac{d^2V_n}{dr^2} + \frac{1}{r}\frac{dV_n}{dr} - \left[\frac{\left(n+\frac{1}{2}\right)^2}{r^2} + \frac{s}{a^2}\right]V_n = \frac{2U_0(-1)^n}{\pi a^2\left(n+\frac{1}{2}\right)}. \tag{4.5.127}$$

Solving Equation 4.5.127 by Hankel transforms via

$$V_n(r,s) = \int_0^\infty A(k) J_{n+\frac{1}{2}}(kr)\, k\, dk, \tag{4.5.128}$$

we find that

$$U(r,\theta,s) = U_0\left[\frac{1}{s} - \frac{2}{\pi}\sum_{n=0}^{\infty}(-1)^n \frac{\cos\left[\left(n+\frac{1}{2}\right)\theta\right]}{s+a^2k^2} J_{n+\frac{1}{2}}(kr)\,\frac{dk}{k}\right]. \quad \textbf{(4.5.129)}$$

Straightforward inversion yields the final solution:

$$u(r,\theta,t) = U_0\left[1 - \frac{2}{\pi}\sum_{n=0}^{\infty}(-1)^n \cos\left[\left(n+\tfrac{1}{2}\right)\theta\right]\int_0^\infty e^{-a^2k^2t} J_{n+\frac{1}{2}}(kr)\,\frac{dk}{k}\right]. \quad \textbf{(4.5.130)}$$

Now[112]

$$\int_0^\infty e^{-a^2k^2t} J_{n+\frac{1}{2}}(kr)\,\frac{dk}{k} = \frac{\Gamma\left(\frac{n}{2}+\frac{1}{4}\right)}{2\Gamma\left(n+\frac{3}{2}\right)}\left(\frac{r^2}{4a^2t}\right)^{\frac{n}{2}+\frac{1}{4}} {}_1F_1\left(\frac{n}{2}+\frac{1}{4}; n+\frac{3}{2}; \frac{r^2}{4a^2t}\right), \quad \textbf{(4.5.131)}$$

where ${}_1F_1(a;b;z)$ is the Kummer function. Reexpressing the Kummer function in terms of Bessel functions,[113] Equation 3.5.130 can be written as

$$\frac{u(r,\theta,t)}{U_0} = 1 - \frac{4}{\pi}\sum_{n=0}^{\infty}\frac{(-1)^n}{2n+1} B_n\!\left(\frac{r^2}{8a^2t}\right)\cos\left[\left(n+\tfrac{1}{2}\right)\theta\right], \quad \textbf{(4.5.132)}$$

where

$$B_n(z) = \sqrt{\frac{\pi z}{2}}\, e^{-z}\left[I_{\frac{n}{2}-\frac{1}{4}}(z) + I_{\frac{n}{2}+\frac{3}{4}}(z)\right]. \quad \textbf{(4.5.133)}$$

Figure 4.5.7 illustrates Equation 4.5.132 when $a^2t = 1$. In Example 5.2.1 we will show how we could have solved this problem via the Wiener-Hopf method.

Problems

1. Let us solve a problem[114] similar to the one that we examined in Example 4.5.2, namely

$$\frac{\partial^2 u}{\partial r^2} + \frac{1}{r}\frac{\partial u}{\partial r} + \frac{\partial^2 u}{\partial z^2} = 0, \qquad 0 \le r < c, \quad 0 \le z < \infty,$$

subject to the boundary conditions

$$\lim_{r\to 0}|u(r,z)| < \infty, \quad u(c,z) = 0, \qquad 0 \le z < \infty,$$

[112] Gradshteyn and Ryzhik, op. cit., Formula 6.631.

[113] Abramowitz, M., and I. A. Segun, 1968: *Handbook of Mathematical Functions*. Dover Publications, 1046 pp. See Formula 13.3.6.

[114] Reprinted from *Int. J. Engng. Sci.*, **6**, B. R. Das, Thermal stresses in a long cylinder containing a penny-shaped crack, 497–516, ©1968, with permission of Elsevier.

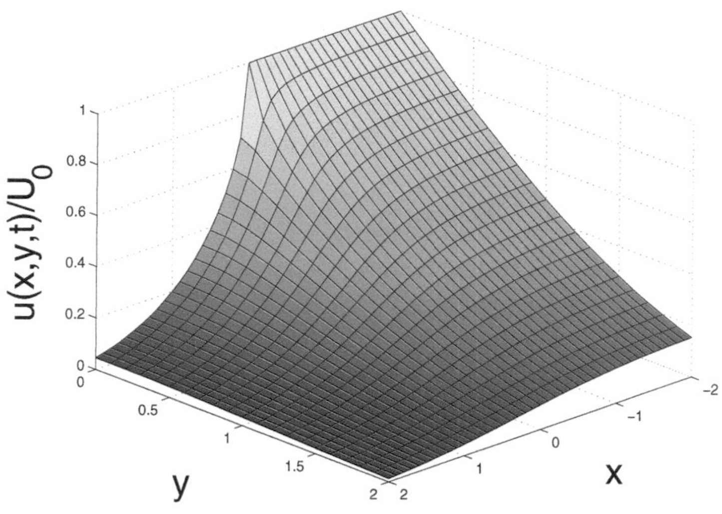

Figure 4.5.7: The solution $u(x, y, t)$ to the mixed boundary value problem governed by Equation 4.5.110 through Equation 4.5.114 with $a^2t = 1$.

$$\lim_{z\to\infty} u(r, z) \to 0, \qquad 0 \le r < c,$$

and

$$\begin{cases} u(r,0) = f(r), & 0 \le r < a, \\ u_z(r,0) = 0, & a < r < c, \end{cases}$$

where $c > a$.

Step 1: Using separation of variables or transform methods, show that the general solution to the problem is

$$u(r, z) = \int_0^\infty A(k)e^{-kz} J_0(kr) \frac{dk}{k} + \int_0^\infty B(k)I_0(kr) \cos(kz) \frac{dk}{k}.$$

Step 2: Show that $A(k)$ and $B(k)$ satisfy the following integral equations:

$$\int_0^\infty A(k)J_0(kr) \frac{dk}{k} + \int_0^\infty B(k)I_0(kr) \frac{dk}{k} = f(r), \qquad 0 \le r < a, \qquad (1)$$

and

$$\int_0^\infty A(k)J_0(kr)\, dk = 0, \qquad a < r < c. \qquad (2)$$

Step 3: Show that by defining

$$A(k) = k \int_0^a g(t) \cos(kt)\, dt = g(a) \sin(ka) - \int_0^a g'(t) \sin(kt)\, dt,$$

Equation (2) is satisfied.

Step 4: From Equation (1), show that $g(t)$ is given by

$$g(t) + \frac{2}{\pi} \int_0^\infty B(k) \cosh(kt)\, \frac{dk}{k} = k(t),$$

where

$$k(t) = \frac{2}{\pi} \frac{d}{dt} \left[\int_0^t \frac{r f(r)}{\sqrt{t^2 - r^2}}\, dr \right].$$

Step 5: From the boundary condition at $r = c$, show that

$$\int_0^\infty A(k) e^{-kz} J_0(kc)\, \frac{dk}{k} + \int_0^\infty B(k) I_0(kc) \cos(kz)\, \frac{dk}{k} = 0.$$

Step 6: Recognizing that $B(k)$ is the Fourier coefficient in a Fourier cosine transform, show that

$$\begin{aligned} B(k) \frac{I_0(kc)}{k} &= -\frac{2}{\pi} \int_0^\infty \cos(kz) \left[\int_0^\infty A(\xi) e^{-\xi z} J_0(\xi c)\, \frac{d\xi}{\xi} \right] dz \\ &= -\frac{2}{\pi} \int_0^\infty A(\xi) \frac{J_0(\xi c)}{k^2 + \xi^2}\, d\xi. \end{aligned}$$

Step 7: Using

$$\int_0^\infty \frac{\xi}{k^2 + \xi^2} \cos(\xi t) J_0(\xi c)\, d\xi = \cosh(kt) K_0(kc),$$

show that

$$\int_0^a g(\tau) \left[\int_0^\infty \frac{\xi}{k^2 + \xi^2} \cos(\xi \tau) J_0(\xi c)\, d\xi \right] d\tau = K_0(kc) \int_0^a g(\tau) \cosh(k\tau)\, d\tau,$$

and

$$B(k) = -\frac{2k K_0(kc)}{\pi I_0(kc)} \int_0^a g(\tau) \cosh(k\tau)\, d\tau.$$

Step 8: Using the results from Step 7, show that

$$g(t) - \int_0^a g(\tau) K(t, \tau)\, d\tau = k(t),$$

where

$$K(t, \tau) = \frac{4}{\pi^2} \int_0^\infty \frac{K_0(kc)}{I_0(kc)} \cosh(k\tau) \cosh(kt)\, dk.$$

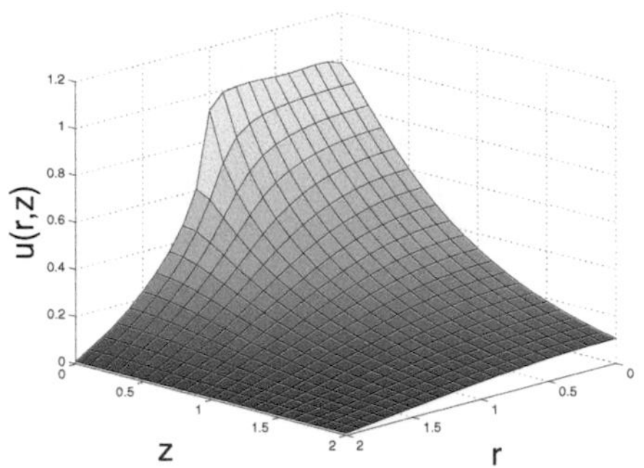

Problem 1

The figure labeled Problem 1 illustrates the solution when $f(r) = 1$, $a = 1$ and $c = 2$.

2. Given $a > 1$ and denoting a non-negative integer by n, let us solve[115]

$$\frac{\partial^2 u}{\partial r^2} + \frac{1}{r}\frac{\partial u}{\partial r} - \frac{n^2}{r^2}u + \frac{\partial^2 u}{\partial z^2} = 0, \qquad 0 \le r < a, \quad 0 \le z < \infty,$$

subject to the boundary conditions

$$\lim_{r\to 0} |u(r,z)| < \infty, \quad u(a,z) = g(z), \qquad 0 \le z < \infty,$$

$$\lim_{z\to\infty} u(r,z) \to 0, \qquad 0 \le r < a,$$

and

$$\begin{cases} u(r,0) = f(r), & 0 \le r < 1, \\ u_z(r,0) = 0, & 1 < r < a. \end{cases}$$

Step 1: Using separation of variables or transform methods, show that the general solution to the problem is

$$u(r,z) = \int_0^\infty A(k)e^{-kz} J_n(kr)\,\frac{dk}{k} + \int_0^\infty B(k) I_n(kr)\cos(kz)\,dk.$$

Step 2: Show that $A(k)$ and $B(k)$ satisfy the following integral equations:

$$\int_0^\infty A(k) J_n(kr)\,\frac{dk}{k} + \int_0^\infty B(k) I_n(kr)\,dk = f(r), \qquad 0 \le r < 1, \qquad (1)$$

[115] Suggested by Sneddon, I. N., 1962: Note on an electrified circular disk situated inside a coaxial infinite hollow cylinder. *Proc. Cambridge Philos. Soc.*, **58**, 621–624. ©1962 Cambridge Philosophical Society. Reprinted with the permission of Cambridge University Press.

$$\int_0^\infty A(k) J_n(kr)\, dk = 0, \qquad 1 < r < a, \tag{2}$$

and

$$\int_0^\infty A(k) e^{-kz} J_n(ka)\, \frac{dk}{k} + \int_0^\infty B(k) I_n(ka) \cos(kz)\, dk = g(z), \qquad 0 \leq z < \infty. \tag{3}$$

Step 3: Using Equation (3) and recognizing the $B(k)$ is the coefficient of a Fourier cosine transform, show that

$$I_n(ka)B(k) = \frac{2}{\pi} \int_0^\infty g(z) \cos(kz)\, dz - \frac{2}{\pi} \int_0^\infty \frac{A(t) J_n(at)}{k^2 + t^2}\, dt.$$

Step 4: Defining

$$A(k) = \sqrt{\frac{\pi k^3}{2}} \int_0^1 \sqrt{t}\, h(t) J_{n-\frac{1}{2}}(kt)\, dt,$$

show that Equation (2) is identically satisfied.

Step 5: Using the results from Step 3 and Step 4 and following Equation 4.5.42 through Equation 4.5.46, show that

$$I_n(ka)B(k) = \frac{2}{\pi} \int_0^\infty g(z) \cos(kz)\, dz - \sqrt{\frac{2}{\pi}} K_n(ka) \int_0^1 \sqrt{k\xi}\, I_{n-\frac{1}{2}}(k\xi) h(\xi)\, d\xi.$$

Hint:

$$\int_0^\infty \xi^{3/2} J_{n-\frac{1}{2}}(t\xi) J_n(a\xi)\, \frac{d\xi}{k^2 + \xi^2} = \sqrt{k}\, I_{n-\frac{1}{2}}(kt) K_n(ka).$$

Step 6: Using Equation (1) and substituting for $A(k)$, show that

$$\int_0^r \frac{t^n\, h(t)}{\sqrt{r^2 - t^2}}\, dt = r^n \left[f(r) - \int_0^\infty B(k) I_n(kr)\, dk \right].$$

Step 7: Solve the integral equation in Step 6 and show that

$$t^n h(t) = \frac{2}{\pi} \frac{d}{dt} \left[\int_0^t \frac{r^{n+1} f(r)}{\sqrt{t^2 - r^2}}\, dr \right] - \frac{2}{\pi} \frac{d}{dt} \left\{ \int_0^\infty B(k) \left[\int_0^t \frac{r^{n+1}}{\sqrt{t^2 - r^2}} I_n(kr)\, dr \right] dk \right\}.$$

Step 8: Using the results from Step 5 to eliminate $B(k)$ in Step 7, show that $h(t)$ is given by the integral equation

$$h(t) = \chi(t) + \int_0^1 h(\xi) K(t, \xi)\, d\xi,$$

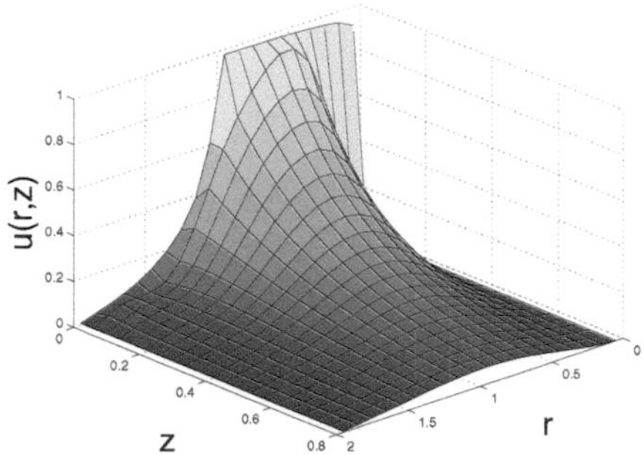

Problem 2

where

$$K(t,\eta) = \frac{2}{\pi}\sqrt{t\eta}\int_0^\infty \frac{\xi K_n(\xi a)}{I_n(\xi a)} I_{n-\frac{1}{2}}(t\xi) I_{n-\frac{1}{2}}(\eta\xi)\, d\xi,$$

and

$$\begin{aligned}\chi(t) = {} & \frac{2}{\pi}t^{-n}\frac{d}{dt}\left[\int_0^t \frac{r^{n+1} f(r)}{\sqrt{t^2-r^2}}\, dr\right] \\ & - \frac{2}{\pi}\int_0^\infty \frac{\sqrt{\xi t}\, I_{n-\frac{1}{2}}(\xi t)}{I_n(\xi a)} \left[\sqrt{\frac{2}{\pi}}\int_0^\infty g(z)\cos(\xi z)\, dz\right] d\xi.\end{aligned}$$

The figure labeled Problem 2 illustrates the potential when $a = 2$, $n = 2$, $g(z) = 0$, and $f(z) = 1$.

3. Given $a > 1$ and denoting a non-negative integer by n, let us solve[116]

$$\frac{\partial^2 u}{\partial r^2} + \frac{1}{r}\frac{\partial u}{\partial r} - \frac{n^2}{r^2}u + \frac{\partial^2 u}{\partial z^2} = 0, \qquad 1 \le r < \infty, \quad 0 \le z < \infty,$$

subject to the boundary conditions

$$u(1,z) = 0, \quad \lim_{r\to\infty} u(r,z) \to 0, \qquad 0 \le z < \infty,$$

$$\lim_{z\to\infty} u(r,z) \to 0, \qquad 1 \le r < \infty,$$

[116] See Rusia, K. C., 1967: Some asymmetric mixed boundary value problems for a half-space with a cylindrical cavity. *Indian J. Pure Appl. Phys.*, **5**, 419–421. Rusia's contribution was to simplify the solution of this problem first posed by Narain, P., 1965: A note on an asymmetric mixed boundary value problem for a half space with a cylindrical cavity. *Glasgow Math. Assoc. Proc.*, **7**, 45–47. Earlier Srivastav (Srivastav, R. P., 1964: An axisymmetric mixed boundary value problem for a half-space with a cylindrical cavity. *J. Math. Mech.*, **13**, 385–393.) solved this problem for the special case $n = 0$.

and

$$\begin{cases} u(r,0) = 0, & 1 \le r < a, \\ u_z(r,0) = -f(r), & a < r < \infty. \end{cases}$$

Step 1: Using separation of variables or transform methods, show that the general solution to the problem is

$$u(r,z) = \int_0^\infty A(k) e^{-kz} J_n(kr)\, dk + \int_0^\infty B(k) K_n(kr) \sin(kz)\, dk.$$

Step 2: Show that $A(k)$ and $B(k)$ satisfy the following integral equations:

$$\int_0^\infty A(k) J_n(kr)\, dk = 0, \qquad 1 < r < a, \tag{1}$$

$$\int_0^\infty kA(k) J_n(kr)\, dk - \int_0^\infty kB(k) K_n(kr)\, dk = f(r), \qquad a < r < \infty, \tag{2}$$

and

$$\int_0^\infty A(k) e^{-kz} J_n(k)\, dk + \int_0^\infty B(k) K_n(k) \sin(kz)\, dk = 0, \qquad 0 \le z < \infty. \tag{3}$$

Step 3: Defining

$$A(k) = \sqrt{k} \int_a^\infty t^{n+\frac{1}{2}} h(t) J_{n-\frac{1}{2}}(kt)\, dt,$$

show that Equation (1) is identically satisfied.

Step 4: Integrating by parts $A(k)$ defined in Step 3, show that

$$A(k) = -\frac{h(a) a^{n+\frac{1}{2}}}{\sqrt{k}} J_{n+\frac{1}{2}}(ka) - \frac{1}{\sqrt{k}} \int_a^\infty h'(t) t^{n+\frac{1}{2}} J_{n+\frac{1}{2}}(kt)\, dt.$$

Step 5: Using Equation (2) and substituting the $A(k)$ given in Step 4, show that

$$\int_r^\infty \frac{h'(t)}{\sqrt{t^2 - r^2}}\, dt = -\sqrt{\frac{\pi}{2}} r^{-n} \left[f(r) + \int_0^\infty kB(k) K_n(kr)\, dk \right], \quad a < r < \infty.$$

Step 6: Solve the integral equation in Step 5 and show that

$$h'(t) = \sqrt{\frac{2}{\pi}} \frac{d}{dt} \left\{ \int_t^\infty \frac{r^{1-n}}{\sqrt{r^2 - t^2}} \left[f(r) + \int_0^\infty kB(k) K_n(kr)\, dk \right] dr \right\},$$

or

$$
\begin{aligned}
h(t) &= \sqrt{\frac{2}{\pi}} \int_t^\infty \frac{r^{1-n}}{\sqrt{r^2-t^2}} \left[f(r) + \int_0^\infty kB(k)K_n(kr)\,dk \right] dr \\
&= \sqrt{\frac{2}{\pi}} \int_t^\infty \frac{r^{1-n} f(r)}{\sqrt{r^2-t^2}}\,dr + \sqrt{\frac{2}{\pi}} \int_0^\infty kB(k) \left[\int_t^\infty \frac{r^{1-n} K_n(kr)}{\sqrt{r^2-t^2}}\,dr \right] dk \\
&= \sqrt{\frac{2}{\pi}} \int_t^\infty \frac{r^{1-n} f(r)}{\sqrt{r^2-t^2}}\,dr + \int_0^\infty \sqrt{k}\, B(k) t^{\frac{1}{2}-n} K_{\frac{1}{2}-n}(kt)\,dk.
\end{aligned}
$$

Hint:

$$
\int_t^\infty \frac{r^{1-n} K_n(kr)}{\sqrt{r^2-t^2}}\,dr = \sqrt{\frac{\pi}{2k}}\, t^{\frac{1}{2}-n} K_{\frac{1}{2}-n}(kt).
$$

Step 7: Recognizing that $K_n(k)B(k)$ is the coefficient of a Fourier sine transform given by Equation (3), show that

$$
\begin{aligned}
K_n(k)B(k) &= -\frac{2}{\pi} \int_0^\infty \left[\int_0^\infty A(t) J_n(t) e^{-tz}\,dt \right] \sin(kz)\,dz \\
&= -\frac{2k}{\pi} \int_0^\infty \frac{A(t) J_n(t)}{t^2+k^2}\,dt.
\end{aligned}
$$

Step 8: Substituting the $A(k)$ given in Step 3 into the results given in Step 7, show that

$$
K_n(k)B(k) = -\frac{2\sqrt{k}}{\pi} I_n(k) \int_a^\infty t^{n+\frac{1}{2}} h(t) K_{n-\frac{1}{2}}(kt)\,dt.
$$

Step 9: Using the results from Step 8 to eliminate $B(k)$ in Step 6, show that $h(t)$ is given by

$$
h(t) = \chi(t) - \int_a^\infty h(\xi) K(t,\xi)\,d\xi,
$$

where

$$
K(t,\xi) = \frac{2t}{\pi} \left(\frac{\xi}{t} \right)^{n+\frac{1}{2}} \int_0^\infty \frac{kI_n(k)}{K_n(k)} K_{n-\frac{1}{2}}(k\xi) K_{\frac{1}{2}-n}(kt)\,dk,
$$

and

$$
\chi(t) = \sqrt{\frac{2}{\pi}} \int_t^\infty \frac{r^{1-n} f(r)}{\sqrt{r^2-t^2}}\,dr.
$$

The figure labeled Problem 3 illustrates this potential when $a=2$ and $f(z) = e^{a-r}$.

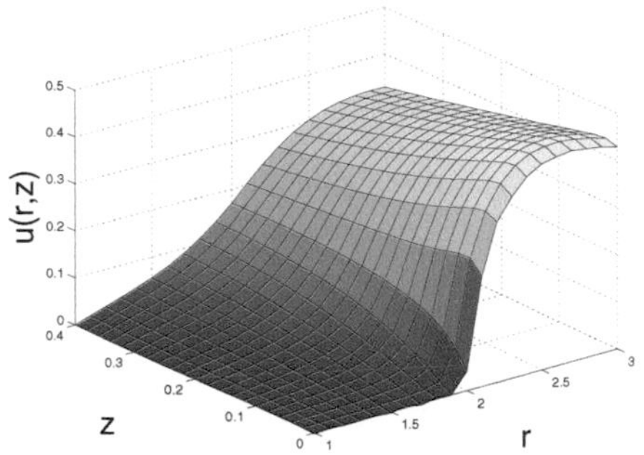

Problem 3

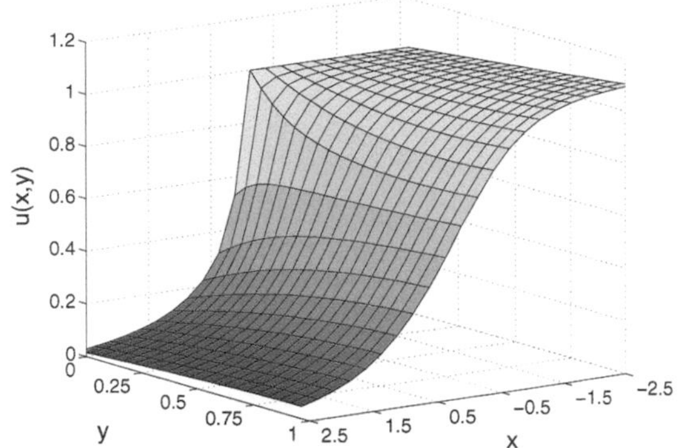

Chapter 5

The Wiener-Hopf Technique

In Example 1.1.3 we posed a mixed boundary value problem that we could not solve by using conventional Fourier transforms because the solution does not vanish as we approach infinity at *both* ends. The Wiener-Hopf technique is a popular method that avoids this problem by defining Fourier transforms over certain regions and then uses function-theoretic analysis to piece together the complete solution.

Although Wiener and Hopf[1] first devised this method to solve singular integral equations of the form

$$f(x) = \int_0^\infty K(x-y) f(y)\, dy + \varphi(x), \qquad 0 < x < \infty, \tag{5.0.1}$$

that had arisen in Hopf's 1928 work on the Milne–Schwarzschild equation,[2] it has been in boundary value problems that it has found its greatest applicability. For example, this technique reduces the problem of diffraction

[1] Wiener, N., and E. Hopf, 1931: Über eine Klasse singulärer Integralgleichungen. *Sitz. Ber. Preuss. Akad. Wiss., Phys.-Math. Kl.*, 696–706.

[2] Shore, S. N., 2002: The evolution of radiative transfer theory from atmospheres to nuclear reactors. *Hist. Math.*, **29**, 463–489.

Figure 5.0.1: One of the great mathematicians of the twentieth century, Norbert Wiener (1894–1964) graduated from high school at the age of 11 and Tufts at 14. Obtaining a doctorate in mathematical logic at 18, he repeatedly traveled to Europe for further education. His work extends over an extremely wide range from stochastic processes to harmonic analysis to cybernetics. (Photo courtesy of the MIT Museum.)

by a semi-infinite plate to the solution of a singular integral equation.[3] The Wiener-Hopf technique[4] then yields the classic result given by Sommerfeld.[5]

Since its original formulation, the Wiener-Hopf technique has undergone simplification by formulating the problem in terms of dual integral equations.[6] The essence of this technique is the process of *factorization* of the Fourier transform of the kernel function into the product of two other Fourier transforms which are analytic and nonzero in certain half planes.

Before we plunge into the use of the Wiener-Hopf technique for solving partial differential equations, let us focus our attention on the mechanics of the method itself. To this end, let us solve the integral equation

$$f(x) = g(x) + \frac{i}{2\kappa_<}\left(\kappa_>^2 - \kappa_<^2\right)\int_0^\infty e^{i\kappa_<|x-\xi|} f(\xi)\,d\xi, \tag{5.0.2}$$

---

[3] Magnus, W., 1941: Über die Beugung electromagnetischer Wellen an einer Halbebene. *Z. Phys.*, **117**, 168–179.

[4] Copson, E. T., 1946: On an integral equation arising in the theory of diffraction. *Quart. J. Math.*, **17**, 19–34.

[5] Sommerfeld, A., 1896: Mathematische Theorie der Diffraction. *Math. Ann.*, **47**, 317–374.

[6] Kaup, S. N., 1950: Wiener-Hopf techniques and mixed boundary value problems. *Comm. Pure Appl. Math.*, **3**, 411–426; Clemmow, P. C., 1951: A method for the exact solution of a class of two-dimensional diffraction problems. *Proc. R. Soc. London, Ser. A*, **205**, 286–308. See Noble, B., 1958: *Methods Based on the Wiener-Hopf Technique for the Solution of Partial Differential Equations*. Pergamon Press, 246 pp.

Figure 5.0.2: Primarily known for his work on topology and ergodic theory, Eberhard Frederich Ferdinand Hopf (1902–1983) received his formal education in Germany. It was during an extended visit to the United States that he worked with Norbert Wiener on what we now know as the "Wiener-Hopf technique." Returning to Germany in 1936, he would eventually become an American citizen (1949) and a professor at Indiana University. (Photo courtesy of the MIT Museum.)

where

$$g(x) = \begin{cases} \dfrac{i}{2\kappa_<} e^{-i\kappa_< x}, & x < 0, \\ \left(x + \dfrac{i}{2\kappa_<} \right) e^{i\kappa_< x}, & x > 0, \end{cases} \tag{5.0.3}$$

and $\Im(\kappa_>)$, $\Im(\kappa_<) \geq \delta > 0$. This integral equation was constructed by Grzesik and Lee[7] to illustrate how various transform methods can be applied to electromagnetic scattering problems.

We intend to solve Equation 5.0.2 via Fourier transforms. An important aspect of the Wiener-Hopf method is the splitting of the Fourier transform into two parts: $F(k) = F_+(k) + F_-(k)$, where

$$F_-(k) = \int_0^\infty f(x) e^{-ikx}\, dx \tag{5.0.4}$$

[7] Grzesik, J. A., and S. C. Lee, 1995: The dielectric half space as a test bed for transform methods. *Radio Sci.*, **30**, 853–862. ©1995 American Geophysical Union. Reproduced/modified by permission of the American Geophysical Union.

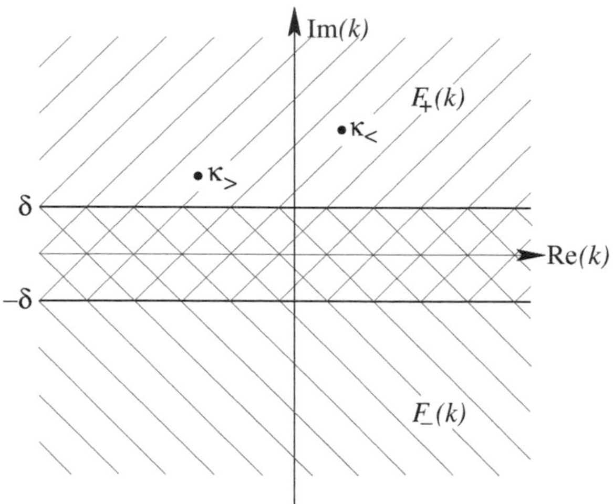

Figure 5.0.3: The location of the half-planes $F_+(k)$ and $F_-(k)$, as well as $\kappa_<$ and $\kappa_>$, in the complex k-plane used in solving Equation 5.0.2 by the Wiener-Hopf technique.

and

$$F_+(k) = \int_{-\infty}^{0} f(x) e^{-ikx}\, dx. \qquad \textbf{(5.0.5)}$$

Because the integral in Equation 5.0.4 converges only if $\Im(k) < 0^+$ when $x > 0$, we have added the subscript "−" to denote its analyticity in the half-space below $\Im(k) < 0^+$ in the k-plane. Similarly, the integral in Equation 5.0.5 converges only where $\Im(k) > 0^-$ and the plus sign denotes its analyticity in the half-space above $\Im(k) > 0^-$ in the k-plane. We will refine these definitions shortly.

Direct computation of $G(k)$ gives

$$G(k) = \frac{1}{2\kappa_<(k-\kappa_<)} - \frac{1}{2\kappa_<(k+\kappa_<)} - \frac{1}{(k+\kappa_<)^2}. \qquad \textbf{(5.0.6)}$$

Note that this transform is analytic in the strip $|\Im(k)| < \delta$. Next, taking the Fourier transform of Equation 5.0.2, we find that

$$F_+(k) + F_-(k) = G(k) + \frac{\kappa_>^2 - \kappa_<^2}{k^2 - \kappa_<^2} F_-(k), \qquad \textbf{(5.0.7)}$$

or in the more symmetrical form:

$$\frac{k+\kappa_<}{k+\kappa_>} F_+(k) + \frac{k-\kappa_>}{k-\kappa_<} F_-(k) = \frac{k+\kappa_<}{k+\kappa_>} G(k). \qquad \textbf{(5.0.8)}$$

This is permissible as long as $F_+(k)$ is analytic in the half-space $\Im(k) > -\delta$, and $F_-(k)$ is analytic in the half-space $\Im(k) < \delta$. See Figure 5.0.3.

Let us focus on Equation 5.0.8. A quick check shows that the first term on the left side is analytic in the same half-plane as $F_+(k)$. Similarly, the second term on the left side is analytic in the same half-plane as $F_-(k)$. This suggests that it might be advantageous to split the right side into two parts where one term is analytic in the same half plane as $F_+(k)$ while the other part is analytic in the half-plane $F_-(k)$. Our first attempt is

$$\frac{k+\kappa_<}{k+\kappa_>}\,G(k) = -\frac{1}{2\kappa_<(k+\kappa_>)} + \widetilde{G}(k), \tag{5.0.9}$$

where

$$\widetilde{G}(k) = \left(\frac{k+\kappa_<}{k+\kappa_>}\right)\left[\frac{1}{2\kappa_<(k-\kappa_<)} - \frac{1}{(k-\kappa_<)^2}\right]. \tag{5.0.10}$$

The first term on the right side of Equation 5.0.9 is analytic in the same half-plane as $F_+(k)$. However, $\widetilde{G}(k)$ remains unsplit because of the simple pole at $k = -\kappa_>$; otherwise, it would be analytic in the same half-plane as $F_-(k)$. To circumvent this difficulty, we add and subtract out the troublesome singularity:

$$\widetilde{G}(k) = \left\{\widetilde{G}(k) - \frac{\operatorname{Res}\left[\widetilde{G}(k); -\kappa_>\right]}{k+\kappa_>}\right\} + \frac{\operatorname{Res}\left[\widetilde{G}(k); -\kappa_>\right]}{k+\kappa_>} = \widetilde{G}_-(k) + \widetilde{G}_+(k), \tag{5.0.11}$$

where

$$\widetilde{G}_+(k) = \left(\frac{\kappa_> - \kappa_<}{k+\kappa_>}\right)\left[\frac{1}{2\kappa_<(\kappa_>+\kappa_<)} + \frac{1}{(\kappa_>+\kappa_<)^2}\right]. \tag{5.0.12}$$

Substituting Equation 5.0.11 into Equation 5.0.9, we can then rewrite Equation 5.0.8 as

$$\frac{k+\kappa_<}{k+\kappa_>}F_+(k) - \widetilde{G}_+(k) + \frac{1}{2\kappa_<(k+\kappa_>)} = -\left(\frac{k-\kappa_>}{k-\kappa_<}\right)F_-(k) + \widetilde{G}_-(k). \tag{5.0.13}$$

Why have we undertaken such an elaborate analysis to obtain Equation 5.0.13? The function on the left side of Equation 5.0.13 is analytic in the half-plane $\Im(k) > -\delta$, while the function on the right side is analytic in the half-plane $\Im(k) < \delta$. By virtue of the principle of analytic continuation, these functions are equal to some function, say $H(k)$, that is analytic over the entire k-plane. We can determine the form of $H(k)$ from the known asymptotic properties of the transform. Here, $H(k)$ must vanish at infinity because 1) $G(k)$ does and 2) both $F_+(k)$ and $F_-(k)$ vanish by the Riemann–Lebesque theorem. Once we have the asymptotic behavior, we can apply Liouville's theorem.

Liouville's theorem:[8] *If $f(z)$ is analytic for all finite value of z, and as $|z| \to \infty$, $f(z) = O(|z|^m)$, then $f(z)$ is a polynomial of degree $\leq m$.*

Here, for example, because $H(k)$ goes to zero as $|k| \to \infty$, $m = 0$ and $H(k) = 0$. Thus, both sides of Equation 5.0.13 vanish.

The only remaining task is to invert the Fourier transforms and to obtain $f(x)$. Because $F(k) = F_+(k) + F_-(k)$,

$$f(x) = \frac{1}{2\pi} \int_{-\infty}^{\infty} \left[F_+(k) + F_-(k)\right] e^{ikx}\, dk. \tag{5.0.14}$$

We will evaluate Equation 5.0.14 using the residue theorem. Since

$$F_+(k) = \frac{(k + \kappa_>)\widetilde{G}_+(k)}{k + \kappa_<} - \frac{1}{2\kappa_<(k + \kappa_<)} \tag{5.0.15}$$

and

$$F_-(k) = \frac{2\kappa_< (k - \kappa_> - 2\kappa_<)}{(k - \kappa_>)(k - \kappa_<)(\kappa_> + \kappa_<)^2}, \tag{5.0.16}$$

we find that

$$f(x) = \frac{2i\kappa_<}{(\kappa_> + \kappa_<)^2} e^{-i\kappa_< x} \tag{5.0.17}$$

if $x < 0$, while for $x > 0$,

$$f(x) = \frac{2\kappa_<}{i\left(\kappa_>^2 - \kappa_<^2\right)} \left(\frac{2\kappa_<}{\kappa_> + \kappa_<} e^{i\kappa_> x} - e^{i\kappa_< x} \right). \tag{5.0.18}$$

Having outlined the mechanics behind solving a Wiener-Hopf problem, we are ready to see how this method is used to solve mixed boundary value problems. The crucial step in the procedure is the ability to break all of the Fourier transforms into two functions, one part is analytic on some upper half-plane while the other is analytic on some lower half-plane. For example, we split $G(k)$ into $G_+(k)$ plus $G_-(k)$. Most often, this factor is in the form of a product: $G(k) = G_+(k)G_-(k)$. In the next two sections, we will show various problems that illustrate various types of factorization.

Problems

1. Consider the following equation that appeared in a Wiener-Hopf analysis by Lehner et al.:[9]

$$\sqrt{\omega^2 + \lambda^2}\, U_+(\omega) = \frac{1}{\omega - \kappa} - T_-(\omega), \tag{1}$$

---

[8] See Titchmarsh, E. C., 1939: *The Theory of Functions*. 2nd Edition. Oxford University Press, Section 2.52.

[9] Lehner, F. K., V. C. Li, and J. R. Rice, 1981: Stress diffusion along rupturing plate boundaries. *J. Geophys. Res.*, **86**, 6155–6169. ©1981 American Geophysical Union. Reproduced/modified by permission of the American Geophysical Union.

where $0 < \lambda$, $\kappa < 0$, $U_+(\omega)$ is analytic in the upper complex ω-plane $0 < \Im(\omega)$ and $T_-(\omega)$ is analytic in the lower half-plane $\Im(\omega) < \tau$, $0 < \tau < \lambda$.

Step 1: Factoring the square root as $\sqrt{\omega + \lambda i}\ \sqrt{\omega - \lambda i}$ so that

$$\sqrt{\omega + \lambda i}\, U_+(\omega) = \frac{1}{(\omega - \kappa)\sqrt{\omega - \lambda i}} - \frac{T_-(\omega)}{\sqrt{\omega - \lambda i}}, \tag{2}$$

show that the left side of (2) is analytic in the upper half-plane $0 < \Im(\omega)$, while the second term on the right side is analytic in the lower half-plane $\Im(\omega) < \tau$.

Step 2: Show that the first term on the right side of (2) is neither analytic on the half-plane $0 < \Im(\omega)$ nor on the lower half-plane $\Im(\omega) < \tau$ due to a simple pole that lies in the lower half-plane $\Im(\omega) < 0$.

Step 3: Show that we can split this troublesome term as follows:

$$\frac{1}{(\omega - \kappa)\sqrt{\omega - \lambda i}} = \frac{1}{(\omega - \kappa)\sqrt{\kappa - \lambda i}} + \left(\frac{1}{\sqrt{\omega - \lambda i}} - \frac{1}{\sqrt{\kappa - \lambda i}} \right) \frac{1}{\omega - \kappa}, \tag{3}$$

where the first term on the right side of (3) is analytic in the half-plane $0 < \Im(\omega)$, while the second term is analytic in the half-plane $\Im(\omega) < \tau$.

Step 4: Show that the factorization of (1) is

$$\sqrt{\omega + \lambda i}\ U_+(\omega) - \frac{1}{(\omega - \kappa)\sqrt{\kappa - \lambda i}} = \left(\frac{1}{\sqrt{\omega - \lambda i}} - \frac{1}{\sqrt{\kappa - \lambda i}} \right) \frac{1}{\omega - \kappa} - \frac{T_-(\omega)}{\sqrt{\omega - \lambda i}}.$$

5.1 THE WIENER–HOPF TECHNIQUE WHEN THE FACTORIZATION CONTAINS NO BRANCH POINTS

In the previous section we sketched out the essence of the Wiener-Hopf technique. An important aspect of this technique was the process of factorization. There, we reexpressed several functions as a sum of two parts; one part is analytic in some lower half-plane, while the other part is analytic in some upper half-plane. Both of these half-planes share some common region. More commonly, the splitting occurs as the *product* of two functions. In this section we illustrate how this factorization arises and how the splitting is accomplished during the solution of a mixed boundary value problem.

• **Example 5.1.1**

Given that $h, \beta > 0$, let us solve the partial differential equation[10]

$$\frac{\partial^2 u}{\partial x^2} + \frac{\partial^2 u}{\partial y^2} - \beta^2 u = 0, \qquad -\infty < x < \infty, \quad 0 < y < 1, \tag{5.1.1}$$

with the boundary conditions

$$u_y(x, 1) - \beta u(x, 1) = 0, \qquad -\infty < x < \infty, \tag{5.1.2}$$

and

$$\begin{cases} u(x, 0) = 1, & x < 0, \\ u_y(x, 0) - (h + \beta)u(x, 0) = 0, & 0 < x. \end{cases} \tag{5.1.3}$$

We begin by defining

$$U(k, y) = \int_{-\infty}^{\infty} u(x, y)e^{ikx}\, dx \quad \text{and} \quad u(x, y) = \frac{1}{2\pi}\int_{-\infty}^{\infty} U(k, y)e^{-ikx}\, dk. \tag{5.1.4}$$

Taking the Fourier transform of Equation 5.1.1, we obtain

$$\frac{d^2U(k, y)}{dy^2} - m^2 U(k, y) = 0, \qquad 0 < y < 1, \tag{5.1.5}$$

where $m^2 = k^2 + \beta^2$. The solution to this differential equation is

$$U(k, y) = A(k)\cosh(my) + B(k)\sinh(my). \tag{5.1.6}$$

Substituting Equation 5.1.6 into Equation 5.1.2 after its Fourier transform has been taken, we find that

$$m\,[A(k)\sinh(m) + B(k)\cosh(m)] - \beta\,[A(k)\cosh(m) + B(k)\sinh(m)] = 0. \tag{5.1.7}$$

The Fourier transform of Equation 5.1.3 is

$$A(k) = \frac{1}{ik} + M_+(k); \tag{5.1.8}$$

and

$$\begin{aligned} \int_{-\infty}^{\infty} \Big[u_y(x, 0) - (h + \beta)u(x, 0)\Big] e^{ikx}\, dx &= \int_{-\infty}^{0} [u_y(x, 0) - (h + \beta)u(x, 0)]\, e^{ikx}\, dx \\ &+ \int_{0}^{\infty} [u_y(x, 0) - (h + \beta)u(x, 0)]\, e^{ikx}\, dx, \end{aligned} \tag{5.1.9}$$

---

[10] Taken from V. T. Buchwald and F. Viera, Linearized evaporation from a soil of finite depth near a wetted region, *Quart. J. Mech. Appl. Math.*, 1996, **49(1)**, 49–64 by permission of Oxford University Press.

or

$$mB - (h + \beta)A = L_-(k), \tag{5.1.10}$$

where

$$M_+(k) = \int_0^\infty u(x,0)e^{ikx}\,dx, \tag{5.1.11}$$

and

$$L_-(k) = \int_{-\infty}^0 \left[u_y(x,0) - (h+\beta)u(x,0)\right]e^{ikx}\,dx. \tag{5.1.12}$$

Here, we have assumed that $|u(x,0)|$ is bounded by $e^{-\epsilon x}$ as $x \to \infty$ with $0 < \epsilon \ll 1$. Consequently, $M_+(k)$ is an analytic function in the half-space $\Im(k) > -\epsilon$. Similarly, $L_-(k)$ is analytic in the half-space $\Im(k) < 0$. We used Equation 5.1.3 to simplify the right side of Equation 5.1.9.

Eliminating $A(k)$ from Equation 5.1.10,

$$mB = L_-(k) + (h+\beta)\left[\frac{1}{ik} + M_+(k)\right]. \tag{5.1.13}$$

Combining Equation 5.1.7, Equation 5.1.8 and Equation 5.1.13, we have that

$$\left[\frac{1}{ik} + M_+(k)\right]\left[hm\cosh(m) - (\beta^2 + h\beta - m^2)\sinh(m)\right] + [m\cosh(m) - \beta\sinh(m)]L_-(k) = 0. \tag{5.1.14}$$

With Equation 5.1.14, we reached the point where we must rewrite it so that it is analytic in the half-plane $\Im(k) < 0$ on one side, while the other side is analytic in the half-plane $\Im(k) > -\epsilon$. The difficulty arises from the terms $hm\cosh(m) - (\beta^2 + h\beta - m^2)\sinh(m)$ and $[m\cosh(m) - \beta\sinh(m)]$. How can we rewrite them so that we can accomplish our splitting? To do this, we now introduce the *infinite product theorem*:

Infinite Product Theorem:[11] *If $f(z)$ is an entire function of z with simple zeros at $z_1, z_2, \ldots$, then*

$$f(z) = f(0)\exp\left[zf'(0)/f(0)\right]\prod_{n=1}^{\infty}\left(1 - \frac{z}{z_n}\right)e^{z/z_n}. \tag{5.1.15}$$

Let us apply this theorem to $\cosh(m) - \beta\sinh(m)/m$. We find that

$$\cosh(m) + \frac{\beta}{m}\sinh(m) = e^{-\beta}F(k)F(-k), \tag{5.1.16}$$

[11] See Titchmarsh, op. cit., Section 3.23.

where

$$F(k) = e^{-\gamma ki/\pi} \prod_{n=1}^{\infty} \left(1 - \frac{ki}{\lambda_n}\right) e^{ki/(n\pi)}, \qquad \textbf{(5.1.17)}$$

γ is Euler's constant, and $\lambda_n > 0$ is the nth root of $\beta \tan(\lambda) = \lambda$. The reason why Equation 5.1.16 and Equation 5.1.17 are useful lies in the fact that $F(k)$ is analytic and nonzero in the half-plane $\Im(k) > -\epsilon$, while $F(-k)$ is analytic and nonzero in the lower half-plane $\Im(k) < 0$. In a similar vein,

$$\cosh(m) - \frac{\beta^2 + h\beta - m^2}{hm} \sinh(m) = e^{-\beta} G(k)G(-k), \qquad \textbf{(5.1.18)}$$

where

$$G(k) = e^{-\gamma ki/\pi} \prod_{n=1}^{\infty} \left(1 - \frac{ki}{\rho_n}\right) e^{ki/(n\pi)}, \qquad \textbf{(5.1.19)}$$

and ρ_n is the nth root of $\tan(\rho) = h\rho/(\beta^2 + h\beta + \rho^2)$. Here, $G(k)$ is analytic in the half-plane $\Im(k) > -\epsilon$ while $G(-k)$ is analytic in the half-plane $\Im(k) < 0$. Substituting Equation 5.1.16 and Equation 5.1.18 into Equation 5.1.14, we obtain

$$\frac{hG(k)M_+(k)}{F(k)} + \frac{F(-k)L_-(k)}{G(-k)} = -\frac{hG(k)}{ikF(k)} = \frac{h}{ik}\left[1 - \frac{G(k)}{F(k)}\right] - \frac{h}{ik}. \qquad \textbf{(5.1.20)}$$

We observe that the first term on the right side of Equation 5.1.20 is analytic in the upper half-space $\Im(k) > -\epsilon$, while the second term is analytic in the lower half-plane $\Im(k) < 0$. We now rewrite Equation 5.1.20 so that its right side is analytic in the upper half-plane, while its left side is analytic in the lower half-plane:

$$\frac{hG(k)M_+(k)}{F(k)} - \frac{h}{ik}\left[1 - \frac{G(k)}{F(k)}\right] = -\frac{F(-k)L_-(k)}{G(-k)} - \frac{h}{ik}. \qquad \textbf{(5.1.21)}$$

At this point we must explore the behavior of both sides of Equation 5.1.21 as $|k| \to \infty$. Applying asymptotic analysis, Buchwald and Viera showed that $G(k)/F(k) \sim k^{1/2}$. Since $M_+(k) \sim k^{-1}$, the first term on the right side of Equation 5.1.20 behaves as $k^{-1/2}$. Because $L_-(k) \sim k^{-1/2}$, the second term behaves as k^{-1}. From Liouville's theorem, both sides of Equation 5.1.21 must equal zero, yielding

$$hM_+(k) = \frac{hF(k)}{ikG(k)}\left[1 - \frac{G(k)}{F(k)}\right], \qquad \textbf{(5.1.22)}$$

and

$$L_-(k) = -\frac{hG(-k)}{ikF(-k)}. \qquad \textbf{(5.1.23)}$$

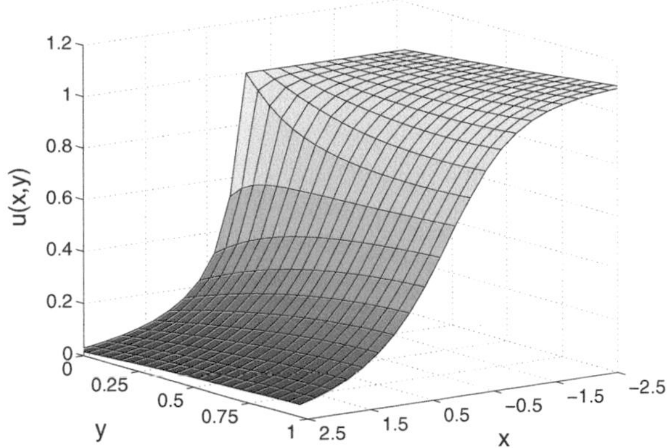

Figure 5.1.1: The solution to Equation 5.1.1 through Equation 5.1.3 obtained via the Wiener-Hopf technique when $h = 2$ and $\beta = 0.1$.

Therefore, from Equation 5.1.4, we have

$$u(x, y) = \frac{1}{2\pi} \int_{-\infty-\epsilon i}^{\infty-\epsilon i} [A(k)\sinh(mz) + B(k)\cosh(mz)]e^{-ikx}\, dk, \qquad \textbf{(5.1.24)}$$

where $A(k)$ is given by a combination of Equation 5.1.8 and Equation 5.1.22, while $B(k)$ follows from Equation 5.1.13, Equation 5.1.22 and Equation 5.1.23. Consequently,

$$ikmA(k) = \frac{F(k)}{G(k)} \quad \text{and} \quad ikmB(k) = (h+\beta)\frac{F(k)}{G(k)} - h\frac{G(-k)}{F(-k)}. \qquad \textbf{(5.1.25)}$$

Finally, we apply the residue theorem to evaluate Equation 5.1.24 and find that

$$u(x, y) = e^{\beta y} + he^{-\beta} \sum_{n=1}^{\infty} \frac{\mu_n^2 G(-i\lambda_n)F(i\lambda_n)}{\lambda_n^2(\lambda_n^2 - \beta)\sin(\mu_n)} \sin(\mu_n y)e^{\lambda_n x}, \qquad x < 0, \qquad \textbf{(5.1.26)}$$

where $\mu_n^2 = \lambda_n^2 - \beta^2$, and

$$u(x, y) = he^{-\beta} \sum_{n=1}^{\infty} B_n[(h+\beta)\sin(\sigma_n y) + \sigma_n \cos(\sigma_n y)]e^{-\rho_n x}, \qquad 0 < x, \qquad \textbf{(5.1.27)}$$

where

$$B_n = \frac{\sigma_n(\rho_n^2 + h\beta)F(-i\rho_n)G(i\rho_n)}{\rho_n^2[(\rho_n^2 + h\beta)(\rho_n^2 + h\beta - h) + h(h+2)\sigma_n^2]\cos(\sigma_n)} \qquad \textbf{(5.1.28)}$$

and $\sigma_n^2 = \rho_n^2 - \beta^2$. Figure 5.1.1 illustrates this solution when $h = 2$ and $\beta = 0.1$.

• Example 5.1.2

In the previous example we used Liouville's theorem to solve the Wiener-Hopf problem. In the following three examples, we illustrate an alternative approach developed by N. N. Lebedev. Here the factorization follows from trial and error.

Presently we solve[12] Laplace's equation

$$\frac{\partial^2 u}{\partial x^2} + \frac{\partial^2 u}{\partial y^2} = 0, \qquad -\infty < x < \infty, \quad 0 < y < \infty, \tag{5.1.29}$$

with the boundary conditions

$$\lim_{|x|\to\infty} u(x,y) \to 0, \qquad 0 < y < \infty, \tag{5.1.30}$$

$$\lim_{y\to\infty} u(x,y) \to 0, \qquad -\infty < x < \infty, \tag{5.1.31}$$

$$u(x,0) = 0, \qquad -\infty < x < \infty, \tag{5.1.32}$$

and

$$\begin{cases} u(x,h^-) = u(x,h^+), \quad \epsilon_1 u_y(x,h^-) = \epsilon_2 u_y(x,h^+), & -\infty < x < 0, \\ \qquad u(x,h^-) = u(x,h^+) = Ve^{-\alpha x}, & 0 < x < \infty, \end{cases} \tag{5.1.33}$$

with $\alpha > 0$.

We begin by noting that a solution to Equation 5.1.29 is

$$u(x,y) = \frac{1}{2\pi}\int_{-\infty}^{\infty} A(k)\frac{\sinh(ky)}{\sinh(kh)} e^{ikx}\, dk, \qquad 0 \le y < h, \tag{5.1.34}$$

and

$$u(x,y) = \frac{1}{2\pi}\int_{-\infty}^{\infty} A(k)e^{-|k|(y-h)} e^{ikx}\, dk, \qquad h < y < \infty. \tag{5.1.35}$$

Note that Equation 5.1.34 and Equation 5.1.35 satisfy not only Laplace's equation, but also the boundary conditions as $|x| \to \infty$, $y \to \infty$, and $u(x,0) = 0$. Because

$$e^{-\alpha x} H(x) = \frac{1}{2\pi i}\int_{-\infty}^{\infty} \frac{e^{ikx}}{k - i\alpha}\, dk, \tag{5.1.36}$$

the boundary condition given by Equation 5.1.33 yields the dual integral equations

$$\int_{-\infty}^{\infty} A(k)K(k)e^{ikx}\, dk = 0, \qquad -\infty < x < 0, \tag{5.1.37}$$

[12] Taken from Lebedev, N. N., 1958: The electric field at the edge of a plane condenser containing a dielectric. *Sov. Tech. Phys.*, **3**, 1234–1243.

and

$$\int_{-\infty}^{\infty} \left[A(k) - \frac{V}{\alpha + ik} \right] e^{ikx} \, dk = 0, \qquad 0 < x < \infty, \tag{5.1.38}$$

where

$$K(k) = \frac{khe^{|k|h}}{\sinh(kh)} \left(1 + \kappa e^{-2|k|h}\right), \qquad \kappa = \frac{\epsilon_1 - \epsilon_2}{\epsilon_1 + \epsilon_2}. \tag{5.1.39}$$

At this point Lebedev introduced the factorization

$$K(w) = K_+(w)K_-(w) = \frac{whe^{\pm wh}}{\sinh(wh)} \left(1 + \kappa e^{\mp 2wh}\right), \tag{5.1.40}$$

where the upper sign corresponds to w with $\Re(w) > 0$, while the lower sign holds when $\Re(w) < 0$,

$$K_+(w) = \Gamma\left(1 - \frac{iwh}{\pi}\right) \exp\left[\frac{iwh}{\pi} \log\left(-\frac{iwh}{\pi}\right) - \frac{iwh}{\pi} + f\left(-\frac{iwh}{\pi}\right)\right], \tag{5.1.41}$$

for $-\pi/2 < \arg(w) < 3\pi/2$, $K_-(w) = K_+(-w)$ with $-3\pi/2 < \arg(w) < \pi/2$; $\Gamma(\cdot)$ is the gamma function and the logarithm takes its principal value. Here,

$$f(z) = \frac{1}{\pi} \int_0^{\infty} \arctan\left[\frac{\kappa \sin(2\pi\eta)}{1 + \kappa \cos(2\pi\eta)}\right] \frac{d\eta}{\eta + z}, \qquad -\pi < \arg(z) < \pi. \tag{5.1.42}$$

The function $f(z)$ is analytic on the complex z-plane cut along the negative real axis, approaches zero as $|z| \to \infty$, and

$$f\left(-\frac{iwh}{\pi}\right) + f\left(\frac{iwh}{\pi}\right) = \log\left(1 + \kappa e^{\mp 2wh}\right), \tag{5.1.43}$$

where the upper and lower signs are taken according to whether $\Re(w) < 0$ or > 0, respectively. Finally, asymptotic analysis reveals that

$$K_+(w) \approx \sqrt{-2iwh}, \qquad |w| \to \infty, \quad -\frac{\pi}{2} < \arg(w) < \frac{3\pi}{2}, \tag{5.1.44}$$

and

$$K_-(w) \approx \sqrt{2iwh}, \qquad |w| \to \infty, \quad -\frac{3\pi}{2} < \arg(w) < \frac{\pi}{2}. \tag{5.1.45}$$

To show that this factorization of $K(w)$ is a correct one, we use the facts that

$$\Gamma\left(1 - \frac{iwh}{h}\right) \Gamma\left(1 + \frac{iwh}{h}\right) = \frac{wh}{\sinh(wh)}, \tag{5.1.46}$$

and

$$\log\left(-\frac{iwh}{\pi}\right) + \log\left(-\frac{iwh}{\pi}\right) = \mp\pi i, \tag{5.1.47}$$

where the upper sign corresponds to $\Re(w) > 0$ while the lower sign corresponds to $\Re(w) < 0$, as well as Equation 5.1.43.

We now use our knowledge of $K(k)$ to construct a solution to Equation 5.1.37 and Equation 5.1.38. Consider Equation 5.1.37 first. A common technique for evaluating Fourier integrals consists of closing the line integral along the real axis with an arc of infinite radius as dictated by Jordan's lemma and then using the residue theorem. Because $x < 0$ here, this closed contour must be a semi-circle of infinite radius in the lower half of the complex k-plane, i.e., $\Im(k) \le 0$. If $K(k)A(k)$ is analytic within this closed contour, then Equation 5.1.37 is satisfied by the Cauchy-Goursat theorem. A quick check confirms that if

$$A(k) = \frac{VK_+(\alpha i)}{(\alpha + ik)K_+(k)}, \tag{5.1.48}$$

then

$$K(k)A(k) = \frac{VK_+(\alpha i)K_-(k)}{\alpha + ik} \tag{5.1.49}$$

is analytic in the lower half-plane $\Im(k) \le 0$. Using similar arguments for Equation 5.1.38, because the expression

$$A(k) - \frac{V}{ik + \alpha} = \frac{V}{ik + \alpha}\left[\frac{K_+(\alpha i)}{K_+(k)} - 1\right] \tag{5.1.50}$$

is analytic in the upper half-plane $\Im(k) \ge 0$, Equation 5.1.38 is satisfied.

Substituting Equation 5.1.48 into Equation 5.1.34 and Equation 5.1.35, we have that

$$u(x, y) = \frac{V}{2\pi i}\int_{-\infty}^{\infty} \frac{K_+(\alpha i)\sinh(ky)}{K_+(k)(k - \alpha i)\sinh(kh)} e^{ikx}\, dk \tag{5.1.51}$$

for $0 \le y \le h$; and

$$u(x, y) = \frac{V}{2\pi i}\int_{-\infty}^{\infty} \frac{K_+(i\alpha)e^{-|k|(y-h)}}{K_+(k)(k - \alpha i)} e^{ikx}\, dk \tag{5.1.52}$$

for $h \le y \le \infty$. Applying the residue theorem to these equations, we find that

$$\begin{aligned}\frac{u(x, y)}{V} = e^{-\alpha y}\frac{\sinh(\alpha y)}{\sinh(\alpha h)} + \frac{1}{\pi}\sum_{n=1}^{\infty} \frac{(-1)^n}{n - \alpha h/\pi}\frac{\Gamma(1 + \alpha h/\pi)}{\Gamma(1 + n)} \\ \times \exp\left[\varphi\left(\frac{\alpha h}{\pi}\right) - \varphi(n) - \frac{n\pi x}{h}\right]\sin\left(\frac{n\pi y}{h}\right)\end{aligned} \tag{5.1.53}$$

for $0 \leq x < \infty$, $0 \leq y \leq h$, where $\varphi(z) = f(z) - z\log(z) + z$. On the other hand,

$$\frac{u(x,y)}{V} = \frac{1-\kappa}{\pi}\Gamma\left(1+\frac{\alpha h}{\pi}\right) e^{\varphi(\alpha h/\pi)} \times \int_0^\infty \frac{e^{\varphi(\eta)} e^{\pi\eta x/h} \sin(\pi\eta y/h)}{\Gamma(1-\eta)(\eta + ah/\pi)[1+2\kappa\cos(2\pi\eta)+\kappa^2]}\,d\eta \tag{5.1.54}$$

for $-\infty < x \leq 0$, $0 \leq y \leq h$;

$$\frac{u(x,y)}{V} = \frac{1}{\pi}\Gamma\left(1+\frac{\alpha h}{\pi}\right) e^{\varphi(\alpha h/\pi)} \times \int_0^\infty \frac{e^{\varphi(\eta)} e^{\pi\eta x/h}\{\sin(\pi\eta y/h) + \kappa\sin[\pi\eta(y-2h)/h]\}}{\Gamma(1-\eta)(\eta+\alpha h/\pi)[1+2\kappa\cos(2h\eta)+\kappa^2]}\,d\eta \tag{5.1.55}$$

for $-\infty < x \leq 0$, $h \leq y < \infty$; and

$$\frac{u(x,y)}{V} = e^{-\alpha x}\cos[\alpha(y-h)] - \frac{1}{\pi}\Gamma\left(1+\frac{\alpha h}{\pi}\right) e^{\varphi(\alpha h/\pi)} \times PV\left\{\int_0^\infty \frac{e^{-\varphi(\eta)-\pi\eta x/\pi}\sin[\pi\eta(y-h)/h]}{\Gamma(1+\eta)(\eta-\alpha h/\pi)}\,d\eta\right\} \tag{5.1.56}$$

for $0 \leq x < \infty$, $h \leq y < \infty$. Equation 5.1.54 through Equation 5.1.56 follow from deforming the line integration along the real axis to one along the imaginary axis. See Lebedev's paper for details. In the limit $\alpha \to 0$, Equation 5.1.53 through Equation 5.1.56 simplify to

$$\frac{u(x,y)}{V} = \frac{y}{h} + \frac{\sqrt{1+\kappa}}{\pi}\sum_{n=1}^\infty \frac{(-1)^n}{n\,n!} e^{-\varphi(n)-n\pi x/h}\sin\left(\frac{n\pi y}{h}\right) \tag{5.1.57}$$

for $0 \leq x \leq \infty$ and $0 \leq y \leq h$;

$$\frac{u(x,y)}{V} = \frac{\sqrt{1+\kappa}}{\pi}(1-\kappa)\int_0^\infty \frac{e^{\varphi(\eta)+\pi\eta x/h}\sin(\pi\eta y/h)}{\eta\Gamma(1-\eta)[1+2\kappa\cos(2\pi\eta)+\kappa^2]}\,d\eta, \tag{5.1.58}$$

for $-\infty < x \leq 0$ and $0 \leq y \leq h$;

$$\frac{u(x,y)}{V} = \frac{\sqrt{1+\kappa}}{\pi}\int_0^\infty \frac{e^{\varphi(\eta)+\pi\eta x/h}\{\sin(\pi\eta y/h) + \kappa\sin[\pi\eta(y-2h)/h]\}}{\eta\Gamma(1-\eta)[1+2\kappa\cos(2\pi\eta)+\kappa^2]}\,d\eta, \tag{5.1.59}$$

for $-\infty < x \leq 0$ and $h \leq y < \infty$; and

$$\frac{u(x,y)}{V} = 1 - \frac{\sqrt{1+\kappa}}{\pi}\int_0^\infty \frac{e^{-\varphi(\eta)-\pi\eta x/h}\sin[\pi\eta(y-h)/h]}{\eta\Gamma(1+\eta)}\,d\eta, \tag{5.1.60}$$

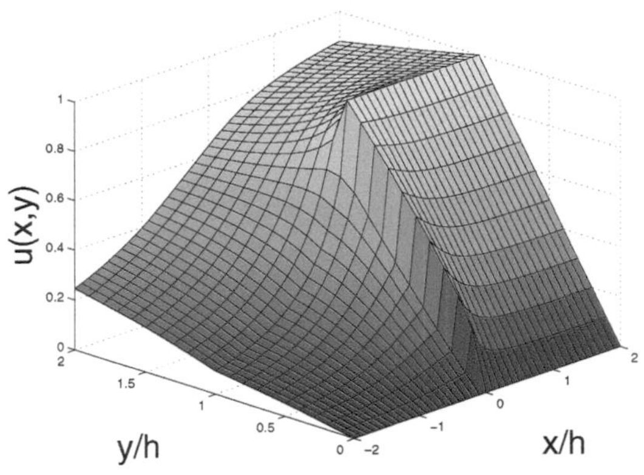

Figure 5.1.2: A plot of the solution to Equation 5.1.29 through Equation 5.1.33 in the limit as $\alpha \to 0$. Here $\epsilon_1 = 3\epsilon_2 = 3$.

for $0 \le x < \infty$ and $h \le y < \infty$. Figure 5.1.2 illustrates Equation 5.1.57 through Equation 5.1.60.

• Example 5.1.3

For the next example, let us solve Laplace's equation[13]

$$\frac{1}{r}\frac{\partial}{\partial r}\left(\frac{\partial u}{\partial r}\right) + \frac{\partial^2 u}{\partial z^2} = 0, \qquad 0 \le r < \infty, \quad -\infty < z < \infty, \tag{5.1.61}$$

with the boundary conditions

$$\lim_{r\to 0} |u(r,z)| < \infty, \quad \lim_{r\to\infty} u(r,z) \to 0, \qquad -\infty < z < \infty, \tag{5.1.62}$$

$$\lim_{|z|\to\infty} u(r,z) \to 0, \qquad 0 \le r < \infty, \tag{5.1.63}$$

and

$$\begin{cases} u_r(a^-,z) = u_r(a^+,z), & -\infty < z < 0, \\ u(a^-,z) = u(a^+,z) = e^{-i\alpha z}, & 0 < z < \infty, \end{cases} \tag{5.1.64}$$

with $\Im(\alpha) < 0$.

We begin by observing that the solution to Equation 5.1.61 is

$$u(r,z) = \frac{1}{2\pi}\int_{-\infty}^{\infty} A(k)\frac{I_0(kr)}{I_0(ka)} e^{ikz}\, dk, \qquad 0 \le r \le a, \tag{5.1.65}$$

---

[13] Adapted from Lebedev, N. N., and I. P. Skal'skaia, 1958: Axially-symmetric electrostatic problem for a thin-walled conductor in the form of a half-infinite tube. *Sov. Tech. Phys.*, **3**, 740–748.

and

$$u(r, z) = \frac{1}{2\pi} \int_{-\infty}^{\infty} A(k) \frac{K_0(|k|r)}{K_0(|k|a)} e^{ikz}\, dk, \qquad a \le r < \infty. \tag{5.1.66}$$

Note that Equation 5.1.65 and Equation 5.1.66 satisfy not only Laplace's equation, but also the boundary conditions as $|z| \to \infty$ and $r \to \infty$. Because

$$e^{-i\alpha z} H(z) = \frac{1}{2\pi i} \int_{-\infty}^{\infty} \frac{e^{ikz}}{k+\alpha}\, dk, \tag{5.1.67}$$

the boundary condition given by Equation 5.1.64 yields the dual integral equations

$$\int_{-\infty}^{\infty} A(k) K(k) e^{ikz}\, dk = 0, \qquad -\infty < z < 0, \tag{5.1.68}$$

and

$$\int_{-\infty}^{\infty} \left[A(k) + \frac{i}{k+\alpha} \right] e^{ikz}\, dk = 0, \qquad 0 < z < \infty, \tag{5.1.69}$$

where $K(k) = I_0(ka)/K_0(|k|a)$.

The difficulty in factoring $K(k)$ lies with the presence of $K_0(z)$ which possesses a branch point at $z = 0$. Our goal remains the same: we wish to factor $K(k)$ as

$$K(k) = K_+(k) K_-(k), \tag{5.1.70}$$

with the properties:

- The function $K_+(k)$ is analytic and has no zeros in a k-plane cut along the negative imaginary axis; the function $K_-(k)$ is analytic and has no zeros in a plane cut along the positive imaginary axis.
- Both $K_+(k)$ and $K_-(k)$ have algebraic growth at infinity, namely

$$K_+(k) \approx \sqrt{-2ika}, \qquad |k| \to \infty, \quad -\frac{\pi}{2} < \arg(k) < \frac{3\pi}{2}, \tag{5.1.71}$$

and

$$K_-(k) \approx \sqrt{2ika}, \qquad |k| \to \infty, \quad -\frac{3\pi}{2} < \arg(k) < \frac{\pi}{2}. \tag{5.1.72}$$

Under these constraints, Lebedev and Skal'skaia showed that

$$K_+(w) = \left(-\frac{2iwa}{\pi} \right)^{1/4} \exp\left[-f\left(-\frac{iwa}{\pi} \right) \right] \times \frac{\exp\left\{ -\frac{iwa}{\pi} \left[1 - \gamma - \log\left(-\frac{iwa}{\pi} \right) - \pi \sum_{n=1}^{\infty} \left(\frac{1}{\gamma_n} - \frac{1}{n\pi} \right) \right] \right\}}{\prod_{n=1}^{\infty} (1 - iwa/\gamma_n) e^{iwa/\gamma_n}}, \tag{5.1.73}$$

and $K_-(w) = K_+(-w)$, where γ is Euler's constant, γ_n is the nth positive root of $J_0(\cdot)$ and

$$f(z) = \frac{1}{\pi}\int_0^\infty \left\{1 - \frac{2}{\pi\eta\left[J_0^2(\eta) + Y_0^2(\eta)\right]}\right\}\log\left(1 + \frac{\eta}{\pi z}\right)\,d\eta. \qquad \textbf{(5.1.74)}$$

We now use these properties of $K(k)$ to construct a solution to Equation 5.1.68 and Equation 5.1.69. Our argument is identical to the one given in the previous example. If we set

$$A(k) = -i\frac{K_+(-\alpha)}{(k+\alpha)K_+(k)}, \qquad \textbf{(5.1.75)}$$

then

$$K(k)A(k) = -i\frac{K_+(-\alpha)K_-(k)}{k+\alpha} \qquad \textbf{(5.1.76)}$$

is clearly analytic in the lower half-plane $\Im(k) \le 0$ and Equation 5.1.68 is satisfied. Furthermore,

$$A(k) + \frac{i}{k+\alpha} = \frac{i}{k+\alpha}\left[1 - \frac{K_+(-\alpha)}{K_+(k)}\right] \qquad \textbf{(5.1.77)}$$

is analytic in the upper half-plane $\Im(k) \ge 0$ and Equation 5.1.69 is satisfied.

Substituting Equation 5.1.75 into Equation 5.1.65 and Equation 5.1.66, we have that

$$u(r,z) = \frac{K_+(-\alpha)}{2\pi i}\int_{-\infty}^\infty \frac{I_0(kr)}{I_0(ka)}\frac{e^{ikz}}{K_+(k)(k+\alpha)}\,dk \qquad \textbf{(5.1.78)}$$

for $0 \le r \le a$; and

$$u(x,y) = \frac{K_+(-\alpha)}{2\pi i}\int_{-\infty}^\infty \frac{K_0(|k|r)}{K_0(|k|a)}\frac{e^{ikz}}{K_+(k)(k+\alpha)}\,dk \qquad \textbf{(5.1.79)}$$

for $a \le r < \infty$. Applying the residue theorem to these equations, we find that

$$u(r,z) = \frac{I_0(\alpha r)}{I_0(\alpha a)}e^{-i\alpha z} + \sum_{n=1}^\infty \frac{K_-(\alpha)J_0(\gamma_n r/a)e^{-\gamma_n z/a}}{aiK_-(-i\gamma_n/a)J_1(\gamma_n)(\alpha + i\gamma_n/a)} \qquad \textbf{(5.1.80)}$$

for $0 \le r \le a$, $0 < z < \infty$. On the other hand, if we deform the original contour that runs along the real axis so that now also runs along the imaginary axis, we obtain

$$u(r,z) = \frac{K_-(\alpha)}{2ai}\int_0^\infty K_+\left(\frac{i\eta}{a}\right)e^{\eta z/a}J_0(\eta)J_0\left(\frac{\eta r}{a}\right)\frac{d\eta}{\alpha - i\eta/a} \qquad \textbf{(5.1.81)}$$

for $0 \le r < \infty$, $-\infty < z < 0$; as well as

$$u(r,z) = \frac{K_0(\pm\alpha r)}{K_0(\pm\alpha r)}e^{-i\alpha z} \qquad \textbf{(5.1.82)}$$

$$- \frac{K_-(\alpha)}{\pi a i}\int_0^\infty \frac{e^{-\eta z/a}\left[J_0(\eta r/a)Y_0(\eta) - J_0(\eta)Y_0(\eta r/a)\right]}{K_-(-i\eta/a)\left[J_0^2(\eta) + Y_0^2(\eta)\right](\alpha + i\eta/a)}\,d\eta$$

for $a \le r < \infty$, $0 < z < \infty$. Figure 5.1.3 illustrates the solution when $\alpha a = 3 - 0.01\,i$.

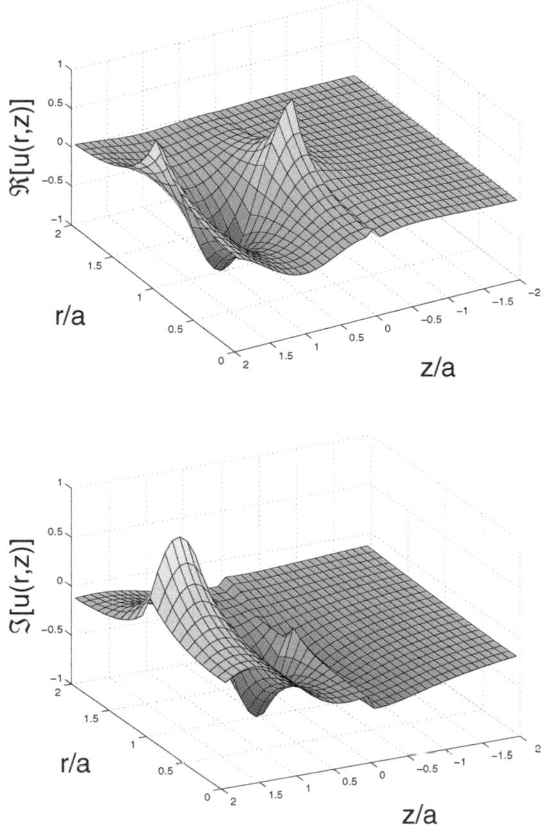

Figure 5.1.3: A plot of the solution to Equation 5.1.61 through Equation 5.1.64 when $\alpha a = 3 - 0.01\,i$.

• Example 5.1.4

Let us solve Laplace's equation[14]

$$\frac{1}{r}\frac{\partial}{\partial r}\left(r\frac{\partial u}{\partial r}\right) + \frac{\partial^2 u}{\partial z^2} = 0, \qquad 0 \le r < b, \quad -\infty < z < \infty, \tag{5.1.83}$$

with the boundary conditions

$$\lim_{z\to-\infty} u(r,z) \to 0, \qquad 0 \le r < b, \tag{5.1.84}$$

$$\lim_{z\to\infty} u(r,z) \to 0, \qquad 0 \le r < b, \tag{5.1.85}$$

[14] Adapted from Lebedev, N. N., and I. P. Skal'skaya, 1960: Electrostatic field of an electron lens consisting of two coaxial cylinders. *Sov. Tech. Phys.*, **5**, 443–450.

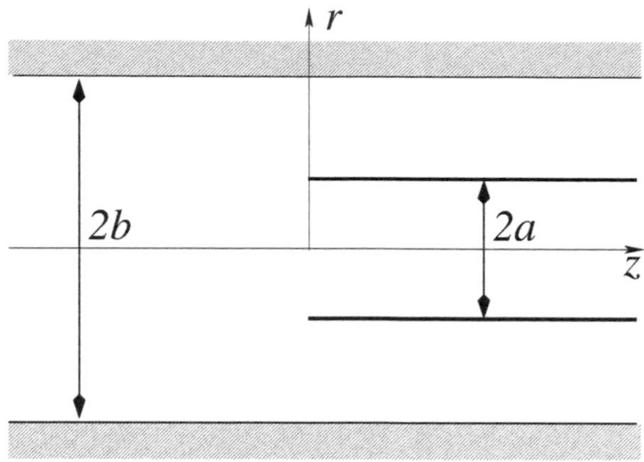

Figure 5.1.4: Schematic of the geometry in Example 5.1.4.

$$u(b,z) = 0, \quad u(a^-,z) = u(a^+,z), \qquad -\infty < z < \infty, \tag{5.1.86}$$

and

$$\begin{cases} u_r(a^-,z) = u_r(a^+,z), & -\infty < z < 0, \\ u(a,z) = Ve^{-\alpha z}, & 0 < z < \infty, \end{cases} \tag{5.1.87}$$

with $\alpha > 0$ and $b > a$. Figure 5.1.4 illustrates the geometry of our problem.

We begin by observing that the solution to Equation 5.1.83 is

$$u(r,z) = \frac{1}{2\pi}\int_{-\infty}^{\infty} A(k)\frac{I_0(kr)}{I_0(ka)}e^{ikz}\,dk, \qquad 0 \le r < a, \tag{5.1.88}$$

and

$$u(r,z) = \frac{1}{2\pi}\int_{-\infty}^{\infty} A(k)\frac{I_0(kr)K_0(kb) - I_0(kb)K_0(kr)}{I_0(ka)K_0(kb) - I_0(kb)K_0(ka)}e^{ikz}\,dk, \qquad a < r < b. \tag{5.1.89}$$

Note that Equation 5.1.88 and Equation 5.1.89 satisfy not only Laplace's equation, but also the boundary conditions as $|z| \to \infty$ and $u(b,z) = 0$. Because

$$e^{-\alpha z}H(z) = \frac{1}{2\pi i}\int_{-\infty}^{\infty} \frac{e^{ikz}}{k - i\alpha}\,dk, \tag{5.1.90}$$

the boundary condition given by Equation 5.1.87 yields the dual integral equations

$$\int_{-\infty}^{\infty} A(k)K(k)e^{ikz}\,dk = 0, \qquad -\infty < z < 0, \tag{5.1.91}$$

and

$$\int_{-\infty}^{\infty} \left[A(k) - \frac{V}{\alpha + ik}\right] e^{ikz}\,dk = 0, \qquad 0 < z < \infty, \tag{5.1.92}$$

where

$$K(k) = \frac{\ln(a/b)\, I_0(kb)}{I_0(ka)\,[I_0(ka)K_0(kb) - I_0(kb)K_0(ka)]}. \tag{5.1.93}$$

In the derivation of Equation 5.1.91 we used the Wronskian[15] involving $I_0(\cdot)$ and $K_0(\cdot)$. We also multiplied Equation 5.1.93 by $\ln(a/b)$ so that $K(0) = 1$.

Before we solve the integral equations, Equation 5.1.91 and Equation 5.1.92, let us examine the singularities in $K(k)$. There are two sources: the zeros of $I_0(ka)$ and $I_0(ka)K_0(kb) - I_0(kb)K_0(ka)$. If we denote the nth zero by $i\gamma_n/a$ and $i\delta_n/(b-a)$, respectively, then γ_n and δ_n are given by the nth root of $J_0(\gamma) = 0$ and

$$J_0\left(\frac{a\delta}{b-a}\right) Y_0\left(\frac{b\delta}{b-a}\right) - J_0\left(\frac{b\delta}{b-a}\right) Y_0\left(\frac{a\delta}{b-a}\right) = 0. \tag{5.1.94}$$

Asymptotic analysis reveals that for large n,

$$\gamma_n = n\pi - \frac{\pi}{4} + O(n^{-1}), \quad \text{and} \quad \delta_n = n\pi + O(n^{-1}). \tag{5.1.95}$$

Let us now turn to the factorization of $K(k)$. A straightforward application of the infinite product theorem yields

$$K(w) = K_+(w)K_-(w), \tag{5.1.96}$$

where

$$K_+(w) = \frac{\displaystyle\prod_{n=1}^{\infty}(1 - iwb/\gamma_n)e^{iwb/\gamma_n}}{\displaystyle\prod_{n=1}^{\infty}(1 - iwa/\gamma_n)e^{iwa/\gamma_n}\prod_{n=1}^{\infty}[1 - iw(b-a)/\delta_n]e^{iw(b-a)/\delta_n}}, \tag{5.1.97}$$

and $K_-(w) = K_+(-w)$. An alternative factorization is

$$K_+(w) = \frac{\displaystyle\prod_{n=1}^{\infty}(1 - iwb/\gamma_n)e^{iwb/\gamma_n}\exp\left\{\frac{iw}{\pi}\left[c + \pi(b-a)\sum_{n=1}^{\infty}\left(\frac{1}{\delta_n} - \frac{1}{\gamma_n}\right)\right]\right\}}{\displaystyle\prod_{n=1}^{\infty}(1 - iwa/\gamma_n)e^{iwa/\gamma_n}\prod_{n=1}^{\infty}[1 - iw(b-a)/\delta_n]e^{iw(b-a)/\delta_n}}, \tag{5.1.98}$$

where $c = (b-a)\ln(b-a) + a\ln(a) - b\ln(b)$. The advantage of using Equation 5.1.98 over Equation 5.1.97 is that $K_\pm(w)$ increases algebraically toward infinity:

$$K_\pm(w) \approx \sqrt{\pm 2ai\ln(b/a)w}, \qquad |w| \to \infty. \tag{5.1.99}$$

---

15 Gradshteyn, I. S., and I. M. Ryzhik, 1965: *Table of Integrals, Series and Products*. Academic Press, Formula 8.474.

We now use this knowledge of $K(k)$ to construct a solution to Equation 5.1.91 and Equation 5.1.92. Following the same reasoning given in Example 5.1.2, $K(k)A(k)$ must be analytic in the lower half-plane $\Im(k) \le 0$. Similarly, the expression $A(k) - V/(\alpha + ik)$ must be analytic in the upper half-plane $\Im(k) \ge 0$. If we set

$$A(k) = \frac{VK_+(\alpha i)}{(\alpha + ik)K_+(k)}, \tag{5.1.100}$$

then

$$K(k)A(k) = \frac{VK_+(\alpha i)K_-(k)}{\alpha + ik} \tag{5.1.101}$$

is clearly analytic in the lower half-plane $\Im(k) \le 0$. Indeed, it decays as $1/\sqrt{w}$ as $|w| \to \infty$. Furthermore,

$$A(k) + \frac{V}{ik + \alpha} = \frac{V}{ik + \alpha}\left[1 - \frac{K_+(\alpha i)}{K_+(k)}\right] \tag{5.1.102}$$

is analytic in the upper half-plane $\Im(k) \ge 0$ and decays as $1/w$ as $|w| \to \infty$. Substituting Equation 5.1.100 into Equation 5.1.88 and Equation 5.1.89, we have that

$$u(r,z) = \frac{VK_+(i\alpha)}{2\pi i} \int_{-\infty}^{\infty} \frac{I_0(kr)}{I_0(ka)} \frac{e^{ikz}}{K_+(k)(k - i\alpha)}\, dk \tag{5.1.103}$$

for $0 \le r \le a$; and

$$u(r,z) = \frac{VK_+(\alpha i)}{2\pi i} \int_{-\infty}^{\infty} \frac{I_0(kr)K_0(kb) - I_0(kb)K_0(kr)}{I_0(ka)K_0(kb) - I_0(kb)K_0(ka)} \frac{e^{ikz}}{K_+(k)(k - i\alpha)}\, dk \tag{5.1.104}$$

for $a \le r \le b$. Applying the residue theorem to Equation 5.1.103 and Equation 5.1.104, we find for the special case of $\alpha = 0$ that

$$\frac{u(r,z)}{V} = 1 - \sum_{n=1}^{\infty} \frac{J_0(\gamma_n r/a) e^{-\gamma_n z/a}}{\gamma_n J_1(\gamma_n) K_+(i\gamma_n/a)} \tag{5.1.105}$$

for $0 \le r \le a$ and $0 < z < \infty$,

$$\frac{u(r,z)}{V} = \frac{1}{\ln(b/a)} \sum_{n=1}^{\infty} \frac{K_+(i\gamma_n/b) e^{\gamma_n z/b}}{\gamma_n^2 J_1^2(\gamma_n)} J_0\left(\frac{\gamma_n a}{b}\right) J_0\left(\frac{\gamma_n r}{b}\right) \tag{5.1.106}$$

for $0 \le r \le b$ and $-\infty < z < 0$, and

$$\begin{aligned}\frac{u(r,z)}{V} &= \frac{\ln(b/r)}{\ln(b/a)} - \frac{\pi}{2} \sum_{n=1}^{\infty} \frac{J_0[\delta_n a/(b-a)] J_0[\delta_n b/(b-a)]}{K_+[i\delta_n/(b-a)]} e^{-\delta_n z/(b-a)} \\ &\quad \times \frac{J_0[\delta_n r/(b-a)] Y_0[\delta_n b/(b-a)] - J_0[\delta_n b/(b-a)] Y_0[\delta_n r/(b-a)]}{J_0^2[\delta_n b/(b-a)] - J_0^2[\delta_n a/(b-a)]}\end{aligned} \tag{5.1.107}$$

for $a \leq r \leq b$ and $0 < z < \infty$.

The MATLAB® code for computing this Wiener-Hopf solution begins by calculating some useful constants:

```
c = (b-a)*log(b-a) + a*log(a) - b*log(b);
euler = -psi(1); S = 0.5 - 3*log(2)/pi;
```

Next we compute γ_n and δ_n using Newton's method. The first guess is provided by Equation 5.1.95. The MATLAB code is

```
for m = 1:iprod
k = m*pi - 0.25*pi;
for n = 0:100
F_prime = - besselj(1,k); F = besselj(0,k);
k = k - F / F_prime;
end
gamma_n(m) = k;
k = m*pi; f1 = a/(b-a); f2 = b/(b-a);
for n = 0:100
F_prime = - f1*besselj(1,f1*k)*bessely(0,f2*k) ...
          - f2*besselj(0,f1*k)*bessely(1,f2*k) ...
          + f2*besselj(1,f1*k)*bessely(0,f2*k) ...
          + f1*besselj(0,f1*k)*bessely(1,f2*k) ;
F = besselj(0,f1*k)*bessely(0,f2*k) ...
  - besselj(0,f2*k)*bessely(0,f1*k) ;
k = k - F / F_prime;
end
delta_n(m) = k;
end
```

Once we find δ_n and γ_n, we turn to $K_+(\cdot)$. Because we cannot compute any expression with an infinite number of multiplications, we truncate it to just `iprod` terms. Following Lebedev and Skal'skaya, we rewrite the various $K_+(\cdot)$ in terms of an universal function $f(x)$ as follows:

$$K_+\left(\frac{i\gamma_m}{a}\right) = f\left(\frac{b-a}{a}\gamma_m\right), \tag{5.1.108}$$

$$K_+\left(\frac{i\gamma_m}{b}\right) = f\left(\frac{b-a}{b}\gamma_m\right), \tag{5.1.109}$$

and

$$K_+\left(\frac{i\delta_m}{b-a}\right) = f(\delta_m), \tag{5.1.110}$$

where

$$f(x) = \frac{P[bx/(b-a)]\exp\{-x[a/(b-a) + \pi S]/\pi\}}{P[ax/(b-a)]Q(x)}, \tag{5.1.111}$$

$$P(x) = \prod_{n=1}^{\infty} \left(1 + \frac{x}{\gamma_n}\right) e^{-x/\gamma_n}, \tag{5.1.112}$$

$$Q(x) = \prod_{n=1}^{\infty} \left(1 + \frac{x}{\delta_n}\right) e^{-x/\delta_n}, \tag{5.1.113}$$

and

$$S = \sum_{n=1}^{\infty} \left(\frac{1}{\delta_n} - \frac{1}{\gamma_n}\right). \tag{5.1.114}$$

However, instead of computing $P(x)$, $Q(x)$ and S from Equation 5.1.112 through Equation 5.1.114, we use the properties of logarithms to reexpress these quantities as

$$\begin{aligned} \ln[P(x)] = \ln\left[\frac{\Gamma(3/4)}{\Gamma(3/4 + x/\pi)}\right] - \frac{x}{\pi}\left[\gamma - \frac{\pi}{2} + 3\ln(2)\right] \\ + \sum_{n=1}^{\infty}\left[\ln\left(1 + \frac{x}{\gamma_n}\right) - \frac{x}{\gamma_n} - \ln\left(1 + \frac{x}{\gamma_n'}\right) + \frac{x}{\gamma_n'}\right], \end{aligned} \tag{5.1.115}$$

$$\begin{aligned} \ln[Q(x)] = -\frac{\gamma x}{\pi} - \ln\left[\Gamma\left(1 + \frac{x}{\pi}\right)\right] \\ + \sum_{n=1}^{\infty}\left[\ln\left(1 + \frac{x}{\delta_n}\right) - \frac{x}{\delta_n} - \ln\left(1 + \frac{x}{n\pi}\right) + \frac{x}{n\pi}\right], \end{aligned} \tag{5.1.116}$$

$$S = \frac{1}{2} - \frac{3}{\pi}\ln(2) + \sum_{n=1}^{\infty}\left(\frac{1}{\delta_n} - \frac{1}{n\pi} - \frac{1}{\gamma_n} + \frac{1}{\gamma_n'}\right), \tag{5.1.117}$$

and $\gamma_n' = n\pi - \pi/4$. The corresponding MATLAB code is

```
% Compute S. Used in computing K_+(k)
for m = 1:iprod
S = S + 1/delta_n(m) - 1/(m*pi)
  - 1/gamma_n(m) + 1/(m*pi-0.25*pi);
end

%          Compute K_+(i gamma_n/a).  Call it K1(m).
for m = 1:iprod
x1 = b*gamma_n(m)/a; x2 = gamma_n(m); x3 = (b-a)*gamma_n(m)/a;
lnP1 = log(gamma(0.75)/gamma(0.75+x1/pi)) ...
```

```
     - x1*(euler-0.5*pi+3*log(2))/pi;
lnP2 = log(gamma(0.75)/gamma(0.75+x2/pi)) ...
     - x2*(euler-0.5*pi+3*log(2))/pi;
lnQ = - euler*x3/pi-log(gamma(1+x3/pi));
for n = 1:iprod
gamma_p = n*pi - 0.25*pi;
lnP1 = lnP1 + log(1+x1/gamma_n(n))-x1/gamma_n(n) ...
     - log(1+x1/gamma_p)+x1/gamma_p;
lnP2 = lnP2 + log(1+x2/gamma_n(n))-x2/gamma_n(n) ...
     - log(1+x2/gamma_p)+x2/gamma_p;
lnQ = lnQ + log(1+x3/delta_n(n))-x3/delta_n(n) ...
    - log(1+x3/(n*pi))+x3/(n*pi);
end
K1(m) = exp(lnP1)*exp(-x3*(c/(b-a)+pi*S)/pi) ...
    / (exp(lnP2)*exp(lnQ));

% ***********************************************************
%         Compute $K_+(i\gamma_n/b)$.  Call it K2(m)
% ***********************************************************

x1 = gamma_n(m); x2 =a*gamma_n(m)/b; x3 = (b-a)*gamma_n(m)/b;
lnP1 = log(gamma(0.75)/gamma(0.75+x1/pi)) ...
     - x1*(euler-0.5*pi+3*log(2))/pi;
lnP2 = log(gamma(0.75)/gamma(0.75+x2/pi)) ...
     - x2*(euler-0.5*pi+3*log(2))/pi;
lnQ = - euler*x3/pi - log(gamma(1+x3/pi));

for n = 1:iprod
gamma_p = n*pi - 0.25*pi;
lnP1 = lnP1 + log(1+x1/gamma_n(n)) - x1/gamma_n(n) ...
     - log(1+x1/gamma_p) + x1/gamma_p;
lnP2 = lnP2 + log(1+x2/gamma_n(n)) - x2/gamma_n(n) ...
     - log(1+x2/gamma_p) + x2/gamma_p;
lnQ = lnQ + log(1+x3/delta_n(n)) - x3/delta_n(n) ...
    - log(1+x3/(n*pi)) + x3/(n*pi);
end

K2(m) = exp(lnP1)*exp(-x3*(c/(b-a)+pi*S)/pi) ...
      / (exp(lnP2)*exp(lnQ));

% ***********************************************************
%         Compute $K_+(i\delta_n/(b-a))$.  Call it K3(m)
% ***********************************************************

x1 = b*delta_n(m)/(b-a); x2 = a*delta_n(m)/(b-a);
```

```
x3 = delta_n(m);
lnP1 = log(gamma(0.75)/gamma(0.75+x1/pi)) ...
     - x1*(euler - 0.5*pi + 3*log(2))/pi;
lnP2 = log(gamma(0.75)/gamma(0.75+x2/pi)) ...
     - x2*(euler - 0.5*pi + 3*log(2))/pi;
lnQ = - euler*x3/pi - log(gamma(1+x3/pi));

for n = 1:iprod
gamma_p = n*pi - 0.25*pi;
lnP1 = lnP1 + log(1+x1/gamma_n(n)) - x1/gamma_n(n) ...
     - log(1+x1/gamma_p) + x1/gamma_p;
lnP2 = lnP2 + log(1+x2/gamma_n(n)) - x2/gamma_n(n) ...
     - log(1+x2/gamma_p) + x2/gamma_p;
lnQ = lnQ + log(1+x3/delta_n(n)) - x3/delta_n(n) ...
    - log(1+x3/(n*pi)) + x3/(n*pi);
end
K3(m) = exp(lnP1)*exp(-x3*(c/(b-a)+pi*S)/pi) ...
      / (exp(lnP2)*exp(lnQ));
end
```

For a given r and z, the solution $u(r, z)$ is computed using the MATLAB code

```
% ***********************************************************
%                  Equation 5.1.105
% ***********************************************************

if ( (z>0) & (r<=1))
u(ii,jj) = 1;
for n = 1:iprod
num = besselj(0,gamma_n(n)*r)*exp(-gamma_n(n)*z);
denom = gamma_n(n)*besselj(1,gamma_n(n))*K1(n);
u(ii,jj) = u(ii,jj) - num/denom;
end; end

% ***********************************************************
%                  Equation 5.1.106
% ***********************************************************

if (z<0)
u(ii,jj) = 0;
for n = 1:iprod
num = K2(n)*exp(a*gamma_n(n)*z/b)*besselj(0,a*gamma_n(n)/b) ...
    * besselj(0,a*r*gamma_n(n)/b);
denom = gamma_n(n)*gamma_n(n)*besselj(1,gamma_n(n))^2;
u(ii,jj) = u(ii,jj) + num/denom;
```

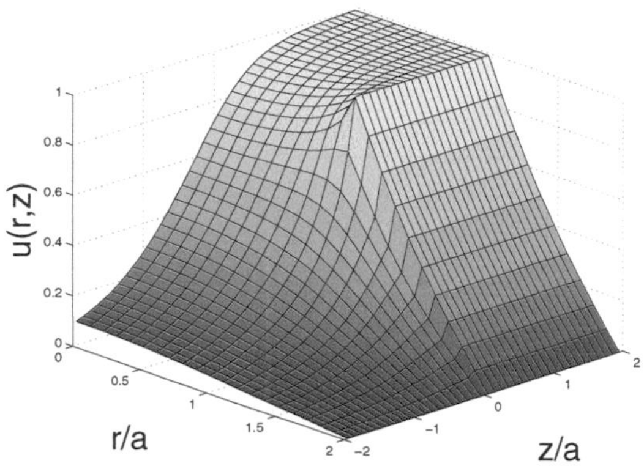

Figure 5.1.5: The solution of Laplace's equation with the mixed boundary conditions given by Equation 5.1.84 through Equation 5.1.87 when $b/a = 2$.

```
end
u(ii,jj) = u(ii,jj)/log(b/a);
end

% ***********************************************************
%                  Equation 5.1.107
% ***********************************************************

if ( (z>0) & (1<=r))
u(ii,jj) = log(b/r)/log(b/a);
for n = 1:iprod
num = besselj(0,a*delta_n(n)/(b-a)) ...
    * besselj(0,b*delta_n(n)/(b-a)) ...
    * (besselj(0,a*delta_n(n)*r/(b-a)) ...
    * bessely(0,b*delta_n(n)/(b-a)) ...
    - besselj(0,b*delta_n(n)/(b-a)) ...
    * bessely(0,a*r*delta_n(n)/(b-a)));
denom = K3(n)*(besselj(0,b*delta_n(n)/(b-a))^2 ...
      -besselj(0,a*delta_n(n)/(b-a))^2);
u(ii,jj) = u(ii,jj) - 0.5*pi*num*exp(-a*delta_n(n)/(b-a))/denom;
end;end
```

Figure 5.1.5 illustrates this solution.

• Example 5.1.5

The Wiener-Hopf technique is often applied to diffraction problems. To illustrate this in a relatively simple form, consider an infinitely long channel

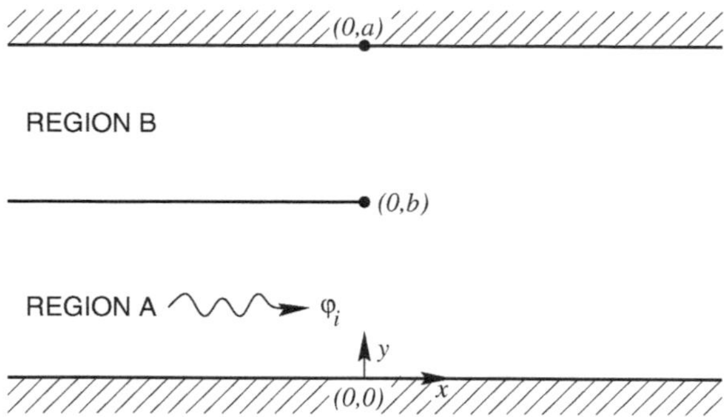

Figure 5.1.6: Schematic of the rotating channel in which Kelvin waves are diffracted.

$-\infty < x < \infty$, $0 < y < a$ rotating on a flat plate[16] filled with an inviscid, homogeneous fluid of uniform depth H. Within the channel, we have another plate of infinitesimal thickness located at $x < 0$ and $y = b$. See Figure 5.1.6. The shallow-water equations govern the motion of the fluid:

$$-i\omega u - fv = -\frac{\partial h}{\partial x}, \tag{5.1.118}$$

$$-i\omega v + fu = -\frac{\partial h}{\partial y}, \tag{5.1.119}$$

and

$$-i\omega h + gH\left(\frac{\partial u}{\partial x} + \frac{\partial v}{\partial y}\right) = 0, \tag{5.1.120}$$

where u and v are the velocities in the x and y directions, respectively, h is the deviation of the free surface from its average height H, g is the gravitational acceleration and f is one-half of the angular velocity at which it rotates. All motions within the fluid behave as $e^{-i\omega t}$.

A little algebra shows that we can eliminate u and v and obtain the Helmholtz equation:

$$\frac{\partial^2 h}{\partial x^2} + \frac{\partial^2 h}{\partial y^2} + k^2 h = 0, \tag{5.1.121}$$

where

$$(\omega^2 - f^2)u = -i\omega\frac{\partial h}{\partial x} + f\frac{\partial h}{\partial y}, \tag{5.1.122}$$

$$(\omega^2 - f^2)v = -i\omega\frac{\partial h}{\partial y} - f\frac{\partial h}{\partial x}, \tag{5.1.123}$$

---

[16] See Kapoulitsas, G. M., 1980: Scattering of long waves in a rotating bifurcated channel. *Int. J. Theoret. Phys.*, **19**, 773–788.

and $k^2 = (\omega^2 - f^2)/(gH)$ assuming that $\omega > f$. We solve the problem when an incident wave of the form

$$h = \phi_i = \exp[(i\omega x - fy)/c] \tag{5.1.124}$$

$$= \exp\{k[ix\cosh(\beta) - y\sinh(\beta)]\}, \tag{5.1.125}$$

a so-called "Kelvin wave," propagates toward the origin from $-\infty$ within the lower channel A. See Figure 5.1.6. We have introduced β such that $f = kc\sinh(\beta)$, $\omega = kc\cosh(\beta)$ and $c^2 = gH$.

The first step in the Wiener-Hopf technique is to write the solution as a sum of the incident wave plus some correction ϕ that represents the reflected and transmitted waves. For example, in Region A, h consists of $\phi_i + \phi$ whereas in Region B we have only ϕ. Because ϕ_i satisfies Equation 5.1.121, so must ϕ. Furthermore, ϕ must satisfy certain boundary conditions. Because of the rigid walls, v must vanish along them; this yields

$$\frac{\partial \phi}{\partial y} - i\tanh(\beta)\frac{\partial \phi}{\partial x} = 0 \tag{5.1.126}$$

along $-\infty < x < \infty$, $y = 0, a$ and $x < 0$, $y = b^{\pm}$. Furthermore, because the partition separating Region A from Region B is infinitesimally thin, we must have continuity of v across that boundary; this gives

$$\left[\frac{\partial \phi}{\partial y} - i\tanh(\beta)\frac{\partial \phi}{\partial x}\right]_{y=b^-} = \left[\frac{\partial \phi}{\partial y} - i\tanh(\beta)\frac{\partial \phi}{\partial x}\right]_{y=b^+} \tag{5.1.127}$$

for $-\infty < x < \infty$. Finally, to prevent infinite velocities in the right half of the channel, h must be continuous at $z = b$, or

$$\phi(x, b^-) + \phi_i(x, b) = \phi(x, b^+) \tag{5.1.128}$$

for $x > 0$.

An important assumption in the Wiener-Hopf technique concerns the so-called "edge conditions" at the edge point $(0, b)$; namely, that

$$\phi = O(1) \quad \text{as} \quad x \to 0^{\pm} \quad \text{and} \quad y = b, \tag{5.1.129}$$

and

$$\frac{\partial \phi}{\partial y} = O(x^{-1/2}) \quad \text{as} \quad x \to 0^{\pm} \quad \text{and} \quad y = b. \tag{5.1.130}$$

These conditions are necessary to guarantee the uniqueness of the solution because the edge point is a geometric singularity. Another assumption introduces dissipation by allowing ω to have a small, positive imaginary part. We can either view this as merely reflecting reality or ensuring that we satisfy the Sommerfeld radiation condition that energy must radiate to infinity. As a result of this complex form of ω, k must also be complex with a small, positive imaginary part k_2.

We solve Equation 5.1.121, as well as Equation 5.1.126 through Equation 5.1.128, by Fourier transforms. Let us define the double-sided Fourier transform of $\phi(x, y)$ by

$$\Phi(\alpha, y) = \int_{-\infty}^{\infty} \phi(x, y) e^{i\alpha x}\, dx, \qquad |\tau| < \tau_0, \tag{5.1.131}$$

as well as the one-sided Fourier transforms

$$\Phi_-(\alpha, y) = \int_{-\infty}^{0} \phi(x, y) e^{i\alpha x}\, dx, \qquad \tau < \tau_0, \tag{5.1.132}$$

and

$$\Phi_+(\alpha, y) = \int_{0}^{\infty} \phi(x, y) e^{i\alpha x}\, dx, \qquad -\tau_0 < \tau, \tag{5.1.133}$$

where $\alpha = \sigma + i\tau$ and σ, τ are real. Clearly,

$$\Phi_-(\alpha, y) + \Phi_+(\alpha, y) = \Phi(\alpha, y). \tag{5.1.134}$$

Note that Equation 5.1.131 through Equation 5.1.133 are analytic in a common strip in the complex α-plane.

Taking the double-sided Fourier transform of Equation 5.1.121,

$$\frac{d^2\Phi}{dy^2} - \gamma^2 \Phi = 0, \tag{5.1.135}$$

where $\gamma = \sqrt{\alpha^2 - k^2}$. In general,

$$\Phi(\alpha, y) = \begin{cases} A(\alpha) e^{-\gamma y} + B(\alpha) e^{\gamma y}, & \text{if} \quad 0 \le y \le b, \\ C(\alpha) e^{-\gamma y} + D(\alpha) e^{\gamma y}, & \text{if} \quad b \le y \le a. \end{cases} \tag{5.1.136}$$

Taking the double-sided Fourier transform of Equation 5.1.126 for the conditions on $y = 0, a$,

$$\Phi'(\alpha, 0) - \alpha \tanh(\beta) \Phi(\alpha, 0) = 0 \tag{5.1.137}$$

and

$$\Phi'(\alpha, a) - \alpha \tanh(\beta) \Phi(\alpha, a) = 0. \tag{5.1.138}$$

Similarly, from Equation 5.1.127,

$$\Phi'(\alpha, b^-) - \alpha \tanh(\beta) \Phi(\alpha, b^-) = \Phi'(\alpha, b^+) - \alpha \tanh(\beta) \Phi(\alpha, b^+). \tag{5.1.139}$$

From the boundary conditions given by Equation 5.1.137 through Equation 5.1.139,

$$B = \lambda A, \tag{5.1.140}$$

$$C = -A\frac{\sinh(\gamma b)}{\sinh[\gamma(a-b)]}e^{\gamma a}, \qquad (\mathbf{5.1.141})$$

and

$$D = -\lambda A\frac{\sinh(\gamma b)}{\sinh[\gamma(a-b)]}e^{-\gamma a}, \qquad (\mathbf{5.1.142})$$

where

$$\lambda = \frac{\gamma + \alpha\tanh(\beta)}{\gamma - \alpha\tanh(\beta)}. \qquad (\mathbf{5.1.143})$$

By taking the one-sided Fourier transform of Equation 5.1.126 (from $-\infty$ to 0) for the condition $x < 0$, we obtain two equations:

$$\Phi'_-(\alpha, b^+) - \alpha\tanh(\beta)\Phi_-(\alpha, b^+) = i\tanh(\beta)\phi(0, b), \qquad (\mathbf{5.1.144})$$

and

$$\Phi'_-(\alpha, b^-) - \alpha\tanh(\beta)\Phi_-(\alpha, b^-) = i\tanh(\beta)\phi(0, b), \qquad (\mathbf{5.1.145})$$

because $\phi(0, b^-) = \phi(0, b^+) = \phi(0, b)$. Using Equation 5.1.144 and Equation 5.1.145 in conjunction with Equation 5.1.139,

$$\begin{aligned}\Phi'_+(\alpha, b^-) &- \alpha\tanh(\beta)\Phi_+(\alpha, b^-) + i\tanh(\beta)\phi(0, b)\\ &= \Phi'_+(\alpha, b^+) - \alpha\tanh(\beta)\Phi_+(\alpha, b^+) + i\tanh(\beta)\phi(0, b)\\ &= P_+(\alpha). \qquad (\mathbf{5.1.146})\end{aligned}$$

Finally, the one-sided Fourier transform of Equation 5.1.128 (from 0 to ∞) yields

$$\Phi_+(\alpha, b^-) + \frac{i\exp[-kb\sinh(\beta)]}{\alpha + k\cosh(\beta)} = \Phi_+(\alpha, b^+). \qquad (\mathbf{5.1.147})$$

Then, by Equation 5.1.134, Equation 5.1.136, Equation 5.1.144 through Equation 5.1.146 and the definition of B,

$$P_+(\alpha) = 2A[\gamma + \alpha\tanh(\beta)]\sinh(\gamma b). \qquad (\mathbf{5.1.148})$$

Let us now introduce the function $Q_-(\alpha)$ defined by

$$Q_-(\alpha) = \tfrac{1}{2}[\Phi_-(\alpha, b^-) - \Phi(\alpha, b^+)]. \qquad (\mathbf{5.1.149})$$

From Equation 5.1.136, Equation 5.1.139, Equation 5.1.141, Equation 5.1.143, Equation 5.1.147 and $D = \lambda C e^{-2\gamma a}$,

$$2Q_-(\alpha) - \frac{i\exp[-kb\sinh(\beta)]}{\alpha + k\cosh(\beta)} = \frac{2A\gamma\sinh(\gamma a)}{[\gamma - \alpha\tanh(\beta)]\sinh[\gamma(a-b)]}. \qquad (\mathbf{5.1.150})$$

Eliminating A between Equation 5.1.148 and Equation 5.1.150, we finally obtain a functional equation of the Wiener-Hopf type:

$$2Q_-(\alpha) - \frac{i\exp[-kb\sinh(\beta)]}{\alpha + k\cosh(\beta)} = \frac{P_+(\alpha)}{\gamma^2 - \alpha^2\tanh(\beta)} \times \frac{\gamma\sinh(\gamma a)}{\sinh(\gamma b)\sinh[\gamma(a-b)]}. \quad \mathbf{(5.1.151)}$$

What makes Equation 5.1.151 a functional equation of the Wiener-Hopf type? Note that $Q_-(\alpha)$ is analytic for $\tau < \tau_0$ while $P_+(\alpha)$ is analytic for $\tau > -\tau_0$. In order for Equation 5.1.151 to be true, we must restrict ourselves to the strip $|\tau| < \tau_0$. Thus, the Wiener-Hopf equation contains complex Fourier transforms which are analytic over the common interval of $\tau_- < \tau < \tau_+$, where $Q_-(\alpha)$ is analytic for $\tau < \tau_+$ and $\tau_- < \tau$.

A crucial step in solving the Wiener-Hopf equation, Equation 5.1.151, is the process of factorization. In the previous section, we did this by adding and subtracting out a particular singularity. Here, we will rewrite $M(\alpha)$ in terms of the product $M_+(\alpha)M_-(\alpha)$, where $M_+(\alpha)$ and $M_-(\alpha)$ are analytic and free of zeros in an upper and lower half-planes, respectively. These half-planes share a certain strip of the α-plane in common. Applying the infinite product theorem separately to the numerator and denominator of

$$M(\alpha) = \frac{\sinh(\gamma b)\sinh[\gamma(a-b)]}{\gamma\sinh(\gamma a)}, \quad \mathbf{(5.1.152)}$$

we immediately find that

$$\begin{aligned} M_+(\alpha) &= M_-(-\alpha) \\ &= \left\{\frac{\sin(kb)\sin[k(a-b)]}{k\sin(ka)}\right\}^{1/2} \exp\left\{\frac{\alpha i}{\pi}\left[b\ln\left(\frac{a}{b}\right) + (a-b)\ln\left(\frac{a}{a-b}\right)\right]\right\} \\ &\times \prod_{n=1}^{\infty}\left(1+\frac{\alpha}{\alpha_{nb}}\right)e^{ib\alpha/(n\pi)} \prod_{n=1}^{\infty}\left[1+\frac{\alpha}{\alpha_{n(a-b)}}\right]e^{i(a-b)\alpha/(n\pi)} \\ &\Big/ \prod_{n=1}^{\infty}\left(1+\frac{\alpha}{\alpha_{na}}\right)e^{ia\alpha/(n\pi)}, \end{aligned} \quad \mathbf{(5.1.153)}$$

where $\alpha_{n\ell} = i\sqrt{n^2\pi^2/\ell^2 - k^2}$ and $\ell = a$ or b or $(a-b)$. The square root has a positive real part or a negative imaginary part. Note that in this factorization $M_+(\alpha)$ is analytic and nonzero in the upper half of the α-plane $(-k_2 < \tau)$ while $M_-(\alpha)$ is analytic and nonzero in the lower half of the α-plane $(\tau < k_2)$.

Substituting this factorization into Equation 5.1.151,

$$\begin{aligned} 2[\alpha - k\cosh(\beta)]M_-(\alpha)Q_-(\alpha) &- \frac{ie^{-kb\sinh(\beta)}[\alpha - k\cosh(\beta)]M_-(\alpha)}{\alpha + k\cosh(\beta)} \\ &= \frac{\cosh^2(\beta)}{[\alpha + k\cosh(\beta)]M_+(\alpha)}P_+(\alpha). \end{aligned} \quad \mathbf{(5.1.154)}$$

Next, we note that

$$\frac{[\alpha - k\cosh(\beta)]M_-(\alpha)}{\alpha + k\cosh(\beta)} = \frac{[\alpha - k\cosh(\beta)]M_-(\alpha) + 2k\cosh(\beta)M_-[-k\cosh(\beta)]}{\alpha + k\cosh(\beta)} - \frac{2k\cosh(\beta)M_-[-k\cosh(\beta)]}{\alpha + k\cosh(\beta)}. \quad \textbf{(5.1.155)}$$

Therefore, Equation 5.1.154 becomes

$$\begin{aligned} 2[\alpha - k\cosh(\beta)]M_-(\alpha)Q_-(\alpha) &- i\exp[-kb\sinh(\beta)] \\ &\times \left\{ \frac{[\alpha - k\cosh(\beta)]M_-(\alpha) + 2k\cosh(\beta)M_-[-k\cosh(\beta)]}{\alpha + k\cosh(\beta)} \right\} \\ &= \frac{\cosh^2(\beta)}{[\alpha + k\cosh(\beta)]M_+(\alpha)} P_+(\alpha) \\ &\quad - i\exp[-kb\sinh(\beta)] \frac{2k\cosh(\beta)M_-[-k\cosh(\beta)]}{\alpha + k\cosh(\beta)}. \quad \textbf{(5.1.156)} \end{aligned}$$

The fundamental reason for the factorization and the subsequent algebraic manipulation is the fact that the left side of Equation 5.1.156 is analytic in $-\tau_0 < \tau$, while the right side is analytic in $\tau < \tau_0$. Hence, both sides are analytic on the strip $|\tau| < \tau_0$. Then by analytic continuation it follows that Equation 5.1.156 is defined in the entire α-plane and both sides equal an entire function $p(\alpha)$. To determine $p(\alpha)$, we examine the asymptotic value of Equation 5.1.156 as $|\alpha| \to \infty$ as well as using the edge conditions, Equation 5.1.129 and Equation 5.1.130. Applying Liouville's theorem, $p(\alpha)$ is a constant. Because in the limit of $|\alpha| \to \infty$, $p(\alpha) \to 0$, then $p(\alpha) = 0$. Therefore, from Equation 5.1.156,

$$P_+(\alpha) = \frac{2ikM_+[k\cosh(\beta)]\exp[-kb\sinh(\beta)]}{\cosh(\beta)} M_+(\alpha). \quad \textbf{(5.1.157)}$$

Knowing $P_+(\alpha)$, we find from Equation 5.1.140 through Equation 5.1.144 that

$$A = \frac{EM_+(\alpha)}{[\gamma + \alpha\tanh(\beta)]\sin(\gamma b)}, \quad \textbf{(5.1.158)}$$

$$B = \frac{EM_+(\alpha)}{[\gamma - \alpha\tanh(\beta)]\sin(\gamma b)}, \quad \textbf{(5.1.159)}$$

$$C = -\frac{EM_+(\alpha)e^{\gamma a}}{[\gamma + \alpha\tanh(\beta)]\sin[\gamma(a-b)]}, \quad \textbf{(5.1.160)}$$

and

$$D = -\frac{EM_+(\alpha)e^{-\gamma a}}{[\gamma - \alpha\tanh(\beta)]\sin[\gamma(a-b)]}, \quad \textbf{(5.1.161)}$$

where

$$E = \frac{ikM_+[k\cosh(\beta)]\exp[-kb\sinh(\beta)]}{\cosh(\beta)}. \tag{5.1.162}$$

With these values of A, B, C and D, we have found $\Phi(\alpha, y)$. Therefore, $\phi(x, y)$ follows from the inversion of $\Phi(\alpha, y)$. For example, for $-\infty < x < \infty, 0 \le y \le b$,

$$\phi(x,y) = \frac{E}{2\pi}\int_{-\infty-\epsilon i}^{\infty-\epsilon i} \frac{M_+(\alpha)}{\sinh(\gamma b)}\left[\frac{e^{-\gamma y}}{\gamma + \alpha\tanh(\beta)} + \frac{e^{\gamma y}}{\gamma - \alpha\tanh(\beta)}\right] e^{-i\alpha x}\,d\alpha. \tag{5.1.163}$$

For $x < 0$ we evaluate Equation 5.1.163 by closing the integration along the real axis with an infinite semicircle in the upper half of the α-plane by Jordan's lemma and using the residue theorem. The integrand of Equation 5.1.163 has simple poles at $\gamma b = n\pi$, where $n = \pm1, \pm2, \ldots$ and the zeros of $\gamma \pm \alpha\tanh(\beta)$. Upon applying the residue theorem,

$$\begin{aligned}\phi(x,y) = &-\frac{k\sinh(\beta)M_+^2[k\cosh(\beta)]}{\sinh[kb\sinh(\beta)]}e^{k[-ix\cosh(\beta)+(y-b)\sinh(\beta)]}\\ &+\frac{2\pi iE}{b^2}\sum_{n=1}^{\infty}\frac{(-1)^n nM_+(-\alpha_n)}{\alpha_{nb}[(n\pi/b)^2+\alpha_{nb}^2\tanh^2(\beta)]}\\ &\times\left[\frac{n\pi}{b}\cos\left(\frac{n\pi y}{b}\right)+\alpha_{nb}\tanh(\beta)\sin\left(\frac{n\pi y}{b}\right)\right]e^{-i\alpha_{nb}x},\end{aligned} \tag{5.1.164}$$

where $\alpha_{nb} = i\sqrt{n^2\pi^2/b^2 - k^2}$. The first term of the right side of Equation 5.1.164 represents the reflected Kelvin wave traveling in the channel $(0 \le y \le b, x < 0)$ to the left. The infinite series represents attenuated, stationary modes.

In a similar manner, we apply the residue theorem to obtain the solution in the remaining domains. They are

$$\begin{aligned}\phi(x,y) = &-\frac{\sinh[k(a-b)\sinh(\beta)]}{\sinh[ka\sinh(\beta)]}e^{k[ix\cosh(\beta)-(y+b)\sinh(\beta)]}\\ &-\frac{2iE}{a}\sum_{n=1}^{\infty}\frac{\sin(n\pi b/a)}{\alpha_{na}M_-(\alpha_{na})[(n\pi/a)^2+\alpha_{na}^2\tanh^2(\beta)]}\\ &\times\left[\frac{n\pi}{a}\cos\left(\frac{n\pi y}{a}\right)-\alpha_{na}\tanh(\beta)\sin\left(\frac{n\pi y}{a}\right)\right]e^{i\alpha_{na}x}\end{aligned} \tag{5.1.165}$$

for $0 \le y \le b, 0 < x$, and

$$\begin{aligned}\phi(x,y) = &\frac{k\sinh(\beta)M_+^2[k\cosh(\beta)]}{\sinh[kd\sinh(\beta)]}e^{k[-ix\cosh(\beta)+(y-a-b)\sinh(\beta)]}\\ &-\frac{2\pi iE}{d^2}\sum_{n=1}^{\infty}\frac{(-1)^n nM_+(\alpha_{nd})}{[(n\pi/d)^2+\alpha_{nd}^2\tanh^2(\beta)]}\\ &\times\left\{\frac{n\pi}{d}\cos\left[\frac{n\pi(y-a)}{d}\right]+\alpha_{nd}\tanh(\beta)\sin\left[\frac{n\pi(y-a)}{d}\right]\right\}e^{-i\alpha_{nd}x}\end{aligned} \tag{5.1.166}$$

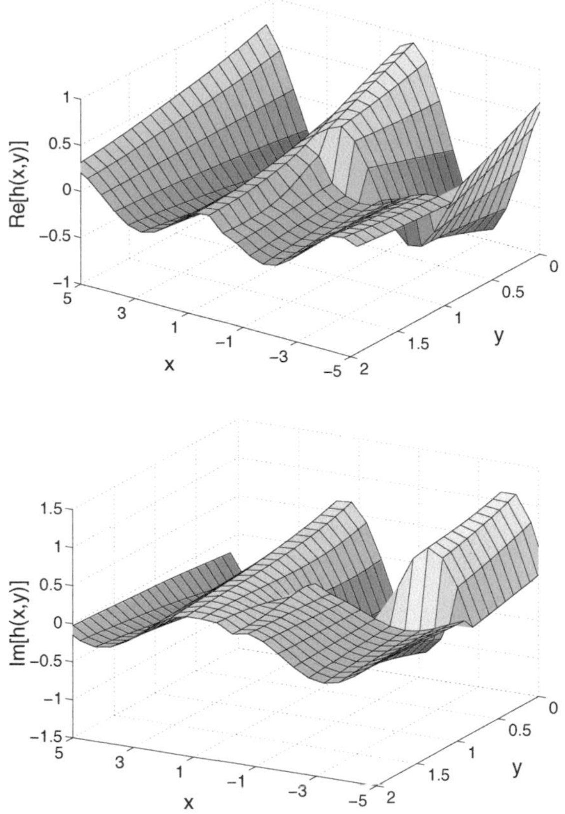

Figure 5.1.7: The real and imaginary parts of the solution to Equation 5.1.121 subject to the boundary conditions given by Equation 5.1.126 through Equation 5.1.128 obtained via the Wiener-Hopf technique when $a = 2$, $b = 1$, $k = 1$ and $\beta = 0.5$.

for $b \le y \le a, x < 0$, where $d = a - b$. Finally, for $b \le y \le a, 0 < x$, $\phi(x, y)$ is given by the sum of $\phi_i(x, y)$ and the solution is given by Equation 5.1.165.

Figure 5.1.7 illustrates the real and imaginary parts of this solution when $a = 2$, $b = 1$, $k = 1$ and $\beta = 0.5$.

• **Example 5.1.6**

Let us solve the biharmonic equation[17]

$$\nabla^4 u = \frac{\partial^4 u}{\partial x^4} + 2\frac{\partial^4 u}{\partial x^2 \partial y^2} + \frac{\partial^4 u}{\partial y^4} = 0, \qquad -\infty < x < \infty, \quad 0 < y < 1, \tag{5.1.167}$$

[17] Taken from Jeong, J.-T., 2001: Slow viscous flow in a partitioned channel. *Phys. Fluids*, **13**, 1577–1582. See also Kim, M.-U., and M. K. Chung, 1984: Two-dimensional slow viscous flow past a plate midway between an infinite channel. *J. Phys. Soc. Japan*, **53**, 156–166.

subject to the boundary conditions

$$u(x,1) = 1, \quad u_y(x,1) = 0, \qquad -\infty < x < \infty, \tag{5.1.168}$$

$$u_y(x,0) = 0, \qquad -\infty < x < \infty, \tag{5.1.169}$$

and

$$\begin{cases} u_{yyy}(x,0) = 0, & -\infty < x < 0, \\ u(x,0) = 0, & 0 < x < \infty. \end{cases} \tag{5.1.170}$$

We begin our analysis by introducing the Fourier transform

$$u(x,y) = 1 + \int_{-\infty}^{\infty} U(k,y)e^{ikx}\,dk \tag{5.1.171}$$

$$= 1 + \int_{-\infty}^{\infty} \Big[A(k)\sinh(ky) + B(k)\cosh(ky) + C(k)y\sinh(ky) + D(y)y\cosh(ky)\Big] e^{ikx}\,dk. \tag{5.1.172}$$

Substituting Equation 5.1.172 into Equation 5.1.168 and Equation 5.1.169, we find that

$$u(x,y) = 1 + \int_{-\infty}^{\infty} \left[\sinh(ky) - \frac{\sinh^2(k) - k^2}{\sinh(k)\cosh(k) + k}\cosh(ky) - ky\cosh(ky) + \frac{ky\,\sinh^2(k)}{\sinh(k)\cosh(k) + k}\sinh(ky)\right] A(k)e^{ikx}\,dk \tag{5.1.173}$$

for $0 < y < 1$. If we now substitute this solution into Equation 5.1.170, we obtain the dual integral equations

$$\int_{-\infty}^{\infty} K(k)A(k)e^{ikx}\,dk = 1, \qquad 0 < x < \infty, \tag{5.1.174}$$

and

$$\int_{-\infty}^{\infty} k^3 A(k)e^{ikx}\,dk = 0, \qquad -\infty < x < 0, \tag{5.1.175}$$

where

$$K(k) = \frac{\sinh^2(k) - k^2}{\sinh(k)\cosh(k) + k} = 2\frac{\sinh^2(k) - k^2}{\sinh(2k) + 2k}. \tag{5.1.176}$$

To solve the dual integral equations, Equation 5.1.174 and Equation 5.1.175, we rewrite them

$$\int_{-\infty}^{\infty} K(k)A(k)e^{ikx}\,dk = \begin{cases} f(x), & -\infty < x < 0, \\ 1, & 0 < x < \infty, \end{cases} \tag{5.1.177}$$

and

$$\int_{-\infty}^{\infty} k^3 A(k) e^{ikx}\, dk = \begin{cases} 0, & -\infty < x < 0, \\ g(x), & 0 < x < \infty, \end{cases} \qquad \textbf{(5.1.178)}$$

where $f(x)$ and $g(x)$ are unknown functions. Taking the Fourier transform of Equation 5.1.177 and Equation 5.1.178 and eliminating $A(k)$ between them, we find that

$$\frac{K(k)}{k^3} G_-(k) = F_+(k) + \frac{1}{2\pi i k}, \qquad \textbf{(5.1.179)}$$

where

$$G_-(k) = \frac{1}{2\pi} \int_0^{\infty} g(x) e^{-ikx}\, dx = k^3 A(k), \qquad \textbf{(5.1.180)}$$

and

$$F_+(k) = \frac{1}{2\pi} \int_{-\infty}^{0} f(x) e^{-ikx}\, dx. \qquad \textbf{(5.1.181)}$$

Here, $F_+(k)$ is an analytic function in the half-plane $\Im(k) > -\epsilon$, where $\epsilon > 0$, while $G_-(k)$ is an analytic function in the half-plane $\Im(k) < 0$.

We begin our solution of Equation 5.1.179 by the Wiener-Hopf technique by noting that we can factor $K(k)$ as follows:

$$K(k) = \frac{k^3}{6} K_+(k) K_-(k), \qquad \textbf{(5.1.182)}$$

where

$$K_+(k) = \prod_{n=1}^{\infty} \frac{(1 + k/k_n)(1 - k/k_n^*)}{(1 + 2k/k_{2n-1})(1 - 2k/k_{2n-1}^*)} = K_-(-k), \qquad \textbf{(5.1.183)}$$

k_n is the nth root of $\sinh^2(k) = k^2$ with $\Re(k_n) > 0$ and $0 < \Im(k_1) < \Im(k_2) < \cdots$. Observe that if k_n is a root, then so are $-k_n$, k_n^* and $-k_n^*$. For $n \gg 1$,

$$k_n \approx \left(n + \tfrac{1}{2}\right) \pi i + \ln\left[\left(n + \tfrac{1}{2}\right) \pi + \sqrt{\left(n + \tfrac{1}{2}\right)^2 \pi^2 - 1}\,\right]. \qquad \textbf{(5.1.184)}$$

Finally, $K_\pm(k) \sim \sqrt{6}(\mp k)^{-3/2}$ as $|k| \to \infty$. Substituting Equation 5.1.182 into Equation 5.1.179 and dividing the resulting equation by $K_+(k)$, we find that Equation 5.1.179 can be written

$$\frac{K_-(k)}{6} G_-(k) - \frac{1}{2\pi i k K_+(0)} = \frac{F_+(k)}{K_+(k)} + \frac{1}{2\pi i k}\left[\frac{1}{K_+(k)} - \frac{1}{K_+(0)}\right]. \qquad \textbf{(5.1.185)}$$

Why have we rewritten Equation 5.1.179 in the form given by Equation 5.1.185? We observe that the left side of Equation 5.1.185 is analytic in the half-plane $\Im(k) < 0$, while the right side of Equation 5.1.185 is analytic in the half-range $\Im(k) > -\epsilon$. Thus, both sides of Equation 5.1.185 are analytic continuations of some entire function $E(k)$. The asymptotic analysis of both sides of Equation 5.1.185 shows that $E(k) \to 0$ as $|k| \to \infty$. Therefore, by Liouville's theorem, $E(k) = 0$ and

$$F_+(k) = \frac{1}{2\pi ik}\left[\frac{K_+(k)}{K_+(0)} - 1\right], \tag{5.1.186}$$

and

$$G_-(k) = \frac{3}{\pi ik K_+(0)K_-(k)}. \tag{5.1.187}$$

Therefore, because $G_-(k) = k^3A(k)$ and defining

$$\begin{aligned}\Psi(k,y) = {} & [\sinh(ky) - ky\cosh(ky)][\sinh(k)\cosh(k) + k] \\ & + ky\sinh^2(k)\sinh(ky) - [\sinh^2(k) - k^2]\cosh(ky),\end{aligned} \tag{5.1.188}$$

we have from Equation 5.1.173 that

$$u(x,y) = 1 + \frac{1}{2\pi i}\int_{-\infty-\epsilon i}^{\infty-\epsilon i} \frac{K_+(k)\Psi(k,y)e^{ikx}}{K_+(0)k[\sinh^2(k) - k^2]}\,dk \tag{5.1.189}$$

$$= 1 + \frac{3}{\pi i}\int_{-\infty-\epsilon i}^{\infty-\epsilon i} \frac{K(k)\Psi(k,y)e^{ikx}}{k^4K_+(0)K_-(k)[\sinh(k)\cosh(k) + k]}\,dk. \tag{5.1.190}$$

The integrals given by Equation 5.1.189 and Equation 5.1.190 can be evaluated by the residue theorem. For $x > 0$, we close the line integral given in Equation 5.1.189 with a semicircle of infinite radius in the upper half-plane and apply the residue theorem. This yields

$$u(x,y) = (3-2y)y^2 + \Re\left\{\sum_{n=1}^{\infty} \frac{K_+(k_n)\Psi(k_n,y)e^{ik_nx}}{K_+(0)k_n[\sinh(k_n)\cosh(k_n) - k_n]}\right\}, \tag{5.1.191}$$

where k_n is the nth zero of $\sinh^2(k) = k^2$. On the other hand, if $x < 0$, we use Equation 5.1.190 and close the line integral with a semicircle of infinite radius in the lower half-plane. Applying the residue theorem,

$$\begin{aligned}u(x,y) = 1 - 6\Re\Bigg\{\sum_{n=1}^{\infty} & \frac{\eta_n y\sinh^2(\eta_n)\sinh(\eta_n y) - [\sinh^2(\eta_n) - \eta_n^2]\cosh(\eta_n y)}{K_+(0)K_-(-\eta_n)\eta_n^4\cosh^2(\eta_n)} \\ & \times e^{-i\eta_n x}\Bigg\},\end{aligned} \tag{5.1.192}$$

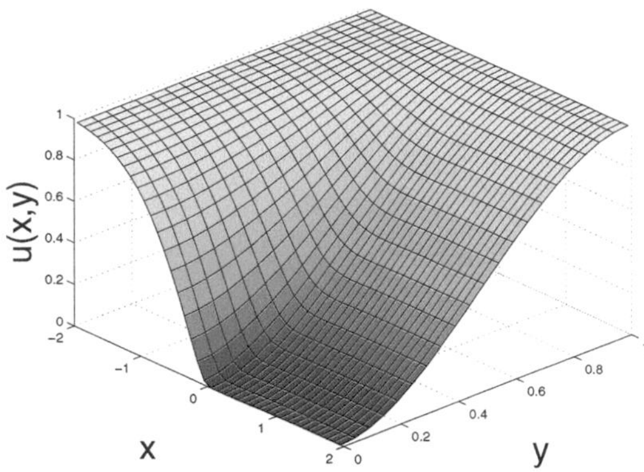

Figure 5.1.8: The solution to the biharmonic equation subject to the boundary conditions given by Equation 5.1.168 through Equation 5.1.170 obtained via the Wiener-Hopf technique.

where η_n is the nth zero of $\sinh(\eta)\cosh(\eta) + \eta = 0$ with positive real and imaginary parts. Figure 5.1.8 illustrates this solution.

Problems

1. Use the Wiener-Hopf technique[18] to solve the mixed boundary value problem

$$\frac{\partial^2 u}{\partial x^2} + \frac{\partial^2 u}{\partial y^2} = 0, \qquad -\infty < x < \infty, \quad 0 < y < 1,$$

with the boundary conditions

$$\lim_{x\to-\infty} u(x,y) \to b - ax, \qquad \lim_{x\to\infty} u(x,y) \to 0, \qquad 0 < y < 1,$$

$$\begin{cases} u_y(x,0) = 0, & x < 0, \\ u(x,0) = 0, & 0 < x, \end{cases}$$

and

$$u_y(x,1) = 0, \qquad -\infty < x < \infty.$$

Step 1: Setting $u(x,y) = b - ax + v(x,y)$, show that the problem becomes

$$\frac{\partial^2 v}{\partial x^2} + \frac{\partial^2 v}{\partial y^2} = 0, \qquad -\infty < x < \infty, \quad 0 < y < 1,$$

---

[18] Adapted from Jeong, J.-T., 2001: Slip boundary condition on an idealized porous wall. *Phys. Fluids*, **13**, 1884–1890.

with the boundary conditions

$$\lim_{x\to-\infty} v(x,y) \to 0, \quad \lim_{x\to\infty} v(x,y) \to ax - b, \qquad 0 < y < 1, \tag{1}$$

$$\begin{cases} v_y(x,0) = 0, & x < 0, \\ v(x,0) = ax - b, & 0 < x, \end{cases} \tag{2}$$

and

$$v_y(x,1) = 0, \qquad -\infty < x < \infty. \tag{3}$$

Step 2: Show that

$$v(x,y) = \int_{-\infty}^{\infty} A(k) \left[\cosh(ky) - \tanh(k)\sinh(ky)\right] e^{ikx}\, dk$$

satisfies the partial differential equation and boundary conditions (1) and (3) in Step 1 if $-\epsilon < \Im(k) < 0$ where $\epsilon > 0$.

Step 3: Using boundary condition (2), show that

$$\int_{-\infty}^{\infty} A(k)e^{ikx}\, dk = ax - b, \qquad 0 < x < \infty,$$

and

$$\int_{-\infty}^{\infty} k\tanh(k)A(k)e^{ikx}\, dk = 0, \qquad -\infty < x < 0.$$

Step 4: By introducing

$$\int_{-\infty}^{\infty} A(k)e^{ikx}\, dk = \begin{cases} ax - b, & 0 < x < \infty, \\ f(x), & -\infty < x < 0, \end{cases}$$

and

$$\int_{-\infty}^{\infty} k\tanh(k)A(k)e^{ikx}\, dk = \begin{cases} g(x), & 0 < x < \infty, \\ 0, & -\infty < x < 0, \end{cases}$$

where $f(x)$ and $g(x)$ are unknown functions, show that

$$A(k) = \frac{bi}{2\pi k} - \frac{a}{2\pi k^2} + F_+(k),$$

and

$$k\tanh(k)A(k) = \frac{1}{2\pi}\int_0^{\infty} g(x)e^{-ikx}\, dx = G_-(k),$$

or

$$K(k)G_-(k) = \frac{bi}{2\pi k} - \frac{a}{2\pi k^2} + F_+(k),$$

where

$$F_+(k) = \frac{1}{2\pi} \int_{-\infty}^{0} f(x)e^{-ikx}\,dx, \qquad \text{and} \qquad K(k) = \frac{\cosh(k)}{k\,\sinh(k)}.$$

Note that $F_+(k)$ is analytic in the half-plane $\Im(k) > -\epsilon$ while $G_-(k)$ is analytic in the half-plane $\Im(k) < 0$.

Step 5: Using the infinite product representation[19] for sinh and cosh, show that

$$K(k) = \frac{K_+(k)K_-(k)}{k^2},$$

where

$$K_+(k) = K_-(-k) = \prod_{n=1}^{\infty} \frac{1 + 2k/[(2n-1)\pi i]}{1 + k/(n\pi i)} = \sqrt{\pi}\frac{\Gamma[1 + k/(\pi i)]}{\Gamma[\frac{1}{2} + k/(\pi i)]},$$

and $\Gamma(\cdot)$ is the gamma function. Note that $K_\pm(k) \sim \sqrt{\mp ki}$ as $|k| \to \infty$.

Step 6: Use the results from Step 5 and show that

$$K_-(k)G_-(k) = \frac{k^2}{K_+(k)}\left[F_+(k) - \frac{a}{2\pi k^2} - \frac{b}{2\pi ik}\right].$$

Note that the right side of the equation is analytic in the upper half-plane $\Im(k) > -\epsilon$, while the left side of the equation is analytic in the lower half-plane $\Im(k) < 0$.

Step 7: Show that each side of the equation in Step 6 is an analytic continuation of some entire function $E(k)$. Use Liouville's theorem to show that $E(k) = -a/[2\pi K_+(0)]$. Therefore,

$$G_-(k) = -\frac{a}{2\pi K_+(0)K_-(k)},$$

and

$$F_+(k) = \frac{a}{2\pi k^2} - \frac{bi}{2\pi k} - \frac{aK_+(k)}{2\pi k^2 K_+(0)}.$$

Step 8: Use the inversion integral and show that

$$\begin{aligned}
u(x,y) &= b - ax \\
&\quad - \frac{a}{2\pi K_+(0)} \int_{-\infty-\epsilon i}^{\infty-\epsilon i} K_+(k)\frac{\cosh(k)\cosh(ky) - \sinh(k)\sinh(ky)}{k^2\cosh(k)} e^{ikx}\,dk \\
&= b - ax \\
&\quad - \frac{a}{2\pi K_+(0)} \int_{-\infty-\epsilon i}^{\infty-\epsilon i} \frac{\cosh(k)\cosh(ky) - \sinh(k)\sinh(ky)}{K_-(k)\,k\,\sinh(k)} e^{ikx}\,dk.
\end{aligned}$$

---

19 See, for example, Gradshteyn and Ryzhik, op. cit., Formulas 1.431.2 and 1.431.4.

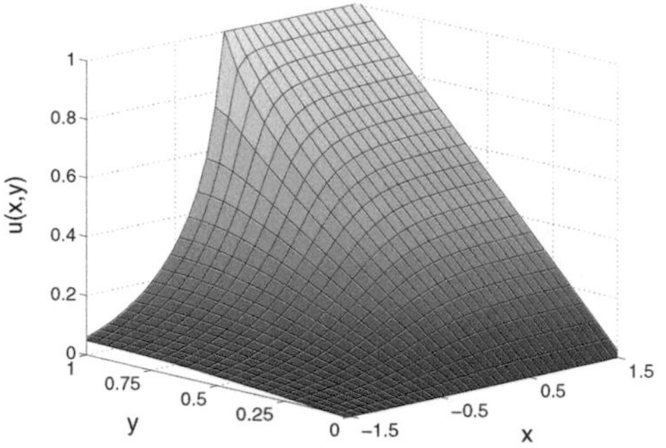

Problem 1

The equation on the first line should be used to find the solution when $x > 0$, while the equation on the second line gives the solution for $x < 0$.

Step 9: Use the residue theorem and show that

$$u(x,y) = \frac{a}{\pi^{3/2}} \sum_{n=1}^{\infty} \frac{\Gamma(n-\frac{1}{2})}{(n-\frac{1}{2})\,\Gamma(n)} \sin\left[\left(n-\tfrac{1}{2}\right)\pi y\right] \exp\left[-\left(n-\tfrac{1}{2}\right)\pi x\right]$$

if $0 < x$, while

$$u(x,y) = b - ax - \frac{a}{\pi^{3/2}} \sum_{n=1}^{\infty} \frac{\Gamma(n+\frac{1}{2})}{n\Gamma(n+1)} \cos(n\pi y) e^{n\pi x}$$

for $x < 0$. Hint: $\Gamma(z+1) = z\Gamma(z)$. The figure labeled Problem 1 illustrates this solution $u(x,y)$ when $b/a = \ln(4)/\pi$. This corresponds to the case where $F_+(k)$ is analytic at $k = 0$.

2. Use the Wiener-Hopf technique to solve the mixed boundary value problem

$$\frac{\partial^2 u}{\partial x^2} + \frac{\partial^2 u}{\partial y^2} = 0, \qquad -\infty < x < \infty, \quad 0 < y < 1,$$

with the boundary conditions

$$u(x,0) = 0, \qquad -\infty < x < \infty,$$

and

$$\begin{cases} u_y(x,1) = 0, & x < 0, \\ u(x,1) = 1, & 0 < x. \end{cases}$$

Step 1: Assuming that $|u(x,1)|$ is bounded by $e^{\epsilon x}$, $0 < \epsilon \ll 1$, as $x \to -\infty$, let us define the following Fourier transforms:

$$U(k,y) = \int_{-\infty}^{\infty} u(x,y) e^{ikx}\,dx, \qquad U_+(k,y) = \int_{0}^{\infty} u(x,y) e^{ikx}\,dx,$$

and

$$U_-(k,y) = \int_{-\infty}^{0} u(x,y)e^{ikx}\,dx,$$

so that $U(k,y) = U_+(k,y) + U_-(k,y)$. Here, $U_+(k,y)$ is analytic in the half-space $\Im(k) > 0$, while $U_-(k,y)$ is analytic in the half-space $\Im(k) < \epsilon$. Take the Fourier transform of the partial differential equation and the first boundary condition and show that it becomes the boundary value problem

$$\frac{d^2U}{dy^2} - k^2U = 0, \qquad 0 < y < 1,$$

with $U(k,0) = 0$.

Step 2: Show that the solution to the boundary value problem is $U(k,y) = A(k)\sinh(ky)$.

Step 3: From the boundary conditions along $y = 1$, show that

$$\sinh(k)A(k) = L_-(k) + \frac{i}{k}, \qquad \text{and} \qquad k\cosh(k)A(k) = M_+(k),$$

where

$$L_-(k) = \int_{-\infty}^{0} u(x,1)e^{ikx}\,dx \quad \text{and} \quad M_+(k) = \int_{0}^{\infty} u_y(x,1)e^{ikx}\,dx.$$

Step 4: By eliminating $A(k)$ from the equations in Step 3, show that we can factor the resulting equation as

$$L_-(k) + \frac{i}{k} = K(k)M_+(k), \tag{1}$$

where $K(k) = \sinh(k)/[k\cosh(k)]$.

Step 5: Using the infinite product representation[20] for sinh and cosh, show that $K(k) = K_+(k)K_-(k)$, where

$$K_+(k) = K_-(-k) = \prod_{n=1}^{\infty} \frac{1 - ik/(n\pi)}{1 - 2ik/[(2n-1)\pi]}.$$

Step 6: Use the results from Step 5 and show that (1) can be rewritten

$$K_+(k)M_+(k) - \frac{i}{K_-(0)k} = \frac{L_-(k)}{K_-(k)} + \frac{i}{kK_-(k)} - \frac{i}{K_-(0)k}.$$

---

[20] See, for example, Gradshteyn and Ryzhik, op. cit., Formulas 1.431.2 and 1.431.4.

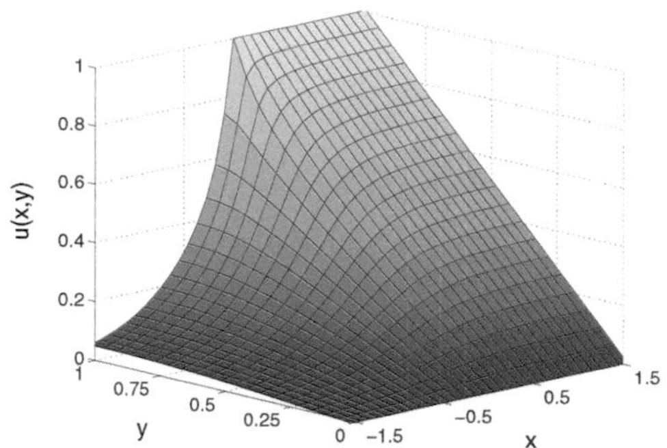

Problem 2

Note that the left side of the equation is analytic in the upper half-plane $\Im(k) > 0$, while the right side of the equation is analytic in the lower half-plane $\Im(k) < \epsilon$.

Step 7: Use Liouville's theorem to show that each side of the equation in Step 6 equals zero. Therefore,

$$L_-(k) = \frac{iK_-(k)}{kK_-(0)} - \frac{i}{k},$$

and

$$U(k, y) = \frac{iK_-(k)\sinh(ky)}{k\sinh(k)}.$$

Step 8: Use the inversion integral and show that

$$u(x, y) = \frac{i}{2\pi}\int_{-\infty+\epsilon i}^{\infty+\epsilon i} \frac{K_-(k)\sinh(ky)}{k\sinh(k)} e^{-ikx}\, dk,$$

or

$$u(x, y) = \frac{i}{2\pi}\int_{-\infty+\epsilon i}^{\infty+\epsilon i} \frac{\sinh(ky)}{k^2\, K_+(k)\cosh(k)} e^{-ikx}\, dk.$$

The first integral is best for finding the solution when $x > 0$, while the second integral is best for $x < 0$.

Step 9: Use the residue theorem and show that

$$u(x, y) = y + \sum_{n=1}^{\infty} \frac{(-1)^n}{n\pi} K_-(-n\pi i)\sin(n\pi y)e^{-n\pi x}$$

if $0 < x$, while

$$u(x, y) = \frac{4}{\pi^2}\sum_{n=1}^{\infty} \frac{(-1)^{n+1}}{(2n-1)^2} \frac{\sin[(2n-1)\pi y/2]}{K_+[(2n-1)\pi i/2]} e^{(2n-1)\pi x/2}$$

for $x < 0$. The figure labeled Problem 2 illustrates this solution $u(x, y)$.

3. Use the Wiener-Hopf technique[21] to solve the mixed boundary value problem

$$\frac{\partial^2 u}{\partial x^2} + \frac{\partial^2 u}{\partial z^2} - u = 0, \qquad -\infty < x < \infty, \quad 0 < z < 1,$$

with the boundary conditions $\lim_{|x|\to\infty} u(x,z) \to 0$,

$$\frac{\partial u(x,0)}{\partial z} = 0, \qquad -\infty < x < \infty,$$

$$\begin{cases} u_z(x,1) = 0, & x < 0, \\ u(x,1) = e^{-x}, & 0 < x. \end{cases}$$

Step 1: Because $u(x,1) = e^{-x}$, we can define the following Fourier transforms:

$$U(k,z) = \int_{-\infty}^{\infty} u(x,z)e^{ikx}\,dx, \qquad U_+(k,z) = \int_0^{\infty} u(x,z)e^{ikx}\,dx,$$

and

$$U_-(k,z) = \int_{-\infty}^{0} u(x,z)e^{ikx}\,dx,$$

so that $U(k,z) = U_+(k,z) + U_-(k,z)$. Therefore, $U_+(k,z)$ is analytic in the half-plane $\Im(k) > -1$, while $U_-(k)$ is analytic in the half-plane $\Im(k) < 0$. Show that we can write the partial differential equation and boundary conditions

$$\frac{d^2U}{dz^2} - m^2 U = 0, \qquad 0 < z < 1,$$

with $U'(k,0) = 0$, $U_+(k,1) = 1/(1-ki)$ and $U'_-(k,1) = 0$, where $m^2 = k^2 + 1$.

Step 2: Show that we can write the solution to Step 1 as

$$U(k,z) = A(k)\frac{\cosh(mz)}{\cosh(m)},$$

with $U'_+(k,1) = m\tanh(m)A(k)$ and $A(k) = U_-(k,1) + i/(k+i)$.

Step 3: By eliminating $A(k)$ from the last two equations in Step 2, show that we can factor the resulting equation as

$$K_+(k)U'_+(k,1) = \frac{m^2\,A(k)}{K_-(k)} = \frac{(k-i)[(k+i)U_-(k,1)+i]}{K_-(k)} = J,$$

where $m\coth(m) = K_+(k)K_-(k)$. Note that the left side of the equation is analytic in the upper half-plane $\Im(k) > -1$, while the right side of the equation is analytic in the lower half-plane $\Im(k) < 0$.

---

[21] See Horvay, G., 1961: Temperature distribution in a slab moving from a chamber at one temperature to a chamber at another temperature. *J. Heat Transfer*, **83**, 391–402.

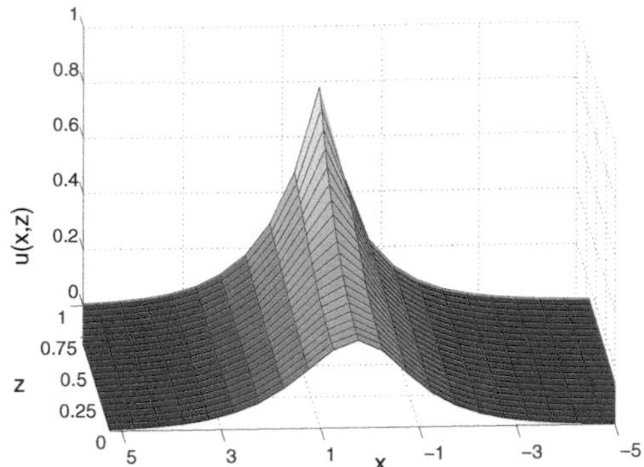

Problem 3

Step 4: It can be shown that $K_-(k) \sim |k|^{1/2}$ as $|k| \to \infty$. Show that $m^2 A(k)/K_-(k)$ cannot increase faster than $|k|^{1/2}$. Then use Liouville's theorem to show that each side equals a constant value J.

Step 5: Use the results from Step 4 to show that $J = 2/K_-(-i)$.

Step 6: From the infinite product theorem we have $K_+(-k) = K_-(k) = \Omega(ik)$, where

$$\Omega(z) = \prod_{n=1}^{\infty} \frac{\dfrac{z}{(n-1/2)\pi} + \sqrt{1 + \dfrac{1}{(n-1/2)^2\pi^2}}}{\dfrac{z}{n\pi} + \sqrt{1 + \dfrac{1}{n^2\pi^2}}}.$$

Use this result and show that

$$u(x,z) = \frac{1}{\pi\Omega(1)} \int_{-\infty-i}^{\infty-i} \frac{\Omega(ik)\cosh(mz)}{(k^2+1)\cosh(m)} e^{-ikx}\, dk.$$

Step 7: Use the residue theorem and show that

$$u(x,z) = e^{-x} - \frac{2i}{\Omega(1)} \sum_{n=0}^{\infty} \frac{\Omega(i\alpha_n)\cosh(\mu_n z)}{\mu_n \alpha_n \sinh(\mu_n)} e^{-i\alpha_n x},$$

where $0 < x$, $\mu_n = (n+1/2)\pi i$ and $\alpha_n = -i\sqrt{1+(2n+1)^2\pi^2/4}$, and

$$u(x,z) = \frac{e^x}{\Omega^2(1)} + \frac{2i}{\Omega(1)} \sum_{n=1}^{\infty} \frac{\cosh(\mu_n z) e^{-i\alpha_n x}}{\alpha_n \Omega(-i\alpha_n)\cosh(\mu_n)},$$

where $x < 0$, $\mu_n = n\pi i$ and $\alpha_n = i\sqrt{1+n^2\pi^2}$. The figure labeled Problem 3 illustrates this solution $u(x,z)$.

4. Use the Wiener-Hopf technique to solve the mixed boundary value problem

$$\frac{\partial^2 u}{\partial x^2} + \frac{\partial^2 u}{\partial y^2} - \beta^2 u = 0, \qquad -\infty < x < \infty, \quad 0 < y < 1,$$

with the boundary conditions

$$u(x, 1) = 0, \qquad -\infty < x < \infty,$$

$$\begin{cases} u(x, 0) = 0, & x < 0, \\ u_y(x, 0) - hu(x, 0) = 1, & 0 < x, \end{cases}$$

where $h > 0$.

Step 1: Assuming that $|u(x, 0)|$ is bounded by $e^{\epsilon x}$, $0 < \epsilon \ll 1$, as $x \to -\infty$, let us introduce the Fourier transforms

$$U(k, y) = \int_{-\infty}^{\infty} u(x, y) e^{ikx}\, dx, \qquad U_+(k, y) = \int_0^{\infty} u(x, y) e^{ikx}\, dx,$$

and

$$U_-(k, y) = \int_{-\infty}^{0} u(x, y) e^{ikx}\, dx,$$

so that $U(k, y) = U_+(k, y) + U_-(k, y)$. Here $U_+(k, y)$ is analytic in the half-plane $\Im(k) > 0$, while $U_-(k, y)$ is analytic in the half-plane $\Im(k) < \epsilon$. Show that we can write the partial differential equation and boundary conditions as the boundary value problem

$$\frac{d^2 U}{dy^2} - m^2 U = 0, \qquad 0 < y < 1,$$

with $U'(k, 1) = 0$, $U'_+(k, 0) - hU_+(k, 0) = i/k$ and $U'_-(k, 0) = 0$, where $m^2 = k^2 + \beta^2$.

Step 2: Show that the solution to Step 1 is $U(k, y) = A(k) \sinh[m(1 - y)]$,

$$U'_+(k, 0) = \sinh(m) A(k), \tag{1}$$

and

$$-\left[m \cosh(m) + h \sinh(m)\right] A(k) = U'_-(k, 0) + \frac{i}{k}, \tag{2}$$

where

$$U'_-(k, 0) = \int_{-\infty}^{0} u_y(x, 0) e^{ikx}\, dx.$$

Step 3: It can be shown[22] that $m\coth(m) + h$ can be factorized as follows: $m\coth(m) + h = K(0)P(k)P(-k)$, where

$$P(k) = \prod_{n=1}^{\infty}\left(1 - \frac{ik}{\rho_n}\right)\left(1 - \frac{ik}{\sqrt{n^2\pi^2 + \beta^2}}\right)^{-1},$$

$K(0) = h + \beta\coth(\beta)$, $\rho_n = \sqrt{\beta^2 + \lambda_n^2}$ and λ_n is the nth root of $\lambda + h\tan(\lambda) = 0$. Note that $P(k)$ is analytic in the half-plane $\Im(k) > 0$, while $P(-k)$ is analytic in the half-plane $\Im(k) < \epsilon$. By eliminating $A(k)$ from (1) and (2) in Step 2 and using this factorization, show that we have the Wiener-Hopf equation

$$K(0)P(k)U_+'(k,0) + \frac{i}{k} = \frac{i}{k}\left[1 - \frac{1}{P(-k)}\right] - \frac{U_-'(k,0)}{P(-k)}. \tag{3}$$

Note that the left side of (3) is analytic in the upper half-plane $\Im(k) > 0$, while the right side is analytic in the lower half-plane $\Im(k) < \epsilon$.

Step 4: It can be shown that $P(k) \sim |k|^{1/2}$. Show that $U_+(k,0) \sim k^{-1}$ and $U_-'(k,0) \sim \ln(k)$ as $|k| \to \infty$. Then use Liouville's theorem to show that each side of (3) equals zero.

Step 5: Use the results from Step 4 to show that

$$K(0)U_+(k,0) = \frac{1}{ki\,P(k)}.$$

Therefore,

$$\begin{aligned} u(x,y) &= \frac{1}{2\pi i K(0)}\int_{-\infty+\epsilon i}^{\infty+\epsilon i} \frac{\sinh[m(1-y)]}{k\sinh(m)P(k)} e^{-ikx}\,dk \\ &= \frac{1}{2\pi i}\int_{-\infty+\epsilon i}^{\infty+\epsilon i} \frac{P(-k)\sinh[m(1-y)]}{k[m\cosh(m) + h\sinh(m)]} e^{-ikx}\,dk. \end{aligned}$$

The first integral is best for computations when $x < 0$, while the second integral is best for $x > 0$.

Step 6: Use the results from Step 5 to show that

$$u(x,y) = -\frac{\pi}{K(0)}\sum_{n=1}^{\infty} \frac{n\,\sin(n\pi y)}{\xi_n^2 P(i\xi_n)} e^{\xi_n x}, \qquad x < 0,$$

---

[22] See Appendix A in Buchwald, V. T., and F. Viera, 1998: Linearized evaporation from a soil of finite depth above a water table. *Austral. Math. Soc., Ser. B*, **39**, 557–576.

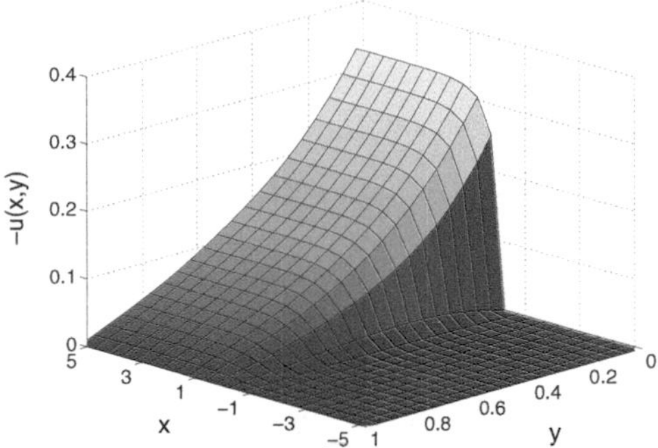

Problem 4

where $\xi_n = \sqrt{n^2\pi^2 + \beta^2}$, and

$$u(x,y) = \frac{e^{-\beta(1-y)} - e^{\beta(1-y)}}{(\beta+h)e^{\beta} + (\beta-h)e^{-\beta}} + \sum_{n=1}^{\infty} \frac{\lambda_n^2 P(i\rho_n)\sin[\lambda_n(1-y)]}{\rho_n^2[h(1+h)+\lambda_n^2]\sin(\lambda_n)} e^{-\lambda_n x},$$

for $0 < x$. The figure labeled Problem 4 illustrates $u(x,y)$ when $h = 1$ and $\beta = 2$.

5. Use the Wiener-Hopf technique to solve the mixed boundary value problem

$$\frac{\partial^2 u}{\partial x^2} + \frac{\partial^2 u}{\partial y^2} - u = 0, \qquad -\infty < x < \infty, \quad 0 < y < 1,$$

with the boundary conditions

$$\frac{\partial u(x,1)}{\partial y} = 0, \qquad -\infty < x < \infty,$$

$$\begin{cases} u(x,0) = 1, & x < 0, \\ u_y(x,0) = 0, & 0 < x. \end{cases}$$

Step 1: Assuming that $|u(x,0)|$ is bounded by $e^{-\epsilon x}$, $0 < \epsilon \ll 1$, as $x \to \infty$, let us define the following Fourier transforms:

$$U(k,y) = \int_{-\infty}^{\infty} u(x,y)e^{ikx}\,dx, \qquad U_+(k,y) = \int_0^{\infty} u(x,y)e^{ikx}\,dx,$$

and

$$U_-(k,y) = \int_{-\infty}^{0} u(x,y)e^{ikx}\,dx,$$

so that $U(k,y) = U_+(k,y) + U_-(k,y)$. Here, $U_+(k,y)$ is analytic in the half-space $\Im(k) > -\epsilon$, while $U_-(k,y)$ is analytic in the half-space $\Im(k) < 0$. Then show that the partial differential equation becomes

$$\frac{d^2U}{dy^2} - m^2 U = 0, \qquad 0 < y < 1,$$

with $U'(k,1) = 0$, where $m^2 = k^2 + 1$.

Step 2: Show that the solution to Step 1 is $U(k,y) = A(k)\cosh[m(1-y)]$.

Step 3: From the boundary conditions along $y = 0$, show that

$$\cosh(m)A(k) = M_+(k) - \frac{i}{k} \qquad \text{and} \qquad -m\sinh(m)A(k) = L_-(k),$$

where

$$M_+(k) = \int_0^\infty u(x,0)e^{ikx}\,dx \quad \text{and} \quad L_-(k) = \int_{-\infty}^0 u_y(x,0)e^{ikx}\,dx.$$

Step 4: By eliminating $A(k)$ from the equations in Step 3, show that we can factor the resulting equation as

$$-m^2\left[M_+(k) - \frac{i}{k}\right] = m\coth(m)L_-(k).$$

Step 5: Using the results that $m\coth(m) = K_+(k)K_-(k)$, where $K_+(k)$ and $K_-(k)$ are defined in Step 6 of Problem 3, show that

$$\frac{i}{K_+(k)} + \frac{K_+(k) - K_+(0)}{k\,K_+(0)K_+(k)} - \frac{(k+i)M_+(k)}{K_+(k)} = \frac{K_-(k)L_-(k)}{k-i} + \frac{1}{k\,K_+(0)}.$$

Note that the left side of the equation is analytic in the upper half-plane $\Im(k) > -\epsilon$, while the right side of the equation is analytic in the lower half-plane $\Im(k) < 0$.

Step 6: Use Liouville's theorem to show that each side of the equation in Step 5 equals zero. Therefore,

$$M_+(k) = \frac{i}{k} + \frac{K_+(k)}{k(k+i)K_+(0)},$$

and

$$U(k,y) = \frac{K_+(k)\cosh[m(1-y)]}{k(k+i)\,K_+(0)\cosh(m)}.$$

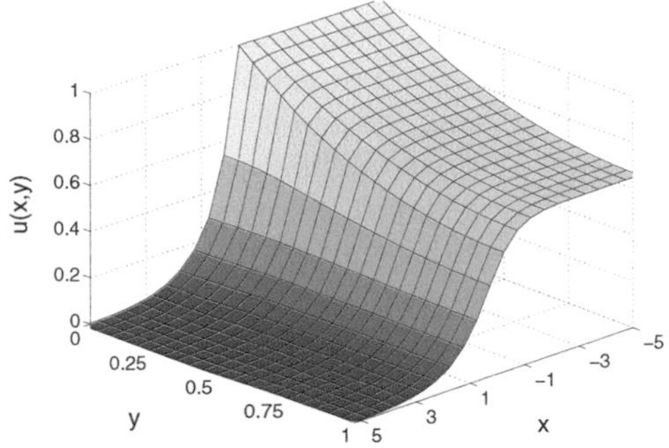

Problem 5

Step 7: Use the inversion integral and show that

$$u(x, y) = \frac{1}{2\pi} \int_{-\infty-\epsilon i}^{\infty-\epsilon i} \frac{K_+(k) \cosh[m(1-y)]}{k(k+i)\, K_+(0) \cosh(m)} e^{-ikx}\, dk,$$

or

$$u(x, y) = \frac{K_-(0)}{2\pi \coth(1)} \int_{-\infty-\epsilon i}^{\infty-\epsilon i} \frac{m \cosh[m(1-y)]}{k(k+i) \sinh(m)} e^{-ikx}\, dk.$$

The first integral is best for finding the solution when $x < 0$, while the second integral is best for $0 < x$.

Step 8: Use the residue theorem and show that

$$u(x, y) = \tanh(1) \frac{K_-(0)}{K_-(-i)} e^{-x} + \tanh(1) \sum_{n=1}^{\infty} \frac{(1+\kappa_n)\, K_-(0)}{\kappa_n^2\, K_-(-i\kappa_n)} \cos(n\pi y) e^{-\kappa_n x}$$

if $0 < x$, where $\kappa_n = \sqrt{1 + n^2\pi^2}$, while

$$u(x, y) = \frac{\cosh(1-y)}{\cosh(1)} + \sum_{n=1}^{\infty} \frac{(1-k_n) K_+(ik_n)}{m_n k_n^2 K_+(0)} \sin(m_n y) e^{k_n x}$$

for $x < 0$, where $k_n = \sqrt{1 + (2n-1)^2\pi^2/4}$ and $m_n = (2n-1)\pi/2$. The figure labeled Problem 5 illustrates this solution $u(x, y)$.

5.2 THE WIENER–HOPF TECHNIQUE WHEN THE FACTORIZATION CONTAINS BRANCH POINTS

In the previous section, the product factors $K_+(k)$ and $K_-(k)$ were always meromorphic, resulting in a solution that consisted of a sum of residues. This occurred because $K(k)$ contained terms such as $m \sinh(m)$ and $\cosh(m)$, whose power series expansion consists only of powers of m^2, and there were no branch points. The form of $K(k)$ was due, in turn, to the presence of a finite domain in one of the spatial domains.

In this section we consider infinite or semi-infinite domains where $K(k)$ will become multivalued. As one might expect, the sum of residues becomes a branch cut integral just as it did in the case of Fourier transforms. There we found that single-valued Fourier transform yielded inverses that were a sum of residues, whereas the inverses of multivalued Fourier transforms contained branch cut integrals.

• Example 5.2.1

An insightful example arises from a heat conduction problem[23] in the upper half-plane $y > 0$:

$$\frac{\partial u}{\partial t} = \frac{\partial^2 u}{\partial x^2} + \frac{\partial^2 u}{\partial y^2}, \qquad -\infty < x < \infty, \quad 0 < t, y, \tag{5.2.1}$$

with the boundary conditions

$$\begin{cases} u(x,0,t) = e^{-\epsilon x}, & 0 < \epsilon, x, \\ u_y(x,0,t) = 0, & x < 0, \end{cases} \tag{5.2.2}$$

and

$$\lim_{y\to\infty} u(x,y,t) \to 0, \tag{5.2.3}$$

while the initial condition is $u(x,y,0) = 0$. Eventually we will consider the limit $\epsilon \to 0$.

What makes this problem particularly interesting is the boundary condition that we specify along $y = 0$; it changes from a Dirichlet condition when $x < 0$, to a Neumann boundary condition when $x > 0$. The Wiener-Hopf technique is commonly used to solve these types of boundary value problems where the nature of the boundary condition changes along a given boundary — the so-called *mixed boundary value problem.*

We begin by introducing the Laplace transform in time

$$U(x,y,s) = \int_0^\infty u(x,y,t)e^{-st}\,dt, \tag{5.2.4}$$

and the Fourier transform in the x-direction

$$\overline{U}_+(k,y,s) = \int_0^\infty U(x,y,s)e^{ikx}\,dx, \tag{5.2.5}$$

---

[23] Simplified version of a problem solved by Huang, S. C., 1985: Unsteady-state heat conduction in semi-infinite regions with mixed-type boundary conditions. *J. Heat Transfer*, **107**, 489–491.

$$\overline{U}_-(k,y,s) = \int_{-\infty}^{0} U(x,y,s)e^{ikx}\,dx, \tag{5.2.6}$$

and

$$\overline{U}(k,y,s) = \overline{U}_+(k,y,s) + \overline{U}_-(k,y,s) = \int_{-\infty}^{\infty} U(x,y,s)e^{ikx}dx. \tag{5.2.7}$$

Here, we have assumed that $|u(x,y,t)|$ is bounded by $e^{-\epsilon x}$ as $x \to \infty$, while $|u(x,y,t)|$ is $O(1)$ as $x \to -\infty$. For this reason, the subscripts "+" and "−" denote that $\overline{U}_+$ is analytic in the upper half-plane $\Im(k) > -\epsilon$, while $\overline{U}_-$ is analytic in the lower half-plane $\Im(k) < 0$.

Taking the joint transform of Equation 5.2.1, we find that

$$\frac{d^2\overline{U}(k,y,s)}{dy^2} - (k^2+s)\overline{U}(k,y,s) = 0, \qquad 0 < y < \infty, \tag{5.2.8}$$

with the transformed boundary conditions

$$\overline{U}_+(k,0,s) = \frac{1}{s(\epsilon - ki)}, \qquad \frac{d\overline{U}_-(k,0,s)}{dy} = 0, \tag{5.2.9}$$

and $\lim_{y\to\infty} \overline{U}(k,y,s) \to 0$. The general solution to Equation 5.2.8 is

$$\overline{U}(k,y,s) = A(k,s)e^{-y\sqrt{k^2+s}}. \tag{5.2.10}$$

Consequently,

$$A(k,s) = \frac{1}{s(\epsilon - ki)} + \overline{U}_-(k,0,s) \tag{5.2.11}$$

and

$$-\sqrt{k^2+s}\;A(k,s) = \frac{d\overline{U}_+(k,0,s)}{dy}. \tag{5.2.12}$$

Note that we have a multivalued function $\sqrt{k^2+s}$ with branch points $k = \pm\sqrt{s}\,i$. Eliminating $A(k,s)$ between Equation 5.2.11 and Equation 5.2.12, we obtain the Wiener-Hopf equation:

$$\frac{d\overline{U}_+(k,0,s)}{dy} = -\sqrt{k^2+s}\left[\overline{U}_-(k,0,s) + \frac{1}{s(\epsilon - ki)}\right]. \tag{5.2.13}$$

Our next goal is to rewrite Equation 5.2.13 so that the left side is analytic in the upper half-plane $\Im(k) > -\epsilon$, while the right side is analytic in the lower half-plane $\Im(k) < 0$. We begin by factoring $\sqrt{k^2+s} = \sqrt{k-i\sqrt{s}}\sqrt{k+i\sqrt{s}}$, where the branch cuts lie along the imaginary axis in the k-plane from $(-\infty i, -\sqrt{s}i]$ and $[\sqrt{s}i, \infty i)$. Equation 5.2.13 can then be rewritten

$$\frac{1}{\sqrt{k+i\sqrt{s}}}\frac{d\overline{U}_+(k,0,s)}{dy} = -\sqrt{k-i\sqrt{s}}\left[\overline{U}_-(k,0,s) + \frac{1}{s(\epsilon - ki)}\right]. \tag{5.2.14}$$

The left side of Equation 5.2.14 is what we want; the same is true of the first term on the right side. However, the second term on the right side falls short. At this point we note that

$$\frac{\sqrt{k-i\sqrt{s}}}{\epsilon-ki}=\frac{\sqrt{k-i\sqrt{s}}-\sqrt{-i\epsilon-i\sqrt{s}}}{\epsilon-ki}+\frac{\sqrt{-i\epsilon-i\sqrt{s}}}{\epsilon-ki}. \tag{5.2.15}$$

Substituting Equation 5.2.15 into Equation 5.2.14 and rearranging terms, we obtain

$$\frac{1}{\sqrt{k+i\sqrt{s}}}\frac{d\overline{U}_{+}(k,0,s)}{dy}+\frac{\sqrt{-i\epsilon-i\sqrt{s}}}{s(\epsilon-ki)}$$
$$=-\sqrt{k-i\sqrt{s}}\,\overline{U}_{-}(k,0,s)-\frac{\sqrt{k-i\sqrt{s}}-\sqrt{-i\epsilon-i\sqrt{s}}}{s(\epsilon-ki)}. \tag{5.2.16}$$

In this form, the right side of Equation 5.2.16 is analytic in the lower half-plane $\Im(k)<0$, while the left side is analytic in the upper half-plane $\Im(k)>-\epsilon$. Since they share a common strip of analyticity $-\epsilon<\Im(k)<0$, they are analytic continuations of each other and equal some entire function. Using Liouville's theorem and taking the limit as $|k|\to\infty$, we see that both sides of Equation 5.2.16 equal zero. Therefore,

$$\overline{U}_{-}(k,0,s)=-\frac{\sqrt{k-i\sqrt{s}}-\sqrt{-i\epsilon-i\sqrt{s}}}{s(\epsilon-ki)\sqrt{k-i\sqrt{s}}}, \tag{5.2.17}$$

$$A(k,s)=\frac{\sqrt{-i\epsilon-i\sqrt{s}}}{s(\epsilon-ki)\sqrt{k-i\sqrt{s}}}, \tag{5.2.18}$$

and

$$U(x,y,s)=\frac{\sqrt{-i\epsilon-i\sqrt{s}}}{2\pi s}\int_{-\infty}^{\infty}\frac{\exp\left(-ikx-y\sqrt{k^2+s}\right)}{(\epsilon-ki)\sqrt{k-i\sqrt{s}}}\,dk. \tag{5.2.19}$$

Upon taking the limit $\epsilon\to 0$ and introducing $x=r\cos(\theta)$, $y=r\sin(\theta)$ and $k=\sqrt{s}\,\eta$, Equation 5.2.19 becomes

$$U(r,\theta,s)=\frac{\sqrt{i}}{2\pi s}\int_{-\infty}^{\infty}\frac{\exp\left\{-r\sqrt{s}\left[i\eta\cos(\theta)+\sin(\theta)\sqrt{\eta^2+1}\right]\right\}}{\eta\sqrt{\eta-i}}\,d\eta. \tag{5.2.20}$$

To invert Equation 5.2.20, we introduce

$$\cosh(\tau)=i\eta\cos(\theta)+\sin(\theta)\sqrt{\eta^2+1}. \tag{5.2.21}$$

Solving for η, we find that $\eta=\sin(\theta)\sinh(\tau)-i\cos(\theta)\cosh(\tau)=-i\cos(\theta-i\tau)$. We now deform the original contour along the real axis to the one defined by

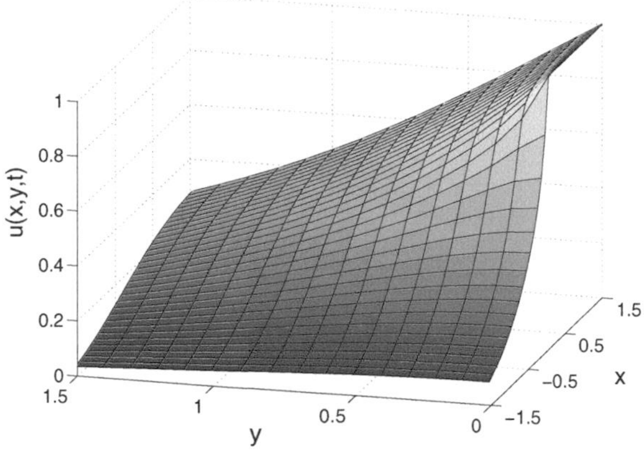

Figure 5.2.1: The solution to Equation 5.2.1 through Equation 5.2.3 obtained via the Wiener-Hopf technique.

η. Particular care must be exercised in the case of $x > 0$ as we deform the contour into the lower half of the k-plane since $0 < \theta < \pi/2$. During this deformation, we pass over the singularity at $k = 0^-$ and must consequently add its contribution to the inverse. Thus, the Laplace transform of the solution now reads

$$U(r,\theta,s) = \frac{e^{-y\sqrt{s}}}{s} H(x) - \frac{1}{\sqrt{2}\,\pi s} \int_{-\infty}^{\infty} \frac{\sin[(\theta - i\tau)/2]}{\cos(\theta - i\tau)} e^{-r\sqrt{s}\cosh(\tau)}\, d\tau. \tag{5.2.22}$$

Taking the inverse Laplace transform of Equation 5.2.22, we obtain

$$u(r,\theta,t) = \operatorname{erfc}\left(\frac{y}{\sqrt{4t}}\right) H(x) - \frac{\sqrt{2}}{\pi} \int_0^{\infty} \Re\left\{ \frac{\sin[(\theta - i\tau)/2]}{\cos(\theta - i\tau)} \right\} \operatorname{erfc}\left[\frac{r\,\cosh(\tau)}{\sqrt{4t}}\right] d\tau. \tag{5.2.23}$$

Figure 5.2.1 illustrates this solution when $t = 1$.

• Example 5.2.2

In Example 1.1.3, we posed the question of how to solve the mixed boundary value problem

$$\frac{\partial^2 u}{\partial x^2} + \frac{\partial^2 u}{\partial y^2} - \alpha^2 u = 0, \qquad -\infty < x < \infty, \quad 0 < y, \tag{5.2.24}$$

with the boundary conditions $\lim_{y\to\infty} u(x,y) \to 0$, and

$$\begin{cases} u(x,0) = 1, & x < 0, \\ u(x,0) = 1 + \lambda u_y(x,0), & 0 < x, \end{cases} \tag{5.2.25}$$

where $0 < \alpha, \lambda$. We showed there that some simplification occurs if we introduce the transformation

$$u(x,y) = \frac{e^{-\alpha y}}{1+\alpha\lambda} + v(x,y), \qquad \textbf{(5.2.26)}$$

so that the problem becomes

$$\frac{\partial^2 v}{\partial x^2} + \frac{\partial^2 v}{\partial y^2} - \alpha^2 v = 0, \qquad -\infty < x < \infty, \quad 0 < y, \qquad \textbf{(5.2.27)}$$

with the boundary conditions $\lim_{y\to\infty} v(x,y) \to 0$ and

$$v(x,0) = \frac{\alpha\lambda}{1+\alpha\lambda}, \quad x < 0, \qquad v(x,0) = \lambda v_y(x,0), \quad 0 < x. \qquad \textbf{(5.2.28)}$$

In spite of this transformation, we showed in Example 1.1.3 that we could not solve Equation 5.2.27 and Equation 5.2.28 by using conventional Fourier transforms. If we assume that $|v(x,y)|$ is bounded by $e^{-\epsilon x}$, $0 < \epsilon \ll 1$, as $x \to \infty$ while $|v(x,y)|$ is $O(1)$ as $x \to -\infty$, can we use the Wiener-Hopf technique here?

We begin by defining the following Fourier transforms in the x-direction:

$$V_+(k,y) = \int_0^\infty v(x,y)e^{ikx}\,dx, \qquad \textbf{(5.2.29)}$$

$$V_-(k,y) = \int_{-\infty}^0 v(x,y)e^{ikx}\,dx, \qquad \textbf{(5.2.30)}$$

and

$$V(k,y) = V_+(k,y) + V_-(k,y) = \int_{-\infty}^\infty v(x,y)e^{ikx}\,dx. \qquad \textbf{(5.2.31)}$$

The subscripts "+" and "−" denote the fact that V_+ is analytic in the upper half-space $\Im(k) > -\epsilon$, while V_- is analytic in the lower half-space $\Im(k) < 0$.

Taking the Fourier transform of Equation 5.2.27, we find that

$$\frac{d^2V(k,y)}{dy^2} - m^2V(k,y) = 0, \qquad m^2 = k^2 + \alpha^2, \qquad 0 < y < \infty, \qquad \textbf{(5.2.32)}$$

along with the transformed boundary condition $\lim_{y\to\infty} V(k,y) \to 0$. The general solution to Equation 5.2.32 is $V(k,y) = A(k)e^{-my}$. Note that m is multivalued with branch points $k = \pm\alpha i$.

Turning to the boundary conditions given by Equation 5.2.28, we obtain

$$A(k) = \frac{\alpha\lambda}{ik(1+\alpha\lambda)} + M_+(k) \quad \text{and} \quad A(k) + m\lambda A(k) = L_-(k), \qquad \textbf{(5.2.33)}$$

where

$$M_+(k) = \int_0^\infty v(x,0)e^{ikx}\,dx \text{ and } L_-(k) = \int_{-\infty}^0 \left[v(x,0) - \lambda v_y(x,0)\right] e^{ikx}\,dx. \tag{5.2.34}$$

Eliminating $A(k)$ in Equation 5.2.33, we obtain the Wiener-Hopf equation:

$$\frac{\alpha\lambda}{ik(1+\alpha\lambda)} + M_+(k) = \frac{L_-(k)}{1+m\lambda}. \tag{5.2.35}$$

Our next goal is to rewrite Equation 5.2.35 so that the left side is analytic in the upper half-plane $\Im(k) > -\epsilon$, while the right side is analytic in the lower half-plane $\Im(k) < 0$. We begin by factoring $P(k) = 1 + \lambda m = P_+(k)P_-(k)$, where $P_+(k)$ and $P_-(k)$ are analytic in the upper and lower half-planes, respectively. We will determine them shortly. Equation 5.2.35 can then be rewritten

$$P_+(k)M_+(k) + \frac{\alpha\lambda P_+(k)}{ik(1+\alpha\lambda)} = \frac{L_-(k)}{P_-(k)}. \tag{5.2.36}$$

The right side of Equation 5.2.36 is what we want; it is analytic in the half-plane $\Im(k) < 0$. The first term on the left side is analytic in the half-plane $\Im(k) > -\epsilon$. The second term, unfortunately, is not analytic in either half-planes. However, we note that

$$\frac{P_+(k)}{ik} = \frac{P_+(k)}{ik} - \frac{P_+(0)}{ik} + \frac{P_+(0)}{ik}. \tag{5.2.37}$$

Substituting Equation 5.2.37 into Equation 5.2.36 and rearranging terms, we obtain

$$P_+(k)M_+(k) + \frac{\alpha\lambda}{1+\alpha\lambda}\left[\frac{P_+(k)}{ik} - \frac{P_+(0)}{ik}\right] = \frac{L_-(k)}{P_-(k)} - \frac{\alpha\lambda P_+(0)}{ik(1+\alpha\lambda)}. \tag{5.2.38}$$

In this form, the left side of Equation 5.2.38 is analytic in the upper half-plane $\Im(k) > -\epsilon$, while the right side is analytic in the lower half-plane $\Im(k) < 0$. Since both sides share a common strip of analyticity $-\epsilon < \Im(k) < 0$, they are analytic continuations of each other and equal some entire function. Using Liouville's theorem and taking the limit as $|k| \to \infty$, we see that both sides of Equation 5.2.38 equal zero. Therefore,

$$M_+(k) = \frac{\alpha\lambda P_+(0)}{ik(1+\alpha\lambda)P_+(k)} - \frac{\alpha\lambda}{ik(1+\alpha\lambda)}, \tag{5.2.39}$$

$$A(k) = \frac{\alpha\lambda P_+(0)}{ik(1+\alpha\lambda)P_+(k)}, \tag{5.2.40}$$

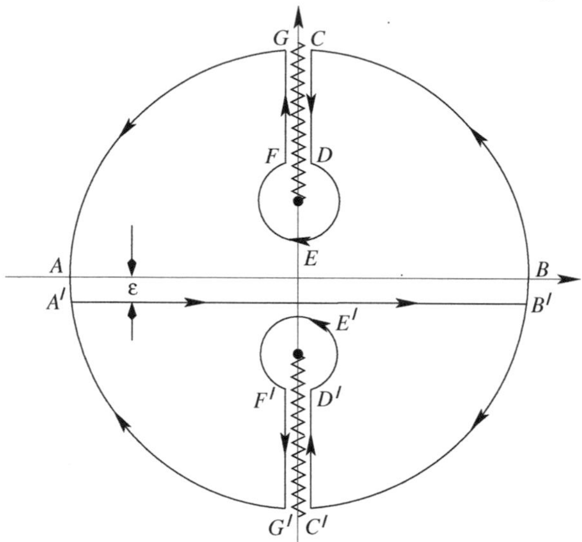

Figure 5.2.2: The contours in inverting the integral in Equation 5.2.41.

and

$$u(x,y) = \frac{e^{-\alpha y}}{1+\alpha\lambda} + \frac{\alpha\lambda P_+(0)}{2\pi i(1+\alpha\lambda)} \int_{-\infty-\epsilon i}^{\infty-\epsilon i} \frac{e^{-my-ikx}}{k\,P_+(k)}\,dk \qquad \textbf{(5.2.41)}$$

$$= \frac{e^{-\alpha y}}{1+\alpha\lambda} + \frac{\alpha\lambda}{2\pi i} \int_{-\infty-\epsilon i}^{\infty-\epsilon i} \frac{P_-(k)}{P_-(0)} \frac{e^{-my-ikx}}{k(1+\lambda m)}\,dk. \qquad \textbf{(5.2.42)}$$

Equation 5.2.41 is best for computing $u(x,y)$ when $x<0$. Using the contour shown in Figure 5.2.2, we deform the original contour to the contour $AGFEDCB$ shown there. During the deformation, we cross the simple pole at $k=0$ and must add its contribution. Therefore,

$$u(x,y) = \frac{e^{-\alpha y}}{1+\alpha\lambda} + \frac{\alpha\lambda\, e^{-\alpha y}}{1+\alpha\lambda} + \frac{\alpha\lambda\, P_+(0)}{2\pi i(1+\alpha\lambda)} \int_{\infty}^{\alpha} \frac{e^{iy\sqrt{\eta^2-\alpha^2}+\eta x}}{P_+(i\eta)} \frac{d\eta}{\eta}$$

$$+ \frac{\alpha\lambda\, P_+(0)}{2\pi i(1+\alpha\lambda)} \int_{\alpha}^{\infty} \frac{e^{-iy\sqrt{\eta^2-\alpha^2}+\eta x}}{P_+(i\eta)} \frac{d\eta}{\eta} \qquad \textbf{(5.2.43)}$$

$$= e^{-\alpha y} - \frac{\alpha\lambda\, P_+(0)}{\pi(1+\alpha\lambda)} \int_{\alpha}^{\infty} e^{x\eta} \sin\left(y\sqrt{\eta^2-\alpha^2}\right) \frac{d\eta}{\eta\, P_+(i\eta)} \qquad \textbf{(5.2.44)}$$

$$= e^{-\alpha y} - \frac{\alpha\lambda\, P_+(0)}{\pi(1+\alpha\lambda)} \int_{1}^{\infty} e^{\alpha x\xi} \sin\left(\alpha y\sqrt{\xi^2-1}\right) \frac{d\xi}{\xi\, P_+(i\alpha\xi)}. \qquad \textbf{(5.2.45)}$$

The contribution from the arcs GA and BC vanish as the radius of the semicircle in the upper half-plane becomes infinite.

Turning to the case $x>0$, we deform the original contour to the contour $A'G'F'E'D'C'B'$ in Figure 5.2.2. During the deformation we do not cross any singularities. Therefore,

$$u(x,y) = \frac{e^{-\alpha y}}{1+\alpha\lambda} + \frac{\alpha\lambda}{2\pi i}\int_{\infty}^{\alpha} \frac{P_-(-i\eta)}{P_-(0)} \frac{e^{-iy\sqrt{\eta^2-\alpha^2}-\eta x}}{1+i\lambda\sqrt{\eta^2-\alpha^2}}\frac{d\eta}{\eta}$$

$$+ \frac{\alpha\lambda}{2\pi i}\int_{\alpha}^{\infty} \frac{P_-(-i\eta)}{P_-(0)} \frac{e^{iy\sqrt{\eta^2-\alpha^2}-\eta x}}{1-i\lambda\sqrt{\eta^2-\alpha^2}}\frac{d\eta}{\eta} \tag{5.2.46}$$

$$u(x,y) = \frac{e^{-\alpha y}}{1+\alpha\lambda} + \frac{\alpha\lambda}{\pi}\int_{\alpha}^{\infty} \frac{P_-(-i\eta)}{P_-(0)} e^{-x\eta}\frac{d\eta}{\eta} \tag{5.2.47}$$

$$\times \frac{\sin\left(y\sqrt{\eta^2-\alpha^2}\right) + \lambda\sqrt{\eta^2-\alpha^2}\cos\left(y\sqrt{\eta^2-\alpha^2}\right)}{1+\lambda^2(\eta^2-\alpha^2)}$$

$$= \frac{e^{-\alpha y}}{1+\alpha\lambda} + \frac{\alpha\lambda}{\pi}\int_{1}^{\infty} \frac{P_-(-i\alpha\xi)}{P_-(0)} e^{-\alpha x\xi}\frac{d\xi}{\xi} \tag{5.2.48}$$

$$\times \frac{\sin\left(\alpha y\sqrt{\xi^2-1}\right) + \alpha\lambda\sqrt{\xi^2-1}\cos\left(\alpha y\sqrt{\xi^2-1}\right)}{1+\alpha^2\lambda^2(\xi^2-1)}.$$

The final task is the factorization. We begin by introducing the functions

$$\varphi(z) = \varphi_+(z) + \varphi_-(z) = \frac{P'(z)}{P(z)} = \frac{z\lambda}{m(1+\lambda m)}, \quad m^2 = z^2 + \alpha^2, \tag{5.2.49}$$

which is analytic in the region $-\alpha < \Im(z) < \alpha$ and $\varphi_\pm(z) = P'_\pm(z)/P_\pm(z)$. By definition, $\varphi_\pm(z)$ is analytic and nonzero in the same half-planes as $P_\pm(z)$. Furthermore,

$$P_\pm(z) = P_\pm(0)\exp\left[\int_0^z \varphi_\pm(\zeta)\,d\zeta\right]. \tag{5.2.50}$$

From Cauchy's integral theorem,

$$\varphi_+(\zeta) = \frac{1}{2\pi i}\int_{-\infty+\epsilon i}^{\infty+\epsilon i} \frac{\varphi(z)}{z-\zeta}\,dz \quad \text{and} \quad \varphi_-(\zeta) = -\frac{1}{2\pi i}\int_{-\infty+\delta i}^{\infty+\delta i} \frac{\varphi(z)}{z-\zeta}\,dz, \tag{5.2.51}$$

where $-\alpha < \epsilon < \delta < \alpha$. We will now evaluate the line integrals in Equation 5.2.51 by converting them into the closed contours shown in Figure 5.2.2. In particular, for φ_+, we employ the closed contour $A'B'C'D'E'F'G'A'$ with the branch cut running from $-\alpha i$ to $-\infty i$, while for the evaluation of φ_-, we use $ABCDEFGA$ with the branch cut running from αi to ∞i. For example,

$$\varphi_+(\zeta) = \frac{\lambda}{2\pi i}\int_{-\infty+\epsilon i}^{\infty+\epsilon i} \frac{z}{m(1+\lambda m)(z-\zeta)}\,dz \tag{5.2.52}$$

$$= \frac{\lambda}{2\pi i}\int_{\infty}^{\alpha} \frac{(-i\eta)(-i\,d\eta)}{\left(i\sqrt{\eta^2-\alpha^2}\right)\left(1+i\lambda\sqrt{\eta^2-\alpha^2}\right)(-i\eta-\zeta)}$$

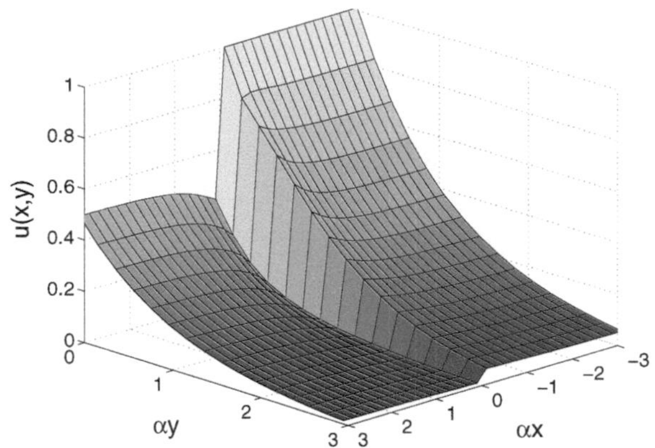

Figure 5.2.3: The solution to Equation 5.2.22 and Equation 5.2.23 obtained via the Wiener-Hopf technique.

$$+\frac{\lambda}{2\pi i}\int_{\alpha}^{\infty}\frac{(-i\eta)(-i\,d\eta)}{\left(-i\sqrt{\eta^2-\alpha^2}\right)\left(1-i\lambda\sqrt{\eta^2-\alpha^2}\right)(-i\eta-\zeta)} \quad \textbf{5.2.53})$$

$$=\frac{\lambda}{\pi}\int_{\alpha}^{\infty}\frac{\eta\,d\eta}{(\zeta+i\eta)\sqrt{\eta^2-\alpha^2}\,[1+\lambda^2(\eta^2-\alpha^2)]} \quad (\textbf{5.2.54})$$

$$=\frac{\rho}{\pi}\int_{0}^{\infty}\frac{ds}{(s^2+\rho^2)\left(\zeta+i\sqrt{s^2+\alpha^2}\right)}. \quad (\textbf{5.2.55})$$

where $s^2 = \eta^2 - \alpha^2$ and $\rho = 1/\lambda$. A similar analysis of $\varphi_-(\zeta)$ shows that $P_-(-\zeta)/P_-(0) = P_+(\zeta)/P_+(0)$. Figure 5.2.3 illustrates $u(x,y)$ when $\alpha\lambda = 1$. The numerical calculations begin with a computation of $\varphi_+(\zeta)$ using Simpson's rule. Then, $P_+(\zeta)/P_+(0)$ is found using the trapezoidal rule. Finally, Equation 5.2.45 and Equation 5.2.48 are evaluated using Simpson's rule.

Problems

1. Use the Wiener-Hopf technique to solve

$$\frac{\partial^2 u}{\partial x^2}+\frac{\partial^2 u}{\partial y^2}-u=-2\rho(x)\delta(y), \qquad -\infty < x, y < \infty,$$

subject to the boundary conditions

$$\lim_{|x|\to\infty} u(x,y)\to 0, \qquad -\infty < y < \infty,$$

$$\lim_{|y|\to\infty} u(x,y)\to 0, \qquad -\infty < x < \infty,$$

and

$$u(x,0)=e^{-x}, \qquad 0 < x < \infty.$$

The function $\rho(x)$ is only nonzero for $0 < x < \infty$.

Step 1: Assuming that $|u(x, y)|$ is bounded by $e^{\epsilon x}$, where $0 < \epsilon \ll 1$, as $x \to -\infty$, let us introduce

$$U(k, y) = \int_{-\infty}^{\infty} u(x, y) e^{-ikx}\, dx \qquad \text{and} \qquad \overline{U}(k, \ell) = \int_{-\infty}^{\infty} U(k, y) e^{-i\ell y}\, dy.$$

Use the differential equation and first two boundary conditions to show that

$$U(k, y) = R(k) \frac{e^{-|y|\sqrt{k^2+1}}}{\sqrt{k^2+1}},$$

where $R(k)$ is the Fourier transform of $\rho(x)$.

Step 2: Taking the Fourier transform of the last boundary condition, show that

$$\frac{R(k)}{\sqrt{k^2+1}} = \frac{1}{1+ik} + F_+(k), \tag{1}$$

where

$$F_+(k) = \int_{-\infty}^{0} u(x, 0) e^{-ikx}\, dx.$$

Note that $R(k)$ is analytic in the lower half-space where $\Im(k) < \epsilon$. Why?

Step 3: Show that (1) can be rewritten

$$\frac{R(k)}{\sqrt{1+ik}} - \frac{\sqrt{2}}{1+ik} = \frac{\sqrt{1-ik} - \sqrt{2}}{1+ik} + \sqrt{1-ik}\, F_+(k).$$

Note that the left side of this equation is analytic in the lower half-plane $\Im(k) < \epsilon$, while the right side is analytic in the upper half-plane $\Im(k) > 0$.

Step 4: Use Liouville's theorem and deduce that $R(k) = \sqrt{2}/\sqrt{1+ik}$.

Step 5: Show that

$$U(k, y) = \frac{\sqrt{2}}{\sqrt{1+ik}} \frac{e^{-|y|\sqrt{k^2+1}}}{\sqrt{k^2+1}}.$$

Step 6: Using integral tables, show that

$$\mathcal{F}^{-1}\left(\frac{e^{-|y|\sqrt{k^2+1}}}{\sqrt{k^2+1}} \right) = \frac{1}{2\pi} \int_{-\infty}^{\infty} \frac{e^{ikx-|y|\sqrt{k^2+1}}}{\sqrt{k^2+1}} dk = \frac{K_0(r)}{\pi},$$

where $r^2 = x^2 + y^2$.

Step 7: Using contour integration, show that

$$\mathcal{F}^{-1}\left(\frac{\sqrt{2}}{\sqrt{1+ik}} \right) = e^{-x} \sqrt{\frac{2}{\pi x}}\, H(x).$$

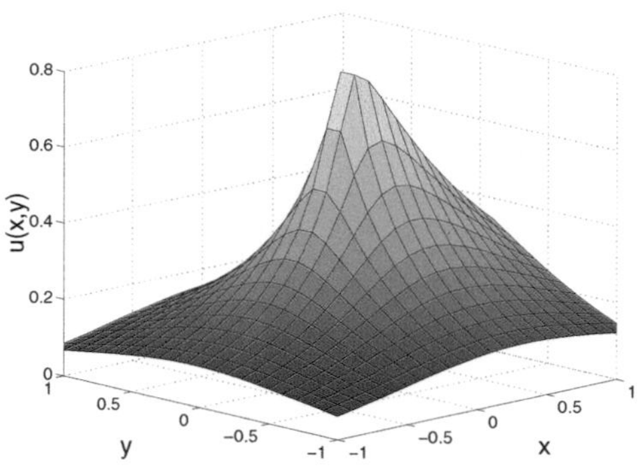

Problem 1

Step 8: Using the results from Step 6 and Step 7 and applying the convolution theorem, show that

$$u(x,y) = e^{-x}\sqrt{\frac{2}{\pi x}}\,H(x) * \frac{K_0(r)}{\pi}.$$

Step 9: Complete the problem and show that

$$u(x,y) = \sqrt{\frac{2}{\pi^3}}\int_0^\infty \frac{e^{-\chi}}{\sqrt{\chi}} K_0\left[\sqrt{(x-\chi)^2+y^2}\,\right]d\chi$$

or

$$u(x,y) = \sqrt{\frac{8}{\pi^3}}\int_0^\infty e^{-\eta^2} K_0\left[\sqrt{(x-\eta^2)^2+y^2}\,\right]d\eta.$$

The figure entitled Problem 1 illustrates this solution.

2. Use the Wiener-Hopf technique to solve

$$\frac{\partial^2 u}{\partial x^2}+\frac{\partial^2 u}{\partial y^2} = u, \qquad -\infty < x < \infty, \quad 0 < y,$$

subject to the boundary conditions

$$\lim_{y\to\infty} u(x,y) \to 0, \qquad -\infty < x < \infty,$$

$$\begin{cases} u_y(x,0) = 0, & x < 0, \\ u(x,0) = e^{-x}, & 0 < x. \end{cases}$$

Step 1: Introducing

$$U(k,y) = \int_{-\infty}^{\infty} u(x,y)e^{ikx}\,dx,$$

use the differential equation and first two boundary conditions to show that $U(k,y) = A(k)e^{-y\sqrt{k^2+1}}$.

Step 2: Taking the Fourier transform of the boundary condition along $y = 0$, show that

$$A(k) = U_-(k) + \frac{1}{1-ki} \qquad \text{and} \qquad U'_+(k) = -\sqrt{k^2+1}\, A(k), \tag{1}$$

where

$$U_-(k) = \int_{-\infty}^{0} u(x,0)e^{ikx}\,dx \qquad \text{and} \qquad U'_+(k) = \int_0^\infty u_y(x,0)e^{ikx}\,dx.$$

Here we have assumed that $|u(x,0)|$ is bounded by $e^{-\epsilon x}$, $0 < \epsilon \ll 1$, as $x \to \infty$ so that U_+ is analytic in the upper half-plane $\Im(k) > -\epsilon$, while U_- is analytic in the lower half-plane $\Im(k) < 0$.

Step 3: Show that (1) can be rewritten

$$-\frac{U_+(k)}{\sqrt{1-ki}} + \frac{\sqrt{2}}{1-ki} = \sqrt{1-ki}\,U_-(k) + \frac{\sqrt{1+ki}-\sqrt{2}}{1-ki}.$$

Note that the right side of this equation is analytic in the lower half-plane $\Im(k) < 0$, while the left side is analytic in the upper half-plane $\Im(k) > -\epsilon$.

Step 4: Use Liouville's theorem and deduce that

$$U_-(k) = \frac{\sqrt{2}}{(1-ki)\sqrt{1+ki}} - \frac{1}{1-ki}.$$

Step 5: Show that

$$U(k,y) = \frac{\sqrt{2}}{\sqrt{1-ki}}\frac{e^{-y\sqrt{k^2+1}}}{\sqrt{k^2+1}} = \frac{1+i}{(k+i)\sqrt{k-i}}\frac{e^{-y\sqrt{k^2+1}}}{\sqrt{k^2+1}}.$$

Step 6: Finish the problem by retracing Step 6 through Step 9 of the previous problem and show that you recover the same solution. Gramberg and van de Ven[24] found an alternative representation

$$u(x,y) = e^{-x} - \frac{\sqrt{2}}{\pi}\int_0^\infty \frac{e^{-x\sqrt{\eta^2+1}}}{\sqrt{\eta^2+1}\sqrt{\sqrt{\eta^2+1}-1}}\sin(\eta y)\,d\eta, \qquad x > 0,$$

---

[24] Gramberg, H. J. J., and A. A. F. van de Ven, 2005: Temperature distribution in a Newtonian fluid injected between two semi-infinite plates. *Eur. J. Mech., Ser. B.*, **24**, 767–787.

and

$$u(x,y) = \frac{\sqrt{2}}{\pi} \int_0^\infty \frac{e^{x\sqrt{\eta^2+1}}}{\sqrt{\eta^2+1}\sqrt{\sqrt{\eta^2+1}+1}} \cos(\eta y)\, d\eta, \qquad x < 0,$$

by evaluating the inverse Fourier transform via contour integration.

3. Use the Wiener-Hopf technique to solve the mixed boundary value problem

$$\frac{\partial^2 u}{\partial x^2} + \frac{\partial^2 u}{\partial y^2} = u, \qquad -\infty < x < \infty, \quad 0 < y,$$

with the boundary conditions $\lim_{y\to\infty} u(x,y) \to 0$,

$$\begin{cases} u(x,0) = 1, & x < 0, \\ u_y(x,0) = 0, & 0 < x. \end{cases}$$

Step 1: Assuming that $|u(x,0)|$ is bounded by $e^{-\epsilon x}$ as $x \to \infty$, where $0 < \epsilon \ll 1$, let us define the following Fourier transforms:

$$U(k,y) = \int_{-\infty}^\infty u(x,y)e^{ikx}\, dx, \qquad U_+(k,y) = \int_0^\infty u(x,y)e^{ikx}\, dx,$$

and

$$U_-(k,y) = \int_{-\infty}^0 u(x,y)e^{ikx}\, dx,$$

so that $U(k,y) = U_+(k,y) + U_-(k,y)$. Here, $U_+(k,y)$ is analytic in the half-space $\Im(k) > -\epsilon$, while $U_-(k,y)$ is analytic in the half-space $\Im(k) < 0$. Then show that the partial differential equation becomes

$$\frac{d^2U}{dy^2} - m^2 U = 0, \qquad 0 < y,$$

with $\lim_{y\to\infty} U(k,y) \to 0$, where $m^2 = k^2 + 1$.

Step 2: Show that the solution to Step 1 is $U(k,y) = A(k)e^{-my}$.

Step 3: From the boundary conditions along $x = 0$, show that

$$A(k) = M_+(k) - \frac{i}{k} \qquad \text{and} \qquad -mA(k) = L_-(k),$$

where

$$M_+(k) = \int_0^\infty u(x,0)e^{ikx}\, dx \quad \text{and} \quad L_-(k) = \int_{-\infty}^0 u_y(x,0)e^{ikx}\, dx.$$

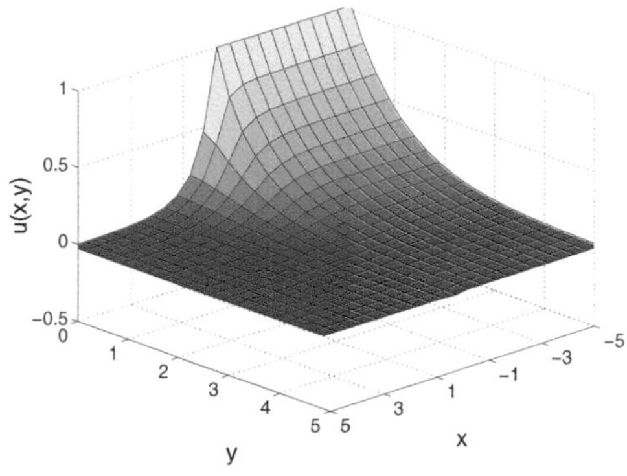

Problem 3

Note that $M_+(k)$ is analytic in the half-space $\Im(k) > -\epsilon$, while $L_-(k)$ is analytic in the half-space $\Im(k) < 0$.

Step 4: By eliminating $A(k)$ from the equations in Step 3, show that we can factor the resulting equation as

$$-\sqrt{k+i}\,M_+(k) + i\frac{\sqrt{k+i}-\sqrt{i}}{k} = \frac{L_-(k)}{\sqrt{k-i}} - \frac{i^{3/2}}{k}.$$

Note that the left side of the equation is analytic in the upper half-plane $\Im(k) > -\epsilon$, while the right side of the equation is analytic in the lower half-plane $\Im(k) < 0$.

Step 5: Use Liouville's theorem to show that each side of the equation in Step 4 equals zero. Therefore,

$$M_+(k) = \frac{i}{k} - \frac{i^{3/2}}{k\sqrt{k+i}} \qquad \text{and} \qquad U(k,y) = -\frac{i^{3/2}}{k\sqrt{k+i}}e^{-y\sqrt{k^2+1}}.$$

Step 6: Use the inversion integral and show that

$$u(x,y) = -\frac{i^{3/2}}{2\pi}\int_{-\infty-\epsilon i}^{\infty-\epsilon i} \frac{\exp\left(-ikx - y\sqrt{k^2+1}\right)}{k\sqrt{k+i}}\,dk.$$

Step 7: Using contour integration and Figure 5.2.2, evaluate the integral in Step 6 and show that

$$u(x,y) = \frac{1}{\pi}\int_1^\infty \cos\left(y\sqrt{\eta^2-1}\right)e^{-x\eta}\frac{d\eta}{\eta\sqrt{\eta-1}}$$

if $x > 0$, and

$$u(x,y) = e^{-y} - \frac{1}{\pi}\int_1^\infty \sin\left(y\sqrt{\eta^2-1}\right)e^{x\eta}\frac{d\eta}{\eta\sqrt{\eta+1}}$$

if $x < 0$. The figure labeled Problem 3 illustrates $u(x,y)$.

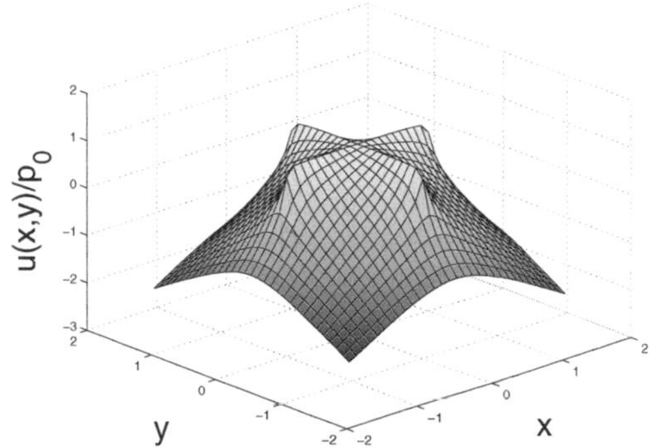

Chapter 6
Green's Function

The use of Green's functions to construct solutions to boundary value problems dates back to nineteenth century electrostatics. In this chapter we first show how to construct a Green's function with mixed boundary conditions. Then we will apply integral representations to a mixed boundary value problem when the kernel is a Green's function. In the last section we specialize to potentials.

6.1 GREEN'S FUNCTION WITH MIXED BOUNDARY VALUE CONDITIONS

We begin our study of Green's function methods by examining how we might construct a Green's function when mixed boundary conditions are present. Consider the rather simple problem[1] of

$$\frac{\partial^2 u}{\partial x^2} + \frac{\partial^2 u}{\partial y^2} = -\delta(x)\delta(y-b), \qquad -\infty < x < \infty, \quad 0 < y < \infty, \tag{6.1.1}$$

subject to the boundary conditions

$$\lim_{|x|\to\infty} u(x,y) \to 0, \qquad 0 < y < \infty, \tag{6.1.2}$$

[1] See Khanzhov, A. D., 1966: A mixed heat conduction boundary problem for a semi-infinite plate. *J. Engng. Phys.*, **11**, 370–371.

$$\lim_{y\to\infty} u(x,y) \to 0, \qquad -\infty < x < \infty, \tag{6.1.3}$$

and

$$\begin{cases} u_y(x,0) = 0, & 0 \le |x| < a, \\ u(x,0) = 0, & a < |x| < \infty. \end{cases} \tag{6.1.4}$$

Using Fourier cosine transforms, the partial differential equation and the boundary conditions given by Equation 6.1.2 and Equation 6.1.3 are satisfied by

$$u(x,y) = \frac{2}{\pi}\int_0^\infty \left[\frac{e^{-k(b-y)}}{4k} + A(k)e^{-k(b-y)}\right]\cos(kx)\,dk, \quad 0 \le y \le b, \tag{6.1.5}$$

and

$$u(x,y) = \frac{2}{\pi}\int_0^\infty \left[\frac{e^{-k(y-b)}}{4k} + A(k)e^{-k(y-b)}\right]\cos(kx)\,dk, \quad b \le y < \infty. \tag{6.1.6}$$

The arbitrary constant $A(k)$ will be used to satisfy the final boundary condition, Equation 6.1.4. Direct substitution of Equation 6.1.5 and Equation 6.1.6 into this boundary condition yields the dual integral equations

$$\int_0^\infty \left[\frac{e^{-kb}}{4} - kA(k)e^{kb}\right]\cos(kx)\,dk = 0, \qquad 0 \le |x| < a, \tag{6.1.7}$$

and

$$\int_0^\infty \left[\frac{e^{-kb}}{4k} + A(k)e^{kb}\right]\cos(kx)\,dk = 0, \qquad a < |x| < \infty. \tag{6.1.8}$$

Noting

$$\cos(kx) = \sqrt{\frac{\pi kx}{2}}J_{-\frac{1}{2}}(kx), \tag{6.1.9}$$

we can reexpress Equation 6.1.7 and Equation 6.1.8 as

$$\int_0^\infty \left[\frac{e^{-kb}}{4} - kA(k)e^{kb}\right]\sqrt{kx}\,J_{-\frac{1}{2}}(kx)\,dk = 0, \qquad 0 \le |x| < a, \tag{6.1.10}$$

and

$$\int_0^\infty \left[\frac{e^{-kb}}{4k} + A(k)e^{kb}\right]\sqrt{kx}\,J_{-\frac{1}{2}}(kx)\,dk = 0, \qquad a < |x| < \infty. \tag{6.1.11}$$

If we introduce $x = a\rho$ and $\eta = ka$, Equation 6.1.10 and Equation 6.1.11 become the nondimensional integral equations

$$\int_0^\infty \left[\frac{e^{-k\eta/a}}{4} - \frac{\eta}{a}A(\eta)e^{b\eta/a}\right]\sqrt{\eta}\,J_{-\frac{1}{2}}(\rho\eta)\,d\eta = 0, \qquad 0 \le |\rho| < 1, \tag{6.1.12}$$

and

$$\int_0^\infty \left[\frac{ae^{-b\eta/a}}{4\eta} + A(\eta)e^{b\eta/a}\right]\sqrt{\eta}\, J_{-\frac{1}{2}}(\rho\eta)\,d\eta = 0, \qquad 1 < |\eta| < \infty. \quad \textbf{(6.1.13)}$$

Let us introduce

$$B(\eta) = \frac{1}{a^2}\sqrt{\frac{\eta}{a}}\left[\frac{a}{4\eta}e^{-b\eta/a} + A(\eta)e^{b\eta/a}\right]. \quad \textbf{(6.1.14)}$$

Then Equation 6.1.12 and Equation 6.1.13 become

$$\int_0^\infty \eta B(\eta) J_{-\frac{1}{2}}(\rho\eta)\,d\eta = h(\rho) = \sqrt{\frac{2}{\pi a\rho}}\,\frac{b}{b^2+a^2\rho^2}, \qquad 0 \le |\rho| < 1, \quad \textbf{(6.1.15)}$$

and

$$\int_0^\infty B(\eta) J_{-\frac{1}{2}}(\rho\eta)\,d\eta = 0, \qquad 1 < |\rho| < \infty. \quad \textbf{(6.1.16)}$$

The solution to these dual integral equations has been given by Titchmarsh[2] with $\nu = -\frac{1}{2}$ and $\alpha = 1$. Solving for $B(\eta)$ we find that

$$B(\eta) = \sqrt{\frac{2\eta}{\pi}}\int_0^1 \mu^{3/2} J_0(\mu\eta)\left[\int_0^1 h(\mu\rho)\sqrt{\frac{\rho}{1-\rho^2}}\,d\rho\right]d\mu \quad \textbf{(6.1.17)}$$

$$= \sqrt{\frac{\eta}{a}}\int_0^1 \frac{\mu\, J_0(\mu\eta)}{\sqrt{b^2+a^2\mu^2}}\,d\mu. \quad \textbf{(6.1.18)}$$

Upon using $B(\eta)$ to find $A(k)$, substituting $A(k)$ into Equation 6.1.5 and Equation 6.1.6, and then evaluating the integrals, we find that

$$u(x,y) = \frac{1}{2\pi}\left\{\frac{1}{2}\ln\left[\frac{(b+y)^2+x^2}{(b-y)^2+x^2}\right]\right. \quad \textbf{(6.1.19)}$$

$$\left. + \ln\left|\frac{R+a^2+b^2+\sqrt{2(a^2+b^2)(y^2-x^2+a^2+R)}}{(b+y)^2+x^2}\right|\right\}$$

where $R = \sqrt{(y^2-x^2+a^2)^2+4x^2y^2}$. Figure 6.1.1 illustrates the solution 6.1.19 when $a/b = 0.5$.

Efimov and Vorob'ev[3] found the Green's function for the three-dimensional Laplace equation in the half-space $z \ge 0$ with the boundary conditions

$$\begin{cases} g_z(x,y,0^+|\xi,\eta,0) = \delta(x-\xi)\delta(y-\eta), & x^2+y^2 < a^2, \\ g(x,y,0^+|\xi,\eta,0) = 0, & x^2+y^2 > a^2. \end{cases} \quad \textbf{(6.1.20)}$$

---

[2] Titchmarsh, E. C., 1948: *Introduction of the Theory of Fourier Integrals.* Oxford, Section 11.16.

[3] Efimov, A. B., and V. N. Vorob'ev, 1975: A mixed boundary value problem for the Laplace equation. *J. Engng. Phys.*, **26**, 664–666.

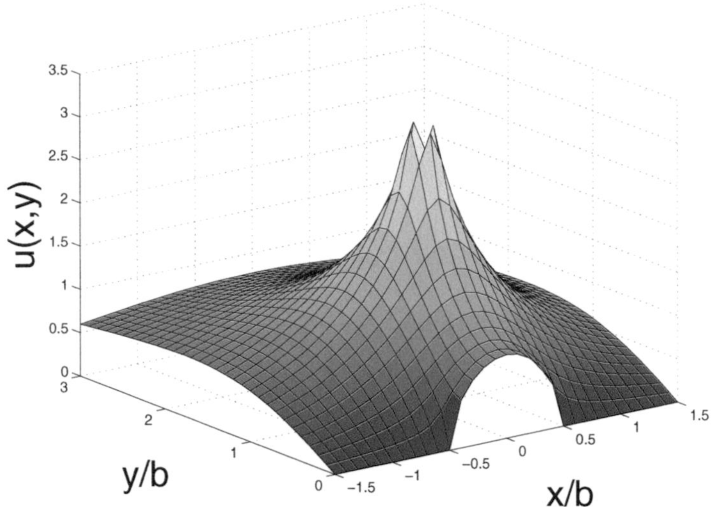

Figure 6.1.1: The Green's function given by Equation 6.1.19 when $a/b = 0.5$.

It is

$$g(x,y,z|\xi,\eta,\zeta) = \frac{2}{\pi r}\arcsin\left\{\frac{\cos[(\theta-\theta_0)/2]}{\cosh(\alpha/2)}\right\}, \tag{6.1.21}$$

where

$$r^2 = (x-\xi)^2 + (y-\eta)^2 + (z-\zeta)^2, \tag{6.1.22}$$

$$\theta = \frac{i}{2}\ln\left[\frac{x^2+y^2+(z-ai)^2}{x^2+y^2+(z+ai)^2}\right], \quad \theta_0 = \frac{i}{2}\ln\left[\frac{\xi^2+\eta^2+(\zeta-ai)^2}{\xi^2+\eta^2+(\zeta+ai)^2}\right], \tag{6.1.23}$$

$$\rho = \frac{1}{2}\ln\left[\frac{\left(\sqrt{x^2+y^2}+a\right)^2+z^2}{\left(\sqrt{x^2+y^2}-a\right)^2+z^2}\right], \quad \rho_0 = \frac{1}{2}\ln\left[\frac{\left(\sqrt{\xi^2+\eta^2}+a\right)^2+\zeta^2}{\left(\sqrt{\xi^2+\eta^2}-a\right)^2+\zeta^2}\right], \tag{6.1.24}$$

$$\varphi = \arctan(y/x), \quad \varphi_0 = \arctan(\eta/\xi), \tag{6.1.25}$$

and

$$\cosh(\alpha) = \cosh(\rho)\cosh(\rho_0) - \sinh(\rho)\sinh(\rho_0)\cos(\varphi-\varphi_0). \tag{6.1.26}$$

Similarly, when the boundary conditions along $z = 0$ read

$$\begin{cases} g(x,y,0^+|\xi,\eta,0) = \delta(x-\xi)\delta(y-\eta), & x^2+y^2 < a^2, \\ g_z(x,y,0^+|\xi,\eta,0) = 0, & x^2+y^2 > a^2, \end{cases} \tag{6.1.27}$$

then the Green's function is

$$g(x, y, z|\xi, \eta, \zeta)$$
$$= \frac{2z}{\pi r_0^3} \arcsin\left[\frac{\sqrt{R - (x^2 + y^2 + z^2 - a^2)}\,\sqrt{a^2 - \xi^2 - \eta^2}}{\sqrt{R(a^2 - \xi^2 - \eta^2) + r_1^2(x, y)r_1^2(\xi, \eta) - 4a^2(x\xi + y\eta)}}\right]$$
$$+ \frac{2\sqrt{2}\,az}{\pi r_0^2\sqrt{(a^2 - \xi^2 - \eta^2)[R - (x^2 + y^2 + z^2 - a^2)]}}, \tag{6.1.28}$$

where

$$r_0^2 = (x - \xi)^2 + (y - \eta)^2 + z^2, \tag{6.1.29}$$

$$r_1^2(x, y) = x^2 + y^2 + a^2, \tag{6.1.30}$$

and

$$R^2 = (x^2 + y^2 + z^2 - a^2)^2 + 4a^2z^2. \tag{6.1.31}$$

6.2 INTEGRAL REPRESENTATIONS INVOLVING GREEN'S FUNCTIONS

Green's functions have long been used to create integral representations for boundary value problems. Here we illustrate how this technique is used in the case of mixed boundary value problems. As before, we will face an integral equation that must solved, usually numerically.

• Example 6.2.1

For our first example, let us complete the solution of

$$\frac{\partial^2 u}{\partial x^2} + \frac{\partial^2 u}{\partial y^2} = 0, \qquad -\infty < x < \infty, \quad 0 < y < L, \tag{6.2.1}$$

subject to the boundary conditions

$$\begin{cases} u_y(x, 0) = -h(x), & |x| < 1, \\ u(x, 0) = 0, & |x| > 1, \end{cases} \tag{6.2.2}$$

$$u(x, L) = 0, \qquad -\infty < x < \infty, \tag{6.2.3}$$

and

$$\lim_{|x|\to\infty} u(x, y) \to 0, \qquad 0 < y < L, \tag{6.2.4}$$

using Green's functions that we began in Example 1.1.4. There we showed that

$$u(x, y) = \frac{1}{2L}\int_{-1}^{1} \frac{f(\xi)\,\sin(\pi y/L)}{\cosh[\pi(x - \xi)/L] - \cos(\pi y/L)}\,d\xi, \tag{6.2.5}$$

where $f(x)$ is given by the integral equation

$$\frac{1}{2L}\int_{-1}^{1} f'(\xi)\coth[\pi(x-\xi)/(2L)]\,d\xi = h(x), \qquad |x|<1. \tag{6.2.6}$$

To solve Equation 6.2.6, we write $h(x)$ as a sum of an even function $h_e(x)$ and an odd function $h_o(x)$. Let us denote by $f_1(\xi)$ that portion of $f(\xi)$ due to the contribution from $h_e(x)$. Then, by integrating Equation 6.2.6 from $-x$ to x, we have that

$$\pi\int_0^x h_e(t)\,dt = \int_0^1 f_1'(\xi)\ln\left|\frac{\sinh[\pi(x-\xi)/(2L)]}{\sinh[\pi(x+\xi)/(2L)]}\right|\,d\xi \tag{6.2.7}$$

$$= \int_0^1 f_1'(\xi)\ln\left|\frac{\tanh[\pi x/(2L)]+\tanh[\pi\xi/(2L)]}{\tanh[\pi x/(2L)]-\tanh[\pi\xi/(2L)]}\right|\,d\xi \tag{6.2.8}$$

for $0\le x\le 1$. From Example 1.2.3, we have that

$$f_1'(x) = \frac{1}{L}\frac{d}{dx}\left[\int_x^1 \frac{\tanh[\pi\xi/(2L)]}{\cosh^2[\pi\xi/(2L)]\sqrt{\tanh^2[\pi\xi/(2L)]-\tanh^2[\pi x/(2L)]}} \times\left\{\int_0^\xi \frac{h_e(\eta)}{\sqrt{\tanh^2[\pi\xi/(2L)]-\tanh^2[\pi\eta/(2L)]}}\,d\eta\right\}d\xi\right], \tag{6.2.9}$$

or

$$f_1(x) = -\frac{1}{L}\int_x^1 \frac{\tanh[\pi\xi/(2L)]}{\cosh^2[\pi\xi/(2L)]\sqrt{\tanh^2[\pi\xi/(2L)]-\tanh^2[\pi x/(2L)]}} \times\left\{\int_0^\xi \frac{h_e(\eta)}{\sqrt{\tanh^2[\pi\xi/(2L)]-\tanh^2[\pi\eta/(2L)]}}\,d\eta\right\}d\xi \tag{6.2.10}$$

for $0\le x\le 1$. Consequently, the portion of the potential $u_1(x,y)$ due to $h_e(\eta)$ can be computed from

$$u_1(x,y) = \frac{\sin(\pi y/L)}{2L}\int_0^1\left[\frac{f_1(\xi)}{\cosh(\pi|x-\xi|/L)-\cos(\pi y/L)} + \frac{f_1(\xi)}{\cosh(\pi|x+\xi|/L)-\cos(\pi y/L)}\right]d\xi, \tag{6.2.11}$$

where $f_1(\xi)$ is given by Equation 6.2.10.

Turning now to finding that portion of $f(\xi)$, $f_2(\xi)$, due to $h_o(x)$, Equation 6.2.6 yields

$$h_o(x) = \frac{1}{2L}\int_0^1 f_2'(\xi)\left\{\coth[\pi(x-\xi)/(2L)]+\coth[\pi(x+\xi)/(2L)]\right\}d\xi. \tag{6.2.12}$$

Integrating Equation 6.2.12 from 0 to x,

$$\pi \int_0^x h_o(t)\,dt = \int_0^1 f_2'(\xi) \ln\left| \frac{\sinh^2[\pi x/(2L)] - \sinh^2[\pi\xi/(2L)]}{\sinh^2[\pi\xi/(2L)]} \right| d\xi \quad \mathbf{(6.2.13)}$$

for $0 \le x \le 1$. From Example 1.2.4,

$$\begin{aligned} f_2'(x) = {} & \frac{1}{2\sinh[\pi x/(2L)]} \frac{d}{dx}\left[\int_x^1 \frac{\sinh(\pi\xi/L)}{\sqrt{\sinh^2[\pi\xi/(2L)] - \sinh^2[\pi x/(2L)]}} \right. \\ & \left. \times \left\{ \int_0^\xi \frac{\sinh[\pi\eta/(2L)]\, h_o(\eta)}{\sqrt{\sinh^2[\pi\xi/(2L)] - \sinh^2[\pi\eta/(2L)]}}\, d\eta \right\} d\xi \right] \\ & + \frac{\pi A}{2L} \frac{\cosh[\pi x/(2L)]}{\sqrt{\sinh^2[\pi/(2L)] - \sinh^2[\pi x/(2L)]}}, \end{aligned} \quad \mathbf{(6.2.14)}$$

where A is an undetermined constant and $0 < x < 1$. Integrating Equation 6.2.14 with respect to x, we obtain

$$\begin{aligned} f_2(x) = {} & -\frac{1}{2L} \int_x^1 \frac{d\chi}{\sinh[\pi\chi/(2L)]} \frac{d}{d\chi}\left[\int_\chi^1 \frac{\sinh(\pi\xi/L)}{\sqrt{\sinh^2[\pi\xi/(2L)] - \sinh^2[\pi\chi/(2L)]}} \right. \\ & \left. \times \left\{ \int_0^\xi \frac{\sinh[\pi\eta/(2L)]\, h_o(\eta)}{\sqrt{\sinh^2[\pi\xi/(2L)] - \sinh^2[\pi\eta/(2L)]}}\, d\eta \right\} d\xi \right] \\ & - A\left[\frac{\pi}{2} - \arcsin\left\{ \frac{\sinh[\pi x/(2L)]}{\sinh[\pi/(2L)]} \right\} \right] \end{aligned} \quad \mathbf{(6.2.15)}$$

for $0 \le x \le 1$. Because $f_2(0) = 0$,

$$\begin{aligned} A = {} & -\frac{1}{\pi L} \int_0^1 \frac{d\chi}{\sinh[\pi\chi/(2L)]} \frac{d}{d\chi}\left[\int_\chi^1 \frac{\sinh(\pi\xi/L)}{\sqrt{\sinh^2[\pi\xi/(2L)] - \sinh^2[\pi\chi/(2L)]}} \right. \\ & \left. \times \left\{ \int_0^\xi \frac{\sinh[\pi\eta/(2L)]\, h_o(\eta)}{\sqrt{\sinh^2[\pi\xi/(2L)] - \sinh^2[\pi\eta/(2L)]}}\, d\eta \right\} d\xi \right]. \end{aligned} \quad \mathbf{(6.2.16)}$$

Consequently, the portion of the potential due to $h_o(x)$ is

$$\begin{aligned} u_2(x,y) = {} & \frac{\sin(\pi y/L)}{2L} \int_0^1 \left[\frac{f_2(\xi)}{\cosh(\pi|x-\xi|/L) - \cos(\pi y/L)} \right. \\ & \left. - \frac{f_2(\xi)}{\cosh(\pi|x+\xi|/L) - \cos(\pi y/L)} \right] d\xi. \end{aligned} \quad \mathbf{(6.2.17)}$$

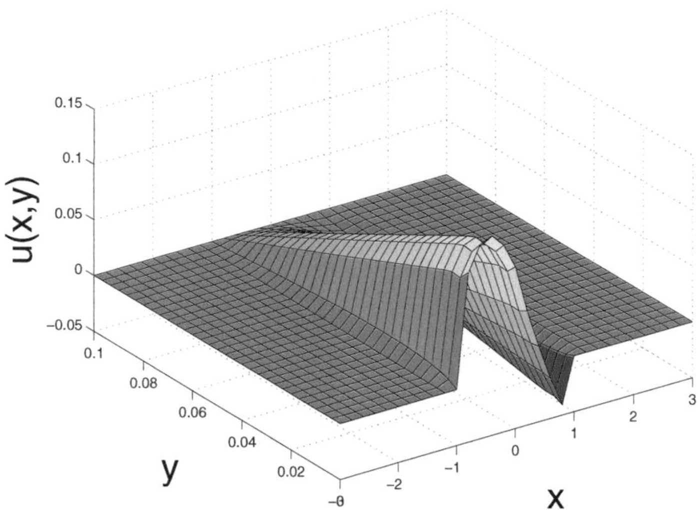

Figure 6.2.1: The solution to Laplace's equation with the boundary conditions given by Equation 6.2.2 through Equation 6.2.4 when $h(x) = (1-x)^2$ and $L = 0.1$.

The potential due to $h(x)$ equals the sum of $u_1(x,y)$ and $u_2(x,y)$. Figure 6.2.1 illustrates this solution when $h(x) = (1-x)^2$ and $L = 0.1$.

Yang et al.[4] solved this problem when they replaced the boundary condition $u(x,L) = 0$ with $u_y(x,L) = 0$. In this case the Green's function becomes

$$\begin{aligned} g(x,y|\xi,\eta) &= \frac{2}{\pi} e^{-\pi|x-\xi|/(2L)} \sin\left(\frac{\pi y}{2L}\right) \sin\left(\frac{\pi\eta}{2L}\right) \\ &+ \frac{1}{4\pi} \ln\left\{ \frac{\cosh[\pi(x-\xi)/(2L)] + \cos[\pi(y+\eta)/(2L)]}{\cosh[\pi(x-\xi)/(2L)] - \cos[\pi(y+\eta)/(2L)]} \right. \\ &\qquad \left. \times \frac{\cosh[\pi(x-\xi)/(2L)] - \cos[\pi(y-\eta)/(2L)]}{\cosh[\pi(x-\xi)/(2L)] + \cos[\pi(y-\eta)/(2L)]} \right\}, \end{aligned} \tag{6.2.18}$$

and Equation 6.2.6 is replaced with

$$h(x) = -\frac{\pi}{4L^2} \int_{-1}^{1} f(\xi) \frac{\cosh[\pi(x-\xi)/(2L)]}{\sinh^2[\pi(x-\xi)/(2L)]} \, d\xi \tag{6.2.19}$$

$$= \frac{1}{2L} \int_{-1}^{1} \frac{f'(\xi)}{\sinh[\pi(x-\eta)/(2L)]} \, d\xi, \qquad |x| < 1. \tag{6.2.20}$$

If we repeat our previous analysis where we set $h(x) = h_e(x) + h_o(x)$, the portion of the potential $u(x,y)$ due to $h_e(x)$ is

[4] Taken with permission from Yang, F., V. Prasad, and I. Kao, 1999: The thermal constriction resistance of a strip contact spot on a thin film. *J. Phys. D: Appl. Phys.*, **32**, 930–936. Published by IOP Publishing Ltd.

$$
\begin{aligned}
u_1(x,y) &= \frac{\sin(\pi y/L)}{4L} \\
&\times \left\{ \int_0^1 \frac{\cosh[\pi x_1/(2L)]\, f_1(\xi)}{\sinh^2[\pi x_1/(2L)]\cos^2[\pi y/(2L)] + \cosh^2[\pi x_1/(2L)]\sin^2[\pi y/(2L)]} \right. \\
&+ \left. \frac{\cosh[\pi x_2/(2L)]\, f_1(\xi)}{\sinh^2[\pi x_2/(2L)]\cos^2[\pi y/(2L)] + \cosh^2[\pi x_2/(2L)]\sin^2[\pi y/(2L)]}\, d\xi \right\},
\end{aligned} \tag{6.2.21}
$$

where $x_1 = |x - \xi|$, $x_2 = |x + \xi|$, and

$$
\begin{aligned}
f_1(x) = -\frac{1}{2L}\int_x^1 & \frac{\sinh(\pi\xi/L)}{\sqrt{\sinh^2[\pi\xi/(2L)] - \sinh^2[\pi x/(2L)]}} \\
& \times \left\{ \int_0^\xi \frac{h_e(\eta)}{\sqrt{\sinh^2[\pi\xi/(2L)] - \sinh^2[\pi\eta/(2L)]}}\, d\eta \right\} d\xi.
\end{aligned} \tag{6.2.22}
$$

On the other hand, the portion of $u(x,y)$ due to $h_o(x)$ is

$$
\begin{aligned}
u_2(x,y) &= \frac{\sin(\pi y/L)}{4L} \\
&\times \left\{ \int_0^1 \frac{\cosh[\pi x_1/(2L)]\, f_2(\xi)}{\sinh^2[\pi x_1/(2L)]\cos^2[\pi y/(2L)] + \cosh^2[\pi x_1/(2L)]\sin^2[\pi y/(2L)]} \right. \\
&- \left. \frac{\cosh[\pi x_2/(2L)]\, f_2(\xi)}{\sinh^2[\pi x_2/(2L)]\cos^2[\pi y/(2L)] + \cosh^2[\pi x_2/(2L)]\sin^2[\pi y/(2L)]}\, d\xi \right\},
\end{aligned} \tag{6.2.23}
$$

where

$$
\begin{aligned}
f_2(x) = -\frac{1}{2L}\int_x^1 & \frac{d\chi}{\sinh[\pi\chi/(2L)]}\frac{d}{d\chi}\left[\int_\chi^1 \frac{\sinh(\pi\xi/L)}{\sqrt{\sinh^2[\pi\xi/(2L)] - \sinh^2[\pi\chi/(2L)]}} \right. \\
& \times \left. \left\{ \int_0^\xi \frac{\sinh[\pi\eta/(2L)]\, h_o(\eta)}{\sqrt{\sinh^2[\pi\xi/(2L)] - \sinh^2[\pi\eta/(2L)]}}\, d\eta \right\} d\xi \right] \\
& - 2A\int_x^1 \frac{d\xi}{\sqrt{\sinh^2[\pi/(2L)] - \sinh^2[\pi\xi/(2L)]}},
\end{aligned} \tag{6.2.24}
$$

and

$$
A = -\frac{1}{4L}\int_0^1 \frac{d\chi}{\sinh[\pi\chi/(2L)]}\frac{d}{d\chi}\left[\int_\chi^1 \frac{\sinh(\pi\xi/L)}{\sqrt{\sinh^2[\pi\xi/(2L)] - \sinh^2[\pi\chi/(2L)]}}\right.
$$

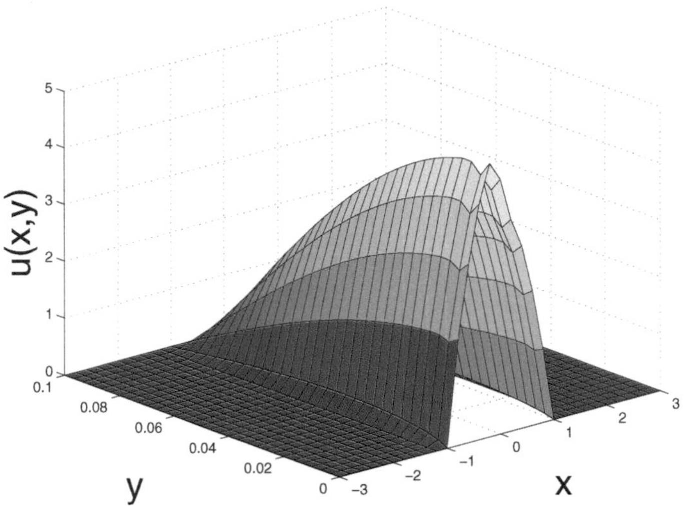

Figure 6.2.2: Same as Figure 6.2.1 except that the boundary condition $u(x, L) = 0$ has been replaced with $u_y(x, L) = 0$.

$$\times \left\{ \int_0^\xi \frac{\sinh[\pi\eta/(2L)]\, h_o(\eta)}{\sqrt{\sinh^2[\pi\xi/(2L)] - \sinh^2[\pi\eta/(2L)]}}\, d\eta \right\} d\xi \Bigg]$$
$$\Bigg/ \left\{ \int_0^1 \frac{d\xi}{\sqrt{\sinh^2[\pi/(2L)] - \sinh^2[\pi\xi/(2L)]}} \right\}. \tag{6.2.25}$$

The potential due to $h(x)$ equals the sum of $u_1(x, y)$ and $u_2(x, y)$. We have illustrated this solution in Figure 6.2.2 when $h(x) = (1 - x)^2$ and $L = 0.1$.

• Example 6.2.2

Green's functions are a powerful technique for solving electrostatic potential problems. Here we illustrate how Lal[5] used this technique to find the electrostatic potential to the mixed boundary value problem:

$$\frac{\partial^2 u}{\partial x^2} + \frac{\partial^2 u}{\partial x^2} = 0, \qquad -\infty < x, y < \infty, \tag{6.2.26}$$

$$u(r, n\pi/2) = p_0, \qquad 0 \le r \le 1, \quad n = 1, 2, 3, 4. \tag{6.2.27}$$

---

[5] Taken with permission from Lal, B., 1978: A note on mixed boundary value problems in electrostatics. *Z. Angew. Math. Mech.*, **58**, 56–58. To see how to solve this problem using conformal mapping, see Homentcovschi, D., 1980: On the mixed boundary value problem for harmonic functions in plane domains. *J. Appl. Math. Phys.*, **31**, 352–366.

We also require that both $u(r, n\pi/2)$ and $u_\theta(r, n\pi/2)$ are continuous if $r > 1$.

From the theory of Green's function,

$$u(r, \theta) = \sum_{n=0}^{3} \int_0^1 A_n(\rho) g(r, \theta|\rho, n\pi/2)\, d\rho, \tag{6.2.28}$$

where

$$g(r, \theta|\rho, \theta_0) = -\frac{1}{4\pi} \ln[r^2 + \rho^2 - 2r\rho\cos(\theta - \theta_0)]. \tag{6.2.29}$$

From symmetry, we have that

$$A_0(\rho) = A_1(\rho) = A_2(\rho) = A_3(\rho). \tag{6.2.30}$$

Upon substituting Equation 6.2.30 into Equation 6.2.28 and applying Equation 6.2.27, we find that

$$\int_0^1 A_0(\rho) \ln|r^4 - \rho^4|\, d\rho = -2\pi p_0, \qquad 0 \le r \le 1. \tag{6.2.31}$$

Let us now introduce $r^2 = \cos(\theta/2)$ and $\rho^2 = \cos(\theta_0/2)$, where $0 \le \theta, \theta_0 \le \pi$, Equation 6.2.31 becomes

$$\int_0^\pi A_0(\rho) \frac{\sin(\theta_0/2)}{\rho} \ln\left|\cos^2(\theta/2) - \cos^2(\theta_0/2)\right|\, d\theta_0 = -8\pi p_0 \tag{6.2.32}$$

$$\int_0^\pi A_0(\rho) \frac{\sin(\theta_0/2)}{\rho} \ln\left|\frac{\cos(\theta) - \cos(\theta_0)}{2}\right|\, d\theta_0 = -8\pi p_0 \tag{6.2.33}$$

$$\int_0^\pi A_0(\rho) \frac{\sin(\theta_0/2)}{\rho} \left\{-2\ln(2) - 2\sum_{n=1}^{\infty} \frac{\cos(n\theta)\cos(n\theta_0)}{n}\right\} d\theta_0 = -8\pi p_0. \tag{6.2.34}$$

By inspection,

$$A_0(\rho) = \frac{4p_0\rho}{\ln(2)\sqrt{1 - \rho^4}}. \tag{6.2.35}$$

Therefore,

$$u(r, \theta) = \frac{4p_0}{\ln(2)} \int_0^1 \frac{\rho}{\sqrt{1 - \rho^4}} \sum_{n=0}^{3} g(r, \theta|\rho, n\pi/2)\, d\rho. \tag{6.2.36}$$

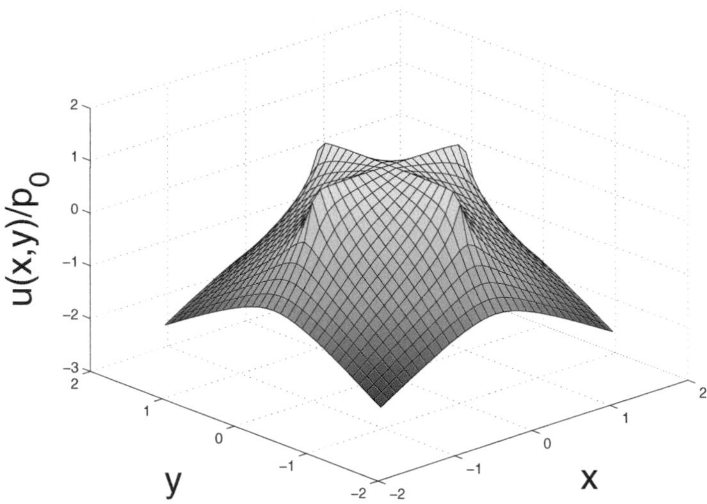

Figure 6.2.3: The electrostatic potential which satisfies the boundary condition $u(r, n\pi/2) = p_0$ when $r < 1$ and $n = 0, 1, 2, 3$.

Figure 6.2.3 illustrates this solution. Lal also found the approximate electrostatic potential for two-cross-shaped charged strips inside a grounded circular cylinder.

• Example 6.2.3: The method of Yang and Yao

Building upon Example 1.1.4 and Example 6.2.1, Yang and Yao[6] developed a general method for finding the potential $u(x, y)$ governed by the nondimensional partial differential equation:

$$\frac{\partial^2 u}{\partial x^2} + \frac{\partial^2 u}{\partial y^2} = 0, \qquad -\infty < x < \infty, \quad 0 < y < L, \tag{6.2.37}$$

subject to the boundary conditions

$$\lim_{|x|\to\infty} u(x, y) \to 0, \qquad 0 < y < L, \tag{6.2.38}$$

$$\begin{cases} u(x,0) = h(x), & |x| \le 1, \\ u_y(x,0) = 0, & |x| \ge 1, \end{cases} \tag{6.2.39}$$

and

$$u(x, L) = 0, \qquad -\infty < x < \infty. \tag{6.2.40}$$

[6] Yang, F.-Q., and R. Yao, 1996: The solution for mixed boundary value problems of two-dimensional potential theory. *Indian J. Pure Appl. Math.*, **27**, 313–322.

Their analysis begins by noting that we can express $u(x, y)$ in terms of the Green's function $g(x, y|\xi, \eta)$ by the integral

$$u(x, y) = \int_{-1}^{1} g(x, y|\xi, 0)\frac{\partial u(\xi, 0)}{\partial \eta}\, d\xi = \int_{-1}^{1} f(\xi)g(x, y|\xi, 0)\, d\xi, \qquad \textbf{(6.2.41)}$$

where $f(\xi)$ is presently unknown and the Green's function $g(x, y|\xi, \eta)$ is given by

$$\frac{\partial^2 g}{\partial x^2} + \frac{\partial^2 g}{\partial y^2} = \delta(x-\xi)\delta(y-\eta), \qquad -\infty < x, \xi < \infty, \quad 0 < y, \eta < L, \ \textbf{(6.2.42)}$$

subject to the boundary conditions

$$\lim_{|x|\to\infty} |g(x, y|\xi, \eta)| < \infty, \qquad 0 < y < L, \qquad \textbf{(6.2.43)}$$

and

$$g_y(x, 0|\xi, \eta) = g(x, L|\xi, \eta) = 0, \qquad -\infty < x < \infty. \qquad \textbf{(6.2.44)}$$

Using standard techniques,[7] the Green's function equals

$$g(x, y|\xi, \eta) = -\sum_{n=1}^{\infty} \frac{2}{(2n-1)\pi} \exp\left[-\frac{(2n-1)\pi|x-\xi|}{2L}\right] \times \cos\left[\frac{(2n-1)\pi y}{2L}\right] \cos\left[\frac{(2n-1)\pi \eta}{2L}\right]. \qquad \textbf{(6.2.45)}$$

Therefore,

$$g(x, y|\xi, 0) = \frac{1}{\pi}\,\Re\left[\ln\left(\frac{\sqrt{r}-1}{\sqrt{r}+1}\right)\right], \qquad \textbf{(6.2.46)}$$

where $r = \exp[-\pi(|x-\xi|-iy)/L\,]$. Substituting Equation 6.2.46 into Equation 6.2.41 and using Equation 6.2.39, we find that

$$h(x) = \frac{1}{\pi}\int_{-1}^{1} f(\xi) \ln\big|\tanh[\pi(x-\xi)/(4L)]\big|\, d\xi. \qquad \textbf{(6.2.47)}$$

At this point, we specialize according to whether $h(x)$ is an even or odd function. Because any function can be written as the sum of an even and odd function, we can first rewrite $h(x)$ as a sum of an even function $h_e(x)$ and an odd function $h_o(x)$. Then we find the potentials for the corresponding $h_e(x)$ and $h_o(x)$. The potential for the given $h(x)$ then equals their sum by the principle of linear superposition.

---

[7] Duffy, D. G., 2001: *Green's Functions with Applications*. Chapman & Hall/CRC, 443 pp. See Section 5.2.

- $h(x)$ is an *even* function

In this case, Equation 6.2.47 can be rewritten

$$\pi h(x) = \int_0^1 f(\xi) \ln\left|\tanh[\pi(x-\xi)/(4L)] \tanh[\pi(x+\xi)/(4L)]\right| d\xi \qquad \mathbf{(6.2.48)}$$

with $0 \le x \le 1$. Taking the x-derivative,

$$h'(x) = \frac{2}{L} \int_0^1 f(\xi) \frac{\sinh[\pi x/(2L)] \cosh[\pi\xi/(2L)]}{\cosh(\pi x/L) - \cosh(\pi\xi/L)} \, d\xi; \qquad \mathbf{(6.2.49)}$$

or,

$$\frac{h'(x)}{\sinh[\pi x/(2L)]} = \frac{1}{L} \int_0^1 f(\xi) \frac{\cosh[\pi\xi/(2L)]}{\sinh^2[\pi x/(2L)] - \sinh^2[\pi\xi/(2L)]} \, d\xi. \qquad \mathbf{(6.2.50)}$$

Using the same techniques outlined in Example 6.2.1, we can solve for $f(x)$ and find that

$$f(x) = \frac{1}{2L \sinh(\pi x/L)} \frac{d}{dx} \left[\int_x^1 \frac{\sinh(\pi\xi/L)}{\sqrt{\sinh^2[\pi\xi/(2L)] - \sinh^2[\pi x/(2L)]}} \times \left\{ \int_0^\xi \frac{h'(\chi) \sinh(\pi\chi/L)}{\sqrt{\sinh^2[\pi\xi/(2L)] - \sinh^2[\pi\chi/(2L)]}} \, d\chi \right\} d\xi \right] + \frac{2A}{\sqrt{\sinh^2[\pi/(2L)] - \sinh^2[\pi x/(2L)]}}, \qquad \mathbf{(6.2.51)}$$

where A is a constant that is determined by $f(0^+)$. Substituting these results into Equation 6.2.41, we obtain the final result.

$$u(x,y) = \frac{1}{\pi} \int_0^1 \left[\ln\left\{ \frac{\sqrt{\sinh^2[\pi|x-\xi|/(2L)] + \sin^2[\pi y/(2L)]}}{\cosh[\pi|x-\xi|/(2L)] + \cos[\pi y/(2L)]} \right\} + \ln\left\{ \frac{\sqrt{\sinh^2[\pi|x+\xi|/(2L)] + \sin^2[\pi y/(2L)]}}{\cosh[\pi|x+\xi|/(2L)] + \cos[\pi y/(2L)]} \right\} \right] f(\xi) \, d\xi. \qquad \mathbf{(6.2.52)}$$

- $h(x)$ is an *odd* function

In the case when $h(x)$ is an odd function, Equation 6.2.47 can be rewritten

$$\pi h(x) = \int_0^1 f(\xi) \ln\left|\frac{\tanh[\pi(x-\xi)/(4L)]}{\tanh[\pi(x+\xi)/(4L)]}\right| d\xi \tag{6.2.53}$$

$$= -\int_0^1 f(\xi) \ln\left|\frac{\sinh[\pi x/(2L)] + \sinh[\pi\xi/(2L)]}{\sinh[\pi x/(2L)] - \sinh[\pi\xi/(2L)]}\right| d\xi \tag{6.2.54}$$

for $0 \le x \le 1$. Again, applying the methods from Example 6.2.1, we find that

$$\begin{aligned} f(x) = \frac{1}{2L}\frac{d}{dx}\Bigg[\int_x^1 & \frac{\sinh(\pi\xi/L)}{\sqrt{\sinh^2[\pi\xi/(2L)] - \sinh^2[\pi x/(2L)]}} \\ & \times \left\{\int_0^\xi \frac{h'(\chi)}{\sqrt{\sinh^2[\pi\xi/(2L)] - \sinh^2[\pi\chi/(2L)]}}\, d\chi\right\} d\xi\Bigg] \\ & - \frac{h(0)\sinh[\pi/(2L)]\mathrm{cotanh}[\pi x/(2L)]}{L\sqrt{\sinh^2[\pi/(2L)] - \sinh^2[\pi x/(2L)]}}. \end{aligned} \tag{6.2.55}$$

Substituting Equation 6.2.55 into Equation 6.2.41, the potential is

$$\begin{aligned} u(x,y) = \frac{1}{\pi}\int_0^1 \Bigg[& \ln\left\{\frac{\sqrt{\sinh^2[\pi|x-\xi|/(2L)] + \sin^2[\pi y/(2L)]}}{\cosh[\pi|x-\xi|/(2L)] + \cos[\pi y/(2L)]}\right\} \\ & - \ln\left\{\frac{\sqrt{\sinh^2[\pi|x+\xi|/(2L)] + \sin^2[\pi y/(2L)]}}{\cosh[\pi|x+\xi|/(2L)] + \cos[\pi y/(2L)]}\right\}\Bigg] f(\xi)\, d\xi \end{aligned} \tag{6.2.56}$$

for $h(x)$ odd.

In a similar manner, the solution to the nondimensional potential problem

$$\frac{\partial^2 u}{\partial x^2} + \frac{\partial^2 u}{\partial y^2} = 0, \qquad -\infty < x < \infty, \quad 0 < y < L, \tag{6.2.57}$$

subject to the boundary conditions

$$\lim_{|x|\to\infty} |u(x,y)| < \infty, \qquad 0 < y < L, \tag{6.2.58}$$

$$\begin{cases} u(x,0) = h(x), & |x| \le 1, \\ u_y(x,0) = 0, & |x| \ge 1, \end{cases} \tag{6.2.59}$$

and

$$u_y(x,L) = 0, \qquad -\infty < x < \infty. \tag{6.2.60}$$

The Green's function is now governed by

$$\frac{\partial^2 g}{\partial x^2}+\frac{\partial^2 g}{\partial y^2}=\delta(x-\xi)\delta(y-\eta), \qquad -\infty<x,\xi<\infty, \quad 0<y,\eta<L, \quad \textbf{(6.2.61)}$$

subject to the boundary conditions

$$\lim_{|x|\to\infty}|g(x,y|\xi,\eta)|<\infty, \qquad 0<y<L, \qquad \textbf{(6.2.62)}$$

and

$$g_y(x,0|\xi,\eta)=g_y(x,L|\xi,\eta)=0, \qquad -\infty<x<\infty. \qquad \textbf{(6.2.63)}$$

The Green's function for this problem is

$$g(x,y|\xi,\eta)=\frac{|x-\xi|}{2L}-\sum_{n=1}^{\infty}\frac{1}{n\pi}\exp\left(-\frac{n\pi|x-\xi|}{L}\right)\cos\left(\frac{n\pi y}{L}\right)\cos\left(\frac{n\pi\eta}{L}\right). \qquad \textbf{(6.2.64)}$$

Therefore,

$$g(x,y|\xi,0)=\frac{|x-\xi|}{2L}+\frac{1}{\pi}\Re[\ln(1-r)], \qquad \textbf{(6.2.65)}$$

where $r=\exp[-\pi(|x-\xi|-iy)/L]$. Upon substituting Equation 6.2.65 into Equation 6.2.41 and using Equation 6.2.39, we obtain an integral equation for $f(\xi)$, namely,

$$h(x)=\frac{1}{\pi}\int_{-1}^{1}f(\xi)\ln\bigl|2\sinh[\pi(x-\xi)/(2L)]\bigr|\,d\xi. \qquad \textbf{(6.2.66)}$$

Once again, we consider the special cases of $h(x)$ as an even or odd function.

- $h(x)$ is an *even* function

In this case, Equation 6.2.66 can be rewritten

$$\pi h(x)=\int_0^1 f(\xi)\ln\bigl|4\sinh[\pi(x-\xi)/(2L)]\sinh[\pi(x+\xi)/(2L)]\bigr|\,d\xi \qquad \textbf{(6.2.67)}$$

$$=\int_0^1 f(\xi)\ln\bigl|2\cosh(\pi x/L)-2\cosh(\pi\xi/L)\bigr|\,d\xi \qquad \textbf{(6.2.68)}$$

for $0\le x\le 1$. Using the techniques shown in Example 6.2.1,

$$f(x)=\frac{1}{L\sqrt{\cosh(\pi x/L)-1}}$$

$$\times \frac{d}{dx}\left\{ \int_x^1 \frac{\sinh(\pi\xi/L)}{\sqrt{2\cosh(\pi\xi/L) - 2\cosh(\pi x/L)}} \right.$$
$$\left. \times \left[\int_0^\xi h'(\chi) \sqrt{\frac{\cosh(\pi\chi/L) - 1}{\cosh(\pi\xi/L) - \cosh(\pi\chi/L)}}\, d\chi \right] d\xi \right\}$$
$$+ \frac{\pi \sinh(\pi x/L)}{L^2 \sqrt{[\cosh(\pi x/L) - 1][\cosh(\pi/L) - \cosh(\pi x/L)]}}$$
$$\times \frac{1}{\ln[\cosh(\pi/L) - 1] - \ln(2)}$$
$$\times \int_0^1 \frac{h(\chi) \sinh(\pi\chi/L)}{\sqrt{[\cosh(\pi\chi/L) - 1][\cosh(\pi/L) - \cosh(\pi\chi/L)]}}\, d\chi. \qquad (\mathbf{6.2.69})$$

Substituting Equation 6.2.69 into Equation 6.2.41, we find that

$$u(x,y) = \frac{1}{2\pi} \int_0^1 \{\ln(4) + \ln[\cosh(\pi|x-\xi|/L) - \cosh(\pi y/L)]$$
$$+ \ln[\cosh(\pi|x+\xi|/L) - \cos(\pi y/L)]\} f(\xi)\, d\xi. \qquad (\mathbf{6.2.70})$$

- $h(x)$ is an *odd* function

Turning to the case when $h(x)$ is an odd function, Equation 6.2.66 can be rewritten

$$\pi h(x) = \int_0^1 f(\xi) \ln\left| \frac{\sinh[\pi(x-\xi)/(2L)]}{\sinh[\pi(x+\xi)/(4L)]} \right| d\xi \qquad (\mathbf{6.2.71})$$
$$= -\int_0^1 f(\xi) \ln\left| \frac{\tanh[\pi x/(2L)] + \tanh[\pi\xi/(2L)]}{\tanh[\pi x/(2L)] - \tanh[\pi\xi/(2L)]} \right| d\xi \qquad (\mathbf{6.2.72})$$

for $0 \le x \le 1$. Using the techniques shown in Example 6.2.1, we find that

$$f(x) = \frac{1}{L}\frac{d}{dx}\left\{ \int_x^1 \frac{\tanh[\pi\xi/(2L)]}{\cosh^2[\pi\xi/(2L)] \sqrt{\tanh^2[\pi\xi/(2L)] - \tanh^2[\pi x/(2L)]}} \right.$$
$$\left. \times \left\{ \int_0^\xi \frac{h'(\chi)}{\sqrt{\tanh^2[\pi\xi/(2L)] - \tanh^2[\pi\chi/(2L)]}}\, d\chi \right\} d\xi \right]$$
$$- \frac{2h(0) \tanh[\pi/(2L)] \operatorname{arcsinh}(\pi x/L)}{L\sqrt{\tanh^2[\pi/(2L)] - \tanh^2[\pi x/(2L)]}}. \qquad (\mathbf{6.2.73})$$

Substituting Equation 6.2.65 and Equation 6.2.73 into Equation 6.2.41, we obtain the final results that

$$u(x,y) = \frac{1}{2\pi}\int_0^1 \ln\left|\frac{\cosh(\pi|x-\xi|/L) - \cos(\pi y/L)}{\cosh(\pi|x+\xi|/L) - \cos(\pi y/L)}\right| f(\xi)\, d\xi \qquad \textbf{(6.2.74)}$$

if $h(x)$ is an odd function.

• **Example 6.2.4: The method of Clements and Love**

In 1974 Clements and Love[8] published a method for finding axisymmetric potentials. Mathematically, this problem is given by

$$\frac{\partial^2 u}{\partial r^2} + \frac{1}{r}\frac{\partial u}{\partial r} + \frac{\partial^2 u}{\partial z^2} = 0, \qquad 0 < r < \infty, \quad 0 < z < \infty, \qquad \textbf{(6.2.75)}$$

subject to the boundary conditions

$$\lim_{r\to 0} |u(r,z)| < \infty, \quad \lim_{r\to\infty} u(r,z) \to 0, \qquad 0 < z < \infty, \qquad \textbf{(6.2.76)}$$

$$\lim_{z\to\infty} u(r,z) \to 0, \qquad 0 < r < \infty, \qquad \textbf{(6.2.77)}$$

and

$$\begin{cases} u(r,0) = U_1(r), & 0 < r < a, \\ u_z(r,0) = -\sigma_0(r), & a < r < b, \\ u(r,0) = U_2(r), & b < r < \infty. \end{cases} \qquad \textbf{(6.2.78)}$$

Clements and Love referred to this problem as a "*Neumann problem*" because of the boundary condition between $a < r < b$.

Clements and Love's method expresses the potential in terms of a Green's function:

$$u(r,z) = \frac{1}{2\pi}\int_0^\infty \sigma(\rho)\left[\int_{-\pi}^{\pi} \frac{d\varphi}{\sqrt{r^2+z^2+\rho^2-2r\rho\cos(\varphi)}}\right]\rho\, d\rho, \qquad \textbf{(6.2.79)}$$

where $\sigma(\rho)$ is presently unknown. The quantity inside of the square brackets is the free-space Green's function. Clements and Love then proved that $\sigma(\rho)$ is given by

$$\rho\,\sigma(\rho) = -\frac{2}{\pi}\frac{d}{d\rho}\left[\int_\rho^a \frac{\xi\, f_1(\xi)}{\sqrt{\xi^2-\rho^2}}\, d\xi\right], \qquad \rho < a, \qquad \textbf{(6.2.80)}$$

$$\sigma(\rho) = \sigma_0(\rho), \qquad a < \rho < b, \qquad \textbf{(6.2.81)}$$

and

$$\rho\,\sigma(\rho) = \frac{2}{\pi}\frac{d}{d\rho}\left[\int_b^\rho \frac{\xi\, f_2(\xi)}{\sqrt{\rho^2-\xi^2}}\, d\xi\right], \qquad b < \rho, \qquad \textbf{(6.2.82)}$$

---

[8] Clements, D. L., and E. R. Love, 1974: Potential problems involving an annulus. *Proc. Cambridge Phil. Soc.*, **76**, 313–325.

where $f_1(\xi)$ and $f_2(\xi)$ are found from the coupled integral equations

$$f_1(r) + \frac{2}{\pi}\int_b^\infty \frac{\xi}{\xi^2 - r^2} f_2(\xi)\, d\xi = g_1(r), \qquad r < a, \tag{6.2.83}$$

$$f_2(r) + \frac{2}{\pi}\int_0^a \frac{r}{r^2 - \xi^2} f_1(\xi)\, d\xi = g_2(r), \qquad r > b, \tag{6.2.84}$$

$$g_1(r) = -\int_a^b \frac{\xi\, \sigma_0(\xi)}{\sqrt{\xi^2 - r^2}}\, d\xi + \frac{d}{dr}\left[\int_0^r \frac{\xi\, U_1(\xi)}{\sqrt{r^2 - \xi^2}}\, d\xi\right], \qquad r < a, \tag{6.2.85}$$

and

$$g_2(r) = -\int_a^b \frac{\xi\, \sigma_0(\xi)}{\sqrt{r^2 - \xi^2}}\, d\xi - \frac{d}{dr}\left[\int_r^\infty \frac{\xi\, U_2(\xi)}{\sqrt{\xi^2 - r^2}}\, d\xi\right], \qquad r > b. \tag{6.2.86}$$

Clements and Love also considered the case when the mixed boundary condition along $z = 0$ reads

$$\begin{cases} u_z(r,0) = -\sigma_1(r), & 0 < r < a, \\ u(r,0) = U_0(r), & a < r < b, \\ u_z(r,0) = -\sigma_2(r), & b < r < \infty. \end{cases} \tag{6.2.87}$$

Clements and Love referred to this problem as a "*Dirichlet problem*" because of the boundary condition between $a < r < b$.

The potential is given by Equation 6.2.79 once again. In the present case,

$$\sigma(\rho) = \begin{cases} \sigma_1(\rho), & \rho < a, \\ \sigma_3(\rho) + \sigma_4(\rho), & a < \rho < b, \\ \sigma_2(\rho), & b < \rho. \end{cases} \tag{6.2.88}$$

The quantities $\sigma_3(\rho)$ and $\sigma_4(\rho)$ are found from

$$\sigma_3(r) = \omega_3(r) + \frac{2b^2(b^2 - a^2)^{3/2}}{\pi\sqrt{b^2 - r^2}} \int_b^\infty \frac{b\tau\, f_2(\tau)}{(b^2 - r^2)\tau^2 + (r^2 - a^2)b^2}\, d\tau, \tag{6.2.89}$$

and

$$\sigma_4(r) = \omega_4(r) + \frac{2b^2(b^2 - a^2)^{3/2}}{\pi\sqrt{r^2 - a^2}} \int_0^a \frac{\tau^2\, f_1(\tau)}{(b^2 - r^2)\tau^2 + (r^2 - a^2)b^2}\, d\tau. \tag{6.2.90}$$

To evaluate Equation 6.2.89 and Equation 6.2.90, we must first compute the quantities $U_3(r)$ and $U_4(r)$ via

$$\begin{aligned} U_3(r) = \tfrac{1}{2}U_0(r) &- \left[\tfrac{1}{2}U_0(a) + A\right]\left(\frac{b-r}{b-a}\right)^2 + A\left(\frac{b-r}{b-a}\right)^3 \\ &+ \left[\tfrac{1}{2}U_0(b) + B\right]\left(\frac{r-a}{b-a}\right)^2 - B\left(\frac{r-a}{b-a}\right)^3, \end{aligned} \tag{6.2.91}$$

and

$$U_4(r) = U_0(r) - U_3(r), \tag{6.2.92}$$

where

$$A = U_0(a) + \tfrac{1}{2}(b-a)U_0'(a) \quad \text{and} \quad B = U_0(b) - \tfrac{1}{2}(b-a)U_0'(b). \tag{6.2.93}$$

Having found $U_3(r)$ and $U_4(r)$, we can compute $\omega_3(r)$ and $\omega_4(r)$ from

$$\begin{aligned}\omega_3(r) = &-\frac{2}{\pi}\int_b^\infty \frac{t}{t^2-r^2}\sqrt{\frac{t^2-b^2}{b^2-r^2}}\,\sigma_2(t)\,dt \\ &-\frac{2}{\pi r}\frac{d}{dr}\left\{\int_r^b \frac{s}{\sqrt{s^2-r^2}}\frac{d}{ds}\left[\int_0^s \frac{t\,U_3(t)}{\sqrt{s^2-t^2}}\,dt\right]ds\right\},\end{aligned} \tag{6.2.94}$$

and

$$\begin{aligned}\omega_4(r) = &-\frac{2}{\pi}\int_0^a \frac{t}{r^2-t^2}\sqrt{\frac{a^2-t^2}{r^2-a^2}}\,\sigma_1(t)\,dt \\ &-\frac{2}{\pi r}\frac{d}{dr}\left\{\int_a^r \frac{s}{\sqrt{r^2-s^2}}\frac{d}{ds}\left[\int_s^\infty \frac{t\,U_4(t)}{\sqrt{t^2-s^2}}\,dt\right]ds\right\}.\end{aligned} \tag{6.2.95}$$

Finally, we find $f_1(\rho)$ and $f_2(\rho)$ from the coupled integral equations

$$f_1(\rho) - \frac{2}{\pi}\int_b^\infty \frac{\tau\,f_2(\tau)}{\tau^2-\rho^2}\,d\tau = g_1(\rho), \qquad \rho < a, \tag{6.2.96}$$

$$\rho\,f_2(\rho) - \frac{2}{\pi}\int_0^a \frac{\tau^2\,f_1(\tau)}{\rho^2-\tau^2}\,d\tau = \rho\,g_2(\rho), \qquad \rho > b, \tag{6.2.97}$$

where

$$g_1(\rho) = \frac{\omega_3(\xi)}{(b^2-\rho^2)^{3/2}}, \qquad \xi = b\sqrt{\frac{a^2-\rho^2}{b^2-\rho^2}}, \qquad \rho < a, \tag{6.2.98}$$

and

$$g_2(\rho) = \frac{\omega_4(\xi)}{(\rho^2-b^2)^{3/2}}, \qquad \xi = b\sqrt{\frac{a^2-\rho^2}{b^2-\rho^2}}, \qquad \rho > b. \tag{6.2.99}$$

Let us illustrate their method by solving[9]

$$\frac{\partial^2 u}{\partial r^2} + \frac{1}{r}\frac{\partial u}{\partial r} + \frac{\partial^2 u}{\partial z^2} = 0, \qquad 0 < r < \infty, \quad 0 < z < \infty, \tag{6.2.100}$$

---

[9] See Yang, F.-Q., and J. C. M. Li, 1995: Impression creep by an annular punch. *Mech. Mater.*, **21**, 89–97.

subject to the boundary conditions

$$\lim_{r\to 0} |u(r,z)| < \infty, \quad \lim_{r\to\infty} u(r,z) \to 0, \qquad 0 < z < \infty, \tag{6.2.101}$$

$$\lim_{z\to\infty} u(r,z) \to 0, \qquad 0 < r < \infty, \tag{6.2.102}$$

and

$$\begin{cases} u(r,0) = 2\sqrt{1-r^2}/\pi, & 0 < r < a, \\ u_z(r,0) = 0 & a < r < 1, \\ u(r,0) = 0, & 1 < r < \infty. \end{cases} \tag{6.2.103}$$

From the nature of the boundary conditions, we use Equation 6.2.79 through Equation 6.2.86. We begin by computing $g_1(r)$ and $g_2(r)$. Substituting $U_1(\xi) = 2\sqrt{1-\xi^2}/\pi$, $U_2(\xi) = 0$, and $\sigma_0(\xi) = 0$ into Equation 6.2.85 and Equation 6.2.86, we find that

$$g_1(r) = \frac{2}{\pi}\frac{d}{dr}\left[\int_0^r \frac{\xi\sqrt{1-\xi^2}}{r^2-\xi^2}\,d\xi\right] = \frac{2}{\pi}\left[1 - \frac{r}{2}\ln\left(\frac{1+r}{1-r}\right)\right] \tag{6.1.104}$$

if $r < a$; and $g_2(r) = 0$ with $r > 1$.

At this point, we must turn to numerical methods to compute $u(r,z)$. Using MATLAB®, we begin our calculations at a given radius **r** and height **z** by solving the coupled integral equations, Equation 6.2.83 and Equation 6.2.84. We do this by introducing `N` nodal point in the region $0 \le \xi \le a$ such that `xi` $=$ `(n` $-$ `1)` $*$ `dr_1`, where `n = 1,2,...,N` and `dr_1 = a/(N-1)`. Similarly, for $1 \le \xi < \infty$, we introduce `M` nodal points such that `xi = 1+(m-1)*dr_2`, where `m = 1,2,...,M` and `dr_2` is the resolution of the grid. Thus, Equation 6.2.83 and Equation 6.2.84 yield N+M equations which we express in matrix notation as $A\mathbf{f} = \mathbf{b}$, where `N` equations arise from Equation 6.2.83 and `M` equations are due to Equation 6.2.84. Because we will evaluate the integrals using Simpson's rule, both `N` and `M` must be odd integers.

The MATLAB code that approximates Equation 6.2.83 is

```
A = zeros(N+M,N+M); % zero out the array A
for n = 1:N
r = (n-1)*dr_1;
b(n) = 1 - 0.5*r*log((1+r)/(1-r));
b(n) = 2*b(n) / pi; % introduce g_1(r) here
A(n,n) = 1;
% evaluate the integral by Simpson's rule
for m = 1:M
xi = 1 + (m-1)*dr_2;
integrand = 2*xi / (pi*(xi*xi-r*r));
if ( (m>1) & (m<M) )
if ( mod(m,2) == 0)
A(n,N+m) = 4*dr_2*integrand/3;
```

```
else
A(n,N+m) = 2*dr_2*integrand/3;
end
else
A(n,N+m) =   dr_2*integrand/3;
end; end; end
```

The MATLAB code that approximates Equation 6.2.84 is

```
for m = 1:M
r = 1 + (m-1)*dr_2;
b(N+m) = 0;
A(N+m,N+m) = 1;
for n = 1:N
xi = (n-1)*dr_1;
integrand = 2*r / (pi*(r*r-xi*xi));
if ( (n>1) & (n<N) )
if ( mod(n,2) == 0 )
A(N+m,n) = 4*dr_1*integrand/3;
else
A(N+m,n) = 2*dr_1*integrand/3;
end
else
A(N+m,n) =   dr_1*integrand/3;
end; end; end
```

Solving these $(N+M)\times(N+M)$ equations, we find `f` which holds $f_1(\xi)$ in its first `N` elements, while $f_2(\xi)$ is given in the remaining `M` elements:

```
f = A\b';
```

Given `f`, we now solve for $\sigma(\rho)$. This is a two-step procedure. First, we evaluate the bracketed terms in Equation 6.2.80 or Equation 6.2.82. For accurate computations in Equation 6.2.80, we note that $\xi\, d\xi/\sqrt{\xi^2-\rho^2} = d\left(\sqrt{\xi^2-\rho^2}\right)$. Equation 6.2.82 employs a similar trick. Then, we evaluate the integral using the trapezoidal rule. Finally, $\sigma(\rho)$ follows from simple finite differences.

```
for m = 1:N-1
t = dr_1*(m-1); bracket_1(m) = 0;
for n = m:N-1
xi_end = n*dr_1; xi_begin = xi_end-dr_1;
f1 = 0.5*(f(n)+f(n+1));
bracket_1(m) = bracket_1(m) + f1*sqrt(xi_end*xi_end-t*t) ...
             - f1*sqrt(xi_begin*xi_begin-t*t);
end;end
bracket_1(N) = 0;

for n = 1:N-1
```

```
rho(n) = (n-0.5)*dr_1;
sigma(n) = 2*(bracket_1(n)-bracket_1(n+1)) / (pi*dr_1*rho(n));
end

bracket_2(1) = 0;
for m = 2:M
t = 1 + dr_2*(m-1); bracket_2(m) = 0;
for n = 1:m-1
xi_end = 1 + dr_2*n; xi_begin = xi_end - dr_2;
f2 = 0.5*(f(N+n)+f(N+n+1));
bracket_2(m) = bracket_2(m) - f2*sqrt(t*t-xi_end*xi_end) ...
               + f2*sqrt(t*t-xi_begin*xi_begin);
end;end

for m = 1:M-1
rho(N-1+m) = 1 + dr_2*(m-0.5);
sigma(N-1+m) = 2*(bracket_2(m+1)-bracket_2(m)) ...
               / (pi*dr_2*rho(N-1+m));
end
```

With $\sigma(\rho)$ we are ready to compute Equation 6.2.79. There are two steps. First, we find the Green's function via Simpson's rule; it is called `green` here. Then we evaluate the outside integral using the midpoint rule. The final solution is called `u(i,j)`.

```
dphi = pi / 10;
for k = 1:(N+M-2)
green = 0;
if (r == 0)
green = 2*pi / sqrt(z*z+rho(k)*rho(k));
else
for ii = 1:21
phi = -pi+(ii-1)*dphi;
denom = r*r+z*z+rho(k)*rho(k)-2*r*rho(k)*cos(phi);
denom = sqrt(denom);
if ( (ii>1) & (ii<21) )
if ( mod(ii,2) == 0)
green = green + 4*dphi/(3*denom);
else
green = green + 2*dphi/(3*denom);
end
else
green = green +   dphi/(3*denom);
end; end; end
if (k < N)
u(i,j) = u(i,j) + sigma(k)*green*rho(k)*dr_1;
else
```

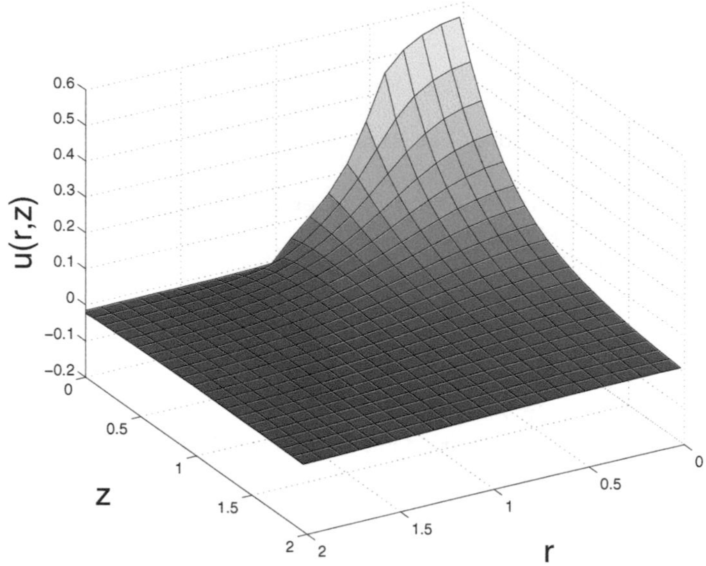

Figure 6.2.4: The solution to Laplace's equation with the boundary conditions given by Equation 6.2.101 through Equation 6.2.103.

```
u(i,j) = u(i,j) + sigma(k)*green*rho(k)*dr_2;
end;end
u(i,j) = u(i,j)/(2*pi);
```

Figure 6.2.4 illustrates this solution when `N = 51`, `M = 401`, `a = 0.5`, and `dr_2 = 0.1`.

6.3 POTENTIAL THEORY

Potentials have long been used to solve Laplace's or Poisson's equation over some region. Here we show how this method has been extended to mixed boundary value problems.

• Example 6.3.1

In this example let us find the solution to Laplace's equation in three dimensions when the boundary condition on the $z = 0$ plane is

$$\begin{cases} u(r,\theta,0) = f(r), & 0 \le r < 1, \\ u_z(r,\theta,0) = 0, & 1 < r < \infty. \end{cases} \tag{6.3.1}$$

Because of the axisymmetric boundary condition, the solution is simply

$$u(r,\vartheta,z) = \int_0^1 \int_0^{2\pi} \frac{\rho\,\sigma(\rho)}{R}\,d\theta\,d\rho, \tag{6.3.2}$$

where

$$R^2 = \rho^2 + r^2 - 2\rho r\cos(\vartheta - \theta) + z^2. \tag{6.3.3}$$

Here R gives the distance from the general point (r, ϑ, z) to the point $(r, \theta, 0)$ on the disk. Because

$$u_z(r, \vartheta, z) = -z \int_0^1 \int_0^{2\pi} \frac{\rho\,\sigma(\rho)}{R^3}\, d\theta\, d\rho, \tag{6.3.4}$$

the boundary condition $u_z(r, \vartheta, 0) = 0$ is satisfied if $r > 1$ because R does not vanish outside of the unit circle. Kanwal and Sachdeva[10] found expressions for $\sigma(\rho)$ in the special case of

$$f(r) = \begin{cases} \lambda, & 0 \le r < \alpha, \\ 1, & \alpha < r < 1. \end{cases} \tag{6.3.5}$$

From Equation 6.3.1, we have that

$$f(\rho) = \int_0^1 \int_0^{2\pi} \frac{\rho\,\sigma(\rho)}{\sqrt{\rho^2 + r^2 - 2\rho r\cos(\vartheta - \theta)}}\, d\theta\, d\rho. \tag{6.3.6}$$

Now,

$$\frac{1}{\sqrt{\rho^2 + r^2 - 2\rho r\cos(\vartheta - \theta)}} = \sum_{n=0}^{\infty} \int_0^{\infty} (2 - \delta_{0n}) J_n(kr) J_n(k\rho) \cos[k(\vartheta - \theta)]\, dk, \tag{6.3.7}$$

$$J_0(k\rho) = \sqrt{\frac{2k}{\pi}} \int_0^{\rho} J_{-\frac{1}{2}}(kx) \frac{\sqrt{x}}{\sqrt{\rho^2 - x^2}}\, dx, \tag{6.3.8}$$

and

$$\int_0^{\infty} J_{-\frac{1}{2}}(kx) J_{-\frac{1}{2}}(ky)\, k\, dk = \frac{\delta(x - y)}{\sqrt{xy}} \tag{6.3.9}$$

where δ_{mn} denotes the Kronecker delta. We can then rewrite Equation 6.3.6 as

$$f(\rho) = 4 \int_0^{\rho} \left[\int_x^1 \frac{t\,\sigma(t)}{\sqrt{t^2 - x^2}}\, dt \right] \frac{dx}{\sqrt{\rho^2 - x^2}}. \tag{6.3.10}$$

If we set

$$S(x) = \int_x^1 \frac{t\,\sigma(t)}{\sqrt{t^2 - x^2}}\, dt, \tag{6.3.11}$$

then

$$\frac{f(\rho)}{4} = \int_0^{\rho} \frac{S(x)}{\sqrt{\rho^2 - x^2}}\, dx. \tag{6.3.12}$$

---

[10] Kanwal, R. P., and B. K. Sachdeva, 1972: Potential due to a double lamina. *J. Appl. Phys.*, **43**, 4821–4822.

We can invert Equation 6.3.12 and find that

$$S(x) = \frac{1}{2\pi}\frac{d}{dx}\left[\int_0^x \frac{\eta f(\eta)}{\sqrt{x^2-\eta^2}}\,d\eta\right], \tag{6.3.13}$$

or

$$S(x) = \begin{cases} \dfrac{\lambda}{2\pi}, & 0 < x < \alpha, \\ \dfrac{\lambda}{2\pi} + \dfrac{(1-\gamma)x}{2\pi\sqrt{x^2-\alpha^2}}, & \alpha < x < 1. \end{cases} \tag{6.3.14}$$

Substituting Equation 6.3.14 into Equation 6.3.11 and inverting,

$$\sigma(t) = \frac{1}{\pi^2}\left[\frac{\lambda}{\sqrt{1-t^2}} - (1-\lambda)\int_\alpha^1 \frac{\eta^2}{\sqrt{\eta^2-\alpha^2}\sqrt{(\eta^2-t^2)^3}}\,d\eta\right], \quad 0 < t < \alpha, \tag{6.3.15}$$

and

$$\sigma(t) = \frac{1}{\pi^2}\left[\frac{\lambda}{\sqrt{1-t^2}} - \frac{1-\lambda}{t}\frac{d}{dt}\left(\int_t^1 \frac{\eta^2}{\sqrt{\eta^2-\alpha^2}\sqrt{\eta^2-t^2}}\,d\eta\right)\right], \quad \alpha < t < 1. \tag{6.3.16}$$

Figure 6.3.1 illustrates the potential $u(x, y, z)$ when $z = 0$ and $z = 0.2$. Here we also selected $\alpha = 0.5$ and $\lambda = 2$. During the numerical evaluation of Equation 6.3.2 the integration with respect to ρ was done first and the time derivative in $t\,\sigma(t)$ was eliminated by an integration by parts.

• Example 6.3.2: Fabrikant's method

In the previous example we found the solution to Laplace's equation in three dimensions in the half-space $z \geq 0$ with the mixed boundary condition given by Equation 6.3.1. During the 1980s Fabrikant[11] generalized this mixed boundary value problem to read

$$\begin{cases} u(r,\theta,0) = f(r,\theta), & 0 \le r < a, \quad 0 \le \theta < 2\pi, \\ u_z(r,\theta,0) = 0, & a < r < \infty, \quad 0 \le \theta < 2\pi. \end{cases} \tag{6.3.17}$$

He showed that the solution to this problem is

$$u(r,\theta,z) = \frac{2}{\pi}\int_0^a \left\{L\left[\frac{\ell_1^2(\eta)}{r\eta^2}\right]\frac{d}{d\eta}\left[\int_0^\eta L(\rho)f(\rho,\theta)\frac{\rho}{\sqrt{\eta^2-\rho^2}}\,d\rho\right]\right\} \times \frac{d\ell_1(\eta)}{\sqrt{r^2-\ell_1^2(\eta)}}, \tag{6.3.18}$$

---

[11] Fabrikant, V. I., 1986: A new approach to some problems in potential theory. *Z. Angew. Math. Mech.*, **66**, 363–368. Quoted with permission.

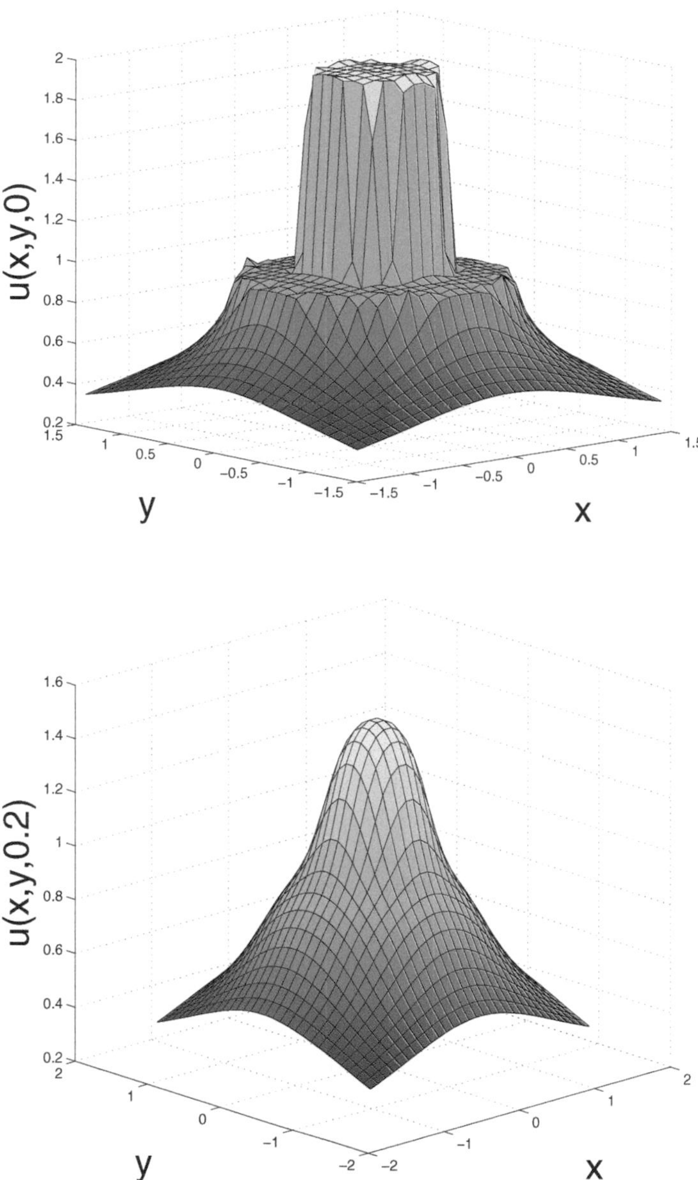

Figure 6.3.1: The solution to three-dimensional Laplace's equation with the boundary conditions given by Equation 6.3.1. The top frame illustrates the solution when $z = 0$, while the lower frame gives the potential at $z = 0.2$. The remaining parameters are $\alpha = 0.5$ and $\lambda = 2$.

or

$$u(r,\theta,z) = \frac{1}{\pi^2}\int_0^a \int_0^{2\pi} \left[\frac{R}{\xi} + \arctan\left(\frac{\xi}{R}\right)\right] \frac{z}{R^3} f(\rho,\vartheta)\, d\vartheta\, \rho\, d\rho, \tag{6.3.19}$$

where $R^2 = r^2 + \rho^2 - 2r\rho\cos(\theta - \vartheta) + z^2$,

$$\ell_{1,2}(x) = \tfrac{1}{2}\sqrt{(x+r)^2 + z^2} \mp \sqrt{(x-r)^2 + z^2}, \tag{6.3.20}$$

the operator $L(k)\sigma(r,\theta)$ is given by

$$L(k)\sigma(r,\theta) = \frac{1}{2\pi}\int_0^{2\pi} \lambda(k, \theta - \vartheta)\sigma(r,\vartheta)\, d\vartheta \tag{6.3.21}$$

$$= \sum_{n=-\infty}^{\infty} k^{|n|} e^{in\theta} \left[\frac{1}{2\pi}\int_0^{2\pi} e^{-in\vartheta}\sigma(r,\vartheta)\, d\vartheta\right], \tag{6.3.22}$$

$$\lambda(k,\vartheta) = \frac{1-k^2}{1+k^2-2k\cos(\vartheta)} = \sum_{n=-\infty}^{\infty} k^{|n|} e^{in\vartheta}, \tag{6.3.23}$$

and

$$\xi = \frac{z\sqrt{a^2-\rho^2}}{\sqrt{\ell_2^2(a) - a^2}} = \frac{\sqrt{\ell_2^2(a) - \ell_1^2(\rho)}\sqrt{\ell_2^2(a) - \ell_2^2(\rho)}}{\ell_2(a)} \tag{6.3.24}$$

$$= \frac{\sqrt{a^2-\rho^2}\sqrt{a^2 - \ell_1^2(a)}}{a} = \frac{\sqrt{a^2-\rho^2}\sqrt{\ell_2^2(a) - r^2}}{\ell_2(a)}. \tag{6.3.25}$$

Equation 6.3.18 is recommended in those cases when the integrals can be evaluated exactly while Equation 6.3.19 is more convenient when the integrals must be computed numerically.

To illustrate Fabrikant's results, consider the case when $f(r,\theta) = w_0$, a constant. In this case,

$$L(\rho)f(\rho,\theta) = \sum_{n=-\infty}^{\infty} \rho^{|n|} e^{in\theta} \left[\frac{1}{2\pi}\int_0^{2\pi} e^{-in\vartheta} w_0\, d\vartheta\right] = w_0 \tag{6.3.26}$$

because all of the terms in the summation vanish except $n = 0$. Therefore,

$$\frac{d}{d\eta}\left[\int_0^{\eta} L(\rho)f(\rho,\theta)\frac{\rho}{\sqrt{\eta^2-\rho^2}}\, d\rho\right] = w_0, \tag{6.3.27}$$

and

$$L\left[\frac{\ell_1^2(\eta)}{r\eta^2}\right]\frac{d}{d\eta}\left[\int_0^{\eta} L(\rho)f(\rho,\theta)\frac{\rho}{\sqrt{\eta^2-\rho^2}}\, d\rho\right] \tag{6.3.28}$$

$$= \sum_{n=-\infty}^{\infty} \left[\frac{\ell_1^2(\eta)}{r\eta^2}\right]^{|n|} e^{in\theta}\left[\frac{1}{2\pi}\int_0^{2\pi} e^{-in\vartheta} w_0\, d\vartheta\right] = w_0.$$

From Equation 6.3.18

$$u(r,\theta,z) = \frac{2w_0}{\pi}\int_0^a \frac{d\ell_1(\eta)}{\sqrt{r^2-\ell_1^2(\eta)}} = \frac{2w_0}{\pi}\arcsin\left[\frac{\ell_1^2(\eta)}{r}\right]\Bigg|_0^a \tag{6.3.29}$$

$$= \frac{2w_0}{\pi}\arcsin\left[\frac{\ell_1^2(a)}{r}\right]. \tag{6.3.30}$$

Fabrikant also considered the case when the mixed boundary condition reads

$$\begin{cases} u_z(r,\theta,0) = \sigma(r,\theta), & 0 \le r < a, \quad 0 \le \theta < 2\pi, \\ u(r,\theta,0) = 0, & a < r < \infty, \quad 0 \le \theta < 2\pi. \end{cases} \tag{6.3.31}$$

In this case he showed that the solution is

$$u(r,\theta,z) = 4C\int_{\ell_2(0)}^{\ell_2(a)}\left[\int_0^{g(x)} \frac{\rho\,d\rho}{\sqrt{g^2(x)-\rho^2}} L\left(\frac{\rho r}{x^2}\right)\sigma(\rho,\theta)\right]\frac{dx}{\sqrt{x^2-r^2}}, \tag{6.3.32}$$

$$u(r,\theta,z) = 4C\int_0^a \left\{\int_0^\eta \frac{\rho\,d\rho}{\sqrt{\eta^2-\rho^2}} L\left[\frac{\rho r}{\ell_2^2(\eta)}\right]\sigma(\rho,\theta)\right\}\frac{d\ell_2(\eta)}{\sqrt{\ell_2^2(\eta)-r^2}}, \tag{6.3.33}$$

or

$$u(r,\theta,z) = \frac{2C}{\pi}\int_0^a\int_0^{2\pi} \arctan\left(\frac{\xi}{R}\right)\frac{\sigma(\rho,\vartheta)}{R}\,d\vartheta\,\rho\,d\rho, \tag{6.3.34}$$

where $g(x) = x\sqrt{1+z^2/(r^2-x^2)}$ and R and ξ have been defined earlier. The constant coefficient C equals $-1/(2\pi)$ in classical potential problems and different values in other applications. Equation 6.3.32 and Equation 6.3.33 are best when the integrals can be evaluated exactly while Equation 6.3.34 should be used otherwise.

To illustrate Equation 6.3.32 through Equation 6.3.34, consider the case when $\sigma(r,\theta) = \sigma_0$, a constant. Then,

$$L\left(\frac{\rho r}{x^2}\right)\sigma(\rho,\theta) = \sum_{n=-\infty}^{\infty}\left(\frac{\rho r}{x^2}\right)^{|n|} e^{in\theta}\left[\frac{1}{2\pi}\int_0^{2\pi} e^{-in\vartheta}\sigma_0\,d\vartheta\right] = \sigma_0, \tag{6.3.35}$$

and

$$\int_0^{g(x)} \frac{\rho\,d\rho}{\sqrt{g^2(x)-\rho^2}} L\left(\frac{\rho r}{x^2}\right)\sigma(\rho,\theta) = \sigma_0 g(x). \tag{6.3.36}$$

Therefore, using Equation 6.3.32,

$$u(r,\theta,z) = 4C\sigma_0\int_{\ell_2(0)}^{\ell_2(a)} g(x)\frac{dx}{\sqrt{x^2-r^2}} \tag{6.3.37}$$

$$= 4C\sigma_0\int_{\ell_2(0)}^{\ell_2(a)} \frac{\sqrt{x^2-r^2-z^2}}{x^2-r^2}x\,dx \tag{6.3.38}$$

$$
\begin{aligned}
&= 2C\sigma_0 \int_{\ell_2(0)}^{\ell_2(a)} \frac{d(x^2 - r^2)}{\sqrt{x^2 - r^2 - z^2}} \\
&- 2C\sigma_0 z^2 \int_{\ell_2(0)}^{\ell_2(a)} \frac{d(x^2 - r^2)}{(x^2 - r^2)\sqrt{x^2 - r^2 - z^2}} \qquad (6.3.39)
\end{aligned}
$$

$$
\begin{aligned}
&= 4C\sigma_0 \sqrt{x^2 - r^2 - z^2}\,\Big|_{\ell_2(0)}^{\ell_2(a)} \\
&- 4C\sigma_0 z \arctan\left(\frac{\sqrt{x^2 - r^2 - z^2}}{z} \right)\Big|_{\ell_2(0)}^{\ell_2(a)} \qquad (6.3.40)
\end{aligned}
$$

$$
= 4C\sigma_0 \sqrt{a^2 - \ell_1^2(a)} - 4C\sigma_0 z \arctan\left[\frac{\sqrt{a^2 - \ell_1^2(a)}}{z} \right], \qquad (6.3.41)
$$

where $\ell_2^2(0) = r^2 + z^2$ and $\ell_2^2(a) + \ell_1^2(a) = a^2 + r^2 + z^2$. Fabrikant has extended his work to spherical coordinates[12] and crack problems.[13]

---

[12] Fabrikant, V. I., 1987: Mixed problems of potential theory in spherical coordinates. *Z. Angew. Math. Mech.*, **67**, 507–518.

[13] Fabrikant, V. I., and E. N. Karapetian, 1994: Elementary exact method for solving mixed boundary value problems of potential theory, with applications to half-plane contact and crack problems. *Quart. J. Mech. Appl. Math.*, **47**, 159–174.

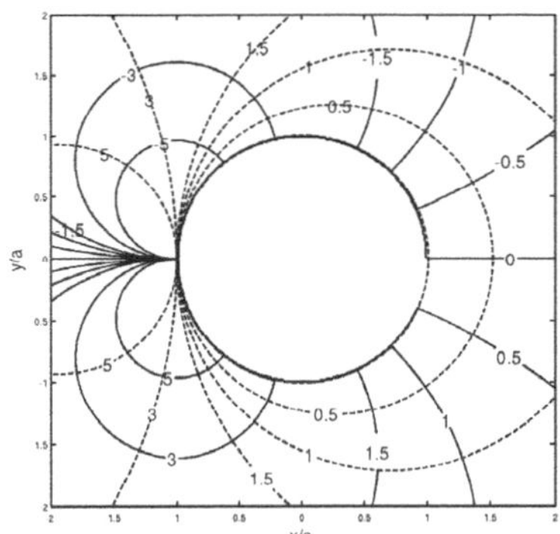

Chapter 7
Conformal Mapping

Conformal mapping is a method from classical mathematical physics for solving Laplace's equation. It is readily shown that an analytic function $w = \xi + i\eta = f(z)$, where $z = x + iy$, transforms Laplace's equation in the xy-plane into Laplace's equation in the $\xi\eta$-plane. The objective here is to choose a mapping so that the solution is easier to obtain in the new domain.

This method has been very popular in fields such as electrostatics and hydrodynamics. In the case of mixed boundary value problems, this technique has enjoyed limited success because the transformed boundary conditions are very complicated and the corresponding solution to Laplace's equation is difficult to find. In this chapter we illustrate some of the successful transformations.

7.1 THE MAPPING $z = w + a\log(w)$

During their study of fringing fields in disc capacitors, Sloggett et al.[1] used this mapping to find the potential in the upper half-plane $y > 0$ where the potential equals V along the line $y = \pi a$ when $-\infty < x < a\ln(a) - a$. Along

[1] Sloggett, G. J., N. G. Barton, and S. J. Spencer, 1986: Fringing fields in disc capacitors. *J. Phys., Ser. A*, **19**, 2725–2736.

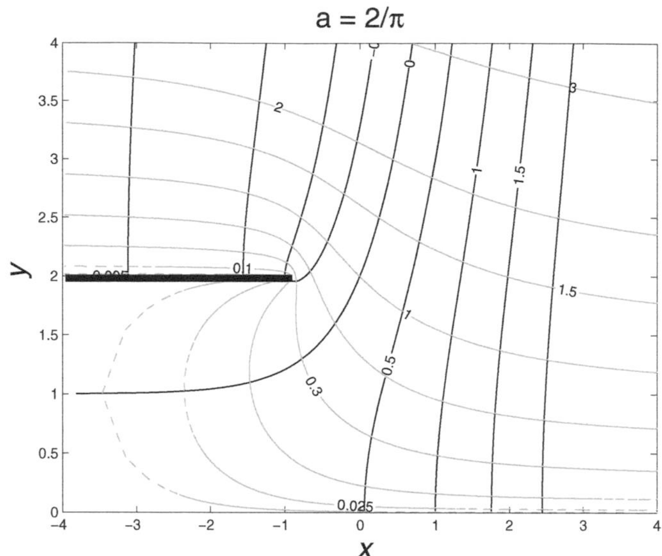

Figure 7.1.1: The conformal mapping $z = w + a\log(w)$ with $a = 2/\pi$. If $w = \xi + i\eta$, the dark (solid) lines are lines of constant ξ while the lighter (dashed) lines give η. The heavy dark line corresponds to the line $-\infty < \xi < 0^-$ and $\eta = 0$.

$y = 0$, $u(x, 0) = 0$. Figure 7.1.1 illustrates this mapping. In particular, the following line segments are mapped along the real axis in the w-plane:

z-plane		w-plane	
$-\infty < x < a\ln(a) - a$	$y = (a\pi)^+$	$-\infty < \xi < -a$	$\eta = 0$
$-\infty < x < a\ln(a) - a$	$y = (a\pi)^-$	$-a < \xi < 0^-$	$\eta = 0$
$-\infty < x < 1$	$y = 0$	$0^+ < \xi < 1$	$\eta = 0$
$1 < x < \infty$	$y = 0$	$1 < \xi < \infty$	$\eta = 0$

Here the $(\cdot)^+$ and $(\cdot)^-$ denote points just above or below $(\cdot)$, respectively.

From Poisson's integral,

$$u(\xi, \eta) = \frac{V}{\pi} \int_0^\infty \frac{\eta}{(\xi - s)^2 + \eta^2}\, ds = \frac{V}{\pi} - \frac{V}{\pi} \arctan\left(\frac{\eta}{\xi}\right). \tag{7.1.1}$$

Therefore, for a given value of ξ and η we can use Equation 7.1.1 to compute the potential. Then the conformal mapping provides the solution for the corresponding x and y. Figure 7.1.2 illustrates this solution. For a given z, Newton's method was used to solve for w. Then the potential follows from Equation 7.1.1.

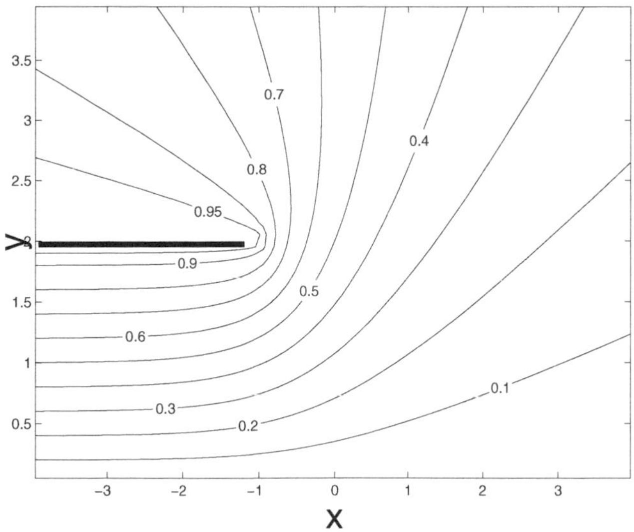

Figure 7.1.2: The solution $u(x,y)$ to a mixed boundary value problem where $u(x,0) = 0$ for $-\infty < x < \infty$ and $u(x,\pi a) = V$ when $-\infty < x < a\ln(a) - a$ with $a = 2/\pi$.

7.2 THE MAPPING $\tanh[\pi z/(2b)] = \mathrm{sn}(w,k)$

In Example 1.1.5 we illustrated how a succession of conformal mappings could solve the mixed boundary value problem

$$\frac{\partial^2 u}{\partial x^2} + \frac{\partial^2 u}{\partial y^2} = 0, \qquad 0 < x < \infty, \quad 0 < y < b, \tag{7.2.1}$$

subject to the boundary conditions

$$u(x,0) = 0, \qquad 0 < x < \infty, \tag{7.2.2}$$

$$\begin{cases} u(x,b) = 1, & 0 < x < a, \\ u_y(x,b) = 0, & a < x < \infty, \end{cases} \tag{7.2.3}$$

and

$$u_x(0,y) = 0, \qquad \lim_{x\to\infty} u(x,y) \to 0, \qquad 0 < y < b. \tag{7.2.4}$$

Here we interchanged x and y and changed the name of some of the parameters. The point here is to show[2] that the conformal mapping

$$\tanh\left(\frac{\pi z}{2b}\right) = \mathrm{sn}(k,w) \tag{7.2.5}$$

---

[2] See Wolfe, P. N., 1962: Capacitance calculations for several simple two-dimensional geometries. *Proc. IRE*, **50**, 2131–2132. See also Gelmont, B., and M. Shur, 1993: Spreading resistance of a round ohmic contact. *Solid-State Electron.*, **36**, 143–146.

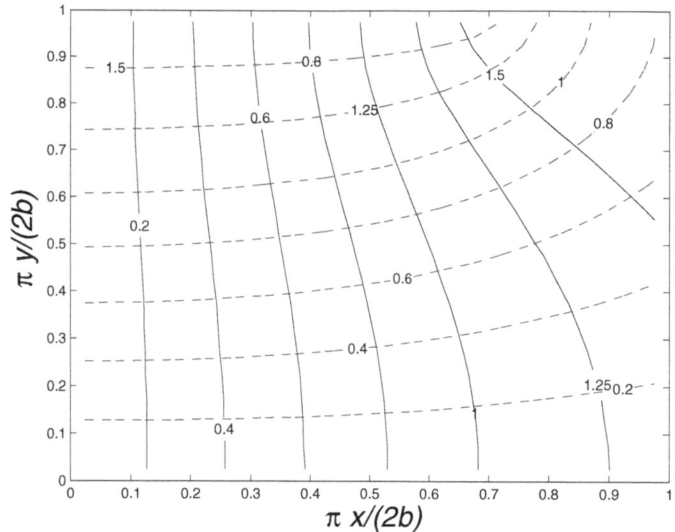

Figure 7.2.1: The conformal mapping $\tanh[\pi z/(2b)] = \text{sn}(w, k)$ when $a/b = \frac{1}{2}$ and $k = \tanh[\pi a/(2b)]$. The solid, dark line gives values of $\Re(w)$ while the dashed, horizontal lines denote $\Im(w)$.

can be used to solve Equation 7.2.1 through Equation 7.2.4, where $z = x + iy$, $w = \xi + i\eta$, $k = \tanh[\pi a/(2b)]$, and $\text{sn}(\cdot, \cdot)$ is one of the Jacobi elliptic functions.

Figure 7.2.1 illustrates lines of constant ξ and η within a portion of the original xy-plane. It shows that the original semi-infinite strip has been mapped into a rectangular region with $0 < \xi < K$ and $0 < \eta < K'$, where K and K' are complete elliptic integrals of the first kind for moduli k and $k' = \sqrt{1 - k^2}$.

Applying the conformal mapping Equation 7.2.5, the problem becomes

$$\frac{\partial^2 u}{\partial \xi^2} + \frac{\partial^2 u}{\partial \eta^2} = 0, \qquad 0 < \xi < K, \quad 0 < \eta < K', \tag{7.2.6}$$

subject to the boundary conditions

$$u(\xi, 0) = 0, \quad u(\xi, K') = 1, \qquad 0 < \xi < K, \tag{7.2.7}$$

and

$$u_\xi(0, \eta) = u_\xi(K, \eta) = 0, \qquad 0 < \eta < K'. \tag{7.2.8}$$

The solution to Equation 7.2.6 through Equation 7.2.8 is simply $u(\xi, \eta) = \eta/K'$. Therefore, lines of constant η/K' give $u(x, y)$ via Equation 7.2.5. Figure 1.1.4 illustrates the solution.

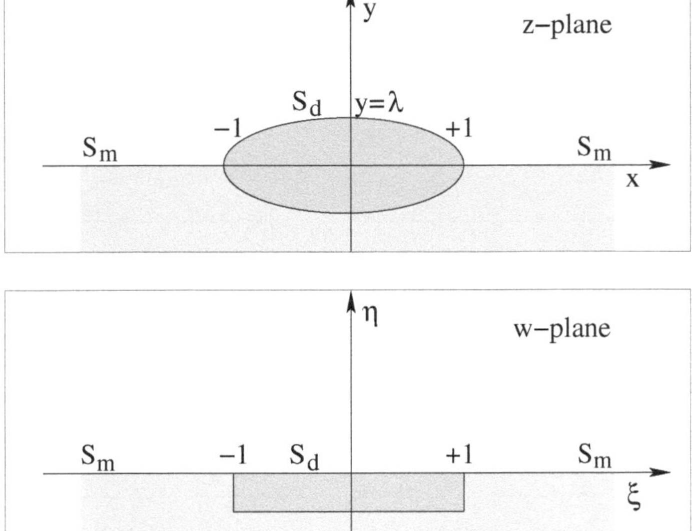

Figure 7.3.1: The conformal mapping used to map the half-plane above the boundary conditions $S_m \cup S_d$ into the half-plane $\eta > 0$ in the w-plane.

7.3 THE MAPPING $z = w + \lambda\sqrt{w^2 - 1}$

This mapping is useful in converting an elliptic shaped domain in the xy-plane into a rectangular one in the $\xi\eta$-plane where $w = \xi + i\eta$. To illustrate this conformal mapping, let us solve Laplace's equation in the half-plane

$$\frac{\partial^2 u}{\partial x^2} + \frac{\partial^2 u}{\partial y^2} = 0, \qquad -\infty < x < \infty, \quad \eta(x,0) < y < \infty, \tag{7.3.1}$$

subject to the boundary conditions

$$\lim_{y\to\infty} u(x,y) \to 0, \qquad -\infty < x < \infty, \tag{7.3.2}$$

and

$$\left.\frac{\partial u}{\partial n}\right|_{S_d} = 0, \qquad \text{and} \qquad u|_{S_m} = 0. \tag{7.3.3}$$

We begin by solving the problem

$$\frac{\partial^2 u}{\partial \xi^2} + \frac{\partial^2 u}{\partial \eta^2} = 0, \qquad -\infty < \xi < \infty, \quad 0 < \eta < \infty, \tag{7.3.4}$$

with the boundary conditions

$$\lim_{|\xi|\to\infty} |u(\xi,\eta)| < \infty, \qquad -\infty < \eta < \infty, \tag{7.3.5}$$

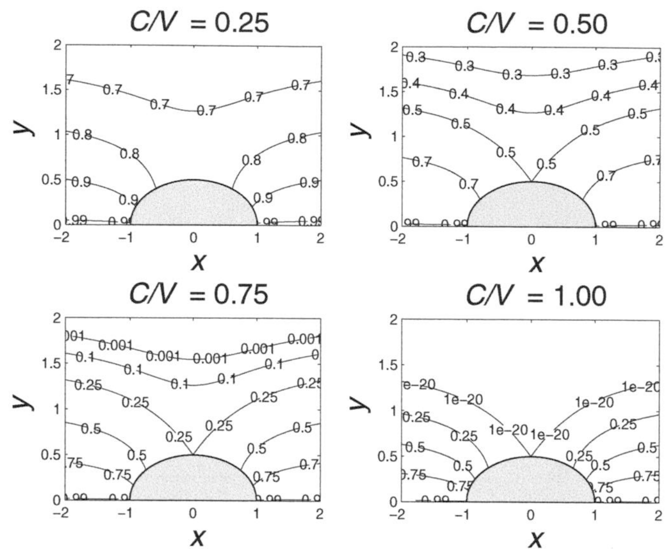

Figure 7.3.2: Plots of the potential $u(x, y)/V$ when $\lambda = 0.5$ for various values of C/V.

$$\lim_{\eta \to \infty} u(\xi, \eta) \to 0, \qquad -\infty < \xi < \infty, \tag{7.3.6}$$

and

$$\begin{cases} u_\eta(\xi, 0) = 0, & |\xi| < 1, \\ u(\xi, 0) = V, & |\xi| > 1. \end{cases} \tag{7.3.7}$$

The solution to this problem is

$$u(\xi, \eta) = V + C\,\Re\left(i\sqrt{w^2 - 1}\right), \tag{7.3.8}$$

because

$$u(\xi, 0) = V + C\,\Re\left(i\sqrt{\xi^2 - 1}\right) = V \tag{7.3.9}$$

if $|\xi| > 1$, and

$$\frac{\partial u(\xi, 0)}{\partial \eta} = C\,\Re\left(-\frac{\xi}{\sqrt{\xi^2 - 1}}\right) = 0 \tag{7.3.10}$$

if $|\xi| < 1$. Therefore,

$$u(x, y) = V + C\,\Re\left(i\frac{-\lambda z + \sqrt{z^2 + \lambda^2 - 1}}{1 - \lambda^2}\right), \tag{7.3.11}$$

where $z = x + iy$ and C is a free parameter.

Figure 7.3.2 illustrates $u(x, y)$ when $\lambda = 0.5$. In the construction of the conformal mapping and solution, it is important to take the branch cut of $\sqrt{w^2 - 1}$ so that it lies along the real axis in the complex w-plane.

7.4 THE MAPPING $w = ai(z-a)/(z+a)$

Let us solve[3] Laplace's equation in a domain exterior to an infinitely long cylinder of radius a

$$\frac{\partial^2 u}{\partial r^2} + \frac{1}{r}\frac{\partial u}{\partial r} + \frac{1}{r^2}\frac{\partial^2 u}{\partial \theta^2} = 0, \qquad a \le r < \infty, \quad 0 < |\theta| < \pi, \tag{7.4.1}$$

subject to the boundary conditions

$$\lim_{r\to\infty} u(r,\theta) \to 0, \qquad 0 \le |\theta| \le \pi, \tag{7.4.2}$$

and

$$\begin{cases} u_r(a,\theta) = 1, & 0 \le |\theta| < \alpha, \\ u(a,\theta) = 0, & \alpha < |\theta| < \pi. \end{cases} \tag{7.4.3}$$

We begin by introducing the conformal mapping:

$$w = \tilde{\xi} + i\tilde{\eta} = ia\frac{z-a}{z+a}, \tag{7.4.4}$$

where $z = x + iy$. Equation 7.4.1 then becomes

$$\frac{\partial^2 u}{\partial \tilde{\xi}^2} + \frac{\partial^2 u}{\partial \tilde{\eta}^2} = 0, \qquad -\infty < \tilde{\xi} < \infty, \quad 0 < \tilde{\eta} < \infty, \tag{7.4.5}$$

with the boundary conditions

$$\lim_{\tilde{\eta}\to\infty} u(\tilde{\xi},\tilde{\eta}) \to 0, \qquad -\infty < \tilde{\xi} < \infty, \tag{7.4.6}$$

and

$$\begin{cases} u_{\tilde{\eta}}(\tilde{\xi},0) = 2a^2/(\tilde{\xi}^2 + a^2), & |\tilde{\xi}| < a\tan(\alpha/2), \\ u(\tilde{\xi},0) = 0, & |\tilde{\xi}| > a\tan(\alpha/2). \end{cases} \tag{7.4.7}$$

We now nondimensionalize $\tilde{\xi}$ and $\tilde{\eta}$ as follows:

$$\xi = \frac{\tilde{\xi}}{a\tan(\alpha/2)} = -\frac{\sin(\theta)}{\tan(\alpha/2)}\frac{2ar}{r^2 + a^2 + 2ar\cos(\theta)}, \tag{7.4.8}$$

and

$$\eta = \frac{\tilde{\eta}}{a\tan(\alpha/2)} = \cot(\alpha/2)\frac{r^2 - a^2}{r^2 + a^2 + 2ar\cos(\theta)}. \tag{7.4.9}$$

Figure 7.4.1 illustrates this conformal mapping. Equation 7.4.5 then becomes

$$\frac{\partial^2 u}{\partial \xi^2} + \frac{\partial^2 u}{\partial \eta^2} = 0, \qquad -\infty < \xi < \infty, \quad 0 < \eta < \infty, \tag{7.4.10}$$

---

[3] See Iossel', Yu. Ya., 1971: A mixed two-dimensional stationary heat-conduction problem for a cylinder. *J. Engng. Phys.*, **21**, 1145–1147.

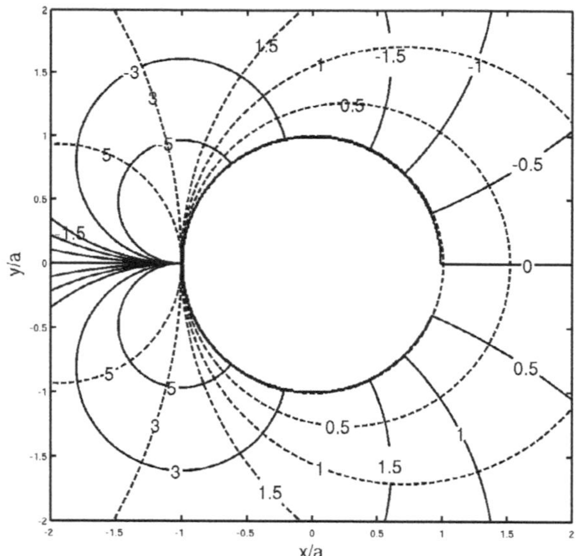

Figure 7.4.1: The conformal mapping given by Equation 7.4.7 through Equation 7.4.9 with $\alpha = \pi/4$. The solid lines are isolines of ξ, while the dashed lines are isolines of η.

with the nondimensional boundary conditions

$$\lim_{\eta\to\infty} u(\xi,\eta) \to 0, \qquad -\infty < \xi < \infty, \tag{7.4.11}$$

and

$$\begin{cases} u_\eta(\xi,0) = 2a\tan(\alpha/2)/[1+\xi^2\tan^2(\alpha/2)], & |\xi|<1, \\ u(\xi,0) = 0, & |\xi|>1. \end{cases} \tag{7.4.12}$$

We now solve Equation 7.4.10 through Equation 7.4.12 using the techniques developed in Section 4.1. Applying Fourier cosine transforms, the solution to Equation 7.4.10 and Equation 7.4.11 is

$$u(\xi,\eta) = \int_0^\infty A(k)e^{-k\eta}\cos(k\xi)\,dk. \tag{7.4.13}$$

Substituting Equation 7.4.13 into the mixed boundary condition Equation 7.4.12, we obtain the dual integral equations

$$\int_0^\infty kA(k)\cos(k\xi)\,dk = -\frac{2a\tan(\alpha/2)}{1+\xi^2\tan^2(\alpha/2)}, \qquad |\xi|<1, \tag{7.4.14}$$

and

$$\int_0^\infty A(k)\cos(k\xi)\,dk = 0, \qquad |\xi|>1. \tag{7.4.15}$$

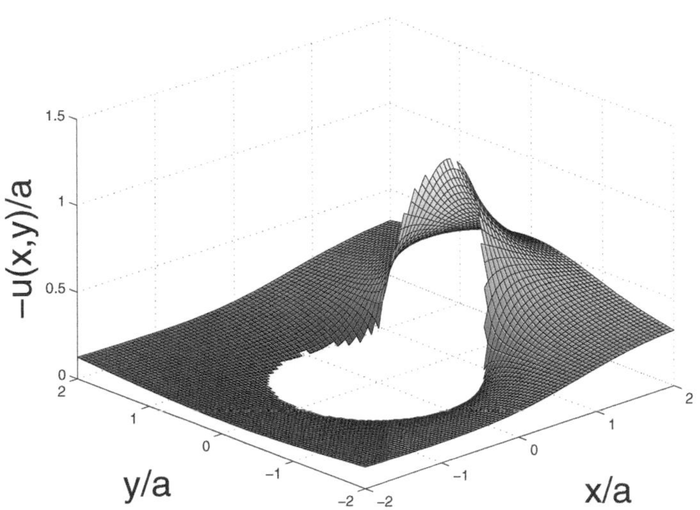

Figure 7.4.2: The solution of Equation 7.4.1 subject to the mixed boundary conditions Equation 7.4.2 and Equation 7.4.3 when $\alpha = \pi/4$.

To solve these dual integral equations, we define $A(k)$ by

$$A(k) = \int_0^1 g(t) J_0(kt)\, dt. \tag{7.4.16}$$

A quick check shows that Equation 7.4.16 satisfies Equation 7.4.15 identically. On the other hand, integrating Equation 7.4.14 with respect to ξ, we find that

$$\int_0^\infty A(k) \sin(k\xi)\, dk = -2a \arctan[\xi \tan(\alpha/2)], \qquad |\xi| < 1. \tag{7.4.17}$$

Next, we substitute Equation 7.4.16 into Equation 7.4.17, interchange the order of integration, then apply Equation 1.4.13 and obtain

$$\int_0^\xi \frac{g(t)}{\sqrt{\xi^2 - t^2}}\, dt = -2a \arctan[\xi \tan(\alpha/2)], \qquad |\xi| < 1. \tag{7.4.18}$$

Applying the results from Equation 1.2.13 and Equation 1.2.14,

$$g(t) = -\frac{4a}{\pi}\frac{d}{dt}\left[\int_0^t \frac{\zeta \arctan[\zeta \tan(\alpha/2)]}{\sqrt{t^2 - \zeta^2}}\, d\zeta\right] \tag{7.4.19}$$

$$= -\frac{2at \tan(\alpha/2)}{\sqrt{1 + t^2 \tan^2(\alpha/2)}}. \tag{7.4.20}$$

Finally, if we substitute Equation 7.4.16 and Equation 7.4.20 into Equation 7.4.13,

$$u(\xi,\eta) = -2a\tan(\alpha/2)\int_0^1 \frac{t}{\sqrt{1+t^2\tan^2(\alpha/2)}}\left[\int_0^\infty J_0(kt)e^{-k\eta}\cos(k\xi)\,dk\right]dt \tag{7.4.21}$$

$$= -\sqrt{2}\,a\tan(\alpha/2) \times \int_0^1 t\sqrt{\frac{t^2+\eta^2+\xi^2+\sqrt{(t^2+\eta^2-\xi^2)^2+4\eta^2\xi^2}}{[(t^2+\eta^2-\xi^2)^2+4\xi^2\eta^2]\,[1+t^2\tan^2(\alpha/2)]}}\,dt \tag{7.4.22}$$

$$= -\sqrt{2}\,a\ln\left[\frac{\tan^2\left(\frac{\alpha}{2}\right)\sqrt{(\beta_2^2-\gamma^2)\tan^2\left(\frac{\alpha}{2}\right)+2\delta\beta_2}+\beta_2\tan^2\left(\frac{\alpha}{2}\right)+\delta}{\tan^2\left(\frac{\alpha}{2}\right)\sqrt{(\beta_1^2-\gamma^2)\tan^2\left(\frac{\alpha}{2}\right)+2\delta\beta_1}+\beta_1\tan^2\left(\frac{\alpha}{2}\right)+\delta}\right], \tag{7.4.23}$$

where $\beta_1 = 2\eta^2$, $\beta_2 = 1+\eta^2-\xi^2+\sqrt{(1+\eta^2-\xi^2)^2+4\xi^2\eta^2}$, $\gamma = 2\xi\eta$, and $\delta = 1+(\xi^2-\eta^2)\tan^2(\alpha/2)$. In Figure 7.4.2 we illustrate the solution when $\alpha = \pi/4$.

7.5 THE MAPPING $z = 2[w - \arctan(w)]/\pi$

Let us solve Laplace's equation in a domain illustrated in Figure 7.5.1

$$\frac{\partial^2 u}{\partial x^2}+\frac{\partial^2 u}{\partial y^2} = 0, \qquad \begin{cases} -b < x < 0, & -\infty < y < \infty, \\ 0 < x < \infty, & 0 < y < \infty, \end{cases} \tag{7.5.1}$$

subject to the boundary conditions

$$\lim_{y\to\infty} |u(x,y)| < \infty, \qquad -b < x < \infty, \tag{7.5.2}$$

$$\lim_{y\to-\infty} |u(x,y)| < \infty, \qquad -b < x < 0, \tag{7.5.3}$$

and

$$\begin{cases} u(-b,y) = T_0, & -\infty < y < \infty, \\ u(0,y) = T_0, & -\infty < y < 0, \\ -u_y(x,0)+hu(x,0) = 0, & 0 < y < \infty. \end{cases} \tag{7.5.4}$$

In the previous problem we used conformal mapping to transform a mixed boundary value problem with Dirichlet and/or Neumann boundary conditions into a simple domain on which we still have Dirichlet and/or Neumann conditions. In the present problem, Strakhov[4] illustrates the difficulties that arise in a mixed boundary value problem where one of the boundary conditions is a Robin condition. Here he suggests a method for solving this problem.

[4] See Strakhov, I. A., 1969: One steady-state heat-conduction problem for a polygonal region with mixed boundary conditions. *J. Engng. Phys.*, **17**, 990–994.

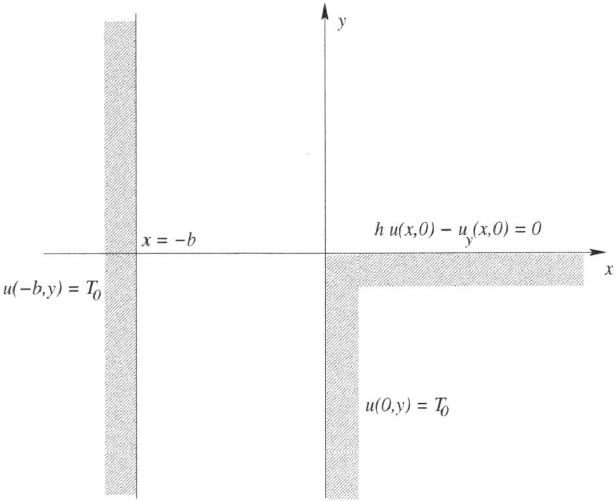

Figure 7.5.1: The domain associated with the problem posed in Section 7.5.

We begin by introducing the conformal mapping:

$$z = \frac{2b}{\pi} \int_0^w \frac{\tau^2}{\tau^2 + 1} \, d\tau = 2b[w - \arctan(w)]/\pi, \tag{7.5.5}$$

where $w = \xi + i\eta$ and $z = x + iy$. Figure 7.5.2 illustrates this conformal mapping. It shows that the original domain is mapped into the first quadrant of the w-plane. In particular, the real semi-axis $0 < x < \infty, y = 0$ on the z-plane becomes the real semi-axis of $0 < \xi < \infty, \eta = 0$ on the w-plane. The boundary $x = 0, -\infty < y < 0$ maps onto the segment $\xi = 0, 0 < \eta < 1$ on the imaginary axis while the boundary $x = -b, -\infty < y < \infty$ maps onto the segment $\xi = 0, 1 < \eta < \infty$.

Upon using the conformal mapping Equation 7.5.5, Equation 7.5.1 becomes

$$\frac{\partial^2 u}{\partial \xi^2} + \frac{\partial^2 u}{\partial \eta^2} = 0, \qquad 0 < \xi < \infty, \quad 0 < \eta < \infty, \tag{7.5.6}$$

with the boundary conditions

$$\lim_{\xi \to \infty} |u(\xi, \eta)| < \infty, \qquad 0 < \eta < \infty, \tag{7.5.7}$$

$$\lim_{\eta \to \infty} |u(\xi, \eta)| < \infty, \qquad 0 < \xi < \infty, \tag{7.5.8}$$

and

$$u(0, \eta) = T_0, \qquad 0 < \eta < \infty, \tag{7.5.9}$$

and

$$-(1 + \xi^2) u_\eta(\xi, 0) + \epsilon \xi^2 u(\eta, 0) = 0, \qquad 0 < \xi < \infty, \tag{7.5.10}$$

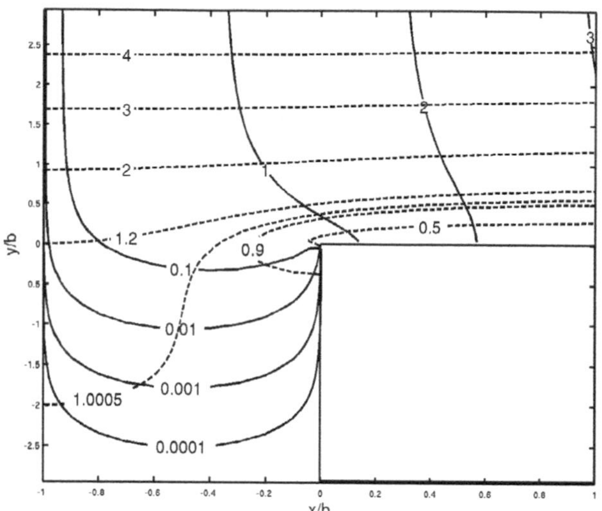

Figure 7.5.2: The conformal mapping given by Equation 7.5.5. The solid lines are lines of constant ξ, while the dashed lines are lines of constant η.

where $\epsilon = 2hb/\pi$.

Motivated by the work of Stokes[5] and Chester,[6] we set

$$u(\xi, \eta) = \Re[\Psi(w)], \tag{7.5.11}$$

where $\Psi(w)$ is an analytic function in any finite portion of the first quadrant of the w-plane. Therefore, the boundary conditions become

$$\Re[\Psi(i\eta)] = T_0, \qquad 0 < \eta < \infty, \tag{7.5.12}$$

and

$$\Re\left[-i(w^2+1)\frac{d\Psi(w)}{dw} + \epsilon w^2 \Psi(w)\right]\Bigg|_{w=\xi} = 0, \qquad 0 < \xi < \infty. \tag{7.5.13}$$

At this point we introduce the function $\chi(w)$ defined by

$$\Psi(w) = \chi(w) - \frac{2iT_0}{\pi}\log(w). \tag{7.5.14}$$

---

[5] Stoker, J. J., 1947: Surface waves in water of variable depth. *Quart. Appl. Math.*, **5**, 1–54.

[6] Chester, C. R., 1961: Reduction of a boundary value problem of the third kind to one of the first kind. *J. Math. Phys. (Cambridge, MA)*, **40**, 68–71.

Why have we introduced $\chi(w)$? Because $\Re[-2iT_0 \log(w)/\pi]$ satisfies Laplace's equation and the boundary conditions Equation 7.5.12 and Equation 7.5.13 as $|w| \to \infty$ in the first quadrant of the w-plane, we anticipate that $\chi(w) = O(1/|w|)$ as $|w| \to \infty$ and $0 \le \arg(w) \le \pi/2$.

Introducing Equation 7.5.14 into Equation 7.5.12 and Equation 7.5.13, we find that

$$\Re[\chi(i\eta)] = 0, \qquad 0 < \eta < \infty, \tag{7.5.15}$$

and

$$\Re\left[-i(w^2+1)\frac{d\chi(w)}{dw} + \epsilon w^2 \chi(w)\right]\Bigg|_{w=\xi} = \frac{2T_0}{\pi}\frac{\xi^2+1}{\xi}, \qquad 0 < \xi < \infty; \tag{7.5.16}$$

or

$$-i(w^2+1)\frac{d\chi(w)}{dw} + \epsilon w^2 \chi(w) = \frac{2T_0}{\pi}\frac{w^2+1}{w} + i\alpha T_0, \tag{7.5.17}$$

where α is a free constant. Integrating Equation 7.5.17,

$$\chi(w) = T_0 \left(\frac{w-i}{w+i}\right)^{\epsilon/2} e^{-i\epsilon w} \int_w^\infty \left(\frac{\zeta-i}{\zeta+i}\right)^{\epsilon/2} e^{i\epsilon\zeta}\left[-\frac{2i}{\pi\zeta} + \frac{\alpha}{\zeta^2+1}\right] d\zeta. \tag{7.5.18}$$

We choose those branches of $(w-i)^{\epsilon/2}$ and $(w+i)^{\epsilon/2}$ that approach $\xi^{\epsilon/2}$ along the positive real axis as $w \to \infty$. The integration occurs over any path in the first quadrant of the w-plane that does not pass through the points $w = 0$ and $w = i$.

We must check and see if Equation 7.5.15 is satisfied. Let $w = i\eta, 1 < \eta < \infty$. Then,

$$\chi(i\eta) = -iT_0 \left(\frac{\eta-1}{\eta+1}\right)^{\epsilon/2} e^{\epsilon\eta} \int_\eta^\infty \left(\frac{\tau+1}{\tau-1}\right)^{\epsilon/2} e^{-\epsilon\tau}\left[\frac{2}{\pi\tau} + \frac{\alpha}{\tau^2-1}\right] d\tau \tag{7.5.19}$$

and $\Re[\chi(i\eta)] = 0$ for $1 < \eta < \infty$. Therefore, Equation 7.5.15 is satisfied.

Consider now $w = i\eta$ with $0 < \eta < 1$. We rewrite Equation 7.5.18 as

$$\begin{aligned}\chi(w) = T_0 \left(\frac{w-i}{w+i}\right)^{\epsilon/2} e^{-i\epsilon w}\Bigg\{B + \frac{2i}{\pi}\Bigg[\int_0^w \left(\left(\frac{\zeta+i}{\zeta-i}\right)^{\epsilon/2} - e^{i\epsilon\pi/2}\right) e^{i\epsilon\zeta}\frac{d\zeta}{\zeta} \\ - e^{i\epsilon\pi/2}\int_w^\infty e^{i\epsilon\zeta}\frac{d\zeta}{\zeta}\Bigg] - \alpha\int_0^w \left(\frac{\zeta+i}{\zeta-i}\right)^{\epsilon/2} e^{i\epsilon\zeta}\frac{d\zeta}{\zeta^2+1}\Bigg\},\end{aligned} \tag{7.5.20}$$

where we have introduced

$$B = -\frac{2i}{\pi}\int_0^\infty \left[\left(\frac{\zeta+i}{\zeta-i}\right)^{\epsilon/2} - e^{i\epsilon\pi/2}\right] e^{i\epsilon\zeta}\frac{d\zeta}{\zeta} + \alpha\int_0^\infty \left(\frac{\zeta+i}{\zeta-i}\right)^{\epsilon/2} e^{i\epsilon\zeta}\frac{d\zeta}{\zeta^2+1}. \tag{7.5.21}$$

In the case of Equation 7.5.21, the integration can occur along any curve in the first quadrant of the w-plane that does not pass through the point $\zeta = i$.

Let us now evaluate Equation 7.5.20 along the segment of the imaginary axis between $w = 0$ and $w = i\eta$ where $0 < \eta < 1$. In this case,

$$\Re[\chi(i\eta)] = T_0 \left(\frac{1-\eta}{1+\eta}\right)^{\epsilon/2} e^{\epsilon\eta} \; \Re\left(Be^{-i\epsilon\pi/2}\right), \qquad 0 < \eta < 1. \qquad \textbf{(7.5.22)}$$

To satisfy Equation 7.5.22,

$$\Re\left(Be^{-i\epsilon\pi/2}\right) = 0. \qquad \textbf{(7.5.23)}$$

Substituting for B and solving for α,

$$\alpha = \frac{1 - \frac{2}{\pi}\int_0^\infty \sin\{\epsilon[\tau - \arctan(\tau)]\}\frac{d\tau}{\tau}}{\int_0^\infty \cos\{\epsilon[\tau - \arctan(\tau)]\}\frac{d\tau}{\tau^2+1}}. \qquad \textbf{(7.5.24)}$$

Noting that $w = i$ is a removable singularity,

$$\chi(i) = \lim_{w\to i} \chi(w) = -\frac{i\alpha T_0}{\epsilon}, \qquad \textbf{(7.5.25)}$$

Equation 7.5.24 now reads

$$\frac{\Psi(w)}{T_0} = \left(\frac{w-i}{w+i}\right)^{\epsilon/2} e^{-i\epsilon w} \int_w^\infty \left(\frac{\zeta+i}{\zeta-i}\right)^{\epsilon/2} e^{i\epsilon\zeta} \left[-\frac{2i}{\pi\zeta} + \frac{\alpha}{\zeta^2+1}\right] d\zeta - \frac{2i}{\pi}\log(w). \qquad \textbf{(7.5.26)}$$

In summary, for a given w, we can compute the corresponding z/b from Equation 7.5.5. The same w is then used to find $\Psi(w)$ and the value of $u(x, y)$. Figure 7.5.3 illustrates the solution when $\epsilon = 2$.

Problems

1. Solve Laplace's equation[7]

$$\frac{\partial^2 u}{\partial x^2} + \frac{\partial^2 u}{\partial y^2} = 0, \qquad -\infty < x < \infty, \quad 0 < y < \infty,$$

subject to the boundary conditions

$$\lim_{|x|\to\infty} u(x, y) \to 0, \qquad 0 < y < \infty,$$

[7] See Karush, W., and G. Young, 1952: Temperature rise in a heat-producing solid behind a surface defect. *J. Appl. Phys.*, **23**, 1191–1193.

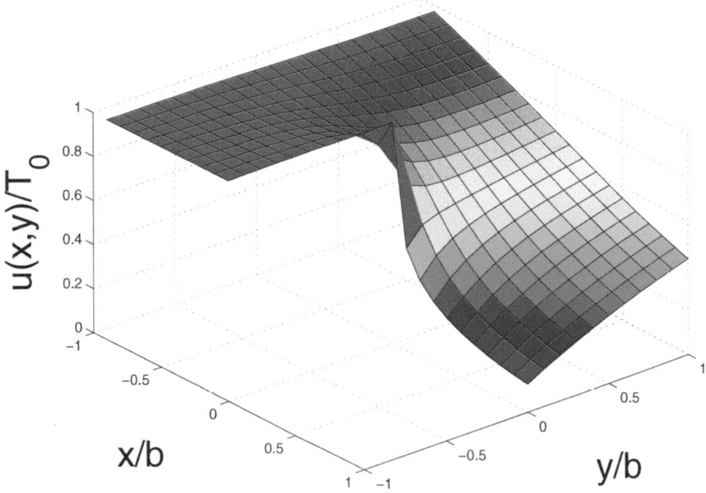

Figure 7.5.3: The solution to the mixed boundary value problem governed by Equation 7.5.1 through Equation 7.5.4 with $\epsilon = 2$.

$$\lim_{y\to\infty} u(x,y) \to 0, \qquad -\infty < x < \infty,$$

and

$$\begin{cases} u_y(x,0) = A, & |x| < a, \\ u(x,0) = 0, & |x| > a. \end{cases}$$

Step 1: Show that conformal map $w = \sqrt{z^2 - a^2}$ maps the upper half of the z-plane into the w-plane as shown in the figure entitled Conformal Mapping $w = \sqrt{z^2 - a^2}$.

Step 2: Show that potential $u(x,y) = -A\,\Im\left(\sqrt{z^2 - a^2} - z\right)$ satisfies Laplace's equation and the boundary conditions along $y = 0$.

Step 3: Using the Taylor expansion for the square root, show that $u(x,y) \to 0$ as $|z| \to \infty$. The figure entitled Problem 1 illustrates this solution.

7.6 THE MAPPING $k_w \operatorname{sn}(w, k_w) = k_z \operatorname{sn}(K_z z/a, k_z)$

In Section 7.2 we illustrated how conformal mapping could be used to solve Laplace's equation on a semi-infinite strip with mixed boundary conditions. Here we solve a similar problem: the solution to Laplace's equation over a rectangular strip with mixed boundary conditions. In particular, we want to

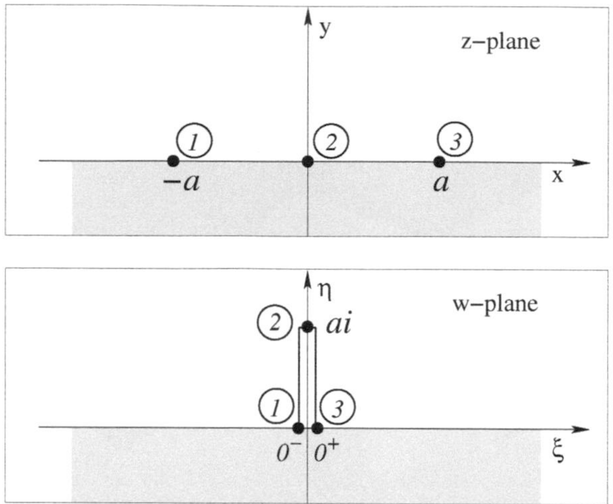

Conformal Mapping $w = \sqrt{z^2 - a^2}$

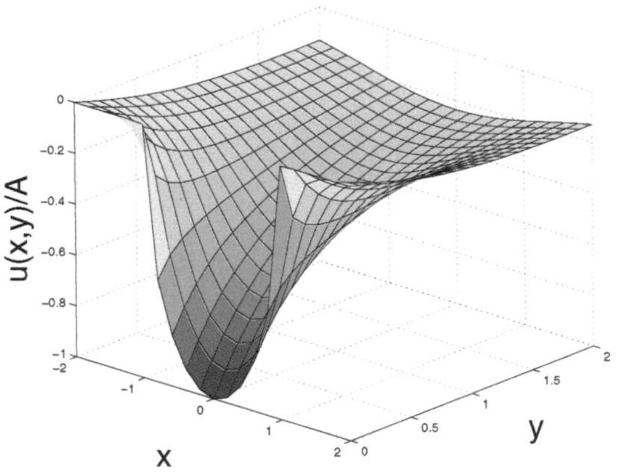

Problem 1

solve Laplace's equation:[8]

$$\frac{\partial^2 u}{\partial x^2} + \frac{\partial^2 u}{\partial y^2} = 0, \qquad 0 < x < a, \quad 0 < y < b, \tag{7.6.1}$$

subject to the boundary conditions

$$u(x, b) = 0, \qquad 0 < x < a, \tag{7.6.2}$$

[8] See Bilotti, A. A., 1974: Static temperature distribution in IC chips with isothermal heat sources. *IEEE Trans. Electron Devices,*, **ED-21**, 217–226.

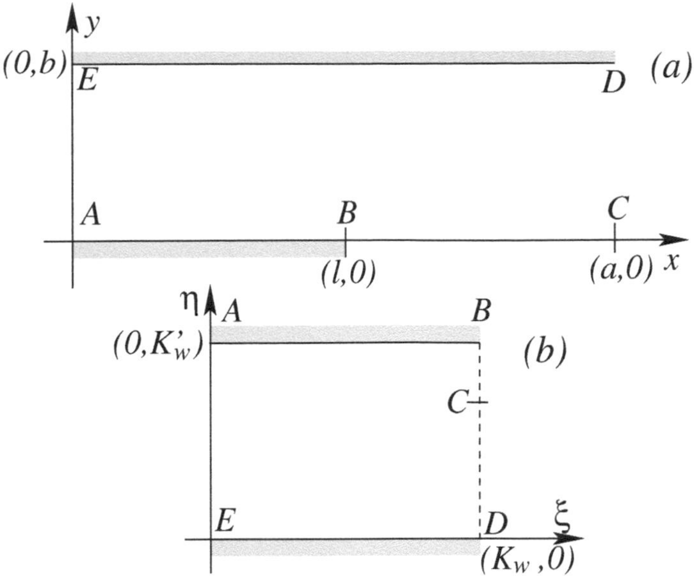

Figure 7.6.1: Schematic of (a) z- and (b) w-planes used in Section 7.6 with the conformal mapping $k_w \text{sn}(w, k_w) = k_z \text{sn}(K_z z/a, k_z)$.

$$\begin{cases} u(x,0) = 1, & 0 < x < \ell, \\ u_y(x,0) = 0, & \ell < x < a, \end{cases} \tag{7.6.3}$$

and

$$u_x(0,y) = u_x(a,y) = 0, \qquad 0 < y < b. \tag{7.6.4}$$

See Figure 7.6.1. In the w-plane, the problem becomes

$$\frac{\partial^2 u}{\partial \xi^2} + \frac{\partial^2 u}{\partial \eta^2} = 0, \qquad 0 < \xi < K_w, \quad 0 < \eta < K'_w, \tag{7.6.5}$$

subject to the boundary conditions

$$u(\xi, K'_w) = 0, \qquad 0 < \xi < K_w, \tag{7.6.6}$$

$$u(\xi, 0) = 1, \qquad 0 < \xi < K_w, \tag{7.6.7}$$

and

$$u_\xi(0,\eta) = u_\xi(K_w,\eta) = 0, \qquad 0 < \eta < K'_w. \tag{7.6.8}$$

The z- and w-planes are related to each other via

$$k_w \,\text{sn}(w, k_w) = k_z \,\text{sn}(K_z z/a, k_z), \tag{7.6.9}$$

where $\text{sn}(\cdot,\cdot)$ is one of the Jacobian elliptic functions, $z = x + iy$, $w = \xi + i\eta$, k_z and k_w are the moduli for the elliptic functions in the z- and w-planes, respectively. In the z-plane, $0 < x/a < K_z$ and $0 < y/b < K'_z$, where K_z and iK'_z are the quarter-periods of the elliptic function. Therefore, we must choose k_z so that $K_z/K'_z = a/b$. Using MATLAB®, this is done as follows:

```
% By guessing k_z, called k, find the closest value
% of K_z/K'_z to a/b.
diff = 10000;
for n = 1:19999
k = 0.00005*n;
k_prime = sqrt(1-k*k);
K = ellipke(k);
K_prime = ellipke(k_prime);
ratio = K/K_prime;
if (abs(ratio-a/b) < diff)
k_z = k; K_z = K; K_prime_z = K_prime;
diff = abs(ratio-a/b);
end; end
```

Once we have k, K_z and K'_z, we can conpute k_w from $k_w = \text{sn}[\ell K_z/a, k_z]$. We also need K_w and K'_w. The corresponding MATLAB code is

```
% Find the corresponding values of k_w, K_w and K'_w.
[sn,cn,dn] = ellipj(ell*K_z/a,k_z);
k_w = k_z*sn;
K_w = ellipke(k_w);
k_prime = sqrt(1-k_w*k_w);
K_prime_w = ellipke(k_prime);
```

Having found k_z, K_z, K'_z and k_w, we are ready to find the values of ξ and η corresponding to a given x and y. This is done in two steps. First we find for a given x and y an approximate value of ξ and η where $0 < \xi < K_w$ and $0 < \eta < K'_w$. Then we use Newton's method to find the exact one-to-one mapping. The MATLAB code is

```
k = k_w; k_1 = 1-k;
for jc = 1:200
zeta_i = K_prime_w*(jc-0.5)/200;
for ic = 1:200
zeta_r = K_w*(ic-0.5)/200;
[s,c,d] = ellipj(zeta_r,k); [s1,c1,d1] = ellipj(zeta_i,k_1);
denom = c1*c1 + k*s*s*s1*s1;
ss_r = s*d1 / denom; ss_i = c*d*s1*c1 / denom;
sn(ic,jc) = ss_r + i*ss_i;
end; end
```

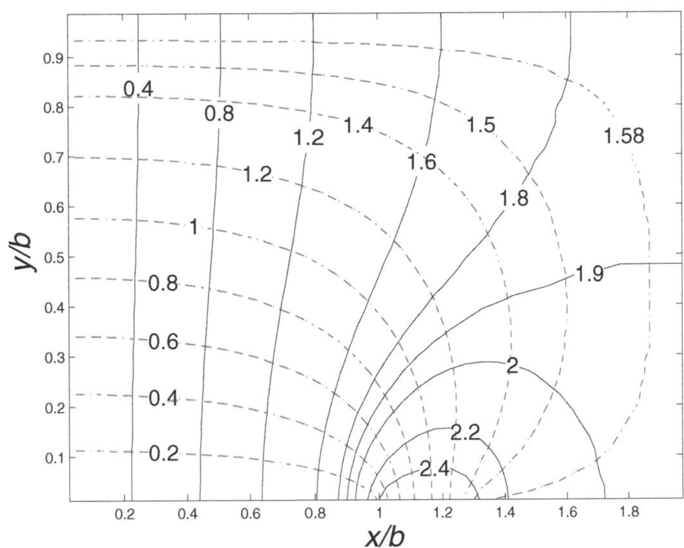

Figure 7.6.2: Lines of constant ξ and η given by the conformal mapping by Equation 7.6.9 when $a = 2$, $b = 1$, and $\ell = 1$.

```
% Next, for a given x and y, find the nearest point
% in the w-plane.

k = k_z; k_1 = 1-k;
for jcount = 1:40
y = b*(jcount-0.5)/40; z_i = K_z*y/a;
for icount = 1:40
x = a*(icount-0.5)/40; z_r = K_z*x/a;
[s,c,d] = ellipj(z_r,k); [s1,c1,d1] = ellipj(z_i,k_1);
denom = c1*c1 + k*s*s*s1*s1;
ss_r = s*d1 / denom; ss_i = c*d*s1*c1 / denom;
rhs = (k_z/k_w)*(ss_r + i*ss_i);
distance = 1000000;
for jc = 1:200
zeta_i = K_prime_w*(jc-0.5)/200;
for ic = 1:200
zeta_r = K_w*(ic-0.5)/200;
F = sn(ic,jc) - rhs;
if (abs(F) < distance)
XX(icount,jcount) = x/b; YY(icount,jcount) = y/b;
UU(icount,jcount) = zeta_r; VV(icount,jcount) = zeta_i;
distance = abs(F);
```

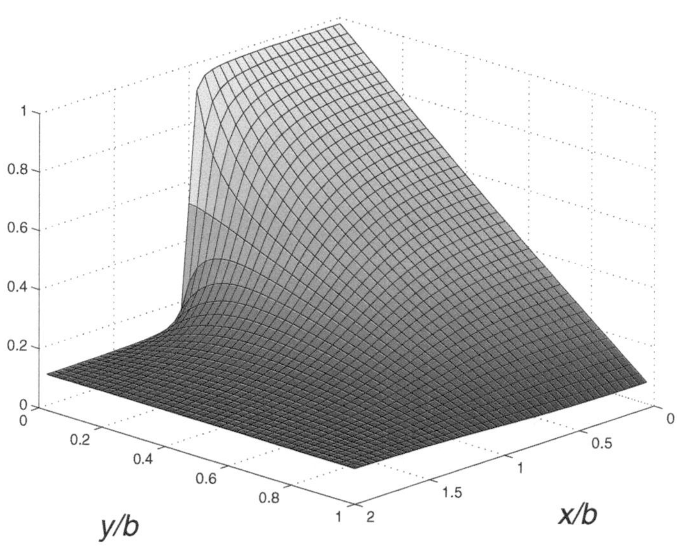

Figure 7.6.3: Solution to Laplace's equation with the mixed boundary value problems given by Equation 7.6.2 through Equation 7.6.4 when $a = 2$, $b = 1$, and $\ell = 1$.

```
end; end; end

% now use Newton's method to get the exact mapping.

zeta_r = UU(icount,jcount); zeta_i = VV(icount,jcount);
for iter = 1:10
[s,c,d] = ellipj(zeta_r,k); [s1,c1,d1] = ellipj(zeta_i,k_1);
denom = c1*c1 + k*s*s*s1*s1;
ssn = (s*d1+i*c*d*s1*c1) / denom;
ccn = (c*c1-i*s*d*s1*d1) / denom;
ddn = (d*c1*d1-i*k*s*c*s1) /denom;
F = ssn - rhs; F_prime = ccn*ddn;
zeta_r = zeta_r - real(F/F_prime);
zeta_i = zeta_i - imag(F/F_prime);
end
% Compute the potential T(x,y)
TT(icount,jcount) = 1 - zeta_i/K_prime_w;
end; end
```

Figure 7.6.2 illustrates lines of constant ξ and η. Having computed the mapping, the solution for Laplace's equation is $u(\xi, \eta) = 1 - \eta/K'_w$. Figure 7.6.3 illustrates the solution after we have transformed back into the xy-plane.

Index

T - #0330 - 071024 - C488 - 234/156/21 - PB - 9780367387587 - Gloss Lamination